OCS EIS/EA
MMS 2004-054

Geological and Geophysical Exploration for Mineral Resources on the Gulf of Mexico Outer Continental Shelf

I0476243

Final Programmatic Environmental Assessment

Prepared by

Continental Shelf Associates, Inc.
759 Parkway Street
Jupiter, Florida 33477

U.S. Department of the Interior
Minerals Management Service
Gulf of Mexico OCS Region

New Orleans
July 2004

FINDING OF NO SIGNIFICANT IMPACT

Programmatic Environmental Assessment
of Geological and Geophysical Exploration for Mineral Resources
on the Gulf of Mexico Outer Continental Shelf

The programmatic environmental assessment (EA) to evaluate the potential environmental impacts of geological and geophysical (G&G) activities in the Gulf of Mexico has been completed. The evaluation encompasses all G&G operations under the regulatory authority of the Minerals Management Service (MMS). The EA has resulted in a Finding of No Significant Impact (FONSI). Based on this EA, we have concluded that the G&G activities evaluated in the EA will not significantly affect the quality of the human environment. Preparation of an environmental impact statement is not required.

The activities analyzed in the EA include seismic surveys, deep-tow side-scan surveys, electromagnetic surveys, geological and geochemical sampling, and remote-sensing surveys. The impact-producing factors considered in the EA include seismic survey noise, vessel and aircraft noise, seafloor disturbance, and space-use conflicts with seismic arrays. Based on established significance criteria, the results of the impact analyses are that G&G activities are not expected to result in significant adverse impacts to any of the potentially affected resources. Potentially adverse but not significant impacts were identified for marine mammals (except the manatee) and commercial and recreational fishing; negligible to potentially adverse but not significant impacts were identified for sea turtles, fish, and benthic communities; and negligible impacts were identified for coastal and marine birds and the manatee.

The MMS currently requires operators engaged in activities on the OCS, including G&G activities, to comply with a number of lease stipulations, Notices to Lessees, and other mitigation measures designed to reduce or eliminate impacts to sensitive environmental resources from impact-producing factors such as vessel or aircraft traffic, anchoring, and trash and debris. These mitigation measures are required under the OCS Lands Act, the Endangered Species Act, and the Marine Mammal Protection Act to ensure environmental protection, consistent environmental policy, and safety. As part of the impact analyses completed in the G&G EA, current protective and mitigation measures were evaluated (Alternative 1). Additional feasible mitigation measures were also considered (Alternatives 2 and 3), as were potential restrictions on concurrent operations within close proximity to one another (Alternative 4), as viable alternatives to further reduce the potential for impacts to marine mammals.

Under the current regulatory *status quo*, the following mitigation measures will be required for seismic surveys in water depths greater than 200 m in the Western and Central Planning Areas and in all water depths in the Eastern Planning Area of the Gulf of Mexico. The need for these mitigations was recognized during development of the EA and these requirements were put into effect by MMS's Notice to Lessees 2004-G01 issued on March 1, 2004.

Mitigation Measures

1. All seismic surveys will include visual monitoring and clearance of a 500-m (radial distance) exclusion zone around the array and in the immediate vicinity of the survey vessel. Visual monitoring will begin no less than 30 minutes prior to the beginning of ramp-up and will continue until seismic operations cease or until sighting conditions do not allow observation of the sea surface (e.g., fog, rain, and darkness). Visual monitoring will be conducted using trained protected species observers. Vessel crews may be used in this capacity provided that the observers are properly trained.

2. All seismic surveys will include ramp-up procedures (i.e., soft start), with ramp-up to be initiated only during daylight hours following clearance of the exclusion zone (i.e., no sightings of whales within the exclusion zone for a minimum of 30 minutes).

3. Continuous (day and night) seismic survey operations will be allowed. However, if a whale is spotted within or transiting towards the exclusion zone surrounding the array and survey vessel, an immediate shutdown of the array will be required. Subsequent restart of the array, using ramp-up and protected species observers, will only be allowed during daylight hours and following clearance of the exclusion zone. Similar restrictions apply to airgun arrays that have been shut down for maintenance.

4. The use of experimental passive acoustic monitoring (to establish the presence and approximate location of vocalizing marine mammals, particularly sperm whales) is encouraged. If an array is shutdown, ramp-up may proceed during periods of poor visibility (e.g., during nighttime or periods of fog or rain, when the 500-m exclusion zone cannot be monitored) if passive acoustic monitoring is employed and no marine mammal vocalizations are evident. Use of a passive acoustic array by an observer proficient in its use will allow ramp-up and the subsequent start of a seismic survey during times of reduced visibility when such ramp-up otherwise would not be permitted using only trained protected species observers.

Dennis Chew
Chief, Environmental Assessment Section
Leasing and Environment, GOM OCS Region

7/1/04
Date

Richard Defenbaugh
Deputy Regional Supervisor
Leasing and Environment, GOM OCS Region

7/1/04
Date

Joseph Christopher
Regional Supervisor
Leasing and Environment, GOM OCS Region

7/1/04
Date

MMS STATEMENT OF RESPONSIBILITY

This Programmatic Environmental Assessment (PEA) has been prepared for the Minerals Management Service (MMS), Gulf of Mexico OCS Region, New Orleans, LA by Continental Shelf Associates, Inc. (CSA), Jupiter, FL, under MMS Contract 1432-01-99-CT-30987. A unique balance of responsibilities is necessary for a private-sector contractor to prepare a National Environmental Policy Act (NEPA) document for a Federal agency. The contractor performs most of the work and is responsible for the technical adequacy and scientific credibility of the analyses, but the Federal agency is responsible for the scope, content, and findings of the document (40 CFR 1506.5).

During the course of this project, MMS has been responsible for

- overall project plan and scope;
- accomplishment of the project, via a competitive procurement, which was won by CSA;
- project performance, including technical and policy reviews of draft documents;
- compliance with NEPA regulations, and with agency-specific NEPA requirements of the U.S. Department of the Interior (USDOI) and of MMS, and with the policies of the MMS and USDOI;
- consultations with other Federal agencies; and
- ultimate "ownership" of the scope, content, and findings of the final PEA.

CSA has been responsible for

- characterizing geological and geophysical (G&G) operations historically conducted, currently operating, or projected to occur in Gulf waters;
- summarizing important characteristics of Gulf resources potentially sensitive to G&G operations, with an emphasis on hearing and sensitivity (to acoustic impact) for marine mammals, sea turtles, and fishes;
- determining whether G&G activities have significant impacts on the Gulf of Mexico's marine, coastal, or human environments, further identifying significant impacts for additional NEPA analysis; and
- describing and evaluating mitigation measures (including the identification of those in current regulations and lease stipulations, plus additional protective measures) that may be applicable to future G&G operations as a means of reducing or eliminating potentially significant impacts.

MMS takes full responsibility for the scope, content, and findings of this PEA.

DISCLAIMER STATEMENT OF
CONTINENTAL SHELF ASSOCIATES, INC.

Continental Shelf Associates, Inc. (CSA), Jupiter, FL, and its team of subcontractors prepared this document for the Minerals Management Service (MMS), Gulf of Mexico OCS Region, New Orleans, LA. The findings presented in this Programmatic Environmental Assessment (PEA) have been determined based solely on a review and summarization of pertinent current scientific knowledge, and the team's understanding of and experience with both geological and geophysical (G&G) operations and the requirements of the National Environmental Policy Act (NEPA) and associated implementing regulations. Neither CSA or any members of its team of subcontractors and independent consultants have financial interests in the G&G industry.

COVER GRAPHICS

Upper right: Schooling barracuda; provided courtesy of Continental Shelf Associates, Inc.

Upper left: Airguns; provided courtesy of WesternGeco.

Lower left: Sperm whale diving, Gulf of Mexico; provided courtesy of W. Lang, Minerals Management Service and K. Mullin, National Marine Fisheries Service.

Lower right: Seismic survey vessel, abstract view; provided courtesy of Continental Shelf Associates, Inc.

Center right: Source directivity plot, inline with survey vessel; provided courtesy of Jack Caldwell, WesternGeco.

CONTENTS

FIGURES

Appendix H

Appendix L

TABLES

Appendix K

Appendix L

ABBREVIATIONS AND ACRONYMS

2C	two component	FR	Federal Register
2D	two dimensional	ft	feet
3D	three dimensional	G	giga
4C	four component	g	gram(s)
4D	four dimensional	G&G	geological and geophysical
μ	micro	GI	generator-injector (airgun)
ABR	auditory brainstem response	GIWW	Gulf Intracoastal Waterway
ATOC	Acoustic Thermometry of Ocean Climate	GMFMC	Gulf of Mexico Fishery Management Council
bbl	barrel(s)	GOM	Gulf of Mexico
BBO	billion barrels of oil	GPS	global positioning system
BO	Biological Opinion	GRN	Gulf Restoration Network
BP	British Petroleum	GulfCet	North-Central and Western Gulf of Mexico Cetacean Study
C°	degree(s) Centigrade		
CATEX	categorical exclusion	h or hr	hour(s)
CEQ	Council on Environmental Quality	HAPC	habitat areas of particular concern
CER	categorical exclusion review	HE	high explosive
CFR	Code of Federal Regulations	HESS	high-energy seismic survey
cm	centimeter(s)	HMS	highly migratory species
CMS	Cetacean Monitoring System	hp	horsepower
COTR	Contracting Officer's Technical Representative	Hz	hertz, cycle per second
		IAGC	International Association of Geophysical Contractors
CSA	Continental Shelf Associates, Inc.		
CZCS	coastal zone color scanner	IFAW	International Fund for Animal Welfare
d	day(s)		
DASAR	Directional Autonomous Seafloor Acoustic Recorders	IMAPS	Integrated Marine Mammal Monitoring and Protection System
dB	decibel(s)		
DCM	deep chlorophyll maximum	in.	inch(es)
DCS	decompression sickness	ITI	Innovative Tranducers, Inc.
DIFAR	Directional Frequency Analysis and Recording	ITS	Incidental Take Statement
		IUSS	Integrated Undersea Surveillance System
DIT	Department of Trade and Industry		
DOD	Department of Defense	JNCC	Joint Nature Conservation Committee
DOE	Department of Energy		
DOI	Department of the Interior	kg	kilogram(s)
DRC	damage risk criteria	kHz	kiloHertz
DWT	dead weight ton(s)	km	kilometer(s)
EA	environmental assessment	kn	knot(s), nautical mile(s)/hour
EEG	electroencephalogram	kw	kilowatt
EEZ	Exclusive Economic Zone	L	liter
EFH	Essential Fish Habitat	LC	Loop Current
EIS	environmental impact statement	LCE	Loop Current eddy
ESA	Endangered Species Act	LDEO	Lamont-Doherty Earth Observatory
ESW	effective strip width		
EWTA	Eglin Water Test Area	LFA	low frequency active
FAA	Federal Aviation Administration	LOA	Letter of Authorization
FEIS	final environmental impact statement	LOOP	Louisiana Offshore Oil Port
FMP	Fishery Management Plan	M	mole
FONSI	finding of no significant impact	m	meter(s)
FPSO	floating production, storage, and offloading	MAR	Mississippi-Atchafalaya River

MAST	Marine Mammal Active Sonar Test	SeaWiFS	Sea viewing Wide-Field of view Scanner
mg	milligram(s)	SEFSC	Southeast Fisheries Science Center
mi	mile(s)	SEL	sound exposure level
min	minute(s)	SHIPS	Seismic Hazard Investigations in Puget Sound
MIT	Massachusetts Institute of Technology	SIO	Scripps Institute of Oceanography
MMBTU	million British thermal units	SGE	surface geochemical exploration
MMC	Marine Mammal Commission	sonar	sound navigation and ranging
MMPA	Marine Mammal Protection Act	SOSUS	sound surveillance system
MMS	Minerals Management Service	SPL	sound pressure level
MOA	memorandum of agreement	SSI	Scientific Solutions, Inc.
MRFSS	Marine Recreational Fisheries Statistics Survey	SST	sea surface temperature
		STAR	*Stenella* Abundance Research
ms	millisecond(s), thousandths of a second	SURTASS LFA	Surveillance Towed Array Sensor System, Low Frequency Active
MSIS	Marine Safety Information System	SWAMP	Sperm Whale Acoustic Monitoring Program
MT	magneto-telluric		
MWA	military warning areas	SWFSC	Southwest Fisheries Science Center
NAAQS	National Ambient Air Quality Standards	SWSS	Sperm Whale Seismic Survey
		TMA	Target Motion Analysis
NAVOCEANO	Naval Oceanographic Office	TMD	theatre missile defense
NEPA	National Environmental Policy Act	TNT	tri-nitro toluene
NGO	non-governmental organization	TTS	temporary threshold shift
NMEA	National Marine Electronics Association	UK	United Kingdom
		ULCC	ultra large crude carrier
NMFS	National Marine Fisheries Service	USCG	U.S. Coast Guard
nm	nanometer	USDOI	U.S. Department of the Interior
nmi	nautical mile(s)	USEPA	U.S. Environmental Protection Agency
NOAA-F	National Oceanic and Atmospheric Administration Fisheries		
		USFWS	U.S. Fish and Wildlife Service
NOIA	National Oceanic Industries Association	VHF	very high frequency
		VLCC	very large crude carrier
NRC	National Research Council	VLF	very low frequency
NTL	Notice to Lessees and Operators	VSP	vertical seismic profile
NYMEX	New York Mercantile Exchange	WHOI	Woods Hole Oceanographic Institution
OCS	outer continental shelf		
ONR	Office of Naval Research	y	year(s)
OPEC	Organization of Petroleum Exporting Countries	YBP	years before present
OSP	optimum sustainable population		
Pa	Pascal		
PBR	potential biological removal		
PEA	Programmatic Environmental Assessment		
PGS	Petroleum Geo-Services		
PMEL	Pacific Marine Environmental Laboratory		
PSIX	Port Site Information Exchange		
PTS	permanent threshold shift		
PVC	poly vinyl chloride		
re 1 μPa	reference level, at 1 microPascal		
rms	root mean squared		
s or sec	second(s)		
SEAMAP	Southeast Area Management and Assessment Program		

SUMMARY

This Programmatic Environmental Assessment (PEA) prepared by the U.S. Department of the Interior's Minerals Management Service (MMS) evaluates environmental impacts of geological and geophysical (G&G) activities in the Gulf of Mexico (GOM). These activities were the subject of a previous Environmental Impact Statement (EIS) prepared by the U.S. Geological Survey in 1976 and a PEA prepared by the MMS in 1984. A new PEA is needed in light of advances in G&G technology, expansion of activities into deep offshore waters, and improved knowledge of acoustic impacts on marine life. Further, those G&G activities that were previously determined to be categorical exclusions (CATEXs), which were exempt from detailed analysis (based on previous environmental documents) need to be revisited and re-evaluated.

Purpose and Need

The action addressed in this PEA encompasses G&G operations under MMS regulatory authority throughout the GOM outer continental shelf (OCS). Further reference to G&G activities in this document refers to those specific to Gulf of Mexico OCS mineral exploration and subject to MMS regulatory authority. G&G activities aimed at OCS mineral exploration are considered a Federal action under the National Environment Policy Act (NEPA).

For the purpose of this analysis, G&G activities include

- seismic surveys (including high-resolution site surveys and various types of seismic exploration and development surveys);
- deep-tow side-scan sonar surveys;
- electromagnetic surveys;
- geological and geochemical sampling; and
- remote sensing (including gravity and magnetic surveys).

Drilling of deep stratigraphic test holes and use of explosives are excluded from consideration because these activities already require preparation of an Environmental Assessment (EA). G&G studies conducted by other Federal agencies and universities for scientific purposes, although similar or identical to operations conducted under MMS G&G permits, do not require an MMS permit and are not specifically addressed in this EA. However, G&G scientific research related to oil, gas, and other mineral resources requires a permit or filing of a notice to MMS.

G&G surveys provide information used by industry and government to evaluate the potential for offshore oil, gas, and methane hydrate resources and geologic hazards. The oil and gas industry needs accurate data on the location, extent, and properties of hydrocarbon resources, as well as information on shallow geologic hazards and seafloor geotechnical properties, in order to explore, develop, produce, and transport hydrocarbons safely and economically. The MMS also needs this information to fulfill its statutory responsibilities to ensure safe operations, support environmental impact analyses, protect benthic resources through avoidance measures, ensure fair market value for leases, make royalty relief determinations, conserve oil and gas resources, and perform other statutory responsibilities. Operations overseen and permitted by the MMS must comply with various environmental laws (e.g., Endangered Species Act [ESA], Marine Mammal Protection Act [MMPA], Magnuson-Stevens Fishery Conservation and Management Act). Information and analysis developed within this PEA will be used by the MMS to support future ESA Section 7 consultations and incidental take authorizations issued under the MMPA. The MMS also uses high-resolution geophysical data in each of its primary mission areas. MMS regulatory staff uses these data to ensure that the proposed site of bottom-founded structures is safe (i.e., via geohazards review) and that the foundations are properly designed (i.e., based on engineering parameters determined from cores), thus ensuring safe operations. MMS environmental staff uses these data to support mitigation measures and decisions to protect benthic and historic archaeological resources, for protection of the environment. MMS resource evaluation staff uses deep seismic data for resources estimation and bid evaluation to ensure that the government receives a fair market value for tracts offered for lease. MMS production and development staff uses 3D data to map reserves and develop conservation evaluations for conservation of resources.

The MMS is mandated to manage the development of OCS oil, gas, and mineral resources, while also ensuring safe operations and protection of the human, marine, and coastal environments. The purpose of the MMS regulatory program is to ensure that the G&G data needed by industry and government are obtained in a technically safe and environmentally sound manner.

Characterization of G&G Activities

A detailed characterization of G&G activities has been developed, in part to provide a basis for identifying impact agents, an important initial step in impact assessment. Among the G&G activities characterized within this PEA, seismic surveys are the main focus because they historically have covered a large area of the Gulf each year and have the greatest potential for impacts on the environment. Further, there are increasing concerns in the regulatory and scientific communities regarding acoustic impacts on marine life, including marine mammals, turtles, and fishes.

MMS permit records are also summarized to determine recent seismic survey activity levels, with survey data (e.g., line kilometers traversed) organized by water depth (i.e., shallow water, <200 m; deeper water, ≥200 m) and by MMS Planning Area.

Based on the technical information compiled in this PEA pertinent to airgun systems being used in the Gulf of Mexico, the distance from an acoustic source (i.e., airgun array) to a predetermined sound pressure level, or target isopleth (e.g., 180 dB re 1 μPa root mean squared [rms]) was calculated. Applying the theory of spherical spreading (i.e., 20log[R]) and considering the potential transition to modified cylindrical spreading and other factors (e.g., array effect, conversion from zero-to-peak to rms), the estimated distance from the source to the 180 dB re 1 μPa (rms) isopleth for a characteristic 240-dB airgun array has been calculated to be ~300 m (~0.16 nmi) for surface and near-surface waters. Within this "impact zone," sound pressure levels would be ≥180 dB, a level at which some studies suggest that auditory impairment of cetaceans has the potential to occur. Similar calculations indicate that the 160 dB re 1 μPa (rms) isopleth extends ~3,000 m (~1.6 nmi) from a typical G&G seismic source. The methodology and rationale employed in these calculations have been detailed in the PEA and supporting technical appendices, concurrent with known limitations and assumptions. For example, the frequency components of a seismic signal have been characterized. However, the differential attenuation of these frequency components has not been thoroughly described in the literature and, therefore, cannot be readily integrated into the distance calculations. Similarly, the biological significance of seismic noise to individual species or species groups remains a research topic of interest and point of discussion among marine biologists and regulators.

Affected Environment and Sensitive Resources

Of the resources considered in detail in this assessment (i.e., marine mammals, sea turtles, coastal and marine birds, fishes, commercial and recreational fisheries, coastal and marine birds, and benthic communities), protected resources are of primary concern - marine mammals, sea turtles, and fishes. Of particular concern are those marine mammal species whose hearing capabilities (based on vocalization characteristics) fall within the low frequencies introduced into the marine environment by some seismic and other geophysical activities. The PEA provides a comprehensive characterization of those biological resources that may be adversely affected by G&G activities. A review of the Gulf's diverse biological resources shows that several species of marine mammals are deemed to be at greater risk of acoustic impact from seismic surveys – sperm, Bryde's, and beaked whales. Therefore, seismic surveys are described in the most detail. However, all remaining G&G activities are also described.

The three species or species groups of most concern – sperm, Bryde's, and beaked whales – are assumed to have some sensitivity to seismic noise (i.e., low frequency), particularly under those circumstances where these species do not move away from higher level exposure. These two species and single species group possess or exhibit traits that prompt particular concerns related to possible acoustic effects from airguns, whether based on low frequency sensitivity (i.e., frequencies of greatest hearing sensitivity that overlap with maximum airgun output) or deep diving habitats (i.e., potentially placing individuals below an operational array, within the zone of highest seismic sound pressure levels). Aggregations of sperm whales are common along the shelf edge in the vicinity of the Mississippi River delta in surface water depths over the 500 to 2,000 m isobath, and studies are continuing in an attempt to determine the group composition (i.e., adult males and females, calves, and immature individuals) of this species within the GOM. Unlike mysticetes, which may remain close to the surface for long periods, sperm whales spend relatively little time at the surface during the course of feeding activity. Sperm whales dive to several hundred meters and beyond for extended periods of time. Such diving activity takes them down to a depth where they could potentially be passed over directly by an operating seismic vessel without visual detection. As airgun arrays are generally configured to produce a maximum, low frequency energy lobe directly downwards towards the seabed, sperm whales may enter a region of increased ensonification relative to more near-surface species.

Bryde's whale is the most frequently sighted baleen species in the Gulf, though considered uncommon. In general, baleen whales are notably absent from waters of the GOM, with the sole exception being Bryde's whale, with a small population present in the northeastern Gulf (off Florida). Strandings and sightings data suggest that this species may be present throughout the year, generally in the northeastern Gulf in surface waters near the 100-m isobath between the Mississippi River delta and southern Florida. Beaked whales that have been classified as strategic stocks in the GOM include three species in the genus Mesoplodon (i.e., Sowerby's, Blainville's, and Gervais' beaked whales) and one in the genus Ziphius (Cuvier's beaked whale). Generally, beaked whales appear to prefer deep water, with sightings routinely in surface waters between the 700 to 2,000 m isobath. In terms of overall sensitivity to G&G activities, baleen whales (e.g., Bryde's) are potentially a relatively "high risk" category amongst the cetacea. Hearing sensitivity at low frequencies down to ~10 Hertz (Hz) is probably good, since many of the vocalizations of baleen whales occur in the low tens to a few hundred Hertz, which implies functional hearing in this range. Similarly, sperm whales also vocalize as low as 100 Hz, with a similar implication regarding functional hearing capabilities.

Beaked whales, as represented by Blainville's beaked whale in the Gulf, may vocalize at <1,000 Hz. In recent years, beaked whales have shown apparent adverse reaction during the use of military sonar (e.g., strandings coincident with North Atlantic Treaty Organization (NATO) and U.S. Navy fleet exercises). While their status as a strategic stock prompts further concern for any acoustic impacts to beaked whales, their deep-diving habits also put this group at risk, similar to sperm whales. As airgun arrays are configured to concentrate their output energy in the lower frequency range (i.e., 10 to 200 Hz), with lower energy levels in the 200 to 500 Hz and 500 to 1,000 Hz range and beyond, there is clearly a good overlap between the expected frequencies of good hearing sensitivity (low threshold) in these species and maximal airgun output at source. Concern for beaked whales is further amplified due to their cryptic nature and associated difficulties locating and identifying individuals in the field.

Alternatives

The alternatives must support the purpose and need as described above. Selecting any one alternative over another does not preclude the MMS from complying with any existing environmental laws. The following alternatives are evaluated in this PEA:

1 – Continuation of the Status Quo (Proposed Action; equivalent to no action);
2 – Addition of Vessel-based Passive Acoustic Monitoring as a Requirement;
3 – Addition of Both Passive and Active Acoustic Monitoring as Requirements; and
4 – Restrict G&G Seismic Survey Activities.

Alternative 1 – Proposed Action (Continuation of the Status Quo)

The Proposed Action addressed in this PEA is continuance of the MMS permitting process for those G&G operations in the GOM subject to MMS regulatory authority, as described in 30 CFR 251.4 and 280.3, and continuance of allowed G&G activities conducted under a lease described in 30 CFR Part 250 Subpart B. For the purpose of this analysis, Alternative 1 includes G&G activities noted previously (i.e., seismic surveys, deep-tow side-scan sonar surveys, electromagnetic surveys, geological and geochemical sampling, and remote sensing). Under the Proposed Action, there would be no change in MMS regulations governing G&G activities. Existing protective measures specified in permit requirements are included in this alternative, as are current requirements applicable to G&G operations in water depths >200 m throughout the GOM and Federal waters <200 m in the Eastern Planning Area (i.e., adherence to the requirements of Notice to Lessees and Operators [NTL] No. 2004-G01 - visual monitoring, ramp-up, shutdown criteria).

The Proposed Action is essentially continuation of the status quo and therefore also serves as a "no action" alternative. In the context of a PEA, "no action" generally means no change from current management direction or level of management intensity.

The Proposed Action meets the underlying purpose and need. The current suite of G&G activities provides the oil and gas industry with sufficiently accurate data on the location, extent, and properties of hydrocarbon resources, as well as information on shallow geologic hazards and seafloor geotechnical properties, in order to explore, develop, produce, and transport hydrocarbons safely and economically. The MMS also needs this information to fulfill its statutory responsibilities to ensure safe operations, to support environmental impact analyses, to protect benthic resources through avoidance measures, to ensure fair market value for leases, to make royalty relief determinations, to conserve oil and gas resources, and to perform other statutory responsibilities.

Alternative 2 – Addition of Vessel-based Passive Acoustic Monitoring as a Requirement

Under this alternative, the existing suite of G&G activities would continue, but with the implementation of one additional mitigation measure – the required use of passive acoustic monitoring. This alternative includes the same requirements as Alternative 1. This alternative is designed to meet the underlying need for G&G data while reducing environmental impacts from seismic surveys. While protective measures are routinely specified in permit requirements (i.e., part of the Proposed Action) and NTL No. 2004-G01 (i.e., ramp-up, visual monitoring, shutdown criteria for seismic operations in water depths >200 m throughout the GOM and Federal waters <200 m in the Eastern Planning Area), the required implementation of an additional, feasible mitigation measure is included in this alternative. To ensure that cetacean species of concern (i.e., the sperm whale, beaked whales, and Bryde's whale) are not present within a predetermined impact zone, this alternative requires the use of vessel-based passive acoustic monitoring.

The rationale for selection of this mitigation measure is based on 1) improved abilities to detect whale vocalizations via passive acoustic methods, and 2) vocalization characteristics of many cetacean species, particularly the deep-diving toothed whale species of concern (the sperm whale and beaked whales) in the northern Gulf of Mexico, and their current status (i.e., Federally listed as an endangered species and strategic stocks, respectively). Based on available sightings data (e.g., from shipboard and aerial surveys, historical records, opportunistic sightings), sperm whales appear to prefer deeper waters over the GOM continental slope. Available sightings data also suggest that Bryde's whales appear to prefer waters of the continental shelf, with regular sightings in the northeastern Gulf near the 100-m isobath.

This mitigation measure is intended to determine presence and location of vocalizing marine mammals; its efficacy is based on whale vocalization and the ability, in real-time, to determine whether a shutdown of the seismic system is warranted. The addition of this possible mitigation to seismic operations would involve additional costs and potential delays to operators in obtaining seismic data. This alternative could slow OCS exploration and development.

Alternative 3 – Addition of Both Passive and Active Acoustic Monitoring as Requirements

Under Alternative 3, existing G&G activities would continue, but with the required implementation of two additional mitigation measures - passive and active acoustic monitoring. This alternative includes the same requirements as Alternative 1 (i.e., application of NTL No. 2004-G01 in water depths >200 m throughout the GOM and Federal waters <200 m in the Eastern Planning Area). All seismic surveys in water depths >200 m will include ramp-up, visual monitoring, and reporting procedures; however, under this alternative, G&G operators would be required to use passive and active acoustic monitoring.

Both passive and active acoustic monitoring have their advantages and disadvantages, which may influence mitigation effectiveness. Additional restrictions on seismic operations would involve additional costs and delays to operators in obtaining seismic data. This alternative could slow OCS exploration and development in similar fashion to those noted under Alternative 2.

Alternative 4 – Restrict G&G Seismic Survey Activities

Under this alternative, existing G&G activities could continue, but with the implementation of additional restrictions on G&G operations. Three restrictions were initially evaluated: 1) a prohibition of simultaneous surveys by more than one G&G seismic survey vessel in areas frequented by sperm whales and Bryde's whales – designated Alternative 4A; 2) cessation of permitting of all types of seismic surveys in the area(s) most frequented by sperm whales – designated Alternative 4B; and 3) seasonal restrictions on G&G activities in the area(s) most frequented by sperm whales – designated Alternative 4C. Each of these alternatives is intended to meet the underlying need for G&G data while reducing potential environmental impacts from seismic surveys. However, on the basis of the initial evaluation, two of the alternatives relating to operational restrictions were eliminated. Alternative 4B does not meet the underlying purpose and need. Alternative 4C was not considered viable due the absence of seasonality trends for species of concern. Alternative 4A (now termed Alternative 4) was considered fully in the focused impact analysis.

Alternative 4 is designed to meet the underlying need for G&G data while reducing environmental impacts from seismic surveys. Under this alternative, G&G surveying operations would be subject to an operational restriction. Specifically, G&G operators would be precluded from conducting simultaneous seismic operations in those portions of the GOM most frequented by sperm whales and Bryde's whales. The purpose of this measure is to remove the potential for simultaneous exposure to seismic noise from concurrent surveys in the same general area (i.e., within approximately a lease block from one another).

While simultaneous seismic survey restrictions (in areas most frequented by sperm and Bryde's whales) may impose limited hardship on the G&G industry, the underlying need will be met. In practice, industry may already limit simultaneous surveys in close proximity to one another, as seismic data collection is adversely affected by extraneous noise sources.

Environmental Impacts

The objectives of the impact analysis are 1) to determine whether G&G activities have significant impacts on the marine, coastal, or human environments of the GOM; and 2) to identify significant impacts for further NEPA analysis. Following development of a comprehensive summary of G&G activities (e.g., vessels, equipment used, survey or sampling protocols, etc.), the impact agents that have the potential to affect each resource were determined, as outlined in **Table S-1**.

Table S-1
Resources of the Affected Environment or Activities and Impact Agents Associated
with G&G Operations

Resource of the Affected Environment	Impact Agent(s)
Marine Mammals	
Sperm whales	Seismic survey noise
Bryde's whales	Seismic survey noise
Beaked whales	Seismic survey noise
Other cetaceans	Seismic survey noise
Manatees	Coastal vessel traffic
Sea Turtles	Seismic survey noise
Fishes	Seismic survey noise
Commercial and Recreational Fisheries	Seismic survey noise; space-use conflicts with seismic arrays
Coastal and Marine Birds	Coastal vessel and aircraft traffic
Benthic Communities	Seafloor disturbance

A preliminary screening was conducted prior to initiation of the formal impact analysis to focus on those G&G activities and resources with potential for non-negligible impacts. The preliminary screening indicates that most G&G activities have negligible impact or no impact on numerous resources of the GOM, including air quality, marine water quality, recreational and commercial diving, marine transportation, cultural resources, military uses, and geology and sediments. The preliminary screening indicates that seismic surveys have potentially adverse impacts on marine mammals, sea turtles, fishes, and commercial and recreational fisheries. Two other resources with negligible impacts (coastal and marine birds, and benthic communities) are also discussed briefly.

The PEA's environmental analysis evaluated the impacts on each of these six resources expected under each alternative separately, including

- Alternative 1 – Proposed Action;
- Alternative 2 – Addition of Vessel-based Passive Acoustic Monitoring as a Requirement (mitigation);
- Alternative 3 – Addition of Both Passive and Active Acoustic Monitoring as Requirements (mitigation); and
- Alternative 4 – Restrict G&G Seismic Survey Activities (operational restriction).

Impacts from potential accidents and cumulative impacts also were determined for each resource.

To complete the impact analysis, resource-specific significance criteria were developed for each resource of the affected environment noted above. The criteria reflect consideration of both the context and intensity of impact (40 CFR 1508.27). Criteria for marine mammals and sea turtles reflect the Federal protected status of all species occurring in the GOM, with significance tailored after a recent EIS prepared for MMS in the GOM. Adverse impacts are classified into one of three levels:

- significant adverse impact (including those that could be mitigated to a non-significant level);
- adverse but not significant; or
- negligible impact.

Impacts also are categorized as direct or indirect. No beneficial impacts (either significant or insignificant) were considered in developing significance criteria because they are not germane in deciding if an EIS is warranted. Cumulative impacts are discussed in a separate section under each resource.

Additional clarification of the significance criteria is warranted. How MMS defines "negligible impact" in this PEA for NEPA purposes is different from "negligible impact," which National Marine Fisheries Service (NMFS) has defined in 50 CFR 216.103 for MMPA. The NMFS definition is "an impact resulting from the specified activity that cannot be reasonably expected to, and is not reasonably likely to, adversely affect the species or stock through effects on annual rates of recruitment or survival." The NMFS definition of negligible impact under MMPA overlaps the adverse but not significant and negligible impacts as defined in this PEA.

The results of the focused impact analysis are summarized in **Table S-2.** This table identifies, on a resource by resource basis, the potential for impacts under each alternative. According to established significance criteria, no significant adverse impacts to any of the Gulf resources are expected. The potential for adverse but not significant impacts was identified for marine mammals (except the manatee) and commercial and recreational fisheries. The potential for impacts ranges from negligible to potentially adverse but not significant for manatees, benthic communities, coastal and marine birds, sea turtles, and fishes.

Reduced potential impacts are noted for each of the alternatives (Alternatives 2, 3, and 4). The use of passive acoustic monitoring (Alternative 2) offers to reduce further the potential exposure to seismic survey noise for those whales that vocalize. The combined use of passive and active acoustic monitoring (Alternative 3) also provides potential impact reductions for whales; however, there are notable limitations for employing active acoustic monitoring systems. Under Alternatives 2 and 3, potential impacts to fishes, commercial and recreational fisheries, sea turtles, coastal and marine birds, and benthic resources remain unchanged relative to Alternative 1. Restricting concurrent seismic operations (Alternative 4) is intended to preclude the potential for simultaneous exposure. Industry practice may already effectively implement this restriction. Limitations identified for alternatives, including potential cost ramifications, are also noted.

Table S-2

Summary and Comparison of Environmental Impacts from Each Alternative on a Resource by Resource Basis

Resource	Alternative 1 – Proposed Action[1]	Alternative 2 – Addition of Vessel-based Passive Acoustic Monitoring as a Requirement	Alternative 3 – Addition of Both Passive and Active Acoustic Monitoring as Requirements	Alternative 4 – Restrict G&G Seismic Survey Operations
Marine Mammals *Sperm whales*	Potentially ***adverse but not significant*** impact from potential exposure to elevated, repetitive, intermittent, and localized noise levels, resulting in possible hearing impairment No mortality or serious injury (i.e, no exceedance of the Potential Biological Removal [PBR] level); no displacement from key habitat; no long-term or permanent displacement from preferred feeding, breeding, or nursery habitats; no substantial or chronic disruption of behavioral patterns that may adversely affect sperm whales through effects on annual rates of recruitment or survival In water depths ≥200 m throughout the GOM and Federal waters <200 m in the Eastern Planning Area, under requirements of Notice to Lessees (NTL) No 2004-G01, visual monitoring of the 500-m exclusion zone reduces the likelihood that whales will be present in close proximity to an array When coupled with ramp-up, these measures may reduce the potential for hearing impairment or other injury to sperm whales from instantaneous start-up of an airgun array Potential for acoustic impact to sperm whales remains, as undetected individuals may enter the zone of maximum ensonification below an array during a dive Limitations: 1) visual monitoring effective during daylight, good visibility/sightability conditions; 2) ramp-up remains unproven as a mitigation measure, although recognized as a common sense measure; 3) species/group specific limitations (detectability); and 4) minor cost ramifications	Potentially ***adverse but not significant.*** Passive acoustic monitoring relies on passive sensing and location of whale vocalizations Potential for impact may be reduced if whales vocalize and are detected Assumed to be effective for sperm whales, as they frequently vocalize Potential for impact to sperm whales is not completely eliminated, impact level remains potentially ***adverse but not significant*** Limitations: Passive acoustic monitoring 1) only works for vocalizing whales; many animals are quiet much of the time, especially when disturbed; 2) difficult to determine the range to the vocalizing animals when using a towed array; 3) cannot readily determine depth to vocalizing animals; 4) for fixed hydrophones (ship or bottom mounted recorders, sonobuoys, ocean bottom cables), area of coverage/detection range may be limited by noise, requiring more sensors to cover a seismic survey area; 5) requires hydrophone arrays be towed behind the survey vessel (or from an additional chase boat); 6) hydrophone performance may be affected by tow speed and the ship's acoustic characteristics, limiting detection range; and 7) cost ramifications	Potentially ***adverse but not significant.*** Passive acoustic monitoring relies on passive sensing and location of whale vocalizations Active acoustic monitoring relies on an active (e g, sonar) search for whales Potential for impact may be reduced Passive acoustic monitoring may be effective for sperm whales, as they frequently vocalize Potential for impact to sperm whales may be reduced if whales vocalize and are detected Advantages of passive acoustic monitoring relative to active acoustic monitoring include: 1) longer ranges can be achieved; 2) omnidirectional; 3) species can be potentially identified by their vocalization signature; 4) no acoustic footprint that could affect the target animals; and 5) more mature and affordable technology Advantages of active acoustic monitoring compared with passive acoustic are as follows: 1) works with non-vocalizing or cryptic whales and those species that exhibit only limited vocalization; 2) can in some cases determine 3D range and bearing, including depth of vocalizing animals; 3) avoids having to stream behind survey vessel if sound source and hydrophone/receiver are hull-mounted; and 4) may involve less bulky equipment, minimizing personnel required for handling and operation Limitations: Active acoustic monitoring has the following disadvantages: 1) active source may be more harmful than the sound source it is being used to mitigate; 2) limited detection ranges depending on power and frequency; 3) inability to identify species based purely on size; 4) limited beam width and associated problems seeing deep-diving whales at close range; 5) active systems could potentially affect the behavior of the animals themselves; 6) towfish would be required, possibly larger than passive acoustic monitoring array; and 7) current costs for development and deployment are higher	Potentially ***adverse but not significant*** Restrictions on concurrent seismic operations will prevent the potential for simultaneous exposure Acoustic impacts to sperm whales would be slightly reduced; however, the potential for acoustic impacts to sperm whales remains Industry practice may already effectively implement this restriction Limitations: None

Table S-2

Summary and Comparison of Environmental Impacts from Each Alternative on a Resource by Resource Basis
(Continued)

Resource	Alternative 1 – Proposed Action[1]	Alternative 2 – Addition of Vessel-based Passive Acoustic Monitoring as a Requirement	Alternative 3 – Addition of Both Passive and Active Acoustic Monitoring as Requirements	Alternative 4 – Restrict G&G Seismic Survey Operations
Bryde's whales	Potentially *adverse but not significant* impact from potential exposure to injurious noise levels (repetitive, intermittent, and localized); same effects as sperm whales, above (e g , no PBR exceedances, etc) Effective visual monitoring when coupled with ramp-up may reduce the potential for hearing impairment or other injury to Bryde's whales from instantaneous start-up of an airgun array Visual monitoring of a pre-determined impact zone reduces the potential for hearing impairment The potential for acoustic impact is reduced Potential for acoustic impacts to Bryde's whales remains Limitations: Ramp-up unproven as a mitigation measure Visual monitoring effective only during daylight, during periods of good visibility; cost ramifications	Potentially *adverse but not significant.* Same as sperm whales above – potential for impact remains Bryde's whales are known to vocalize, but the frequency of their vocalizations may be problematic; passive acoustic effective only when whales vocalize; geographic and depth limits of this species may further limit mitigation effectiveness Limitations: Same as above.	Potentially *adverse but not significant.* Bryde's whales are known to vocalize; frequency of vocalization may be problematic; passive acoustic effective only when whales vocalize; geographic and depth limits of this species may further limit mitigation effectiveness Active acoustic may be problematic Potential for impact remains Limitations: Same as above	Potentially *adverse but not significant.* Same as sperm whales above – potential for impact remains the same Limitations: Same as above
Beaked whales	Potentially *adverse but not significant* impact from potential exposure to injurious noise levels (repetitive, intermittent, and localized); same effects as sperm whales, above (e g , no PBR exceedances, etc)	Potentially *adverse but not significant.* Beaked whales same as Bryde's whales above – potential for impact remains Limitations: Same as above	Potentially *adverse but not significant.* Beaked whales same as Bryde's whales above – potential for impact remains Limitations: Same as above	Potentially *adverse but not significant.* Same as sperm whales above – potential for impact remains Limitations: Same as above
Other cetaceans	Potentially *adverse but not significant* impact from potential exposure to injurious noise levels (repetitive, intermittent, and localized)	Potentially *adverse but not significant.* Same as sperm whales above – potential for impact emains Limitations: Same as above	Potentially *adverse but not significant.* Same as sperm whales above – potential for impact emains Limitations: Same as above	Potentially *adverse but not significant.* Same as sperm whales above – potential for impact remains Limitations: Same as above
Manatees	Negligible impact due to unlikely exposure	*Negligible.* No effect of mitigation; potential for impact remains the same Limitations: Same as above	*Negligible.* No effect of mitigation; potential for impact remains the same Limitations: Same as above	*Negligible.* No effect of mitigation; potential for impact remains the same Limitations: Same as above
Sea Turtles	Impacts primarily *negligible*, but may elevate to potentially *adverse but not significant* from potential exposure to injurious noise levels (repetitive, intermittent, and localized) and vessel traffic; seismic noise may disturb sea turtles and may produce temporary or permanent hearing impairment in some individuals, but is unlikely to cause death or life-threatening injury Seismic surveys and other G&G activities are not expected to cause long-term or permanent displacement from critical habitat/preferred habitat, nor result in destruction or adverse modification of critical habitat	*Negligible* to potentially *adverse but not significant.* No effect of mitigation due to lack of vocalization; potential for impact remains the same Limitations: Same as above No effect of mitigation due to lack of vocalization	*Negligible* to potentially *adverse but not significant.* No effect of mitigation; potential for impact remains the same Limitations: Same as above	*Negligible* to potentially *adverse but not significant.* No effect of mitigation; potential for impact remains the same Limitations: Same as above

Table S-2

Summary and Comparison of Environmental Impacts from Each Alternative on a Resource by Resource Basis

(Continued)

Resource	Alternative 1 – Proposed Action[1]	Alternative 2 – Addition of Vessel-based Passive Acoustic Monitoring as a Requirement	Alternative 3 – Addition of Both Passive and Active Acoustic Monitoring as Requirements	Alternative 4 – Restrict G&G Seismic Survey Operations
Fishes	*Negligible* to potentially *adverse but not significant* impact from seismic survey noise (repetitive, intermittent, and localized); noise may disturb fish and may produce temporary or permanent hearing impairment in some individuals, but is unlikely to cause death or life-threatening injury Seismic surveys are not expected to cause long-term or permanent displacement of any listed species from critical habitat/preferred habitat, nor to result in destruction or adverse modification of critical habitat or essential fish habitat	*Negligible* to potentially *adverse but not significant.* No effect of mitigation; potential for impact remains the same Limitations: Same as above	*Negligible* to potentially *adverse but not significant.* No effect of mitigation; potential for impact remains the same Limitations: Same as above	*Negligible* to potentially *adverse but not significant.* No effect of mitigation; potential for impact remains the same Limitations: Same as above
Commercial and Recreational Fisheries	Potentially *adverse but not significant* impact from space-use conflicts (between seismic surveys and longline fisheries) and seismic survey noise (short-term and localized decreases in catchability, not to an extent that would be expected to result in economic losses)	Potentially *adverse but not significant.* No effect of mitigation; potential for impact remains the same Limitations: Same as above	Potentially *adverse but not significant.* No effect of mitigation; potential for impact remains the same Limitations: Same as above	Potentially *adverse but not significant.* No effect of mitigation; potential for impact remains the same Limitations: Same as above
Coastal and Marine Birds	*Negligible* impact from seismic surveys (repetitive, intermittent, and localized noise) and aircraft and vessel traffic (noise, disturbance)	*Negligible.* No effect of mitigation; potential for impact remains the same Limitations: Same as above	*Negligible.* No effect of mitigation; potential for impact remains the same Limitations: Same as above	*Negligible.* No effect of mitigation; potential for impact remains the same Limitations: Same as above
Benthic Communities	*Negligible* impact from geological and geochemical sampling, anchors, and bottom cables (placement, retrieval) on soft bottom communities, with *negligible* to potentially *adverse but not significant* impacts to sensitive benthic communities (if unidentified prior to bottom-related activities) No seismic related impacts[2]	*Negligible* (see footnote 2) No effect of mitigation; potential for impact remains the same Limitations: Same as above	*Negligible* (see footnote 2) No effect of mitigation; potential for impact remains the same Limitations: Same as above	*Negligible* (see footnote 2) No effect of mitigation; potential for impact remains the same Limitations: Same as above

[1] In water depths ≥200 m throughout the GOM and Federal waters <200 m in the Eastern Planning Area, includes ramp-up and visual monitoring per NTL No 2004-G01; in water depths <200 m elsewhere, no visual monitoring or ramp-up required

[2] Negligible impacts to sensitive benthic resources are expected from bottom-related (i e , seafloor) activities due to existing protective measures and operational restrictions, coupled with proper identification of known sensitive resources; increased impact levels might be realized if sensitive resources remain unidentified (see **PEA Section III.G - Benthic Communities**)

As evident in **Table S-2**, Alternatives 2 and 3 do not result in a reduction in impact (due to operational restrictions) to most of the resources listed. Impacts to those resources remain unchanged relative to Alternative 1.

Among sea turtles, visual monitoring (under Alternative 1) offers limited mitigation against vessel strikes under those conditions where individual turtles may be sighted. Alternatives 2, 3, and 4 offer no reduction in impact level relative to Alternative 1.

Among the marine mammals, there are expected decreases in impacts to all vocalizing marine mammals (with the exception of manatees) under Alternative 2 (passive acoustic monitoring) and Alternative 3 (passive and active acoustic monitoring), in spite of the limitations noted for each mitigation measure. Alternative 2 offers the greatest potential for reduced impacts to vocalizing species (e.g., sperm whales); impacts to non-vocalizing marine mammals remain unchanged. A combination of existing NTL requirements and passive acoustic monitoring, while not completely eliminating the limitations inherent in each individual measure, is expected to provide the greatest degree of assurance that no marine mammals (with the exception of manatees) have ventured into the exclusion zone of an operational seismic array. Under Alternative 3, some marine mammals (i.e., those that vocalize) may realize benefit from passive acoustic monitoring, and some species may be detectable using active acoustic monitoring techniques. However, there are limitations and potential impacts associated with active acoustic monitoring that may outweigh its potential benefits (e.g., increased anthropogenic noise in the environment, use of sound sources whose sound pressure levels may exceed acceptable exposure levels, etc.). Alternative 4 offers an indeterminate reduction in the potential for impact to sperm and Bryde's whales; however, current industry practice may already address the need to avoid concurrent seismic survey activity.

In all cases where impacts are expected to decrease, there has been no reduction in impact designation level as initially determined under Alternative 1, due in part to the limitations inherent in each mitigation. Only the potential for impact has been reduced as a result of the mitigation measure.

For accidents, all of the Gulf resources evaluated could be affected by a spill caused by an accident involving a G&G vessel. Based on the historical occurrence of vessel accidents (e.g., three incidents reported in the GOM during 1996 and 1997 involving "research vessels," inclusive of G&G vessels), the probability of such incidents occurring is quite low, with the potential for a pollution incident even lower. Records of the volumes of released lubricating oil or diesel fuel in documented G&G vessel accidents has generally been low (i.e., two to five minor releases per year in the GOM). An event involving a survey vessel could result in release of diesel fuel, but such an event has an extremely remote probability of occurring. Thus, incidents involving survey vessels are not expected to result in significant impacts on any of the Gulf resources considered in this analysis. In all cases, impacts are negligible.

A cumulative activity scenario was developed that identified major activities occurring in the GOM. Similarly, the noise environment of the Gulf also was described. Major impact producing factors (i.e., coincident with other similar activities) under the cumulative activity scenario include vessel traffic (i.e., cargo, tanker, military, commercial fishing, recreational boating) and its associated noise and shipstrike potential. Analysis of the cumulative scenario (exclusive of the Proposed Action) produced predicted impact levels, by resource, which ranged from negligible to potentially adverse but not significant (i.e., no significant impacts were evident). The incremental impact of the Proposed Action was then compared to the cumulative scenario impact determinations to predict incremental impacts. In terms of vessel activity levels, seismic survey vessel activity represents a very small component of total vessel activity in Gulf waters. For example, oil and gas support vessels account for approximately one quarter of a million transits per year in Gulf waters, with commercial vessels >10,000 dead weight tons (DWT) contributing another 36,000 trips. By comparison, approximately 20 seismic surveys may occur annually in the Gulf, or <0.01% of the total activity from these three sources (i.e., oil and gas support operations, commercial cargo and tanker activity, seismic surveys). Commercial fishing and recreational boating, military operations, and ocean study activities also contribute to the cumulative vessel activity level, further reducing the relative contribution from seismic surveys. Therefore, G&G operations contribute an extremely minor amount of additional vessel activity in the GOM. Using the predictions of future seismic survey activity levels in the Gulf developed as part of this analysis, seismic vessel activity is not expected to produce a significant incremental increase in vessel activity levels. The cumulative incremental impact attributed to G&G vessel traffic is negligible.

Comprehensive measurements of ambient noise levels in the GOM are lacking. On the basis of analysis of the predominant noise sources identified for the GOM and their relative contributions to total noise levels, the conclusion is that seismic surveys represent a relatively minor, intermittent, and non-stationary component of the overall noise environment. Seismic surveys are typically conducted over multiple blocks. As mobile noise sources, seismic surveys produce repetitive, localized, and short-term increases in ambient noise levels, with the period between potential exposure ranging from hours to days (i.e., time between separate passes of a seismic survey vessel). In the near field, within ~300 m or so of an array, received sound levels may reach or exceed 180 dB re 1 μPa (rms); levels of 160 dB re 1 μPa (rms) may extend to ~3,000 m from an array. At even greater distances, sound from a seismic survey is of a similar nature to other commercial vessel activity. Given the current vessel activity and its associated infrastructure, future seismic survey activity is not expected to produce a significant incremental increase in ambient noise levels. Analysis of cumulative noise impacts on marine mammals, sea turtles, fishes, commercial and recreational fisheries, coastal and marine birds, and benthic communities in the Gulf of Mexico suggests that impacts are either negligible or potentially adverse but not significant. The cumulative incremental impact attributed to G&G vessel noise to these resources is negligible.

This impact analysis represents the final determination of potential impact (or range of potential impacts) associated with G&G operations to various sensitive resources present in OCS waters of the GOM. Summary information has been compiled from the best available data sources, with appropriate qualification of recognized data gaps, limitations, or assumptions applied in the impact analysis. Detailed technical information that supports these impact determinations has been placed in a series of appendices, as appropriate. Limitations or recognized data gaps that have been identified in the analysis include adequacy of life history and ecological data for GOM species, adequacy of survey data, assumptions and interpolations applied to the available data, mitigation effectiveness, and significant data gaps (e.g., recognition and avoidance of seismic noise, seismic pulse frequency characteristics, and the range of marine mammal hearing sensitivities, etc.). Where identified limitations or data deficiencies have been noted, a realistic approach or interpretation of available data was employed in the impact analysis.

I. INTRODUCTION

The Outer Continental Shelf Lands Act, as amended, mandates the Secretary of the Interior through the Minerals Management Service (MMS), to manage the development of outer continental shelf (OCS) oil, gas, and mineral resources, while protecting the human, marine, and coastal environments. To meet these objectives, the MMS will use this document, prepared pursuant to the National Environmental Policy Act (NEPA), as a planning and management tool. This programmatic Environmental Assessment (PEA) evaluates environmental impacts of geological and geophysical (G&G) activities in OCS waters of the Gulf of Mexico (GOM). These activities were the subject of a previous Environmental Impact Statement (EIS) by the U.S. Department of the Interior, Geological Survey (1976) and a PEA by the MMS (Johnson et al., 1984). A new PEA is needed in light of advances in G&G technology, expansion of activities into deep offshore waters, and improved knowledge of acoustic impacts on marine life. This PEA revisits and re-evaluates those G&G activities currently being conducted in the GOM and determines whether significant impacts to Gulf resources are occurring.

A. BACKGROUND

The oil and gas industry conducts several types of G&G surveys in the GOM to obtain data on hydrocarbon resources and geologic hazards (see **PEA Section II.B.1. - Types of G&G Activities** for descriptions of survey types). Most of the data come from seismic surveys, in which one or more airguns are towed behind a ship and acoustic energy pulses are directed into the seafloor. The acoustic signals reflect off subsurface sedimentary layers and are recorded by hydrophones, which are towed on streamers behind the ship or spaced along bottom cables or vertical cables. The data gathered by seismic surveys enable industry to assess more accurately potential hydrocarbon reservoirs and help to optimally locate exploration and development wells, maximizing extraction and production from a reservoir. High-resolution seismic site surveys, along with side-scan sonar surveys, are used to detect geohazards, archaeological resources, and certain types of benthic communities. Other types of G&G activities that provide data on hydrocarbon resources and/or geohazards include electromagnetic surveys, bottom sampling, and several remote sensing methods (e.g., radar imaging, aeromagnetic surveys, gravity surveys, and gravity gradiometry).

The most common G&G activities in the Gulf are seismic surveys using either streamers or bottom cables. State-of-the-art computer mapping systems represent the subsurface in three dimensions (3D) and can enhance various aspects of the data set. These 3D seismic data have enabled industry to identify, with greater precision, where the most economical deep-water prospects are located. The 3D technology also is being used in developed areas on the shallower shelf to identify previously overlooked hydrocarbon-bearing zones in currently producing formations and new productive horizons near or below the currently producing formations. However, because 3D modeling requires much denser data coverage (i.e., closer line spacing) than the older 2D seismic surveys, areas already covered using 2D techniques must be resurveyed. Further, 3D surveys may be repeated over producing fields to characterize production reservoirs; these 4D or time-lapse surveys are becoming more frequent as the technology for analyzing the data is developed.

The number of prelease geophysical permits in the GOM has been consistently high over the last several years (see **PEA Section II.B.2 - Level of Activity** and **PEA Table II-4**). As many as five regional surveys may be conducted at any one time in the Gulf, with more than 20 surveys annually. With the implementation of deep-water royalty relief and the record-breaking number of bids in recent lease sales, geophysical surveying companies have committed to large speculative 2D and 3D geophysical surveys in larger areas, and are moving out into progressively deeper water. At present, less than a thousand lease blocks in the Western and Central Planning Areas have not been covered by 3D seismic surveys. These remaining blocks are likely to be surveyed over the next several years. In addition, bottom cable and vertical cable surveys will be done over older 3D surveys in order to collect additional or new information. Time-lapse (4D) surveys also will be done over existing fields. New technologies, increased levels of activity, denser and repeated coverage, and expansion into new areas (e.g., the Gulf's deep-water environment), coupled with increased environmental concerns, are factors that warrant a reassessment of environmental impacts from G&G activities.

B. PURPOSE AND NEED

Action

The action addressed in this PEA is continuance of the MMS permitting process for those G&G operations in the GOM subject to MMS regulatory authority, as described in 30 CFR 251.4 and 280.3, and continuance of G&G activities conducted under a lease described in 30 CFR Part 250 Subpart B. In other words, the "action" being evaluated is a continuation of G&G activities in the GOM under current MMS permit and lease requirements (i.e., continuation of the status quo). Since initiation of this PEA in 2000, the status quo for G&G activities in the GOM has realized a significant evolution concurrent with increased awareness of government, industry, academia, and special interest groups regarding the effects of noise in the marine environment (e.g., see reviews in National Research Council, 2003). Details of the evolution of the status quo for G&G activities are provided in **Appendix A**; given that this assessment has been based on the development of reasonable alternatives, including analysis of the status quo, it is important to recognize that the alternatives considered also have changed over the course of the past several years. For the purpose of this analysis, G&G activities include seismic surveys (including high-resolution site surveys and various types of seismic exploration and development surveys), deep-tow side-scan sonar surveys, electromagnetic surveys, geological and geochemical sampling, and remote sensing. Drilling of deep stratigraphic test holes and use of explosives are excluded from consideration because these activities already require preparation of an Environmental Assessment (EA). G&G activities aimed at OCS mineral exploration are considered a Federal action under the NEPA.

Need

G&G surveys provide information used by industry and government to evaluate the potential for offshore oil, gas, and methane hydrate resources and geologic hazards. The oil and gas industry needs accurate data on the location, extent, and properties of hydrocarbon resources, as well as information on shallow geologic hazards and seafloor geotechnical properties, in order to explore, develop, produce, and transport hydrocarbons safely and economically. The MMS also needs this information to fulfill its statutory responsibilities to ensure safe operations, support environmental impact analyses, protect benthic resources through avoidance measures, ensure fair market value for leases, make royalty relief determinations, conserve oil and gas resources, and perform other statutory responsibilities. For example, the MMS must comply with various environmental laws such as the Endangered Species Act (ESA), the Marine Mammal Protection Act (MMPA), and the Magnuson-Stevens Fishery Conservation and Management Act. Therefore, this information also will be used by the MMS to support future ESA Section 7 consultations and incidental take authorizations issued under the MMPA.

Agency Purpose

The MMS is mandated to manage the development of OCS oil, gas, and mineral resources, while also ensuring safe operations and protection of the human, marine, and coastal environments. The purpose of the MMS regulatory program is to ensure that the G&G data needed by industry and government are obtained in a technically safe and environmentally sound manner. MMS regulations at 30 CFR 251 mandate that G&G activities may not interfere with or endanger operations under any lease or right-of-way, or permit issued, nor may they cause harm or damage to aquatic life, property, or to the marine, coastal, or human environments.

C. REGULATORY FRAMEWORK

The MMS is responsible for overseeing all activities associated with mineral resource development on the Federal OCS, including assessment, leasing, exploration, development, production, and royalty management. G&G activities aimed at mineral exploration are subject to a complex series of permits and notices. The MMS Resource Evaluation Program oversees G&G data acquisition and permitting activities, pursuant to regulations in 30 CFR (Dellagiarino et al., 1997, 1998, 2000). Specifically, these include 1) Part 251 regulating prelease G&G exploratory operations for oil, gas, and sulfur resources; and 2) Part 280 regulating prelease prospecting activities (Fulton, 1998). Postlease G&G activities are governed by 30 CFR 250 regulations and by applicable Notices to Lessees and Operators (NTLs). Other regulations also pertain to one or more of the issues considered in this analysis (e.g., the President's Council on Environmental Quality [CEQ] guidelines for implementation of NEPA; pertinent regulations administered by the National Marine Fisheries Service[1] [NMFS]). Applicable NMFS regulations for protection of marine mammals have been evaluated and are presented in **Appendix B (Section I - Review of Pertinent Regulations)**. MMS formally consulted with NOAA-F on G&G activities in 2002 as part of the consultations for Lease Sale 184 and for Lease Sales 185, 187, 190, 192, 194, 196, 198, 200, and 201 (i.e., the Multi-Sale EIS consultation). NOAA-F incorporated G&G issues into the Biological Opinions for both of these consultations, and MMS has implemented required and recommended mitigation and conservation measures through lease stipulations and NTLs. Subsequent consultations may be required if any of the four reinitiation "triggers" at 50 CFR 402.16 occur. MMS also has petitioned NOAA-F for rulemaking under the MMPA to authorize "small takes" incidental to seismic surveys conducted in the GOM.

D. ENVIRONMENTAL ANALYSES

On the basis of the 1976 U.S. Geological Survey EIS and the 1984 EA prepared by the MMS, it was determined that the vast majority of G&G activities were categorical exclusions (CATEXs) (exceptions noted below). A CATEX is "a category of actions which do not individually or cumulatively have a significant effect on the environment... and for which, therefore, neither an environmental assessment or environmental impact statement is required" (40 CFR 1508.4). The only exceptions were drilling of deep stratigraphic test holes and use of explosives (Appendix 10 of 516 Department Manual 6), both of which require an EA. Explosives were used prior to 1989 but have since been replaced by airgun sources, which generate superior acoustic signals and do not cause the damaging environmental impacts of explosives[2]. It is assumed that no explosives would be used in future seismic surveys permitted by MMS on the OCS. Following the environmental reviews in 1976 and 1984, G&G-related CATEXs were formally adopted as department policy by the U.S. Department of the Interior (USDOI). Concurrence on this determination also was provided by the CEQ. The MMS has conducted an internal environmental review of current G&G technology and evolving environmental concerns, such as underwater acoustic effects on marine organisms. Given the new technology and emerging research, the MMS concluded that G&G activities now listed as CATEXs should be re-evaluated under a PEA to determine whether new technology or environmental information has altered the 1984 determinations.

---

[1] The NMFS was formally renamed National Oceanic and Atmospheric Administration Fisheries (NOAA-F) in 2002, although the NMFS moniker continues to be used by the agency; NMFS and NOAA-F are used interchangeably in the PEA.

[2] Although explosives have been replaced by airguns as a sound source for OCS seismic surveys, they still may be used for seismic surveys in State waters.

The following environmental impact analysis represents a final determination of potential impact (or range of potential impacts) associated with G&G operations to various sensitive resources present in OCS waters of the GOM. Summary information has been compiled from the best available data sources, with appropriate qualification of recognized data gaps, limitations, or assumptions applied in the impact analysis. Detailed technical information that supports these impact determinations has been placed in a series of appendices, as appropriate, and cross-referenced. While this allows the reader to concentrate on the immediate issue of impact assessment without inclusion of an extensive technical discussion, it also allows for a more detailed, in-depth presentation of pertinent technical issues that support the impact determination.

This PEA of G&G operations in the GOM has been prepared pursuant to NEPA to ensure that environmental information is available to decision-makers and the public, and that decisions are based on an understanding of the environmental consequences of the decision at hand. On the basis of this PEA, it has been concluded that G&G activities and the sounds they produce (detailed in **Appendices C** and **D**) do not cause significant environmental impacts. Therefore, G&G operations will be listed by MMS as a category of activities that are generally excluded from the NEPA requirements to prepare an EA or EIS. However, there may be specific G&G proposals that are exceptions to this CATEX. Therefore, MMS will review each subsequent G&G permit request to determine if the proposed activity is categorically excluded or is an exception that requires preparation of an EA. This categorical exclusion review (CER) will apply criteria specified in the Departmental Manual [516 DM 2.3.A(3)] to determine whether or not an EA is required. Permit-specific EAs will summarize, reference, and tier to this PEA, as appropriate. If an EA concludes that significant impacts will occur, then an EIS will be prepared, unless the permit applicant withdraws the application or modifies the application so that significant impacts will not occur. MMS routinely imposes certain environmental mitigation measures on G&G permit approvals to avoid or reduce environmental impacts, including permits for actions that are categorically excluded and do not require preparation of an EA. On a case-by-case basis, MMS may impose additional environmental mitigation measures on future permit applications to further avoid or reduce environmental impacts. Many of these mitigation measures are described in this PEA (see **Appendix E, Section II - Mitigation Measures and Operational Restrictions**). In the future, these mitigation measures may be revised, and others developed, and can be imposed by the MMS without a need for a new PEA.

E.   OBJECTIVES AND SCOPE

The objectives of this PEA are

- to characterize G&G activities being conducted on the GOM OCS;
- to identify any significant impacts from such activities that would require further NEPA analysis; and
- to evaluate feasible mitigation measures, including the identification of any protective measures currently found within existing regulations and lease stipulations.

The PEA encompasses existing and near-future G&G operations in Federal waters of the GOM (**Figure I-1**). Data from 1988 to present (or most recent available) were used to estimate annual levels of activity. Economic, geographic, and technological factors were considered in estimating future activity levels (see **PEA Section II.B.2 - Level of Activity**).

In accordance with CEQ guidance, the PEA focuses on those activities and resources for which the potential for significant impacts exists. The main environmental issue is the impact of noise from seismic operations on marine organisms (especially marine mammals, turtles, and fishes). Other issues were identified by systematically considering the impact agents and potentially affected resources for each type of G&G activity. A preliminary screening is conducted to identify activities having little or no environmental impact, which are not analyzed further (see **PEA Section III.A.2 - Preliminary Screenings**). The balance of **PEA Sections III.B** through **III.H** discuss resource - and activity-specific impacts that remained following preliminary screening.

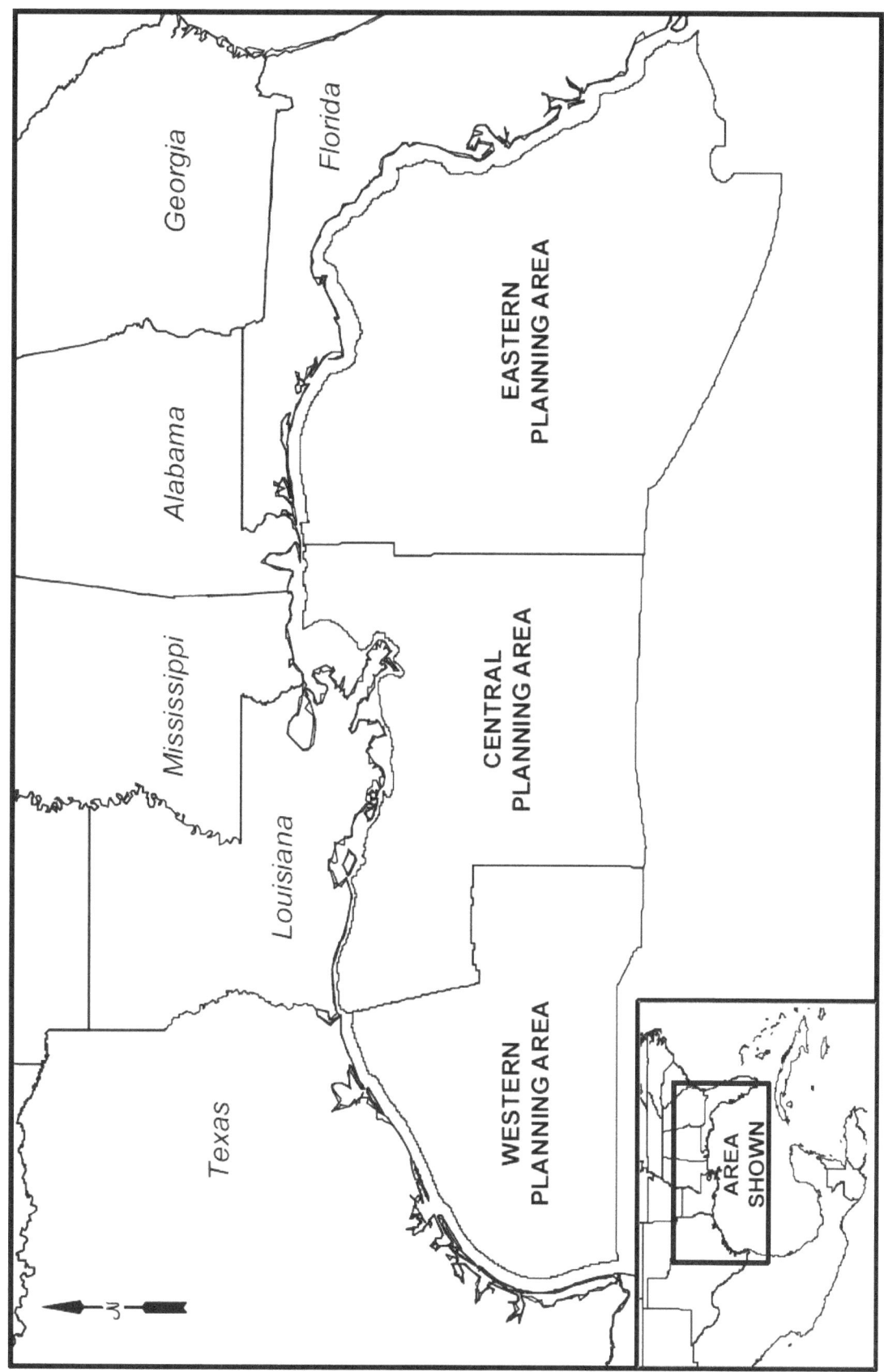

Figure I-1. Gulf of Mexico Outer Continental Shelf Planning Areas.

II. ALTERNATIVES INCLUDING THE PROPOSED ACTION

A. RANGE OF ALTERNATIVES

The alternatives must support the purpose and need as outlined previously in **PEA Section I**. Selecting any one alternative over another does not preclude the MMS from complying with any existing environmental laws. The following alternatives are evaluated in this PEA:

1 – Continuation of the Status Quo (Proposed Action; equivalent to no action);
2 – Addition of Vessel-based Passive Acoustic Monitoring as a Requirement;
3 – Addition of Both Passive and Active Acoustic Monitoring as Requirements; and
4 – Restrict G&G Seismic Survey Activities.

The alternatives represent different ways of addressing the purpose and need as described previously in **PEA Section I**. Other alternatives (i.e., slim-hole drilling, passive seismic) were evaluated and rejected because they failed to address purpose and need adequately, as detailed in **Appendix E (Section I - Alternatives Considered but not Analyzed)**. A brief overview of the alternatives is given below, followed by detailed individual descriptions.

1. Continuation of the Status Quo (Proposed Action, Alternative 1)

The Proposed Action addressed in this PEA is continuance of the MMS permitting process for those G&G operations in the GOM under MMS regulatory authority, as described in 30 CFR 250.201, 30 CFR 251.4, and 30 CFR 280.3. For the purpose of this analysis, G&G activities include seismic surveys (including high-resolution site surveys and various types of seismic exploration and development surveys), deep-tow side-scan sonar surveys, electromagnetic surveys, geological and geochemical sampling, and remote sensing, as detailed in **Appendix D**. *(Note: Drilling of deep stratigraphic test holes and use of explosives are excluded from consideration because these activities already require preparation of an EA.)* Under the Proposed Action, there would be no change in MMS regulations governing G&G activities. Existing protective measures specified in permit requirements are included in this alternative, as are current requirements applicable to G&G operations in water depths >200 m in the GOM and all OCS waters of the Eastern Planning Area (i.e., adherence to the requirements of NTL No. 2004-G01; see **Table II-1**). Mitigation measures commonly applied to G&G permits are described in **Appendix E (Section II - Mitigation Measures and Operational Restrictions)**.

The Proposed Action is essentially continuation of the status quo and therefore also serves as a "no action" alternative. In the context of a PEA, "no action" generally means no change from current management direction or level of management intensity (CEQ, "Forty Most Asked Questions Concerning CEQ's NEPA Regulations" [46 FR 18026, March 23, 1981]).

2. Addition of Vessel-based Passive Acoustic Monitoring as a Requirement (Alternative 2)

Under this alternative, the existing suite of G&G activities would continue, but with the implementation of one additional mitigation measure – the required use of passive acoustic monitoring. This alternative includes the same requirements as Alternative 1. This alternative is designed to meet the underlying need for G&G data while reducing environmental impacts from seismic surveys. While protective measures are routinely specified in permit requirements (i.e., part of the Proposed Action) and NTL No. 2004-G01 (i.e., ramp-up, visual monitoring, shut-down criteria for seismic operations in water depths >200 m and all OCS waters of the Eastern Planning Area), the required implementation of an additional, feasible mitigation measure is included in this alternative. To ensure that cetacean species of concern (i.e., the sperm whale, beaked whales, and Bryde's whale) are not present within a predetermined

Table II-1
Summary of Current Requirements for Seismic Activities Conducted Under Notice to Lessees
and Operators (NTL) No. 2004-G01 in the U.S. Gulf of Mexico[1]

NTL No. 2004-G01 requires geophysical and geological operators conducting seismic operations in all Federal outer continental shelf (OCS) waters >200 m deep in the Gulf of Mexico, and all Federal waters <200 m deep in the Eastern Planning Area of the Gulf of Mexico, to:

- employ ramp-up;
- utilize trained protected species observers; and
- complete Minerals Management Service reporting requirements.

Ramp-up can be initiated:

- only during periods of sufficient visibility when visual observers are able to scan and clear an area (i.e., exclusion zone) 500 m around the seismic array.

Trained protected species observers (visual observers) must:

- clear the exclusion zone at and below the sea surface within a radius of 500 m surrounding the center of an airgun array and the area within the immediate vicinity of the survey vessel;
- fail to observe marine mammals or sea turtles within (or approaching) the exclusion zone for a period of 30 minutes, after which ramp-up operations may begin; and
- continue to monitor the exclusion zone visually (once ramp-up has been completed and the seismic array is operating at full power) until seismic operations cease or sighting conditions do not allow observation of the sea surface (e.g., fog, rain, darkness).

If a whale is sighted either within the exclusion zone or moving towards the exclusion zone, the array must be shut down until the area can be cleared. The seismic array may be powered down to a minimum level of 160 dB re 1 μPa (rms) without reinitiating ramp-up.

This NTL also details the *optional* use of passive acoustic monitoring (for vocalizing marine mammals) during periods of poor visibility (e.g., at night, during periods of fog or limited visibility).

These NTL requirements are applicable to all surface seismic operations; requirements are slightly different for shorter term, site-specific bore hole seismic surveys.

Performance of these mitigation measures will be a condition of approval of applications for geophysical permits, and will be applicable to geophysical activities conducted under lease terms for all seismic survey operations proposed within all Gulf of Mexico OCS waters >200 m deep and all OCS waters <200 m deep in the Eastern Planning Area.

---

[1] See **Appendix A**, pp. A-10 through A-18, for the NTL text in its entirety.

impact zone, this alternative requires the use of vessel-based passive acoustic monitoring. This mitigation measure has been evaluated in **PEA Section III (Environmental Impacts)**.

The rationale for selection of this mitigation measure is based on the 1) improved abilities to detect whale vocalizations via passive acoustic methods, and 2) vocalization characteristics of many cetacean species, particularly the deep-diving toothed whale species of concern (the sperm whale and beaked whales) in the northern GOM, and their current status (i.e., Federally listed as an endangered species and strategic stocks, respectively). According to available sightings data (e.g., from shipboard and aerial surveys, historical records, opportunistic sightings), sperm whales appear to prefer deeper waters over the GOM continental slope. Available sightings data (e.g., from shipboard and aerial surveys, historical records, opportunistic sightings) suggest that Bryde's whales appear to prefer waters of the continental shelf, with regular sightings in the northeastern Gulf near the 100-m isobath. Species account information has been provided in **Appendix F (Section I.A - Biological Environment, Marine Mammals)**.

3.   Addition of Both Passive and Active Acoustic Monitoring as Requirements (Alternative 3)

Under Alternative 3, existing G&G activities would continue, but with the required implementation of two additional mitigation measures - passive and active acoustic monitoring. This alternative includes the same requirements as Alternative 1 (i.e., application of NTL No. 2004-G01 in water depths >200 m throughout the GOM and all OCS waters of the Eastern Planning Area). All seismic surveys in these areas will include ramp-up, visual monitoring, and reporting procedures; however, under this alternative, G&G operators would be required to use passive and active acoustic monitoring.

4.   Restrict G&G Seismic Survey Activities (Alternative 4)

Under this alternative, the existing suite of G&G activities would continue but with the implementation of additional restrictions on G&G seismic operations. This alternative is designed to meet the underlying need for G&G data while reducing environmental impacts from seismic surveys. Under this alternative, G&G surveying operations would be subject to an operational restriction. Specifically, G&G operators would be precluded from conducting simultaneous seismic operations in those portions of the GOM most frequented by sperm whales and Bryde's whales. The purpose of this measure is to remove the potential for simultaneous exposure to seismic noise from concurrent surveys in the same general area. Consideration also was given to 1) cessation of permitting for all types of seismic surveys in areas most frequented by sperm whales, and 2) restricting all types of seismic surveys in those portions of the GOM most frequented by sperm whales on a seasonal basis (i.e., temporal exclusion). However, these alternatives were removed from further consideration, as detailed in **Appendix E (Section II.B.1 - Geographic and Seasonal Restrictions)**.

B.   ALTERNATIVE 1: PROPOSED ACTION

The Proposed Action, in this case representing the "no action" alternative and continuation of the status quo, consists of the various G&G activities currently occurring and estimated to occur in the foreseeable future (i.e., over the next several decades) in the Western, Central, and Eastern Planning Areas of the GOM. This includes both prelease and postlease activities. Drilling of deep stratigraphic test holes and use of explosives for seismic exploration are excluded from consideration because these activities already require preparation of a PEA.

The following sections briefly describe each type of G&G activity and estimated levels of activity on an annual basis. Detailed descriptions of operations and equipment are provided in **Appendix D**.

1.   Types of G&G Activities

Table II-2 summarizes the type of G&G activities occurring in the GOM. These include

- seismic surveys (includes high-resolution site surveys and various types of seismic exploration and development surveys);
- deep-tow side-scan sonar surveys;
- electromagnetic surveys;
- geological and geochemical sampling; and
- remote sensing (including gravity and magnetic surveys).

Seismic surveys are the main focus of this PEA because they historically have covered a large area of the Gulf each year and have the greatest potential for "significant" impacts on the environment. Further, there are increasing concerns in the regulatory and scientific communities regarding acoustic impacts on marine life including marine mammals, turtles, and fishes (see **PEA Section III**). Therefore, seismic surveys are described in the most detail. Other techniques are described briefly, with additional details on all techniques provided in **Appendix D**.[1]

a.   *Seismic Surveys*

Table II-3 summarizes seismic survey operations occurring in the GOM[2]. Two general types of seismic surveys are conducted. *High-resolution site surveys* collect data from near the seafloor surface to several kilometers or more below the seafloor for various reasons – to investigate geohazards and soil conditions, to identify potential benthic biological communities (or habitats) and archaeological resources, or for exploration purposes. Such high resolution data may be used for initial site evaluation for drilling rig emplacement and for platform or pipeline design and emplacement. High-resolution site survey data obtained at greater depths below the seafloor are used for exploration purposes. *Seismic exploration and development surveys* are conducted to obtain data on geological formations from the sediment near-surface to several thousand meters deep (below the sediment surface). This information enables industry accurately to assess potential hydrocarbon reservoirs and helps to optimally locate exploration and development wells, maximizing extraction and production from a reservoir. A survey vessel tows an array (or dual arrays) of high-pressure "airguns," which serve as the sound source. In conventional 2D and 3D seismic surveys, the receivers (hydrophones) are on streamer cables towed behind the survey vessel at a depth of 5 to 10 m below the surface of the water. Individual streamers can be up to 12 km in length or, on rare occasions, even longer. Tail buoys with radar reflectors and strobe lights are used to mark the end of the streamers and make them more visible and/or detectable to other vessels. Alternatively, receivers may be deployed on the seafloor (bottom cable surveys), on vertical cables suspended from the sea surface (vertical cable surveys), or on vertical cables suspended within well bores (vertical seismic profile (VSP) surveys). When conventional 3D ship surveys or bottom cable surveys are repeated at the same location over time, these are referred to as 4D or time-lapse surveys.

Large vessels in the 60- to 90-m class are typical of those used to conduct 2D and 3D seismic exploration and development surveys. Smaller vessels (37 to 47 m) are used for high-resolution site

---

[1] Hydrocarbon "sniffers" have historically been used in the Gulf of Mexico; however, this technique is nearly obsolete. Further discussion is provided in **Appendix D, Section I.D.3 – Hydrocarbon "Sniffers."**

[2] Seismic sources such as sparkers, unibooms, and mini-sleeve exploders are not discussed, as these sources are either not currently in use or their use in the Gulf of Mexico is extremely limited. Hence, this PEA does not cover environmental impacts that would occur if these types of sources were employed in the Gulf of Mexico.

Table II-2
Typical Characteristics of G&G Activities in the Gulf of Mexico

Activity	Prelease or Postlease	Survey Scale	Purpose	Description	Survey Platform	High-Energy Sound Sources	Seafloor Activities
Seismic Surveys							
• High-resolution site surveys	Postlease	Lease block	Locate shallow hazards, obtain engineering data for placement of structures; also used for exploration	Airgun(s) and streamer cable towed back and forth across lease block	Ship	Single or multiple airguns	None
• Seismic exploration surveys	Mainly prelease (speculation)	Multiple lease blocks	Delineate and monitor potential oil and gas reservoirs	Airgun array(s) towed across survey area; location of receivers varies (see below):	(see below)	Multiple airguns (single or dual source array, each consisting of three, 6-airgun subarrays)	(see below)
• Streamer surveys (2D, 3D)				Receivers on streamer cables	Ship	2D: single array; 3D: dual array	None
• Bottom cable surveys				Receivers on bottom cables	Multiple ships	Multiple airguns (dual array)	Cables placed temporarily on seafloor[1]
• Vertical cable surveys	Postlease	Lease block		Receivers on vertical cables	Two ships	Multiple airguns (dual array)	Vertical cables temporarily anchored to seafloor

Table II-2
Typical Characteristics of G&G Activities in the Gulf of Mexico
(Continued)

Activity	Prelease or Postlease	Survey Scale	Purpose	Description	Survey Platform	High-Energy Sound Sources	Seafloor Activities
• Vertical seismic profile surveys	Postlease	Lease block	Correlate geologic data to seismic data	Receivers on vertical cables lowered into a borehole	Crane-suspended from rig; work boat	Single airgun (crane); four to eight airguns, towed array (work boat)	Receivers inserted into borehole; minimal seafloor activity
Deep-Tow Side-Scan Sonar Surveys	Postlease	Lease block	Locate shallow hazards, cultural resources, hard bottom areas	Side-scan sonar "fish" towed about 25 to 30 m above bottom	Ship	None	Chain dragged on seafloor (helps maintain "fish" height above bottom)
Electromagnetic Surveys							
• Magneto-telluric surveys	Prelease	Multiple lease blocks	Delineate potential oil and gas reservoirs	Receivers placed on seafloor to detect natural electrical and magnetic fields	Ship	None	Receivers placed temporarily on seafloor
• Bi-pole surveys	Prelease	Multiple lease blocks	Delineate potential oil and gas reservoirs	Receivers detect electrical signals sent into seafloor	Ship	None	Receivers placed temporarily on seafloor
Geological/Geochemical Sampling							
• Bottom sampling	Mainly postlease	Lease block	Obtain physical and chemical data on surface sediments	Sediment samples collected using a gravity or piston corer, grab, dredge	Ship	None	Removal of bottom sample (few centimeters to several meters)
• Shallow coring	Mainly postlease	Lease block	Obtain physical and chemical data on surface sediments	Conventional rotary drilling from a boat or drilling barge	Ship	None	Removal of bottom sample (several meters)

Table II-2
Typical Characteristics of G&G Activities in the Gulf of Mexico
(Continued)

Activity	Prelease or Postlease	Survey Scale	Purpose	Description	Survey Platform	High-Energy Sound Sources	Seafloor Activities
Remote Sensing							
• Radar imaging	Prelease	Regional/ multiple lease blocks	Detect oil slicks on sea surface	Radar detects dampening of ocean surface capillary waves	Satellite	None	None
• Aeromagnetic surveys	Prelease	Regional/ multiple lease blocks	Delineate potential oil and gas reservoirs	Magnetometer measures earth's magnetic field and/or its vertical gradient	Aircraft (fixed wing)	None	None
• Gravity surveys	Prelease	Multiple lease blocks	Delineate potential oil and gas reservoirs	Gravity meter measures earth's gravitational field	Ship (or rarely, helicopter)	None	None
• Gravity gradiometry	Prelease	Multiple lease blocks	Delineate potential oil and gas reservoirs	Instrument measures earth's gravity gradient	Ship	None	None
• Marine magnetic surveys	Mainly postlease	Regional/ multiple lease blocks	Locate buried pipelines and items of archaeological interest; also used for exploration	Magnetometer measures the earth's magnetic field and/or its vertical gradient	Ship	None	None

[1] Ocean bottom cables may remain on the seafloor for several days. In som cases, cables may be left on the seafloor for several weeks as part of time-lapse (4D) survey efforts.

Table II-3
Typical Characteristics of Seismic Surveys in the Gulf of Mexico

Survey Type	Ship(s)	Ship Speed	Airgun(s)	Source Level (zero-to-peak)[1]	Firing Rate[2]	Streamers	Grid/Line Spacing	Time to Complete
High-resolution site surveys (2D)	Single ship, 37 m	3 to 3.5 kn	Single airgun	229 to 233 dB re 1 μPa at 1 m	7 to 8 s (12.5 m)	Single 600-m streamer towed about 700 m behind ship	300 x 900 m grid size; 129 linear km per block	2 days (lease block)
High-resolution site surveys (3D)	Single ship, 47 m	3 to 3.5 kn	Tri-cluster of airguns	233 dB re 1 μPa at 1 m	7 to 8 s (12.5 m)	Up to six streamers, 100 to 200 m long	66 lines per lease block[3]	5 days (lease block)
Seismic exploration surveys (2D)	Single ship, 60 to 90 m	4.5 kn	Single source array of three 6-gun subarrays	233 to 240 dB re 1 μPa at 1 m	16 s (37.5 m)	Single streamer, 8 to 12 km long	Lines 100-166 km long, about 2 km apart	Days, weeks, or months depending on survey area
Seismic exploration surveys (3D)	Single ship, 80 to 90 m	4.5 kn	Dual source arrays, each consisting of three 6-gun subarrays	233 to 240 dB re 1 μPa at 1 m	16 s (37.5 m); two sources, alternate firings	6 to 12 streamers, 3 to 8 km long and spread out over 600 to 1,500 m	Grid size 24 x 48 m; lines 100-166 km long, about 1 km apart	Days, weeks, or months depending on survey area
Ocean bottom cable surveys	Several smaller ships[4]	4.5 kn (sometimes 6 kn)	Same as for 3D seismic (dual source)	233 to 240 dB re 1 μPa at 1 m	10 s (25 m); two sources, alternate firings	N/A	Parallel lines of cables 50 m apart	Days, weeks, or months depending on survey area

Table II-3
Typical Characteristics of Seismic Surveys in the Gulf of Mexico
(Continued)

Survey Type	Ship(s)[5]	Ship Speed	Airgun(s)	Source Level (zero-to-peak)[1]	Firing Rate[2]	Streamers	Grid/Line Spacing	Time to Complete
Vertical cable surveys	Two ships[5]	4.5 kn	Same as for 3D seismic (dual source)	233 to 240 dB re 1 µPa at 1 m	10 s (25 m); two sources, alternate firings	N/A	Operational area 14 x 20 km; grid size 50 x 80 m	Hours or days
Vertical seismic profile surveys	Single ship, 47 m	3 to 3.5 kn	Tri-cluster of airguns	233 dB re 1 µPa at 1 m	For zero offset surveys: no boats, intermittent firing rates of 4 to 8 times over 20 s followed by quiet times of 10 min, hours to complete, 180-dB airgun. For walk-away surveys: 4 to 8 airguns 180 to 230 dB, repeat walkway for every cable repositioning in borehole, takes hours, use boats. For 3D Surveys: same as 3D seismic survey, but must repeat for every cable lowering, takes days.	N/A	Variable	Days

[1] rms (i.e., root mean square) source levels are approximately 10 dB lower than those listed as zero-to-peak.

[2] Distance traveled, in parentheses, is a function of ship speed and firing rate.

[3] High resolution site surveys conducted on lease blocks identified as having a high probability for the presence of historic archaeological resources (e.g., shipwrecks) are required to use 50-m grid spacing, per NTL No. 98-06.

[4] Vessel requirements: two for cable layout/pickup, one for recording, one for shooting, two for utility.

[5] Both vessels deploy cables, then one becomes the source boat and the other redeploys cables.

surveys. Further detail on survey vessels is provided in **Appendix D** (**Section IV.E - Ships, High Resolution Operations**). The vessels follow precise, pre-plotted lines so that the desired coverage of the seafloor is achieved. With seismic gear deployed, the operating vessel speed is typically in the 3.0 to 3.5 kn range (5.6 to 6.5 km/h) for high-resolution site surveys and 4.5 kn (8.3 km/h) for 2D or 3D seismic exploration and development surveys. An integrated navigational system keeps track of where the airguns are fired, as well as the position and depth of the streamer cables. Field operations are usually conducted 24 h a day.

Details on seismic sources are presented in **Appendix D** (**Section II - Seismic Sources**). A typical marine seismic source is a sleeve-type airgun array, which releases compressed air into the water, creating an acoustical energy pulse that penetrates the seafloor. The airguns are towed 5 to 7 m below the sea surface and release compressed air at intervals of several seconds, creating a regular series of strong acoustic impulses separated by silent periods lasting 10 to 20 s, depending on survey type and depth to the target formations. Individual airguns are suspended in the water from a float system referred to as a subarray. Each subarray contains 6 or 7 individual airguns spaced from 2.5 to 3 m apart, making the total subarray 12.5 to 18 m long. Typically three (sometimes four) subarrays are combined to form an array. When three subarrays are used, the spacing is 8 m between subarrays, and when four are used, the spacing is 12 m. Thus, the overall width of the airgun array is generally 16 to 36 m.

1)   *High-Resolution Site Surveys*

High-resolution site surveys are conducted to investigate the shallow subsurface for geohazards and soil conditions, as well as to identify potential benthic biological communities (or habitats) and archaeological resources in support of review and mitigation measures for OCS exploration and development plans. Information also can be recovered at much greater depths, so that some surveys are used for exploration purposes. A typical operation consists of a ship towing an airgun about 25 m behind the ship and a 600-m streamer cable with a tail buoy. The ship travels at 3 to 3.5 kn (5.6 to 6.5 km/h), and the airgun is fired every 7 to 8 s (or about every 12.5 m). Typical surveys cover one lease block, which is 4.8 km on a side. MMS regulations require information be gathered on a 300- by 900-m grid, which amounts to about 129 line km of data per lease block. If the MMS has identified a block as having a high probability for the presence of historic archaeological resources (i.e., shipwrecks), grid points must be on a 50-m spacing (i.e., pursuant to NTL No. 2001-G01). Including line turns, the time to survey one block is about 36 h; however, streamer and airgun deployment and other operations add to the total survey time.

Recently, 3D high resolution surveys using ships towing multiple streamer cables have become available. Since multiple streamers are towed, the ships tend to be slightly larger (47 m vs. 37 m). Up to six streamers 100 to 200 m long are used with a tri-cluster of airguns. With this system, 66 sail lines are necessary per block, which takes about 5 days to collect.

2)   *2D Seismic Exploration Surveys*

2D seismic exploration surveys are conducted in the GOM by geophysical contractors. When geophysical contractors license data to multiple clients that they acquire on a speculative basis, the surveys are known within the industry as "spec" surveys. When geophysical contractors sell the rights to proprietary data that they shoot for the exclusive use of a single client, the surveys are known as proprietary surveys. "Spec" surveys are conducted over large multi-block areas, whereas proprietary surveys cover only a few blocks. Although the number of 2D surveys is small compared with 3D surveys, they are important as the survey can cover a larger area in less detail, resulting in a lower cost per area covered.

The ships conducting these surveys are generally 60 to 90 m long and tow a single source array 100 to 200 m behind the ship (see **Appendix D**, **Section I.A.2 - Description of Activities, 2D Surveys**). The source array typically consists of three subarrays of six or seven airguns each, and it is about 12.5 to 18 m long and 16 to 36 m wide. Following behind the source array another 100 to 200 m is a single streamer on the order of 8 to 12 km long. The ship tows this apparatus at a speed of about 4.5 kn (8.3 km/h). About every 16 s (i.e., a distance of 37.5 m for a vessel travelling at 4.5 kn), the airgun array is fired (the actual time between firings varies depending on ship speed).

To complete a survey, the ship sails down a track from 12 to 20 h (100 to 166 km), depending upon the size of the survey area. Reaching the end of the track, the ship takes 2 to 3 h to turn around and start down another track. The spacing between tracks is usually on the order of 2 km. This procedure takes place day and night and may continue for days, weeks, or months depending upon the size of the survey area.

3)   *3D Seismic Exploration Surveys*

As with 2D surveys, almost all 3D seismic exploration surveys in the GOM are conducted by geophysical contractors as "spec" surveys, conducted over large, multi-block areas. Proprietary surveys are usually conducted over only a few blocks.

The ships conducting these surveys are generally 80 to 90 m long, or slightly larger than those used in 2D surveys since they are towing more equipment. These ships tow two source arrays (aligned in parallel with one another) 100 to 200 m behind the ship. The two source arrays are identical and are the same as used in the 2D surveys described previously. Following behind the dual source arrays another 100 to 200 m are anywhere from 6 to 12 streamer cables 3 to 8 km long and spread out over a breadth of 600 to 1,500 m (see **Appendix D**, **Section I.A.3 - Description of Activities, 3D Surveys** for details).

The survey ship tows the apparatus at a speed of 4.5 kn (8.3 km/h). About every 16 s (i.e., a distance of 37.5 m for a vessel travelling at 4.5 kn), one of the dual airgun arrays is fired. Sixteen seconds later, the other array is fired. The timing between firings varies depending on ship speed to achieve the desired spacing.

To complete a survey, the ship continues down a track from 12 to 20 h (i.e., a distance of 100 to 166 km at 4.5 kn), depending upon the size of the survey. Reaching the end on the track, the ship takes 2 to 3 h to turn around and start down another track. This procedure takes place day and night and may continue for days, weeks, or months depending upon the size of the survey.

4)   *Ocean Bottom Cable Surveys*

Ocean bottom cable surveys were originally designed to enable seismic surveys in congested areas such as producing fields with their many platforms and producing facilities. Recently, these surveys have been found to be useful for obtaining four-component (4C) data (seismic pressure, as well as vertical and two horizontal motions of the water bottom, or seafloor), yielding more information about the fluids and rock characteristics in the subsurface (e.g., Ebrom, 2000). Standard hydrophones limit the water depth to which surveys can be conducted to about 183 m. However, new technology now allows surveys to be conducted at water depths of up to 2,500 m or more. As systems are deployed at deeper and deeper depths, both the physical demands upon the system and the concern for the integrity of the system increase. Depth limitations are imposed due to the construction of the carrier systems (e.g., pressure seals, strength of cable). Whether the carrier system is comprised of streamer-type or logging-type cables, the question remains as to whether they are capable of sustaining their own weight over the vertical distance between sea surface and the seafloor, as well as stresses placed on these systems during deployment and retrieval.

Ocean bottom cable surveys require the use of multiple ships (usually two ships for cable layout/pickup, one ship for recording, one ship for shooting, and two smaller utility boats). These ships are generally smaller than those used in streamer operations, and the utility boats can be very small. Operations begin by dropping cables off the back of the layout boat. Cable length is typically 4,200 m but can be up to 12 km. Groups of seismic detectors (usually hydrophones and vertical motion geophones) are attached to the cable in intervals of 25 to 50 m. Multiple cables are laid parallel to each other using this layout method with a 50-m interval between cables. When the cable is in place, a ship towing a dual airgun array passes between the cables, firing every 25 m. Sometimes a faster source ship speed of 6 kn instead of the normal 4.5 kn speed is used with an increase in time between airgun firings. After a source line is shot, the source ship takes about 10 to 15 minutes to turn around and pass down between the next two cables. When a cable is no longer needed to record seismic data, it is retrieved by the cable pickup ship and moved to the next recording position. A particular cable can lay on the bottom anywhere from 2 h to several days, depending upon operation conditions. Normally, a cable is left in place about 24 h. In some cases, cables may be left on the bottom for future time-lapse (4D) surveys (see below).

5)   *Time-Lapse (4D) Surveys*

The purpose of time-lapse (4D) surveys is to monitor the depletion of the reservoir and to locate zones of bypassed production in an already discovered oil or gas field. A time-lapse survey requires repeat surveys with highly accurate navigation to ensure the same subsurface points are measured on each repeat survey. Time-lapse surveys are usually repeated every 6 months to a year, but occasionally the repeat interval can be as short as 4 months.

Time-lapse surveys can use either seismic streamer cables or ocean bottom cables to house the seismic detectors. In either case, the procedure closely resembles the ones described previously for 3D and ocean bottom cable surveys. The main difference is in the size of the survey. Since the oil or gas field already has been located, the survey is much smaller, and survey time is much shorter. An average survey takes 2 to 4 weeks and can cover 20 square kilometers.

Although the technique began using streamer cables, the difficulty in locating the sensors with suitable precision led to the use of bottom cables, then to fixed bottom cables. When fixed bottom cables are used, the survey time, after the first survey, is much shorter since all that has to be done is connect the fixed bottom cable to the recording instruments and start shooting.

6)   *Vertical Cable Surveys*

Vertical cable surveys, although uncommon, are similar to ocean bottom cable surveys in that the receivers are deployed and then shot into by a source boat. However, they are substantially different from ocean bottom surveys in that the receivers are located on vertical cables anchored to the ocean bottom. These surveys are normally conducted at water depths up to 1,700 m; however, when specially constructed hydrophones are used, surveys can be conducted in water depths up to 2,500 m.

Two identically configured boats are used during a vertical cable survey. Both boats are used initially to place the cables. During the survey, one boat is used as a source boat and the other to recover and redeploy the cables.

The vertical cables are deployed on two overlapping grids. On each grid, vertical cables are deployed every 2 km. One grid is staggered 1 km to the other such that any one vertical cable is no more than 1.4 km from its closest neighbor. Normally 28 or 32 vertical cables are deployed at any one time. At the bottom of each vertical cable is an anchor composed of 680 kg of steel. The active section of the cable is 375 m long and contains 16 specially constructed hydrophones spaced 25 m apart. At the top are buoyant floats to keep the cable as vertical as possible.

Once the cables are in place, the source boat begins shooting in such a way that each vertical cable receives shots at a distance of 5 km in all directions. This is accomplished by sailing down lines parallel to the grid of vertical cables. Once the shooting boat shoots a line 1 km beyond the first row of vertical cables, that row is recovered and redeployed. Cables may be left in place for hours or days, depending upon the size of the survey and operating conditions. Vessel speed is normally 4.5 kn. The dual airgun array is the same as normally used in 3D streamer surveys (see **Appendix D**, **Section I.A.6 - Description of Activities, Vertical Cable Surveys** for details).

7)   *VSP Surveys*

VSP surveys are surveys where seismic data are recorded from sensors placed in a borehole (i.e., a hole vertical to the ocean surface or seafloor) with seismic sources deployed in various geometries around the vertical array of sensors. VSP surveys are either zero offset or check shot surveys (i.e., when the seismic source is placed very near to the vertical array of sensors) or walk-away surveys (i.e., when the seismic source is placed in a series of positions along a radial line from the vertical array of sensors). Each survey type is designed to acquire specific information. Zero offset and walk-away VSP surveys are by far and away the most common VSP surveys conducted in GOM. Less common are 3D VSP surveys where the source is deployed in an area surrounding the vertical array of sensors.

In all VSP surveys, sensors are lowered down a borehole before production tubing is placed in the well bore or the well is abandoned. The sensors lowered down the borehole can be connected together in strings of 16 to 36 receivers spaced from 15 m to 150 m apart depending upon the survey objective and other variables. After lowering the sensor string to the lowest portion of the borehole to be surveyed, the sensors are temporarily clamped to the side of the well bore and seismic signals recorded. Subsequently, the sensors are repositioned and the next set of seismic signals recorded. Seismic sources used in VSP surveys are the same as those used in conventional seismic surveys. Zero offset surveys are conducted using a small volume single airgun suspended by a crane located on the deck of the drilling rig. Walk-away surveys utilize a work boat with only four to eight airguns. 3D VSP surveys use the same airgun arrays as used for conventional 2D and 3D surveys. These airgun arrays can vary from 1,000 to 5,000 cubic inches, depending upon the depth of the objective. Typical airgun array depths are 7 to 10 m below the surface.

One method used to provide 3D coverage is for the source vessel to travel in a spiral track. The source vessel begins the spiral track at a distance of 200 m from the borehole and keeps the distance between spirals equal to the number of arrays times the array separation (e.g., two arrays spaced 120 m apart equals a 240-m track spacing). First one airgun array will fire, then 12 to 14 s later the other airgun array will fire. At a typical vessel speed of 4.5 to 5 kn, the distance between firings is between 28 and 36 m. The source vessel continues on the spiral out to a distance of up to 9 km. If the borehole sensor string needs to be raised to another level, the whole procedure is repeated.

Total time spent on VSP surveys depends upon the type of survey, the objectives of the survey, the cost of the drilling rig and the equipment used. For a zero offset survey or a walk-away survey, the survey can take less than a day, in the absence of any serious equipment failures. For a zero offset survey, the airguns are fired four to eight times for 20 s, followed by a 5- to 20-min quiet time during which the sensor string is raised; the airguns are fired again for four to eight times for 20 s and so on until the survey is completed. A 3D survey may require up to 10 days to complete, however, 30% of that time may be with the airguns in standby mode.

8)   *Other Variations*

Multi-ship Surveys. Multi-ship surveys are an integral part of ocean bottom surveys and vertical cable surveys. However, in the quest for seismic data recorded at greater distances from the source, multi-ship streamer surveys (where both ships tow streamer cables but only the lead ship fires its seismic source) are becoming more prevalent. Generally, this technique is used to either obtain converted wave data (shear wave) or to penetrate hard seafloor layers. Two sorts of operations are in use – one 2D and the other 3D. The 2D operation places two ships, one behind the other, each towing single streamer cables at a distance apart and a streamer length sufficient to record seismic data at distances of 8 to 12 km from the source. The 3D operation places two ships, one behind the other, towing multiple (up to 12) streamers 2,500 m long a sufficient distance apart to record seismic data at distances of 9 km from the source. In both 2D and 3D surveys, only the lead ship is used as a seismic source boat.

Undershooting. Undershooting, where an obstacle stands between the seismic source and the receiver, is used most commonly in conjunction with ocean bottom surveys. This is because the surveys typically are located in areas of congestion and the cost is low. Operationally the technique is identical to normal ocean bottom operations, except the obstructing structure lies between the source boat and the ocean bottom cable.

Undershooting using 2D or 3D streamer cable ships is rare and is required when an obstacle, most often a production platform, interferes with the routine collection of data in the immediate vicinity of the obstacle. Undershooting is the technique whereby a vessel towing an airgun array, and perhaps streamer cables of its own, passes on one side of the obstacle. Concurrently, another vessel towing streamers passes on the other side of the same obstacle. The energy from the airgun array passes through the subsurface beneath the obstacle and is recorded by the vessel on the other side of the obstacle, hence the name "undershooting." Undershooting using ocean bottom cables as receivers is sometimes necessary when the geometric disposition of multiple obstacles is complex (e.g., a seismic vessel towing streamers may not always be able to maneuver between obstacles), but ocean bottom cable is generally more expensive than using surface streamers as receivers.

b.   *Deep-Tow Side-Scan Sonar Surveys*

Deep-tow side-scan sonar surveys are conducted in the GOM primarily for engineering studies involving the placement of production facilities and pipelines. The surveys provide information about seafloor topography and help to identify the presence of sand flows, hydrates, seeps, and potential hard bottom areas.

Operations are conducted from ships towing data communications cables up to 7 km long, which enables operations in water depths up to 3,000 m deep. Close to the end of the cable is a 30- to 45-m long section of chain to keep the sensor package (fish) tracking at approximately 25 to 30 m above the bottom. The chain drags along the seafloor, cutting a trench approximately 10 cm wide by 15 cm deep (4 in. wide by 6 in. deep). In situations where the chain can become entangled in shipwrecks, well heads, or hard bottom areas, the chain is removed, and the sensor package is kept above the bottom by adjusting the length of the tow cable.

The sensor package housed in the towed "fish" consists of two sonar devices, one mounted on the port side and the other on the starboard side of the fish. The beam width of each of the sonar sources is 1°

to 2°, which combined with the height above the seafloor of 25 to 30 m, yields continuous seafloor coverage. Also included in the sensor package is a pinger for subbottom profiling.

c. *Electromagnetic Surveys*

Electromagnetic surveys are used to help delineate potential oil and gas reservoirs. There are two practical electromagnetic techniques applicable to marine surveys. Both the magneto-telluric (MT) and bi-pole methods have been primarily applied in the marine environment in a research mode. However, the MT technique has seen limited use in the GOM, where about 400 stations have been occupied. There is talk of turning the bi-pole technique into a commercial device, but there has been no action as of this date. A third experimental technique that has not yet been tried is described briefly in **Appendix D, Section I.C - Electromagnetic Surveys**.

In the MT technique, no electrical currents are induced into the earth, but the receiver device detects the natural electrical and magnetic fields present in the earth. Ships are used to deploy and retrieve the recording devices. These devices are about 1.5 m high by 1 m on a side and are attached to a concrete anchor about 60 cm on a side, 15 cm high, and weighing about 136 kg. Also attached to the recording device are four arms sticking out from each side of the box with an electrode on each end. These arms are about 20 m long and made of 5-cm plastic polyvinyl chloride (PVC) pipe. Inside the recording box is a magnetometer and a long-term recording device, which allows the box to remain on the water bottom for days at a time. The recording box is retrieved by using an acoustic pinger that releases the anchor from the recording box, which then floats to the surface.

In the bi-pole technique, two cables (joined together, with the second cable a few hundred feet longer than the first) are towed around by a ship. Attached to the end of each cable is a metal cylinder about 3 m long and 0.3 m in diameter. At regular intervals the ship stops, the cables sink to the bottom, and an electrical signal is input through the cables and into the seafloor. These electrical signals are detected by previously deployed receivers 2 to 10 km away from the source and arranged in a line or profile. The receiver boxes are attached to concrete blocks like those used in the MT technique. Inside the receiver boxes are recording devices that allow for recording for a few days. When the recording is finished, an acoustic pinger releases the recording box from the anchor, and the recording box floats to the surface for retrieval.

d. *Geological and Geochemical Sampling*

Geological and geochemical sampling is conducted to obtain samples of the seafloor for physical and/or chemical analyses. Physical analyses are used in engineering studies for placement of structures such as platforms and pipelines. Chemical analyses (surface geochemical prospecting) are based on the premise that upward migrated petroleum from deep source rocks and reservoirs can be detected in near-surface sediments and are used to evaluate exploration potential. Usually, a program of bottom sampling and shallow coring is conducted simultaneously using a small marine drilling vessel (USDOI, Geological Survey, 1976). Other sampling techniques using towed hydrocarbon "sniffers" are rarely used (see **Appendix D, Section I.D.3 - Hydrocarbon "Sniffers"**).

1) *Bottom Sampling*

Bottom sampling involves devices that penetrate only a few centimeters to several meters below the seafloor. Samples of surficial sediments are typically obtained by dropping a piston core or gravity core ("dart"), essentially a weighted tube, to the ocean floor and recovering it with an attached wire line. Samples also can be obtained using a grab, which is a device with a jaw-like mechanism, or with a dredge, which is a wire cage dragged along the seafloor.

For the last 15 years, almost all the surface geochemical exploration techniques in the northern GOM have involved piston coring to obtain sediment samples for the analysis of upward migrated hydrocarbons (Brooks et al., 1986, 1997; Sassen et al., 1993). The typical piston core is a 6-m long, 7.5-cm diameter pipe with a 910-kg core weight. In some earlier studies, lighter and shorter gravity cores were used. Over the last 20 years, it is estimated that 10,000 to 15,000 piston and gravity cores have been taken in the northern Gulf, primarily in continental slope water depths.

In gravity coring, wire is paid out from the coring winch at a fairly fast speed allowing the corer to hit the bottom with a force proportional to the weight of the corer and the speed at which it is deployed. Penetration into the bottom is limited by the sediment type, friction of the sediment on the outside and inside walls of the core barrel, and the resistance of the water exiting the top of the core barrel. In contrast, a piston corer uses a "free fall" of the coring rig to achieve a greater initial force on impact, and a sliding piston inside the core barrel to reduce inside wall friction with the sediment and to assist in the evacuation of displaced water from the top of the corer. The core barrel dimensions are generally 6-m long by 7.6-cm internal diameter by 9-cm outer diameter. Coring at lengths greater than 6 m is possible but not common.

2)   *Heat Flow Measurements*

Another tool in limited use in deep-water exploration is a heat flow probe. This technique, used primarily in academic circles as a research tool, provides geochemical and geological information that aids in understanding regional scale hydrodynamics and the potential for occurrence of hydrocarbons. Heat flow measurements are conducted with a device that looks much like a piston corer. The device measures both temperature gradient and thermal conductivity *in situ* over subbottom depth intervals of up to 6 m. The footprint and impact on the sediments is almost identical to that of piston or gravity coring.

3)   *Shallow Coring*

Shallow coring is done by conventional rotary drilling equipment from a drilling barge or boat. Penetration is usually limited to the recovery of several meters of consolidated rock.

e.   **Remote Sensing**

1)   *Radar Imaging*

Radar imaging by satellite is currently used to detect oil slicks on the sea surface. This is possible because when the oil molecules reach the sea surface, they form a thin layer that dampens the ocean surface capillary waves. The detection of oil slicks requires quiet water conditions and consequently is limited by sea state as well as satellite position and frequency of coverage. The resolution of the radar images ranges from 8 to 100 m with a swath width range of 50 to 500 km. The radar satellite is in a near polar orbit at an altitude of 798 km. The cycle time for a duplicate orbit is 24 days, but a common spot on the earth can be revisited every 5 days and surveyed with different viewing parameters. The MMS does not permit nor approve radar imaging surveys.

2)   *Aeromagnetic Surveys*

Aeromagnetic surveys are conducted in the GOM to look for deep crustal structure, salt related structure, and intrasedimentary anomalies. The surveys are flown by fixed wing aircraft, with flight lines on the order of 400 km (250 mi) long, at a height of 75 to 150 m above the sea surface, and are flown at speeds of about 220 km/h. Flight line spacing ranges from 500 to 800 m apart with cross lines every

2,000 to 3,000 m. Acquisition rates are on the order of 1,000 to 2,000 km of data per day. The earth's magnetic field is measured by either a proton precision or cesium vapor magnetometer mounted in a "stinger" projection from the tail of the aircraft. On occasion, two magnetometers are used to measure not only the total magnetic field but also the vertical gradient of the field. Magnetometers also can be towed behind a ship. This usually is in conjunction with a seismic survey but can be run as a separate survey.

3)   *Gravity Surveys*

Marine gravity data can be collected with instruments on the seafloor, in boreholes, in ships, or in helicopters. Originally, data were collected on the seafloor, but modern technology has moved the collection point to ships. Marine gravity meters have, in some cases, been housed in a ship while it is conducting a seismic survey. However, the preferred method has been to use dedicated ships in order to acquire more precise data. With the advent of global positioning system (GPS) navigation and larger, more stable seismic ships, it is now possible to achieve the same order of accuracy with meters placed in seismic ships as in dedicated ships. Data grids for gravity surveys range from 1.6 km x 8 km to 9.7 km x 32 km. Gravity data may also be collected using helicopters. However, helicopter surveys are rare in the GOM because of the logistics required to keep the craft in the air for extended periods far from shore.

4)   *Gravity Gradiometry*

Measuring the earth's gravity gradient is now possible with the release of Defense Department technology. The instrument is housed in a box located in the center of a survey ship. In shallow water, the ship sails a 0.25-km by 1-km grid, and in deep water, a 1-km by 2-km grid is used. Typically, a 20-block area is selected for survey, and this can be completed in about 2 days.

5)   *Marine Magnetic Surveys*

Marine magnetic surveys measure the earth's magnetic field for the purpose of determining structure and sedimentary properties of subsurface horizons. These surveys are usually conducted in conjunction with a seismic survey, allowing the navigation information to be used for both surveys. The development of low power digital sensors has allowed the sensor package to be towed behind the seismic source array, which has greatly improved operational efficiency of magnetic surveys. The sensor is housed in a cylindrical package measuring approximately 1 m long and 15 to 20 cm in diameter and weighing about 14 kg. The electronics package inside the case contains about 1 L of chemically inert fluid that is non-toxic. The sensor is towed behind one of the sub-arrays of the seismic source array at distances of 50, 100, or 150 m (behind the array), although 100 m is the most common. The sensor is towed at a depth of 3 m and makes use of depth devices mounted on the cable to maintain a constant depth.

2.   Level of Activity

a.   Characterization of Activity

Although gravity, magnetic, and electromagnetic surveys are conducted in the GOM, by far the greatest activity is seismic data acquisition. The principal seismic technology is 3D seismic, using either streamers or bottom cables. Time-lapse (4D) surveys are becoming more frequent as the technology for analyzing the data are developed. Four-component (4C) seismic surveys conducted with bottom cables and vertical cable surveys are in their infancy.

In developing a level of activity, only the seismic techniques are considered, as they are the main focus of the environmental analysis in this PEA. The number of lease blocks surveyed is one measure of activity. Seismic survey activity is typically measured by the number of active "crews." Each crew is a complete entity with the ability to collect seismic data. Crews can be ships towing streamer cables, or a fleet of ships collecting either bottom cable or vertical cable seismic data. For the purposes of describing the level of activity, the period from 1988 to the present adequately characterizes the recent level of G&G activity in the GOM. Another measure of activity is permits issued by MMS for 3D seismic surveys. While permits do not cover seismic surveys conducted under lease terms, most 3D seismic surveys are conducted under permit. While it is also of interest to project future levels of G&G seismic activity in the GOM, such predictions must be viewed as gross approximations because many factors influence activity levels, as detailed as follows. For the purposes of this analysis, recent trends in permitting levels were evaluated, current driving factors considered, and estimates of future survey activity projected.

b.   *Factors Driving Activity*

1)   *Economics*

The price of oil and natural gas is the predominant driver for exploration activity. Prices for oil are determined on the world market and are primarily set by the Organization of Petroleum Exporting Countries (OPEC). Over the period from August 1985 to May 2003, the closing futures price of Light Sweet Crude Oil varied from a low in 1998 of $11.21 per barrel to a high of $36.09 per barrel in 1990 (Haver Analytics, written comm., 2003). Over a 10-year period from October 1988 to December 1998, the price of oil began at a low of $11.63 per barrel and rose gradually to an average price of about $20 per barrel (with one spike at $36.09 per barrel in October 1990), then declined to another low of $11.21 per barrel in December 1998. Since then, there have been two peaks (i.e., one in November 2000 at $34.16 per barrel, a second in February 2003 at $35.78 per barrel) interspersed with a low of $19.53 per barrel in December 2001. Most recently (May 2004), the price of oil has reached historic high levels of around $40 per barrel. The general trend seems to be prices increasing over the next few years.

The price of gas is less dependent upon the world market, as gas is usually delivered from field to market by gas pipelines. Gas prices have shown a gradual increase during the decade of the 1990's from a price of about $1.80 per million British thermal units (MMBTU) (closing futures price New York Mercantile Exchange [NYMEX]) to a price of about $2.70 per MMBTU in December 2001 (Haver Analytics, written comm., 2003). One price spike occurred during December 2000 when the price rose to $8.32 per MMBTU. Currently (June 2003) prices seem to be on another rise; however, no trend to prices can be discerned. During the projection period, relatively short-term price spikes for natural gas may be expected to occur (e.g., as occurred during winter 2000-2001), and the price can be expected to increase over the levels in the 1990's.

This oil and gas price scenario assumes a relatively stable economic environment over the long-term with regard to oil and gas producing prices. Offsetting this optimism is the current oversupply of marine seismic vessels for exploration for new and exploitation of discovered fields and the economic viability of the seismic contracting business. This has caused some vessels to sit idle and profits for geophysical exploration companies to become elusive. Some reduction in vessel capacity will occur. Indeed, the drastic reduction of crews from 41 in 1999 to 15 in 2000 is indicative of this problem.

2)   *Geography*

Oil and gas exploration on the continental shelf of the northern GOM is in a mature state, although large discoveries are expected in deeper waters. From a seismic exploration view, about 900 blocks in the Western and Central Planning Areas have not yet been surveyed with 3D seismic techniques (R. Brinkman, MMS GOM OCS Region, written comm., 1999, 2000). For the purpose of this analysis, several factors were considered in the development of activity level projections. Annual survey activity levels during the past several years were considered (i.e., 1993-2002), as well as the number of permits issued during the first 6 months of 2003. Based on trends evident in the past several years, the assumption is that the remaining blocks would most likely be surveyed over the next several years, with a gradual tapering off of activity. Resurveying of blocks also has been integrated into the projections. It is assumed that a lower level of new seismic survey activity will occur in the Eastern Planning Area relative to the remaining two Planning Areas (i.e., the vast majority of survey activities are expected in the Central and Western Planning Areas). Industry interest in the Eastern Gulf has historically been limited to the westernmost portions of the planning area (e.g., Destin Dome 56 Unit; Lease Sale 181 area).

3)   *Technology*

The development of new technology has been a key factor in continuing new seismic activity. Technology improvements to the 3D method are expected to cause previously surveyed blocks to be resurveyed. In the past, improvements in technology have caused resurveying at approximately 8-year intervals and reprocessing of data every 5 years. However, current economic conditions have caused seismic contractors to decrease their investments in new technology. Indeed, these contractors are reducing the number of crews in order to reduce cash flow. Because of these events, new technology will take longer to come into the mainstream, and the resurvey time will be on the order of 10 to 14 years. Four-component (4C) bottom cable and vertical cable surveys also will be done over older 3D surveys in order to collect additional or new information. Time lapse surveys will be done over existing fields. On occasion, these time lapse surveys will be 4C.

*c.   **Correlation of Factors to Activity***

The economic factor will be the most important driving factor affecting G&G activity over the next several years, given the relative stability or instability of oil and gas prices and the economic health of the exploration industry. Over the past 10 to 12 years, periods of relative price stability have been interspersed with rapid price fluctuations due to regional and worldwide instabilities and conflicts. The oil and gas pricing scenario assumes a relatively stable economic environment over the long-term. The geographic factor is projected to cause the survey of blocks of interest in the GOM that have not been surveyed. Based on a review of permit history and MMS consultation, the estimation is that 900 blocks remain to be surveyed; such new surveys are expected to occur between now and 2008, with highest activity levels expected in the near term (i.e., 2004-2005). Technology will be the driving factor that will require each block (i.e., each block already surveyed) to be re-shot with a new survey using either streamer cables, water bottom cables with hydrophones and vertical geophones (two component [2C]), water bottom cables with 4C, or vertical cables. Normally, new technology would prompt the resurveying of blocks with an approximate 8-year delay, but with the decrease in investment in new technology that period should expand to approximately 14 years. All blocks without production will likely be resurveyed; those with production will most likely be resurveyed with time lapse surveys. Adding these two numbers together (newly surveyed and resurveyed blocks) gives the number of blocks available for crew activity. This figure varies from 1,041 (in 2003) to 6,520 blocks (in 2011). The numbers of lease blocks surveyed since 1988 and the numbers of seismic crews working are presented in **Table II-4**.

The methodologies used to compile **Table II-4** have been outlined in **Appendix D (Section V.C - Correlation of Factors to Activity)**. Line miles and actual blocks surveys were derived from MMS statistics, and it was estimated that approximately 230 line miles are completed per block. Estimates of blocks remaining to be surveyed (900) were calculated from MMS statistics; it has been assumed that the

Table II-4

Lease Blocks Surveyed[1] and Numbers of Seismic Crews Under the Proposed Action

Year	Line Miles[2]	Actual Blocks Surveyed	Estimated Blocks Surveyed[3]	MMS Blocks Surveyed[4]	Repeat Blocks[5]	Total Blocks	Crews Total[6]	Crews 3D	Crews 2D	Crews 2D (est.)	Crews 3D (est.)	Crews Total (est.)
1988	193,089	--	841	841	--	841	--	--	--	--	--	--
1989	187,402	--	816	816	--	816	31	--	--	--	--	--
1990	369,540	--	1,610	1,610	--	1,610	25	--	--	--	--	--
1991	363,259	--	1,583	1,583	--	1,583	21	--	--	--	--	--
1992	384,435	1,825	--	1,825	--	1,825	23	--	--	--	--	--
1993	641,514	1,913	--	1,913	--	1,913	31	21	10	--	--	--
1994	566,142	2,488	--	2,488	--	2,488	34	29	5	--	--	--
1995	636,824	3,112	--	3,112	--	3,112	32	28	4	--	--	--
1996	637,311	3,045	--	3,045	--	3,045	37	30	7	--	--	--
1997	--	6,520	--	6,520	--	6,520	50	38	9	--	--	--
1998	--	5,183	--	5,183	--	5,183	41	33	8	--	--	--
1999	--	3,612	--	3,612	--	3,612	15	8	7	--	--	--
2000	--	3,003	--	3,003	--	3,003	17	8	9	--	--	--
2001	--	3,648	--	3,648	--	3,648	15	6	9	--	--	--
2002	--	3,698	--	3,698	841	4,539	--	--	--	--	--	--
2003	--	--	--	225	816	1,041	--	--	--	8	8	16
2004	--	--	--	225	1,610	1,835	--	--	--	8	15	23
2005	--	--	--	113	1,583	1,696	--	--	--	8	14	22
2006	--	--	--	113	1,825	1,938	--	--	--	8	16	24
2007	--	--	--	112	1,913	2,025	--	--	--	8	16	24
2008	--	--	--	112	2,488	2,600	--	--	--	8	21	29
2009	--	--	--	--	3,112	3,112	--	--	--	8	25	33
2010	--	--	--	--	3,045	3,045	--	--	--	8	25	33
2011	--	--	--	--	6,520	6,520	--	--	--	8	53	61
2012	--	--	--	--	5,183	3,270	--	--	--	8	42	50
2013	--	--	--	--	3,612	3,612	--	--	--	8	29	37
2014	--	--	--	--	3,003	3,003	--	--	--	8	24	32
2015	--	--	--	--	450	450	--	--	--	8	4	12

[1] Tabular data represent prelease survey activity levels only; data are unavailable for postlease surveys or other surveys (e.g., pipeline rights of way, scientific research).

[2] Records maintained by MMS (Gulf of Mexico OCS Region) during 1988-1991 were in line miles only regardless of whether the survey was 2D or 3D; during 1992-1996, both line miles and number of blocks surveyed were also recorded, allowing for calculation of blocks/mile. After 1996, only blocks are recorded regardless of survey type (2D, 3D). Line miles of data collected are equal to the length of transects surveyed times the number of streamers. For example, a survey of one mile transect distance by a ship towing six streamers results in 6 line miles of data collected.

[3] Estimated blocks surveyed during the period 1988-1991 based on line miles traveled and/or blocks/mile data (for the period 1992-2002).

[4] Projections (beyond 2002) are approximations based on the number of remaining unsurveyed blocks (900).

[5] Repeat blocks represent those blocks to be resurveyed.

[6] The term "crews" refers to a seismic data acquisition operation. In most cases, it represents a ship towing streamer cables, but can represent multiple ships (e.g., bottom cable surveys; vertical cable surveys). Crew statistics courtesy of IHS Energy Group, World Geophysical News.

unsurveyed blocks will be surveyed over a 6-year period. The first repeat survey to be conducted on blocks has been extended to 14 years because the technology has not improved at the rapid pace displayed in the 1990's, and the economic condition of the G&G contractors is poor. Actual crews totals for 2D and 3D surveys were obtained from an industry source. The estimated number of 3D crews was calculated based on the assumption that 3D surveys are 10 times as efficient as 2D crews in covering a given area; this implies that 90% of the blocks surveyed were done by 3D crews. This approach also provided the basis for estimating crew efficiency.

A simplified trend analysis was also used to project the number of seismic surveys expected in the foreseeable future in the GOM. Recent historical survey activity was considered, as were the number of unsurveyed blocks and projected resurvey activities (i.e., estimated 14-year resurvey cycle). Historical activity levels were derived from MMS permit application records (see **Appendix D**, **Figures D-6 through D-11**).

*Note: Tabular data presented in **Table II-4** represent prelease survey activity levels only; data are generally unavailable for postlease surveys or other surveys (e.g., pipeline rights of way, scientific research). Permitted activity for 3D surveys has been summarized in **Appendix D (Section V - Level of Activity)**. Further, it is important to recognize that projections (beyond 2003) are only approximations based on the remaining unsurveyed blocks (900) and the projection of a 14-year resurvey cycle. While **Table II-4** projects activity out to 2015, there is no historical basis or implied significance to this endpoint.*

The activity levels realized in the Gulf in recent years (i.e., 1995 to 2003) are expected to produce a corresponding increase in resurveying activities on a 14-year cycle, assuming other factors remain relatively unchanged. It is apparent from this analysis that a) several factors strongly influence survey activity levels (i.e., projections are only approximations), and b) there is a "general trend" or "cycle" evident, based on remaining unsurveyed blocks and projected resurvey activity tied to an approximate 14-year timeframe. Further, it is unknown how many new, unsurveyed blocks may be identified in the near future (e.g., from future lease sales and leasing activity).

One factor that cannot be easily integrated into these projections is the geology of specific areas of interest. For example, MMS notes that the area seaward of the Sigsbee Escarpment (i.e., Lund and Lund South in the Central GOM Planning Area) contains only a few deeply buried prospects. For the near term in this area, 3D surveys will probably be of limited areal extent and may result in additional blocks being surveyed for the second and third time sooner. Areas such as Mississippi Canyon and northern Atwater, where drilling activity has been high during recent years, contain some blocks that have been surveyed three times during the period 1993 through 2003.

Using data for the period 1994 to 1999, "crew efficiency" was also calculated by dividing the number of 3D crews by the number of blocks surveyed. Crew efficiency varied from 0.004 crews per block to 0.009 crews per block, with a median value of 0.009 (see **Table II-4**). Assuming 0.009 crew efficiency and multiplying by the total number of blocks estimated to be surveyed from 2003 on, the number of active 3D crews is predicted to range from 8 to 53. Dellagiarino et al. (1998, 2000) indicated that the percentage of 3D surveys (relative to all geophysical surveys permitted by the MMS since 1993) has ranged from a low of 40% (in 1998) to a high of 54% (in 1993). MMS notes that 2D permitting activity has remained relatively static over the past several years. The number of 2D crews is assumed to remain constant at eight crews in the foreseeable future. The resulting predictions for total crew and 3D crew activity are listed in **Table II-4**.

In addition to the 2D or 3D seismic surveys (prevalent in the GOM) are the VSP surveys conducted on a much smaller geographic scale (e.g., within a single lease block). Seismic sources used during a VSP survey are the same as those used in conventional seismic surveys. Zero offset surveys (check shot surveys) utilize a single, small volume airgun suspended from the deck of a drilling rig. Walk-away surveys utilize a work boat equipped with four to eight airguns, while 3D VSP surveys use the same airgun arrays as used for conventional 2D and 3D seismic surveys. Check shot surveys are the most common type of borehole seismic survey conducted in the GOM. Geographic distribution of check shot surveys for several 3-year periods is outlined in **Appendix D (Section V.C - Correlation of Factors to Activity)**. Comparisons of 2D and 3D seismic surveys with check shot survey activity levels can only be made with appropriate caveats, given the inherent differences in survey characteristics (i.e., frequency of airgun firing, geographic distribution of survey activity, duration of survey operations). It is evident, however, that check shot survey activity in recent years has followed the general trend evident in the region's exploration and development sectors - increasing movement into deeper waters of the Gulf.

3.   Mitigation Measures Included

The Proposed Action includes mitigation measures that are implemented through regulations governing prelease and postlease G&G activities. Pursuant to 30 CFR 251.4, a permit must be obtained to conduct prelease geological or geophysical exploration for oil, gas, and sulphur resources. (Permits for exploration for other mineral resources are pursuant to 30 CFR 280.3.) Permit applications must be submitted to MMS in accordance with the requirements outlined in 30 CFR 251.5 and 30 CFR 251.6 and explained further in applicable Letters to Permittees. The Letter to Permittees dated January 20, 1989, specifies forms and maps, stipulations, and special provisions applicable to most permit activity. The 30 CFR 251 regulations do not apply to G&G activities conducted by, or on behalf of, a lessee on a leased block. Such G&G activities are governed by 30 CFR 250.201 regulations and by applicable NTLs.

The stipulations applying to both prelease and postlease G&G activities include several environmental protective measures (see **Appendix E, Section II - Mitigation Measures and Operational Restrictions**). Key points include the following:

- Explosives cannot be used except under written authorization from the Regional Supervisor. Further protective measures (including Endangered Species Act Section 7 consultation with the NMFS) apply in the event that explosives are used. Explosives are not covered here because their use automatically requires preparation of a separate EA.

- Several measures are specified to protect Florida manatees from vessel strikes. These include a separate ESA consultation.

- Bottom disturbing activities are prohibited within the No-Activity Zones of certain topographic features, and anchoring is restricted within the "Pinnacle Trend" area of the northeastern GOM. Further protections apply for activities within the Flower Garden Banks National Marine Sanctuary.

- The permittee must report discovery of any archaeological resource (shipwreck/prehistoric site) to the MMS and take precautions to protect the resource from operational activities.

Further, G&G activities in waters >200 m deep in the GOM and all OCS waters of the Eastern Planning Area have recently (i.e., August and October 2002, June 2003, March 2004) been required to comply with a series of additional mitigation measures designed to minimize or eliminate potential impacts to marine mammals. These mitigation measures, most recently encompassed within NTL No. 2004-G01, include the use of ramp-up during daylight hours, use of trained marine mammal observers, initiation of reporting requirements, and the requirement to shutdown an operational array when whales are sighted within or moving towards a predetermined impact zone.

a. *Rationale for Ramp-Up*

Ramp-up (also known as "soft start," "slow start," or "slow build up") entails the gradual increase in intensity of a sound source (e.g., airgun array) over a period of 15 to 30 minutes, until maximum source levels are reached, as detailed in **Appendix E, Section II.B.3, Ramp-Up**. The intent of ramp-up is to either avoid or reduce the potential for instantaneous hearing damage to an animal (from the sudden initiation of an acoustic source) that might be located in close proximity to an airgun array. Increasing sound levels are designed to warn animals of pending seismic operations (at full power) and to allow sufficient time for those animals to leave the immediate area. Increasing sound levels (e.g., from an airgun array) are thought to be annoying or aversive to marine mammals. The effectiveness of ramp-up has yet to be documented, either within U.S waters or overseas. The MMS is considering study designs that may be effective for evaluating ramp-up as it applies to G&G operations in the GOM, indicative of the extremely limited state of knowledge regarding this mitigation measure. Ramp-up has been used as a common sense measure in various international arenas (see **Appendix E, Section II.B.3, Ramp-Up** and summary by Pierson et al., 1998). Under optimal conditions, sensitive individuals are expected to move out of the area, beyond the range where hearing damage might occur. A discussion of the application of ramp-up and its efficacy is found in **Appendix E, Section II.B.3, Ramp-Up**. In spite of the absence of data on its effectiveness, a Gulf-wide requirement for ramp-up is currently in place as part of NTL No. 2004-G01. Under this NTL, ramp-up is used in conjunction with visual monitoring.

b. *Rationale for Visual Monitoring*

The use of visual monitoring (i.e., use of trained observers to scan the ocean surface for signs of whale presence), as required under NTL No. 2004-G01, is intended to establish and maintain a zone around the seismic vessel that is clear of whales, thereby reducing or eliminating the potential for hearing damage. Visual monitoring includes visual clearance of the impact zone immediately prior to and during ramp-up, as well as continual monitoring of the area around the vessel and operating array during seismic operations. In the event one or more whales are sighted within the impact zone, ramp-up or seismic operations are halted until the area is clear. Visual observations have definite limitations. Sea state, sun angle, limited visibility, and observer fatigue can all affect observer sighting success (i.e., perception bias). Routine activities of marine mammals (e.g., diving duration patterns, pod size, overt behaviors) show considerable variability between species, thereby affecting whether or not animals are sighted (i.e., availability bias). During nighttime operations, or during periods of reduced visibility, visual monitoring of the impact zone becomes problematic.

c. *Calculated Impact Zone*

Based on the technical information pertinent to airgun systems provided in **Appendix D, Section II.A - Seismic Sources, Airguns** and the impact zone calculations (and assumptions) outlined in **Appendix E, Section II.B.2 - Impact Zones and Real-Time Monitoring** the distance from an acoustic source (i.e., airgun array) to target isopleths (i.e., 160 and 180 dB re 1 µPa rms) can be estimated. The significance of these isopleths is discussed further in **Appendix B, Section V - Other Factors and Considerations**, along with the current efforts being undertaken by NMFS to establish appropriate species-specific guidelines for acoustic exposure of various marine mammals to man-made sound.

Calculation of an impact zone radius relative to a seismic array must consider the array effect, particularly for surface and near surface waters (i.e., from the horizontal plane to approximately 30° below the horizon; e.g., see **Appendix C, Section III - Seismic Source Levels**). Seismic arrays are designed to direct the sound downward with maximum energy directed towards the seafloor and minimum sound directed parallel to the array. The "array effect" (see **Appendix D, Figure D-4**) reduces sound source levels (in surface and near surface waters) from the array by 20 to 60 dB or more and is frequency dependent. Richardson et al. (1995) note that sound pressure levels may propagate differently in the horizontal direction, citing differing effective source levels perpendicular vs. parallel to the array axis. The array effect may cause the target isopleth to be ellipsoid in shape (i.e., the isopleth is not a perfect circle around the array) with many peaks and troughs superimposed on the ellipsoid. In light of all of these factors, there is no single value that can be assigned to the array effect. Determination of a precise value for the array effect is controversial, and a widely accepted value has neither been determined nor agreed to by experts in the field. However, in order to simplify the analysis presented in this PEA, a conservative estimate for the array effect has been established at 20 dB.

Further, airgun array sound source levels are typically expressed as zero-to-peak. Conversion from zero-to-peak to root mean squared (rms) sound levels reduces the sound source level by approximately 10 dB. Details of this calculation are outlined in **Appendix C, Section III - Seismic Source Levels**.

Calculations under the assumption of free-field spherical spreading and the potential influence of modified cylindrical spreading (i.e., 20log[R] and 15log[R], respectively) are detailed in **Appendix C, Section III - Seismic Source Levels**. The transition from unbound 20log[R] spreading loss conditions to something more complex (e.g., involving sound pulse reverberations from the seabed) is related to water depth (i.e., estimated transition at 1.5 times the water depth). Additional assumptions and limitations evident in these basic calculations also are outlined. Radial distances to the isopleths of interest are as follows:

- *Actual range to 180 dB re 1 μPa (rms) target isopleth* = ~300 m
- *Actual range to 160 dB re 1 μPa (rms) target isopleth* = ~3,000 m

Due to the sound source levels characteristic of G&G seismic systems being used in the GOM, and because of the directional aspect of the airgun sound, marine mammals that may be present beneath an array are at greater risk from sound exposure and its potential effects than those found at the same distance from the array in surface or near the surface waters. The required use of visual monitoring (under NTL No. 2004-G01) and ramp-up provides for establishment of a conservative impact zone. Calculations presented in **Appendix C (Section III - Seismic Source Levels)** utilize an appropriate sound transmission loss model. For G&G seismic operations in Gulf waters that are >200 m deep and all OCS waters of the Eastern Planning Area, surface and near surface waters within a radial distance of 500 m around a seismic array should be kept clear of whales; in consideration of the basic calculations presented above (i.e., range to 180 and 160 dB re 1 μPa [rms] target isopleths of ~300 and ~3,000 m, respectively), whales and other marine mammals outside of the 500-m exclusion zone required under NTL No. 2004-G01 are not likely to be exposed to levels near 180 dB re 1 μPa (rms).

d.   *Observers' Roles*

Trained marine mammal observers placed aboard seismic vessels function in one of several ways. When whales are observed (either within the 500-m impact zone or approaching the impact zone), observers will call for the temporary shutdown of the airgun array. Under these conditions, observational notes (e.g., species, abundance, overt behaviors) will be taken along with salient physical characteristics (e.g., vessel location and direction of travel, array type and acoustic output, etc.). Resumption of survey activities will occur when the whale has left the impact zone. Observers also will serve as monitors and recorders, creating a database of information regarding numbers, types, and activity of marine mammals in the vicinity of seismic vessels.

In addition, observers will be required to monitor the impact zone for 30 min prior to powering up a seismic array. This will ensure that marine mammals are not present within the impact zone, thereby avoiding exposure to acoustic energy in close proximity to the array. Under those circumstances where the impact zone cannot be fully observed for at least 30 min prior to seismic operations (e.g., during nighttime or fog), operations should be suspended until the impact zone is clear.

e.   Species- or Group-Specific Sensitivities

In terms of overall sensitivity to G&G activities, baleen whales (e.g., Bryde's) are possibly a relatively "high risk" category amongst the Cetacea, as discussed in **Appendix G, Section II.A.1 - Mysticetes, Behavioral Effects**. Hearing sensitivity at low frequencies down to ~10 Hz is almost certainly good, and many of the vocalizations of baleen whales occur in the low tens to a few hundred Hertz, which implies functional hearing in this range. Similarly, sperm whales also vocalize as low as 100 Hz, with a similar implication regarding functional hearing capabilities.

Beaked whales, as represented by Blainville's beaked whale, vocalize at <1,000 Hz. While this group may not be as susceptible to low frequency airgun noise as sperm and Bryde's whales, beaked whales are deep divers. Therefore, there is potential for beaked whales undertaking prolonged deep dives to enter the area beneath an array where maximum airgun energy is focused. As outlined in detail in **Appendix G, Section II.C.2 - Odontocetes: Dwarf and Pygmy Sperm Whales, and Beaked Whales, Physical Effects**, beaked whales reportedly have been affected during a series of acoustic-related events (e.g., March 2000 strandings in the Bahamas following possible exposure to high intensity, mid-frequency sonar pulses; September 2002 strandings in the Canary Islands following naval maneuvers). It is noteworthy that seismic pulses and mid-frequency sonar pulses are quite different (i.e., pulses from airgun arrays are broadband with most of the energy below 1 kHz; typical military mid-frequency sonars operate at frequencies of 2 to 10 kHz, generally with a relatively narrow bandwidth at any one time). Given that sonar and seismic sounds have quite different characteristics and duty cycles, it is not appropriate to infer a direct connection between the use of sonar and seismic arrays with subsequent strandings. Evidence suggests, however, that these sound sources may, under certain circumstances, have a potential to lead to hearing damage and disorientation in sensitive species (e.g., beaked whales).

In terms of airgun arrays and their maximal energy output (i.e., in the region of a few tens of Hertz), there is clearly a good overlap between the expected frequencies of good hearing sensitivity (low threshold) in these species and maximal airgun output at source.

4.   Evaluation Relative to Purpose and Need

The Proposed Action meets the underlying need specified in **PEA Section I**. The current suite of G&G activities provides the oil and gas industry with sufficiently accurate data on the location, extent, and properties of hydrocarbon resources, as well as information on shallow geologic hazards and seafloor geotechnical properties, in order to explore, develop, produce, and transport hydrocarbons safely and economically. The current suite of G&G activities also provides the MMS with data needed to fulfill its statutory responsibilities to ensure safe operations, to support environmental impact analyses, to protect benthic resources through avoidance measures, to ensure fair market value for leases, to make royalty relief determinations, to conserve oil and gas resources, and to perform other statutory responsibilities.

C.   ALTERNATIVE 2: ADDITION OF VESSEL-BASED PASSIVE ACOUSTIC MONITORING AS A REQUIREMENT

Under this alternative, the existing suite of G&G activities would continue but with the implementation of passive acoustic monitoring as an additional required mitigation measure. This alternative is designed to meet the underlying need for G&G data while reducing environmental impacts from seismic surveys.

1.   G&G Activities Included

This alternative would include the same suite of G&G activities included in the Proposed Action.

2.   Mitigation Measures Included

Existing mitigation measures (protective measures routinely specified in permit requirements, lease stipulations, and NTLs) described previously for the Proposed Action also would be included in this alternative. The following mitigation measure also would be added in this alternative: use of vessel-based passive acoustic monitoring.

Rationale for Passive Acoustic Monitoring

Passive acoustic monitoring (i.e., no acoustic sources are used, only listening devices) can occur either from a vessel-based system or from a hydrophone or sonobuoy array placed on the seafloor, or both. For example, operations conducted in the Alaskan Beaufort Sea used both approaches, while a sonobuoy array was used in the Santa Barbara Channel and hydrophones were employed in Puget Sound. Mitigation measures implemented as part of recent seismic operations in U.S. waters have been summarized in **Appendix E** (see **Table E-2**).

There are recognizable limitations to passive acoustic monitoring, including: 1) passive systems only work for vocalizing whales (i.e., many animals are quiet much of the time, especially when disturbed); 2) output from passive systems makes it difficult to determine the range to the vocalizing animals when using a towed array; 3) passive systems cannot readily determine depth to vocalizing animals; 4) for fixed hydrophones (ship or bottom mounted recorders, sonobuoys, ocean bottom cables), the area of coverage/detection range may be limited by noise, requiring more sensors to cover a seismic survey area; 5) passive systems require that the hydrophone arrays be towed behind the survey vessel (or from an additional chase boat); and 6) hydrophone performance may be affected by tow speed and the ship's acoustic characteristics, effectively limiting the detection range. Trained personnel also are required to operate the equipment and to interpret acoustic signatures.

As discussed under Alternative 3 (see **PEA Section II.D**), the use of passive acoustics has advantages over that of active acoustics in several ways, including: 1) longer ranges can be achieved; 2) passive systems are omnidirectional; 3) species can be potentially identified using a passive system by their vocalization signature; 4) with passive systems, there are no acoustic footprints that could affect the target animals; and 5) passive systems offer a more mature and affordable, though still developing, technology.

3.   Other Mitigation Measures Evaluated

The following mitigation measures were considered, but not included in this alternative, as discussed in **Appendix E** (**Section II.B - Possible Mitigation Measures and Operational Restrictions**):

- seasonal restrictions;
- reduction of sound source levels; and
- sound baffling (to reduce high-frequency noise produced by airguns).

4.   Evaluation Relative to Purpose and Need

This alternative meets the underlying need specified in **PEA Section I**. The current suite of G&G activities provides the oil and gas industry with sufficiently accurate data on the location, extent, and properties of hydrocarbon resources, as well as information on shallow geologic hazards and seafloor geotechnical properties, in order to explore, develop, produce, and transport hydrocarbons safely and economically. As noted previously, the current suite of G&G activities also provides the MMS with data needed to fulfill its statutory responsibilities to ensure safe operations, to support environmental impact analyses, to protect benthic resources through avoidance measures, to ensure fair market value for leases, to make royalty relief determinations, to conserve oil and gas resources, and to perform other statutory responsibilities.

However, the additional restrictions on seismic operations would involve additional costs and delays to operators in obtaining seismic data. This alternative could slow OCS exploration and development. Requirements for visual and acoustic monitoring in deep-water portions of the central and western Gulf are more likely to affect OCS exploration and development because these are areas with a high level of industry activity.

D.   ALTERNATIVE 3: ADDITION OF BOTH PASSIVE AND ACTIVE ACOUSTIC MONITORING AS REQUIREMENTS

Under this alternative, existing G&G activities would continue, but there would be implementation of two additional requirements - use of both passive and active acoustic monitoring - for seismic operations in the GOM. This alternative is designed to meet the underlying need for G&G data while reducing environmental impacts from seismic surveys.

1.   G&G Activities Included

This alternative would include the same suite of G&G activities included in the Proposed Action.

2.   Mitigation Measures Included

Existing mitigation measures (protective measures routinely specified in permit requirements and lease stipulations) described previously for the Proposed Action also would be included in this alternative. The following mitigation measure also would be added in this alternative: use of both passive and active acoustic monitoring.

Rationale for Passive and Active Acoustic Monitoring

Passive acoustic monitoring does not emit sounds (i.e., no acoustic sources are used, only listening devices) and can occur either from a vessel-based system or from a hydrophone or sonobuoy array, or both. Active acoustic monitoring utilizes sound (e.g., sonar) to locate submerged animals. As noted under Alternative 2, there are several recent examples where this mitigation measure has been applied. Operations conducted in the Alaskan Beaufort Sea used both approaches to passive acoustic monitoring, while a sonobuoy array was used in the Santa Barbara Channel and hydrophones were employed in Puget Sound. Mitigation measures implemented as part of recent seismic operations in U.S. waters have been summarized in **Appendix E** (see **Table E-2**). There are recognizable limitations to passive acoustic monitoring, as noted previously. Determinations of range and bearing (e.g., are the sources within the impact zone, are the sources approaching the impact zone, etc.) may be problematic. Trained personnel are required to operate the equipment and to interpret acoustic signatures. From a biological standpoint, not all marine mammal species vocalize.

The use of passive acoustics has advantages over that of active acoustics in several ways, including: 1) longer ranges can be achieved; 2) passive systems are omnidirectional; 3) species can be potentially identified using a passive system by their vocalization signature; 4) with passive systems, there are no acoustic footprints that could affect the target animals; and 5) passive systems offer a more mature and affordable, though still developing, technology.

The advantages of active acoustic monitoring (when compared to passive acoustic monitoring) include: 1) active systems work with non-vocalizing or cryptic whales and those species that exhibit only limited vocalization; 2) active systems can, in some cases, determine 3D range and bearing, including depth of vocalizing animals; 3) active systems do not have to be placed behind survey vessel if sound source and hydrophone/receiver are hull-mounted; and 4) active systems may involve less bulky equipment, minimizing personnel required for handling and operation.

Active acoustic monitoring has several disadvantages, including 1) the active source may potentially be more harmful than the sound source it is being used to mitigate; 2) active systems have limited detection ranges depending on power and frequency; 3) active systems are unable to identify species based purely on size; 4) active systems have a limited beam width and associated problems seeing deep diving whales at close range; 5) active systems could potentially affect the behavior of the animals themselves; 6) active systems require the use of a towfish, which might be larger than a passive acoustic monitoring array; and 7) the current costs for development and deployment of active systems are higher.

3.   Other Mitigation Measures Evaluated

Other mitigation measures were considered, but not included in this alternative, as discussed in **Appendix E (Section II - Mitigation Measures and Operational Restrictions)**.

4.   Evaluation Relative to Purpose and Need

This alternative meets the underlying need specified in **PEA Section I**. Current G&G activities provide the oil and gas industry with sufficiently accurate data on the location, extent, and properties of hydrocarbon resources, as well as information on shallow geologic hazards and seafloor geotechnical properties, in order to explore, develop, produce, and transport hydrocarbons safely and economically.

However, the additional restrictions on seismic operations would involve additional costs and delays to operators in obtaining seismic data. This alternative could slow OCS exploration and development in similar fashion to those noted under Alternative 2.

E.   ALTERNATIVE 4: RESTRICT G&G SEISMIC SURVEY ACTIVITIES

Under this alternative, the existing suite of G&G activities would continue, but there would be restrictions on seismic surveying operations in those portions of the GOM where sperm whales and Bryde's whales congregate. This alternative is designed to meet the underlying need for G&G data while reducing even further any potential environmental impacts from seismic surveys.

1.   G&G Activities Included

This alternative would include the same suite of G&G activities included in the Proposed Action.

2.   Mitigation Measures Included

Existing mitigation measures (protective measures routinely specified in permit requirements and lease stipulations) described previously for the Proposed Action also would be included in this alternative.

3.   Operational Restrictions Evaluated

While no additional mitigation measures would be added in this alternative, G&G surveying operations would be subject to one or two operational restrictions, including

- Alternative 4A - prohibition of simultaneous seismic surveys by more than one G&G survey vessel in those portions of the GOM most frequented by sperm and Bryde's whales.

Under Alternative 4A, the underlying purpose and need would be met, although limited economic impact might occur under those conditions where two or more G&G operators wished to survey simultaneously in those portions of the Gulf most frequented by sperm and Bryde's whales.

The MMS is mandated to manage the development of OCS oil, gas, and mineral resources. Among other things, MMS has a responsibility to ensure that OCS development occurs in an expeditious and orderly fashion. Upon meeting appropriate Federal requirements, OCS lessees are legally entitled to explore, develop, and produce oil and gas contained within their lease area. The lessee then must proceed with "due diligence" in the exploration and development of their lease. Information gathered from G&G activities is an important tool for the industry in meeting the above due diligence requirement for exploration and development. As described earlier, industry needs this information in order to explore, develop, produce, and transport hydrocarbons. MMS also needs this information in order to evaluate resource potential and carry out its regulatory responsibilities. Therefore, forestalling such activities would not serve the needs of industry or the MMS.

4.   Evaluation Relative to Purpose and Need

While simultaneous seismic survey restrictions in sperm or Bryde's whale congregation areas (Alternative 4A) may impose limited hardship on the G&G industry, the underlying need will be met. In contrast, the cessation of all seismic activities in sperm and Bryde's whale areas (Alternative 4B) or seasonal limitations in sperm whale areas both failed to meet fully the underlying need specified in **PEA Section I**. The current suite of G&G activities provides the oil and gas industry with sufficiently accurate data on the location, extent, and properties of hydrocarbon resources, as well as information on shallow geologic hazards and seafloor geotechnical properties, in order to explore, develop, produce, and transport hydrocarbons safely and economically. Further, the additional restrictions on seismic operations would involve additional costs and delays to operators in obtaining seismic data. This alternative could slow OCS exploration and development. Geographic restrictions would severely limit the ability of geophysical contractors to acquire data, ultimately affecting industry's ability to explore and develop hydrocarbon resources on the GOM OCS. In view of these constraints, this alternative places undesirable limits on collection of data the MMS needs to fulfill its statutory responsibilities to ensure safe operations, to support environmental impact analyses, to protect benthic resources through avoidance measures, to ensure fair market value for leases, to make royalty relief determinations, to conserve oil and gas resources, and to perform other statutory responsibilities.

F. SUMMARY AND COMPARISON OF ALTERNATIVES

A tabular comparison of the advantages and disadvantages of each alternative is presented in **Table II-5**. A more detailed summary and comparison of alternatives is provided in **PEA Section III** (see **Table III-4**) as part of a summary evaluation of potential impacts by resource. In addition, consideration has also been given in **PEA Section III** to possible combinations of mitigation measures and operational restrictions on a Gulf-wide basis, or select application of mitigation measures and operational restrictions on a geographic basis (i.e., hybridizations). The summary discussion of impacts at the end of **PEA Section III** considers further the advantages and disadvantages of such hybridization.

Table II-5
Comparison of Alternatives – Advantages and Disadvantages

Alternative	Comparisons	
	Advantages	Disadvantages
1 – Continuation of the Status Quo (No Action)	Meets Purpose and Need. Under NTL No. 2004-G01, geological and geophysical operators conducting surface seismic surveys[1] in all Gulf of Mexico outer continental shelf (OCS) waters >200 m deep must use visual monitoring and ramp-up; geophysical operators conducting surface seismic surveys in OCS waters <200 m deep in the Eastern Planning Area also must use visual monitoring and ramp-up. Visual monitoring is based on maintaining a 500-m (radial distance) exclusion zone that is clear of whales (all marine mammals in the Gulf of Mexico exclusive of dolphins and manatees). Ramp-up, the gradual increase in acoustic output from a seismic array, provides an opportunity for whales, other marine mammals, and sea turtles to leave the area, avoiding maximum acoustic output and possible physiological damage. If whales are spotted either within the exclusion zone, or heading into the exclusion zone, trained observers call for immediate shutdown or powering down[2] of the system. This provides the basis for immediate system shutdown or powering down of the system until the exclusion zone is clear.	NTL No. 2004-G01 not in effect in OCS waters <200 m deep, except in the Eastern Planning Area. In OCS waters >200 m deep throughout the Gulf and OCS waters <200 m deep in the Eastern Planning Area, potential for minor delays in surveying operations (i.e., 30 min of visual monitoring prior to ramp-up), with associated cost ramifications. Ramp-up is not a proven mitigation measure, only a common sense measure. Visual monitoring effectiveness is limited by perception and availability biases (e.g., daylight hours with good visibility; low [Beaufort] sea state; surface activity by whales, recognition by observers). Requires trained personnel. Acceptable observation periods limited to daylight hours with no fog, low sea state, and limited sun glare. Nighttime vision aids have not proven to be reliable.
2 – Addition of Vessel-Based Passive Acoustic Monitoring as a Requirement	Meets Purpose and Need. Provides additional data regarding whale presence, and a basis for system shutdown or powering down (see footnote 1) until the exclusion zone is clear. Passive acoustic monitoring is based on reception/interpretation of whale vocalizations, regardless of visual monitoring limitations. If a whale is identified (via passive acoustic or visual monitoring methods) either within the exclusion zone, or heading into the exclusion zone, observers call for immediate system shutdown or powering down of the system.	Limited to those species known to vocalize. Intermittent vocalization a problem, as is the determination of bearing and distance to the source. Passive acoustic monitoring cannot be used during surveying, only prior to ramp-up or during vessel turns (if acoustic sources are not being used). Requires trained personnel.

Table II-5

Comparison of Alternatives – Advantages and Disadvantages

(Continued)

Alternative	Comparisons	
	Advantages	Disadvantages
3 – Addition of Both Passive and Active Acoustic Monitoring as Requirements	Meets Purpose and Need. Provides additional data for determination of whale presence, and a basis for system shutdown or powering down of the system (see footnote 1) until the exclusion zone is clear. Passive acoustic monitoring is based on reception/interpretation of whale vocalizations. Active acoustic monitoring is based on active sonar emissions. If a whale is identified (via visual, passive acoustic, or active acoustic monitoring methods) either within the exclusion zone, or heading into the exclusion zone, observers call for immediate system shutdown or powering down of the system.	Passive acoustic monitoring is limited to those species known to vocalize. Intermittent vocalization is a problem, as is the determination of bearing and distance to the source. Passive acoustic monitoring cannot be used during surveying, only prior to ramp-up or during vessel turns (if acoustic sources are not being used). Active acoustic monitoring has not yet been fully tested; there are limitations associated with accurate determination of range and depth of marine mammal species. Active sources may be more harmful than the sound source being mitigated. Requires trained personnel.
4 – Restrict G&G Survey Activities		
4A – No simultaneous surveying in sperm and Bryde's whale areas	Meets Purpose and Need. Reduces or eliminates the possibility of multiple acoustic sources adversely affecting sperm and Bryde's whales.	May produce limited hardship only under those circumstances where two or more operators wish to survey simultaneously.

[1] Borehole seismic surveys differ from surface seismic surveys (i.e., smaller airgun arrays, average survey time of 12-24 h, sound source usually not moving, requires the capability of moving the receiver in the borehole between shots). Altered mitigations apply only to borehole seismic surveys, including a) during daylight hours, when visual observations of the exclusion zone are being performed as required in this NTL, borehole seismic operations will not be required to ramp-up for shutdowns of 30 min or less in duration, as long as no whales, other marine mammals, or sea turtles are observed in the exclusion zone during the shutdown. If a whale, other marine mammal, or sea turtle is sighted in the exclusion zone, ramp-up is required and may begin only after visual surveys confirm that the exclusion zone has been clear for 30 min; b) during nighttime or when conditions prohibit visual observation of the exclusion zone, ramp-up will not be required for shutdowns of 20 min or less in duration. For borehole seismic surveys that utilize passive acoustics during nighttime and periods of poor visibility, ramp-up is not required for shutdowns of 30 min or less; c) nighttime or poor visibility ramp-up is allowed only when passive acoustics are used to ensure that no whales are present in the exclusion zone (as for all other seismic surveys). Operators are strongly encouraged to acquire the survey in daylight hours when possible; d) protected species observers must be used during daylight hours, as required in this NTL, and may be stationed either on the source boat or on the associated drilling rig or platform if a clear view of the sea surface in the exclusion zone and adjacent waters is available; e) all other mitigations and provisions for seismic surveys as set forth in this NTL will apply to borehole seismic surveys.

[2] Under NTL No. 2004-G01, when a whale is sighted within or heading towards the exclusion zone (i.e., within an estimated 500 m of the sound source array), the array must be shut down. The vessel may continue on its course, but all airgun discharges must cease. The vessel operator must comply immediately with such a call by an on-watch visual observer. When no whales are sighted for at least a 30-min period, ramp-up of the source array may begin. Ramp-up cannot begin unless conditions allow the sea surface to be visually inspected for whales for 30 min prior to commencement of ramp-up, unless passive acoustic monitoring is used (i.e., ramp-up cannot begin after dark or in conditions that prohibit visual inspection [fog, rain, etc.] of the exclusion zone). Any shutdown due to a whale sighting within the exclusion zone must be followed by a 30-min all-clear period and then a standard, full ramp-up. Any shutdown for other reasons, including, but not limited to, mechanical or electronic failure, resulting in the cessation of the sound source for a period greater than 20 min, also must be followed by full ramp-up procedures. In recognition of occasional, short periods of the cessation of airgun firing for a variety of reasons, periods of airgun silence not exceeding 20 min in duration will not require ramp-up for the resumption of seismic operations if a) visual surveys are continued diligently throughout the silent period (requiring daylight and reasonable sighting conditions), and b) no whales, other marine mammals, or sea turtles are observed in the exclusion zone. If whales, other marine mammals, or sea turtles are observed in the exclusion zone during the short silent period, resumption of seismic survey operations must be preceded by ramp-up.

III. ENVIRONMENTAL IMPACTS

A. INTRODUCTION

1. Significance Criteria

The objectives of the impact analysis are 1) to determine whether G&G activities have significant impacts on the marine, coastal, or human environments of the GOM; and 2) to identify significant impacts, if any, for further NEPA analysis.

For the impact analysis, resource-specific significance criteria were developed for each category of the affected environment. The criteria reflect consideration of both the context and intensity of impact (40 CFR 1508.27). Criteria for marine mammals and sea turtles reflect the Federal protected status of all species occurring in the GOM. Adverse impacts are classified into one of three levels:

- significant adverse impact (including those that could be mitigated to non-significance);
- adverse but not significant impact; or
- negligible impact.

Significance criteria presented in this analysis, reflecting accepted threshold levels for significance (i.e., thresholds are resource-specific), are based on a recent EIS (USDOI, MMS, 2001b) for proposed floating production, storage and offloading (FPSO) systems being considered in the deep-water regions of the GOM. Impacts are also categorized as direct or indirect. No beneficial impacts (either significant or non-significant) have been identified. Cumulative impacts are discussed in a separate section under each resource.

2. Preliminary Screening

After a review of previous environmental assessments (EAs) and EISs (e.g., USDOI, MMS, 1996, 1997a,b, 1999, 2000, 2001a,b, 2002a,b) and relevant literature pertinent to historic and projected OCS activities (e.g., Baud et al., 2002), the following resources were initially considered for impact analysis:

- marine mammals (including ESA listed species[1] and strategic stocks[2]);
- sea turtles (all are ESA listed species);

---

[1] The NMFS and the U.S. Fish and Wildlife Service (USFWS) are responsible for designating "listed species" – those species formally designated (or under consideration, potentially to be designated) as endangered (E) or threatened (T) (including E or T [S/A] species, for *similarity in appearance*), pursuant to the ESA of 1973, as amended. Once listed, the precarious status of that species is formally recognized, and various requirements are triggered (e.g., consultation with appropriate agencies, restrictions on "take" [see **Appendix B**], recovery plan development).

[2] NMFS is responsible for managing stocks of cetaceans and pinnipeds under the MMPA of 1972. In the MMPA, the term "stock" means a group of marine mammals of the same species or smaller taxa in a common spatial arrangement that interbreed when mature. Some GOM cetacean stocks are classified as *strategic*. Strategic stocks are those that are 1) subject to a level of direct human-caused mortality that exceeds their Potential Biological Removal (PBR) level (defined in **Appendix F, Section I.A.1.5 – Abundance Estimates**); 2) declining and thus likely to be listed as threatened species under the ESA within the foreseeable future; and/or 3) currently listed as endangered or threatened species under the ESA. The latter would include the sperm whale and manatee. Other strategic stocks within the Gulf are four species of beaked whales; the short-finned pilot whale; and bay, sound, and estuarine communities of the bottlenose dolphin.

- fishes (including listed species and ichthyoplankton);
- commercial and recreational fisheries;
- coastal and marine birds (including ESA listed species);
- benthic communities;
- cultural resources;
- military uses;
- recreational and commercial diving;
- marine transportation;
- geology/sediments; and
- air and water quality.

A preliminary screening was conducted to focus the impact analysis on those G&G activities and resources with potential for non-negligible impacts. First, a matrix was prepared to identify impact agents associated with each type of G&G activity (**Table III-1**). The impact agents are 1) airgun noise; 2) sonar noise; 3) seafloor disturbance; 4) vessel traffic; 5) towed streamers; and 6) aircraft traffic. A second matrix was prepared to identify resources potentially affected by each type of G&G activity (**Table III-2**). In this preliminary analysis, the level of impact associated with each interaction was categorized as *no impact* (i.e., no measurable impact to a resource evident), *negligible impact* (i.e., measurable but relatively minor impact to a resource predicted), or *potentially adverse impact* (i.e., measurable impact to a resource predicted).

The preliminary screening indicates that most G&G activities have negligible impact or no impact. The following G&G activities do not require further analysis because they are expected to have a negligible impact (or no impact) on the environment:

- *All of the remote sensing methods* (radar imaging, aeromagnetic surveys, gravity surveys, gravity gradiometry, and marine magnetic surveys). These involve only a small amount of vessel and/or aircraft traffic and are considered to have little or no environmental impact on any resource.

- *Electromagnetic surveys* (MT and bi-pole methods). In addition to small amounts of ship traffic, these surveys involve only minor disturbance to the seafloor (temporary placement of receiver boxes on the bottom). Potential impacts are negligible.

The following resources were determined to have negligible or no impacts and are not considered further:

- *Cultural resources.* Of the G&G activities under consideration, only geological and geochemical sampling (bottom sampling and shallow coring) could adversely affect shipwrecks and prehistoric sites. These sites are most likely to occur inshore of the 45-m isobath (see **Appendix F, Section II.C - Cultural Resources**). Bottom sampling and shallow coring occur both prelease (as an exploratory tool) and postlease after a shallow hazards survey of a lease block has been conducted, which would identify potential archaeological sites. Under lease stipulations, lessees are required to report discovery of any archaeological resource to the MMS and take precautions to protect the resource from operational activities (see **Appendix E, Section II.A.2 - Stipulations and Protective Measures**). This required mitigation has proven effective over many years, and no adverse impacts are anticipated.

Table III-1
Potential Impact Agents Associated with Geological and Geophysical (G&G) Activities

	Airgun Noise	Sonar Noise	Seafloor Disturbance	Vessel Traffic	Towed Streamers	Aircraft Traffic
Seismic Surveys						
High-resolution site surveys	X	--	--	X	X	--
Streamer surveys (high-resolution, 2D, 3D, multi-ship)	X	--	--	X	X	--
Ocean bottom cable surveys	X	--	X	X	--	--
Vertical cable surveys	X	--	X	X	X	--
Vertical seismic profile surveys	X	--	--	X	--	--
Deep-Tow Side-Scan Sonar Surveys	--	X	--	X	--	--
Electromagnetic Surveys						
Magneto-telluric method	--	--	X	X	--	--
Bi-pole method	--	--	X	X	--	--
Geological and Geochemical Sampling						
Bottom sampling	--	--	X	X	--	--
Shallow coring	--	--	X	X	--	--
Remote Sensing						
Radar imaging	--	--	--	--	--	--
Aeromagnetic surveys	--	--	--	--	--	X
Gravity surveys	--	--	--	X	--	X
Gravity gradiometry	--	--	--	X	--	--
Marine magnetic surveys	--	--	--	X	--	--

X – indicates that the G&G operation normally produces the designated impact agent.
-- – indicates that the G&G operation does not normally produce the designated impact agent.

Table III-2
Preliminary Screening of Potential Impacts

Activity	Marine Mammals	Sea Turtles	Fishes	Commercial and Recreational Fisheries	Coastal and Marine Birds	Benthic Communities	Cultural Resources	Military Uses	Recreational and Commercial Diving	Marine Transportation	Geology/Sediments	Air and Water Quality
Seismic Surveys												
Streamer surveys (high-resolution, 2D, 3D, multi-ship)	X	X	X	X	o	--	--	--	o	o	--	o
Ocean bottom cable surveys	X	X	X	X	o	o	--	--	o	o	o	o
Vertical cable surveys	X	X	X	X	o	o	--	--	o	o	o	o
Vertical seismic profile surveys	X	X	X	o	o	--	--	--	o	o	--	o
Deep-Tow Side-Scan Sonar Surveys	o	o	o	o	--	--	--	--	o	--	--	o
Electromagnetic Surveys												
Magneto-telluric method	--	--	--	--	--	o	--	--	--	--	o	o
Bi-pole method	--	--	--	--	--	o	--	--	--	--	o	o
Geological and Geochemical Sampling												
Bottom sampling	--	--	--	--	--	o	o	--	--	--	o	o
Shallow coring	--	--	--	--	--	o	o	--	--	--	o	o
Remote Sensing												
Radar imaging	--	--	--	--	--	--	--	--	--	--	--	--
Aeromagnetic surveys	--	--	--	--	--	--	--	--	--	--	--	o
Gravity surveys	--	--	--	--	--	--	--	--	--	--	--	o
Gravity gradiometry	--	--	--	--	--	--	--	--	--	--	--	o
Marine magnetic surveys	--	--	--	--	--	--	--	--	--	--	--	o

X – indicates potentially adverse impact (i.e., measurable impact to a resource predicted).

o – indicates negligible impact (i.e., measurable but relatively minor impact to a resource predicted).

-- – indicates no impact (i.e., no measurable impact to a resource evident).

- *Military uses*. All military activities in the GOM OCS occur within warning areas designated by the Department of Defense. Lessees and permittees conducting G&G operations are required to coordinate with the appropriate military command. This required multi-use conflict mitigation has proven effective over many years, and no adverse impacts are anticipated.

- *Recreational and commercial diving*. Given current levels of recreational and commercial diving activity[3] in the GOM, G&G activities have the potential to affect commercial and recreational divers via airgun and sonar noise and vessel traffic (**PEA Table II-2**). Several of the acoustic sources used in typical G&G seismic activities exceed the 150 dB re 1 μPa recommended maximum sound level for dive sites at their source[4]. With the exception of limited and localized VSP surveys, seismic surveys are not routinely conducted close to platforms or anchored dive vessels, where their large towed streamer arrays might entangle divers or facility components (e.g., risers, anchor buoys, etc.). Streamers also are equipped with radar reflectors that can be detected and avoided by dive boats in the vicinity. Due to spreading and transmission losses, and the fact that seismic sources are designed to direct their energy downward (towards the seafloor), sound source levels diminish horizontally with increasing distance in the water column. The effects of continuous sound and sound intensity on human divers have been documented[5]. It is highly unlikely that commercial or recreational divers would be close enough to repetitive, intermittent, and localized seismic survey sources to experience adverse impacts. Given these factors (i.e., limited survey activity in close proximity to platforms, sound attenuation with distance, studies of underwater noise and their effects on divers), impacts to divers and dive boats from seismic survey vessels are considered to be negligible.

- *Marine transportation*. All G&G vessel activities are conducted according to the navigation rules of the U.S. Corps of Engineers and the U.S. Coast Guard. G&G vessel traffic uses existing shipping lanes and routes, including the Gulf Intracoastal Waterway. Seismic surveys are not conducted within shipping lanes where large towed streamer arrays might

---

[3] There are nearly 2,500 commercial divers employed in the U.S. (Occupational Employment Statistics, 1999), with the majority working in the GOM. Most commercial diving activity in the Gulf is centered around installation, maintenance, or removal of offshore oil platforms and pipelines. For recreational diving, there are hundreds of dive shops catering to several thousand recreational divers in Gulf states. Recreational diving activity in the eastern Gulf is associated with the reefs and hard bottom areas seen along the west Florida continental shelf. In the central and western Gulf, recreational diving is primarily realized on wrecks, around oil platforms, and on topographic highs (e.g., Flower Garden Banks).

[4] Three main human physiological symptoms can be associated with exposure to continuous, high intensity, low frequency sound sources, including 1) sensory nervous system disruption through the epidermis (i.e., increased vibrotactile sensitivity; the frequency response prompting this condition peaks at ~250 Hz), resulting in tingling and numbness in the diver's extremities; 2) acoustically forced vibration of gas pockets in the gastrointestinal tract, which causes abdominal discomfort; and 3) temporary threshold shift (TTS) in hearing caused by intense or prolonged high sound levels. Other effects such as arterial resonance and lung hemorrhaging are possible, but are rarely seen in instances where explosive devices are not in use (United Nations Educational, Scientific, and Cultural Organization, 1996).

[5] Steevens et al. (1997) summarized studies that showed no indication of adverse effects in divers exposed to continuous, low frequency (240 Hz) underwater sound levels ranging from 130 to 190 dB, with exposure ranging from 4 to 15 min. Seismic operations employ intermittent, repetitive sound pressure levels of 215 to 240 dB re 1 μPa (zero-to-peak), with attenuation to 160 or 180 dB (rms) levels at hundreds to thousands of meters from the sound source (see **Appendix C, Section III - Seismic Source Levels**). Ranges where received sound levels from typical seismic equipment used in the GOM exceed 190 dB are extremely limited, estimated at tens of meters from the seismic array (see **Appendix C, Section III** for calculations of radial distance).

affect shipping traffic. Streamers have a radar reflector that can be detected by other vessels in the vicinity of seismic surveys. Impacts on marine transportation are negligible.

- *Geology and sediments.* Of the G&G activities under consideration, only geological and geochemical sampling (bottom sampling and shallow coring) could adversely affect sediments. Alteration of ambient sediments could be expected via surficial sediment disturbance, resuspension, and creation of minor surficial features (e.g., gouges, holes, depressions, etc.). The total seafloor area disturbed during these activities annually is much less than 1% of the area of a single lease block. Further, geological and geochemical sampling will have no effect on local or regional geology. Other activities such as bottom cable surveys, vertical cable surveys, and electromagnetic surveys involve temporarily placing instruments on the seafloor, but no adverse impacts are anticipated. The impacts on geology and sediments are negligible.

- *Air and water quality.* Ships and aircraft involved in G&G activities produce air pollutant emissions. Survey vessels discharge treated sanitary and domestic wastes from U.S. Coast Guard-approved sanitation units. Impacts on water and air quality are negligible.

The preliminary screening indicates that seismic surveys have potentially adverse impacts on marine mammals, sea turtles, fishes, and commercial and recreational fisheries. The rest of the impact analysis focuses mainly on these categories. Two other resources with negligible impacts (coastal and marine birds, and benthic communities) are also discussed briefly due to the potential for adverse impacts that have been reduced to a non-significant level by existing mitigation measures.

B.   MARINE MAMMALS

1.   Affected Environment

Twenty-nine species of marine mammals are known to occur in the GOM (**Table III-3**; see **Appendix F, Section I.A - Marine Mammals** for further information). There are 28 cetacean species, which include 7 mysticete (baleen whales) and 21 odontocete (toothed whales and dolphins) species; and one sirenian species, the West Indian manatee. Their population status is indicated using the following categories (adapted from Würsig et al., 2000):

- *Common*: a species that is abundant and widespread throughout the region in which it occurs.
- *Uncommon*: a species that does not occur in large numbers, and may or may not be widely distributed throughout the region in which it occurs.
- *Rare*: a species present in such small numbers throughout the region that it is seldom seen.
- *Extralimital*: a species known on the basis of few records that are probably the result of unusual movements of few individuals into the region.

Mysticetes (baleen whales) other than Bryde's whale are considered extralimital or rare in the Gulf (Würsig et al., 2000). Bryde's whale is the most frequently sighted baleen species in the Gulf, though considered uncommon. Strandings and sightings data suggest that this species may be present throughout the year, generally in the northeastern Gulf near the 100-m isobath between the Mississippi River delta and southern Florida (Davis et al., 2000; Würsig et al., 2000).

Table III-3

Marine Mammals of the Gulf of Mexico

Scientific Name	Common Name	Management Status[1]	Population Status[2]	Scientific Name	Common Name	Management Status[1]	Population Status[2]
ORDER CETACEA	**WHALES AND DOLPHINS**			Family Delphinidae	Dolphins (Delphinids)		
SUBORDER MYSTICETI	BALEEN WHALES			*Stenella frontalis*	Atlantic spotted dolphin	none	4
Family Balaenidae	Right whales			*Tursiops truncatus*	Bottlenose dolphin	none	4
Eubalaena glacialis	Northern right whale	E, S	1	*Stenella clymene*	Clymene dolphin	none	4
Family Balaenopteridae	Rorquals			*Pseudorca crassidens*	False killer whale	none	3
Balaenoptera musculus	Blue whale	E, S	1	*Lagenodelphis hosei*	Fraser's dolphin	none	4
Balaenoptera edeni	Bryde's whale	none	3	*Orcinus orca*	Killer whale	none	3
Balaenoptera physalus	Fin whale	E, S	2	*Peponocephala electra*	Melon-headed whale	none	4
Megaptera novaeangliae	Humpback whale	E, S	2	*Stenella attenuata*	Pantropical spotted dolphin	none	4
Balaenoptera acutorostrata	Minke whale	none	2	*Feresa attenuata*	Pygmy killer whale	none	3
Balaenoptera borealis	Sei whale	E, S	2	*Globicephala macrorhynchus*	Short-finned pilot whale	S	4
SUBORDER ODONTOCETI	TOOTHED WHALES/DOLPHINS			*Grampus griseus*	Risso's dolphin	none	4
Family Physeteridae	Sperm whales			*Steno bredanensis*	Rough-toothed dolphin	none	4
Physeter macrocephalus	Sperm whale	E, S	4	*Stenella longirostris*	Spinner dolphin	none	4
Family Kogidae	Pygmy and dwarf sperm whales			*Stenella coeruleoalba*	Striped dolphin	none	4
Kogia breviceps	Pygmy sperm whale	none	4[3]	**ORDER SIRENIA**	**DUGONGS AND MANATEES**		
Kogia simus	Dwarf sperm whale	none	4[3]	Family Trichechidae	Manatees		
Family Ziphidae	Beaked whales			*Trichechus manatus latirostris*	Florida manatee	E	2[4]
Mesoplodon densirostris	Blainville's beaked whale	S	2-4[3]	*Trichechus manatus manatus*	Antillean manatee	E	2
Ziphus cavirostris	Cuvier's beaked whale	S	2-4[3]				
Mesoplodon europaeus	Gervais' beaked whale	S	3				
Mesoplodon bidens	Sowerby's beaked whale	S	1				

[1] Management status: E = endangered under the Endangered Species Act of 1973; S = strategic stock under the Marine Mammal Protection Act of 1972, as indicated by Waring et al. (1999).

[2] Population status: 1 = extralimital; 2 = rare; 3 = uncommon; 4 = common (adapted from Würsig et al. 2000).

[3] Determining the population status of Blainville's and Cuvier's beaked whales and dwarf and pygmy sperm whales (*Kogia*), which occur in the Gulf of Mexico, is problematic. Würsig et al. (2000) classify the presence of Blainville's and Cuvier's beaked whales in the Gulf as rare. The National Marine Fisheries Service (NMFS) notes that beaked whales are difficult to identify to species, they are hard to see, and they occur in small groups. In general, only Cuvier's beaked whales and adult male Blainville's beaked whales can be identified in the field. Nevertheless, NMFS suggests that sightings of beaked whales and *Kogia* in the Gulf are not rare or that uncommon. During all NMFS aerial and ship surveys combined, there have been sightings of about 75 beaked whale groups (15 as Cuvier's beaked whale, 36 as *Mesoplodon* spp., 2 as Blainville's beaked whale, and 22 as unidentified ziphiids). While these sightings are widely distributed in the deep waters of the northern Gulf, because they occur in small groups (usually <4 to 6), the abundance of each beaked whale category is low compared with species with a similar number of sightings that occur in much larger groups. Another factor to consider is the sightability of beaked whales and *Kogia*; they rarely leap out of the water or splash at the surface and are difficult to see unless seas are very calm (Beaufort sea state 0, 1). While a quantitative analysis has not been performed, in general, as the sea state decreases, the number of beaked whale sightings increases. The majority of NMFS surveys have been conducted in sea states that are not optimal for sighting beaked whales. Therefore, NMFS suggests that Blainville's and Cuvier's beaked whales are at least uncommon, and depending on how abundance is viewed (group sightings or number of individuals), may in fact, along with Gervais' beaked whale, be common. Because of the difficulties distinguishing Gervais' and Blainville's beaked whale, it may be that if one species is truly rare, the other is without doubt common or uncommon. On the basis of the frequency of their sightings, the Marine Mammal Commission considers *Kogia* as common in the northern Gulf of Mexico.

[4] Excluding the Florida coast, the Florida manatee is considered rare in the northern Gulf of Mexico.

Most of the odontocetes are considered common. The population levels of certain odontocete species in the Gulf, such as beaked whales and dwarf and pygmy sperm whales, have historically been difficult to assess accurately. Historical sighting records of these species suggest they have widespread distributions. They are believed to be deep-diving animals and therefore may spend extended periods of time below the surface. These species also tend to occur or are commonly sighted as solitary individuals or in relatively small group sizes that are usually not visibly demonstrative while on the surface (e.g., leaping or creating splash). Anecdotal evidence suggests that beaked whales shy away from vessels. Therefore, these species may not always be seen during field survey efforts, especially in conditions of elevated sea states or sunlight glare (reflection) on the sea surface. Consequently, the estimated population densities of these species in the Gulf are likely to be underestimated. Sightings data collected from recent, deep-water surveys in the Gulf, along with historic strandings data, suggest that the status of beaked whales in the Gulf may range from extralimital (in the case of Sowerby's beaked whale) to uncommon or common. Dwarf and pygmy sperm whales are considered uncommon to common in the Gulf (Würsig et al., 1996).[6]

The sperm whale is the only endangered species of marine mammal likely to come in contact with offshore G&G seismic activities in the GOM. Sperm whales are the most common large whale in the GOM. Survey data suggest that they are most concentrated over and around the 1,000-m depth contour south of the Mississippi River delta and at similar depths off southeast Florida, near the Dry Tortugas. Historic sightings and strandings data suggest that sperm whales may be resident (i.e., their occurrence is persistent or recurrent) within the GOM, though details of their seasonal movements remain unclear (Davis et al., 2000; Würsig et al., 2000). Seven mysticete species may occur in the Gulf; however, six species are considered rare or extralimital (i.e., northern right, blue, fin, humpback, minke, and sei whales; Jefferson, 1995; Jefferson and Schiro, 1997). One mysticete species (i.e., Bryde's) is considered to be uncommon (Würsig et al., 2000). One endangered sirenian species, the West Indian manatee (including two subspecies, the Florida and Antillean manatees) inhabits only coastal marine, brackish, and freshwater habitats.

No areas of critical habitat have been designated for the sperm whale or any mysticetes in the GOM. The Florida manatee has two critical habitats in the GOM: Crystal River and southwest Florida from around Tampa Bay south to the southwestern tip of the state (USDOI, USFWS, 1996). In addition, nearshore areas from Crystal River to Apalachicola, and from Sarasota south to the southwestern tip of Florida, are identified as migratory routes for Florida manatees.

2. Impacts of Routine Activities - Alternative 1 (Proposed Action)

Impact agents associated with routine G&G seismic activities that may affect marine mammals include noise from seismic surveys and deep-tow side-scan sonar surveys, and vessel traffic from various types of G&G surveys. The following sections address impacts from seismic surveys, side-scan sonar surveys, and vessel traffic in general.

---

[6] Determining the population status of Blainville's and Cuvier's beaked whales that occur in the GOM is problematic. Würsig et al. (2000) classify their presence in the Gulf as rare. In contrast, NMFS notes that beaked whales are difficult to identify to species, they are hard to see, and they occur in small groups (see **Table III-3** footnote c and **Appendix F, Section I.A – Marine Mammals**). NMFS suggests that Blainville's beaked whale and Cuvier's beaked whale are at least uncommon and depending on how abundance is viewed (group sightings or number of individuals), may in fact, along with Gervais' beaked whale, be common. Würsig et al. (2000) consider *Kogia* to be uncommon, while the Marine Mammal Commission (MMC) considers this group to be common in the northern GOM.

a.   *Seismic Surveys*

1)   *Vessel Noise*

The primary sources of sounds from all power vessel classes include propeller cavitation and "singing," hull noises, and machinery noises. Propeller cavitation is usually the dominant noise source of vessels underway, though these noise levels may be exceeded if the vessel is powered by medium- to high-speed diesel engines (Ross, 1976). Propeller singing is a resonant vibration noise of a moving propeller. Propulsion machinery noises originate from a diverse array of mechanical sources within the hull. In addition, other noises include flow noise from water dragging along the hull and bubbles breaking in the vessel's wake. Larger vessels, like seismic survey vessels, tend to produce stronger and lower-frequency sounds up to approximately 50 Hz. The effects of moving seismic vessels on marine mammals (originating from vessel movement, noise, or both) are difficult to assess with accuracy, based on the wide array of observed behavioral responses to moving vessels, both between species and among species. Generally, it is conservative to suggest that noises associated with seismic survey vessel traffic may occasionally elicit behavioral changes in marine mammals that are in proximity to the vessel. These behavioral changes may include some evasive maneuvers such as diving or changes in swimming direction but are not expected to adversely impact these species. Several species of marine mammals (e.g., dolphins) also are attracted to vessels for bowriding activity. Under these circumstances, vessel noise may be attracting these species. Anecdotal evidence of bowriding during active airgun operations suggests, for such species, that the vessel itself may produce a sound shadow at the surface of the water near the bow, that vessel and seismic noise produce no adverse impact, and/or that the benefits from bowriding exceed any adverse impact associated with noise.

2)   *Seismic Noise*

Seismic surveys were identified as a primary concern at the Marine Protected Species Workshop held by MMS in New Orleans in 1999. Concerns focused on the amount of noise-producing activity that occurs in the Gulf, changes in levels and characteristics of ambient noise, the level of duplication of seismic survey effort, and whether seismic operations have already affected the response, distribution, abundance, or productivity of species or stocks. The concern is illustrated by the observation made during multi-year acoustic surveys conducted under the MMS-sponsored North-Central and Western GOM Cetacean Study (also known as "GulfCet I and II") that the percent of survey effort with seismic signals present increased from 10% to 34% between 1991-1994 and 1996-1997 (Norris et al., 2000). This increase coincidentally corresponds to higher numbers of seismic surveys during the same period, as indicated previously in **PEA Table II-4**. A preliminary analysis of previous seismic survey activity and projections of future activity levels (see **PEA Section II.B.2 - Level of Activity** and **Appendix D, Section V - Level of Activity**) suggests that seismic operations follow a cycle, further suggesting that ensonification (i.e., introduction of anthropogenic noise) of the marine environment from seismic sources is also cyclic.

Potential impacts of seismic surveys on marine mammals have been reviewed by Richardson et al. (1995), Gordon et al. (1998), and Davis et al. (1998). A literature review and discussion of potential impacts is presented in **Appendix G**. A key point is that almost all impacts of seismic surveys have been inferred or assumed by implication rather than observed. There have been no documented instances of deaths, physical injuries, or auditory (physiological) effects on marine mammals from seismic surveys. Behavioral responses have been observed in many instances, primarily in mysticetes. However, the biological importance of such behavioral responses (i.e., to the individual animals and populations involved) has not been determined.

Richardson et al. (1995) defined four zones of potential noise effects on marine mammals (in order of decreasing severity):

- Hearing loss, discomfort, and injury (physical effects);
- Masking;
- Responsiveness (behavioral effects); and
- Audibility.

These categories provide a useful framework for discussing impacts of seismic surveys on GOM marine mammals. Audibility *per se* is not an impact and will not be discussed; however, audibility that leads to responsiveness has been addressed in this analysis. The impact analysis focuses on those species groups most likely to come into contact with seismic operations in the GOM. These are Bryde's whale (the only mysticete likely to occur in the Gulf, although characterized as uncommon), sperm whale, dwarf and pygmy sperm whales, beaked whales, and delphinids. Manatees are excluded from the analysis; because of their coastal habitat, they are highly unlikely to come in contact with seismic survey activities other than coastal vessel traffic (addressed later in a separate section).

3)   *Hearing Loss, Discomfort, and Injury*

Physical impacts of seismic survey noise may range from temporary hearing impairment to gross physical injury. Unlike explosives that produce omnidirectional pressure pulses, sound energy emitted from seismic airgun arrays is focused downward (see **Appendix D, Section II - Seismic Sources**). Studies have shown that gross physical damage, usually in the form of organ injury, is closely correlated with the positive acoustic impulses generated from an underwater explosion (Richardson et al., 1995). Considering the much lesser strength of acoustic impulses and the downward focus of seismic airgun arrays, it is unlikely seismic sources could produce gross physical damage in marine mammals. Such effects, if they were to occur, would be limited to animals positioned in very close proximity[7] to the airgun array, a zone that would most likely be avoided by marine mammals. Only an airgun starting up near full power immediately above a marine mammal would be likely to cause physical (e.g., organ) injury. Therefore, the main concern of auditory-induced physical effects is the potential for temporary or permanent hearing impairment (i.e., physiological change or damage to the auditory mechanism resulting from exposure to acoustic energy). The auditory mechanism experiences temporary and permanent impairments at lower levels of acoustic pressure or energy than other tissues or organs (Ketten, 1995).

At the lowest end of the physical effects scale to the auditory mechanism is a condition known as temporary threshold shift (TTS), a temporary, reversible form of hearing impairment. In TTS, the lower threshold of hearing in the relevant frequency band is increased (i.e., hearing becomes less sensitive) when exposed to a critical combination of sound intensity and duration. Moving up the physical effects scale, a phenomenon known as permanent threshold shift (PTS) occurs. With PTS, the hearing threshold shift is non-recoverable - i.e., permanent damage results to the auditory system. Repeated exposure to TTS levels without sufficient recovery time can lead to PTS. Finally, the most severe effects are physical damage, where organs and tissue structures rupture due to the high and rapid pressure fluctuations. An excellent discussion and summary of TTS and PTS can be found in Department of the Navy (2001b).

---

[7] "Very close proximity" is estimated to encompass that zone located immediately around and beneath an array (e.g., within several tens of meters of each airgun). Precise calculations of injury distance are problematic, variably affected by the species present and its sensitivity to seismic energy, animal orientation in the water, airgun output and array geometry, and physical characteristics of the environment, among others. Airgun characteristics are detailed in **Appendix D** (**Section II - Seismic Sources**), while calculations of radial distance to the isopleth of regulatory interest (i.e., 180 dB re 1 μPa [rms]) are provided in **Appendix C** (**Section III - Seismic Source Levels**).

A recent analysis of marine mammal hearing compiled by Ketten (1998) shows that mysticetes (baleen whales) exhibit inferred hearing thresholds of 10 to 31,000 Hz, with dominant frequencies of 16 to 25,000 Hz. Baleen whale hearing has not been extensively studied. There are no specific data regarding sensitivity, frequency or intensity discrimination, or localization abilities in baleen whales. Baleen whales apparently are more dependent on low frequency sounds than other marine mammals. The lack of specific data on baleen whale hearing abilities remains a major limitation in evaluating the effects of manmade noise on this group. Among the odontocetes, hearing thresholds are highly species-specific, with dominant frequencies of 120 to 60,000 Hz in delphinids (Ketten, 1998). Toothed whales are probably sensitive to sounds above approximately 10 kHz. Below the 10-kHz level, sensitivity deteriorates with decreasing frequency, with the possible exception of the sperm whale (Carder and Ridgway, 1990). The sensitivity of many toothed whale species to high frequency sounds is attributed to their use of high frequency sound pulses in echolocation and moderately high frequency calls for communication. Low frequency hearing has not been studied extensively in toothed whales; however, some species may be able to detect sound frequencies as low as 60 to 105 Hz. Below 1 kHz, where most industrial noise energy is concentrated, toothed whale hearing sensitivity appears to be relatively poor. Toothed whales also possess good frequency and intensity discrimination abilities, as well as good directional localization capabilities. Additional information pertinent to marine mammal hearing can be found in **Appendix G**. There are no documented data on auditory-induced physical effects of underwater seismic noise on mysticetes, sperm whales, dwarf and pygmy sperm whales, or beaked whales. One TTS study has been done on bottlenose dolphins (Ridgway et al., 1997; Schlundt et al., 2000). The study was designed to predict auditory effects of military sonars, and it involved pure tones of 1-s duration, with frequencies of 3, 20, and 75 kHz. Sound pressure levels causing TTS ranged from 192 to 201 dB[8]. Additional experiments have been done by Ridgway and colleagues using simulated acoustic signatures of underwater explosions (Finneran et al., 2000b). Due to the type of sound source and the frequency range, neither of these data sets can be assumed applicable to seismic pulses (i.e., seismic sources are not expected to produce the same type of impact as explosives or sonar).

Richardson et al. (1995) attempted to use human Damage Risk Criteria (DRC) to predict noise levels that could cause TTS and/or PTS in marine mammals. They calculated that a DRC for a marine mammal exposed to 100 seismic pulses might be a received sound pressure level of 178 to 208 dB re 1 µPa. They emphasized that these values are extremely speculative, given the unknown relevance of human in-air data to marine mammals underwater. Richardson et al. (1995) further estimated that, for a typical airgun array operating in arctic waters, marine mammals would have to be within about 1 km to the side of an airgun array to be exposed to the lower end of this range, and probably within 100 m to experience levels of 200 dB or more. In contrast, sound propagation estimates developed for proposed 3D seismic operations in the Santa Barbara Channel (California) indicated that the 180-dB (average pulse pressure level) contour was 316 m from the source (LePage et al., 1995). Quantitative predictions of sound transmission (and sound attenuation) vary on a site-specific basis, according to a series of variables (i.e., source level and spectral characteristics of the sound, the rate of sound attenuation with distance, ambient sound characteristics, and physical conditions of the environment). Output also varies between arrays depending upon airgun and array size and array geometry. In all cases, the animals beneath an array are exposed to the highest energy levels.

There has been some scientific consensus that received sound pressure levels above 180 dB re 1 µPa involve some risk of hearing impairment. At a workshop on marine mammals and low-frequency sound convened by the MMS High-Energy Seismic Survey (HESS) Team in 1997, an expert panel expressed concern about sound levels above 180 dB re 1 µPa with respect to overt

---

[8] All dB values are referenced to 1 microPascal (re 1 µPa) at 1 m. Values should be considered as rms, unless otherwise noted, per **Appendix C** (**Section II.C - rms Levels**).

behavioral, physiological, and hearing effects on marine mammals in general. While the HESS workshop discussions resulted in recommendations regarding sound exposure levels, a formal proceedings document outlining such recommendations has yet to be published. The Department of the Navy (1999a, 2001a) used a received level of 180 dB re 1 µPa as a "conservative" criterion for risk of potential hearing impairment from a single pulse.

Use of a single, broadband sound pressure level such as 180 dB re 1 µPa as a TTS criterion is problematic for several reasons:

1) At onset levels, TTS is highly frequency-dependent. Animals with poor hearing (i.e., high hearing threshold) in the low frequency range are unlikely to be affected. The Department of the Navy (1998, 1999b, 2001b) used a TTS criterion based on whether the energy in any 1/3 octave band exceeded 182 dB re 1 µPa$^2 \cdot$ s.

2) It is not known whether animals would, in all cases, avoid exposure to levels that could cause TTS. There is, however, considerable evidence of marine mammals avoiding seismic pulses at received levels much lower than those considered likely to cause hearing impairment (Richardson et al., 1995; also see **Appendix G**). However, because auditory discomfort may occur at levels higher than those inducing TTS (Yost, 1994), animals could (in some cases) experience TTS without necessarily leaving the area.

3) Not all animals will experience TTS at a given exposure level. Instead, a gradient will occur. The Department of the Navy (1999a, 2001a,b) used a risk-continuum model to predict numbers of marine mammals potentially experiencing hearing impairment.

4) Cetacean ears have specializations that protect them from the animal's own high level vocalizations. However, these would not necessarily protect them from an unanticipated, high-level, externally generated sound (Ketten, 1998).

5) The 180 dB re 1 µPa (rms) criterion does not take into account duration of the sound exposure or the number of intermittent impulses allowed during an exposure. A typical seismic vessel surveys at 4.5 kn and fires a seismic array every 16 s, producing an intermittent, repetitive, and localized noise source.

In terms of overall risk of auditory impacts from seismic surveys, several points are noteworthy: 1) airgun arrays concentrate their output energy in the lower frequency range (i.e., 10 to 200 Hz, with lower energy levels in the 200 to 500 Hz and 500 to 1,000 Hz range, and beyond) and also tend to have more broadband signatures with energy extending well above 1 kHz (Goold and Fish, 1998; Sodal, 1999); and 2) cetaceans exhibit vocalizations (and thus an implication of suggested hearing capabilities) across a broad range of frequencies, suggesting differential susceptibility to seismic noise. **Figure III-1** depicts the vocalization characteristics of various baleen and toothed whales (as summarized by Richardson et al., 1995; Wartzok and Ketten, 1999; and National Research Council, 2003; see **Appendix G**, **Table G-1**) and the overlap with the frequency output from a typical airgun array. In terms of overall risk of auditory impacts, mysticetes in general appear to be at greater risk relative to other cetaceans. Hearing sensitivity at low frequencies down to ~10 Hz is almost certainly good, and many of the vocalizations of mysticetes occur in the low tens to a few hundred Hertz (Thompson et al., 1990; Richardson et al., 1995; Crane and Lashkari, 1996; Rivers, 1997; Stafford et al., 1998, 1999; Wartzok and Ketten, 1999), which implies (but does not prove) functional hearing in this range.

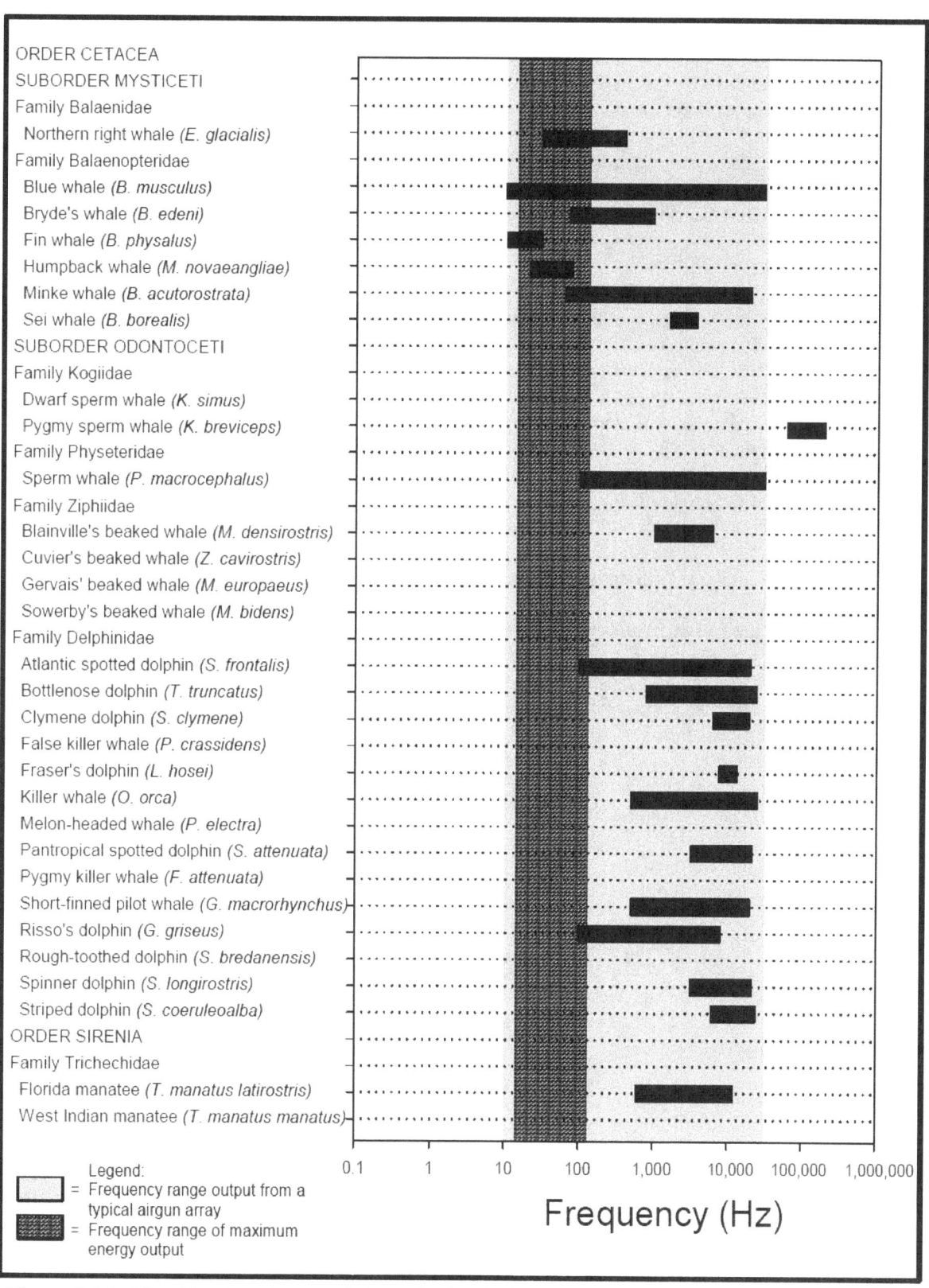

Figure III-1. Vocalization characteristics of select baleen and toothed whales relative to maximum and total output from a typical airgun array.

Bryde's whale is the only mysticete occurring regularly, though uncommonly, in the GOM. Although there are no auditory data for this species, vocalization frequencies have been documented (see **Appendix G, Table G-1**); it is generally considered that the auditory abilities of all mysticete species are broadly similar, based upon vocalization frequencies and ear anatomy (Ketten, 1998).

As airgun arrays are configured to output maximal energy in the region of a few tens of Hertz (see **Appendix D, Section II.A - Airguns**), there is clearly a possible overlap between the expected frequencies of good hearing sensitivity (low threshold) in mysticetes and maximal airgun output at source.

In contrast to the mysticetes discussed previously, odontocetes are probably sensitive to sounds above ~10 kHz (i.e., documented use of high frequency sound pulses in echolocation; moderately high frequency calls for communication), with decreasing frequency sensitivity below this level. Toothed whale hearing sensitivity appears to be relatively poor below 1 kHz. Only the sperm whale (among the odontocetes) represents a possible exception. Odontocetes of interest include delphinids, dwarf and pygmy sperm whales, beaked whales, and the sperm whale; hearing sensitivities of these species or species groups have been detailed in **Appendix G, Section II - Marine Mammal Impact Sensitivity**).

Based on the predominantly high-frequency range of their hearing, most delphinids would be at low risk for TTS and other auditory impacts from seismic surveys. This is not to say that they cannot hear and respond to seismic signals, but rather that most of the energy in the pulses is at frequencies <200 Hz, where delphinid hearing is relatively poor.

Due to their cryptic nature, beaked whales may be underrepresented among the Gulf marine mammal fauna, as noted previously. Given that the vocalization data available for beaked whales suggest the possibility of overlap with seismic sources (i.e., Blainville's beaked whale in the Gulf may vocalize at <1,000 Hz), their strategic stock status prompts further concern for any acoustic impacts to this group. Their deep diving habits also put this group at some potential risk, similar to sperm whales. As airgun arrays are configured to output maximal energy in the region of a few tens of Hertz with lower energies into the 1,000-Hz range and beyond, there also is overlap between the expected frequencies of good hearing sensitivity (low threshold) in these species and maximal airgun output at source.

One odontocete that may potentially be at some risk of auditory impact is the sperm whale. Although there are no audiograms for sperm whales, Norris et al. (2000) suggest that they have good low frequency hearing because of their large size and use of relatively low frequency pulsing. Sperm whales produce clicks with a frequency range from <100 Hz to 30 kHz, with most of the energy at 2 to 4 kHz and 10 to 16 kHz (Richardson et al., 1995). They also have distinct spectral components in their clicks at frequencies as low as 400 Hz (Goold and Jones, 1995). Although the function of these low frequency components is not clear, their presence suggests functionality and implies the ability to perceive them through the auditory system.

4)   *Masking*

Auditory masking occurs when a sound signal that is of importance to a marine mammal (e.g., communication calls, echolocation, environmental sounds cues) is rendered undetectable due to the high noise-to-signal ratio in a relevant frequency band. In the case of seismic surveys, where potential masking noise takes a pulsed form with a low duty cycle (~10%, or a 1-s disturbance in the sound field in every 10 s of ambient noise), the effect of masking is likely to be low relative to continuous sounds such as ship noise. Davis et al. (1998) considered masking to be of little consequence in relation to possible impacts of seismic surveys on the Scotian Shelf, largely due to the low duty cycle of seismic pulses. In contrast, Gordon et al. (1998) pointed out signal duration increases with range from the source and speculated that there is some potential for masking at low frequencies, mainly of consequence to mysticetes. From either viewpoint, masking is unlikely to represent an important impact of seismic sources on GOM odontocete whales and dolphins. It is possible that seismic survey activity may, on occasion, mask mysticete whale vocalizations. However, mysticetes whales are classified as extralimital, rare, or uncommon in the GOM, depending upon the species in question.

5)   *Behavioral Responses*

A number of studies have documented behavioral effects in response to seismic surveys, primarily for mysticetes (Richardson et al., 1995). Studies pertaining to the effects of stress and stress responses in mammals, including studies on marine mammals, have been reviewed by Curry (1999). However, in most cases, the biological importance of stress responses in marine mammals (e.g., effects on energetics, survival, reproduction, population status) remains unknown.[9]

Mysticetes. The mysticetes have been the most studied group of marine mammals in terms of observations of behavioral changes in response to seismic operations. As noted previously, there is clearly a possible overlap between the expected frequencies of good hearing sensitivity (low threshold) in mysticetes and maximal airgun output at source. Bryde's whale is the only mysticete species occurring regularly in the GOM, although its presence is categorized as uncommon. Although there have been no studies of Bryde's whale reactions to seismic surveys, it is generally considered that the auditory abilities of all mysticete species are broadly similar, based upon vocalization frequencies and ear anatomy (Ketten, 1998). Limited data on Bryde's whale reactions to other anthropogenic disturbance suggest little response to slowly approaching boats (Watkins, 1981), and that this species, like others, also appears to be easier to approach when feeding (Gallardo et al., 1983).

Given that no audiograms of mysticetes have been obtained, it is impossible to define what level of sound above hearing threshold may cause behavioral effects. For this reason, observations at sea have concentrated on relating received sound levels to observed behavioral changes (Malme et al., 1983, 1984, 1985, 1986, 1988; Reeves et al., 1984; Richardson et al., 1986; Ljungblad et al., 1988; McDonald et al., 1993; Richardson and Malme, 1993; Richardson, 1998; McCauley et al., 2000). It has been supposed, in the absence of audiogram data, that hearing threshold around 10 Hz may occur at a level of some 80 dB re 1 μPa (Ketten, 1998). If this were the case, exposure to sound at 10 Hz and at a level of, say, 160 dB re 1 μPa might represent a level 80 dB over hearing threshold in a mysticete.

Behavioral reactions (avoidance) have been noted in gray whales in response to received pulse levels of 164 dB re 1 μPa rms (Malme et al., 1984), equivalent to a sound exposure level of 158 dB re 1 μPa (Davis et al., 1998). Bowhead whales have also been shown to exhibit avoidance of an area of seismic surveying where received levels reached approximately 130 dB re 1 μPa rms (W.J. Richardson, oral comm. to J.C. Goold, 1998). Such values stimulating behavioral avoidance are not inconsistent with the notion of a low frequency hearing threshold of some 80 dB in mysticetes.

Sperm whales. Sperm whales have a fundamentally different lifestyle from that of mysticetes. Sperm whales are deep diving, pelagic predators that echolocate at depth using sonar clicks and feed on deep water cephalopods and fishes. Auditory thresholds of adult sperm whales have not been obtained, but it is reasonable to suppose, based on their vocalizations, that they are sensitive to a wide range of frequencies.

Possible sensitivity to low frequency sounds has been reported in sperm whales by Bowles et al. (1994), where sounds at 57 Hz with source levels of 209 to 220 dB re 1 μPa may have caused sperm

---

[9] Pertinent NMFS regulations regarding levels of harassment and "take" are discussed in **Appendix B**.

whales to stop vocalizing and/or to leave the area of ensonification. Sperm whales are a highly vocal species under natural conditions (i.e., they click almost continuously during dives), and interruption or cessation of their vocal activity has often been cited as a reaction to manmade noise. Watkins and Scheville (1975) showed that sperm whales interrupted click production in response to pinger (6- to 13-kHz) sounds. Mate et al. (1994) reported temporarily decreased sperm whale abundance in an area of seismic operations in the northeastern GOM. However, acoustic arrays recorded sperm whales producing click sequences during dives within 4 nmi of an active 3D seismic vessel during surveys conducted in 2001. It was not determined whether these animals were behaviorally affected or not by these activities (S. Swartz, NMFS, Southeast Fisheries Science Center [SEFSC], oral comm., 2001). Watkins et al. (1993) report interruption of vocal activity and immediate submergence by two sperm whales exposed to high level submarine sonar pulses. Andre et al. (1997) showed that 10-kHz pulses, with source levels of 180 dB re 1 μPa, induced startle reactions in sperm whales. Goold (1999) documented an event where six sperm whales in shallow water were actively driven through a narrow channel in a coherent formation, using ship noise and echosounder/fishfinder emissions from a flotilla of 10 vessels; sound characteristics and levels were not determined. One contradictory observation, however, reports no alteration in sperm whale vocal activity when exposed to received levels of 173 dB re 1 μPa rms from 1-g TNT detonators (Madsen and Mohl, 2000).

Sperm whales are most likely acoustically aware of their environment and can exhibit behavioral reactions in a number of ways, including interruption of vocal activity and locomotive avoidance. There are, as yet, insufficient data to assign thresholds for acoustic disturbance to sperm whales.

An additional factor to consider in the case of sperm whales is their deep diving habit. Unlike mysticetes, which may remain close to the surface for long periods, sperm whales spend relatively little time at the surface during the course of feeding activity. This means they would be less likely to receive any surface shielding afforded by refractive effects caused by near-surface hydrographic conditions, which can occur in some instances. In addition, the sperm whale dive takes them down to a depth where they could potentially be passed over directly by an operating seismic vessel without their being visually detected. As airgun arrays are generally configured to produce a maximum, low frequency energy lobe directly downwards towards the seabed, sperm whales may enter a region of increased ensonification relative to more near-surface species.

Dwarf/pygmy sperm whales and beaked whales. Little is known of the acoustic abilities of either of these groups. All species of both groups are deep diving (Willis and Baird, 1998) and almost certainly use echolocation clicks spanning the sonic and low ultrasonic frequency range. Pulsed sounds with peak frequencies below 13 kHz have been recorded from pygmy sperm whales (Caldwell and Caldwell, 1987), and the anatomy and physical properties of the dwarf sperm whale head have been shown to be consistent with production of echolocation clicks (Cranford et al., 1996; Goold and Clarke, 2000). Thompson et al. (1990) report a short duration (0.42 s) frequency modulated "cry" from a captive, stranded adult female, pygmy sperm whale, with start and finish frequencies of 1.36 and 1.48 kHz, respectively. No audiograms have been obtained in pygmy sperm whales, dwarf sperm whales, or beaked whales, and there are insufficient data to ascribe avoidance thresholds (Richardson et al., 1995). It is possible, however, that these species may, as in the case of sperm whales, be sensitive to a wide range of sound frequencies, including those produced by seismic gun arrays. This factor, along with their similar deep-diving habits and relatively widespread distributions in the Gulf, may warrant concerns for these groups (i.e., dwarf/pygmy sperm whales and beaked whales) from seismic survey activities. Beaked whales have been shown to be susceptible to acoustic disturbance (Frantzis, 1998; Balcomb and Claridge, 2000). Clicks and frequency modulated whistles have been reported from beaked whales, with frequencies ranging between 300 Hz and 40 kHz (Lynn and Reiss, 1992; Dawson et al., 1998; Rogers and Brown, 1999).

Delphinidae. The Delphinidae is a diverse group including the true dolphins, killer whales, and pilot whales. There have been few studies of the impact of seismic surveys on members of the Delphinidae; indeed, Richardson et al. (1995) comment on an almost total lack of studies on effects of G&G seismic activities on delphinid species. More recently, however, several studies have been conducted documenting the effects of seismic operations on delphinid species, as detailed below.

Physical Effects – One TTS study has been done on bottlenose dolphins (Ridgway et al., 1997; Schlundt et al., 2000). The study was designed to predict auditory effects of military sonars, and it involved pure tones of 1-s duration, with frequencies of 3, 20, and 75 kHz. The study succeeded in inducing TTS, and the sound pressure levels causing TTS ranged from 192 to 201 dB.

Au et al. (1999) observed 12 to 18 dB TTS in a bottlenose dolphin exposed to 50 min of octave-band noise centered at 7.5 kHz. Finneran et al. (2000a) discuss a behavioral response paradigm to measure masked underwater hearing thresholds in two species of odontocete – bottlenose dolphin and beluga – before and after exposure to seismic pulses from a watergun. The existence of experimental results is alluded to, but no further data are presented.

Behavioral Effects – The TTS study of Ridgway et al. (1997) also produced overt behavioral reactions, not in line with trained behavior, at levels at least 10 dB below those that induced TTS. In terms of effects of G&G vessel traffic itself, this is not expected to cause significant impacts to delphinids in terms of noise signatures. However, there is evidence that vessel traffic can affect the acoustic behavior of dolphins (Van Parijs and Corkeron, 2001). These workers describe how the vocal behavior of Pacific humpback dolphins is modified in response to transiting vessel traffic. Dolphins significantly increased their rate of whistling immediately after a vessel had moved through the study area at ranges of 1.5 km or less. Effects were particularly noticeable for mother-calf groups, and it is suggested that the noise from transiting vessels affects dolphin group cohesion. Mother-calf pairs were the most disturbed and have the greatest need to re-establish vocal contact following noise masking. This finding may have particular relevance for the disturbance of critical habitats (e.g., breeding, nursery areas). Marine mammals may habituate to common sounds and sound sources. Curry (1999) reported that an evaluation of captive beluga whales after exposure to drilling rig noise did not result in stress (determined by measurements of pre- and post-exposure levels of blood catecholamines) to the animals.

Historically, seismic survey operators have considered airguns to be sources of low frequency energy (<200 Hz) only. As such, it was not considered that the emissions would be audible to dolphin species, given their high frequency biased hearing and their relatively poor sensitivity at low frequency. However, recent measurements of airgun sources at sea (Goold and Fish, 1998; Sodal, 1999) have demonstrated that, although airgun arrays are a source of primarily low frequency energy, there also is energy at higher frequencies. These energies encompass the entire audio frequency range of 20 Hz to 20 kHz (Goold and Fish, 1998), and extend well into the ultrasonic range up to 50 kHz (Sodal, 1999). This high frequency energy must be taken into account when considering seismic interactions with members of the Delphinidae. Further, and contrary to early perceptions, the high frequency components of airgun emissions are of sufficient level to exceed the dolphin auditory threshold curve at these low frequencies, even after considerable spreading loss (J.C. Goold, oral comm., 2001; also see Goold and Fish, 1998).

Since the delphinid auditory system has a relatively poor response at the low frequency end (about 110 dB re 1 μPa at 200 Hz) and increases in sensitivity towards the ultrasonic range, it is clear that a gradient of increasing sensitivity exists over a broad frequency range to the frequency of peak sensitivity. Further, although an airgun pulse will have maximal energy at a few tens of Hertz, with energy rolling off towards the higher kHz frequencies, the increase in dolphin hearing sensitivity in this region, in a sense, compensates for this. It may be considered, very generally, that the seismic pulse spectrum and the delphinid audiogram track one another over a bandwidth of tens of kHz.

Goold and Fish (1998) recorded airgun pulses during a 2D seismic survey from a 2,120-cubic inch airgun array, across a frequency range 200 Hz to 22 kHz. It was determined that an equivalent narrow-band seismic signal level at 20 kHz was 133 dB re 1 µPa rms. This figure is some 78 dB above the equivalent point in the dolphin audiogram at 20 kHz and is a figure obtained at 1 km from the airgun array. Observations of common dolphins, during the seismic surveys from which these measurements were taken, revealed that dolphins were able to tolerate seismic pulses at a distance of 1 km from the array (Goold, 1996), so it is assumed that a received level of 133 dB re 1 µPa rms at 20 kHz is tolerable.[10] However, dolphins were not documented at ranges closer than 1 km, so it was considered that this might be representative of an upper level of tolerance.

Seismic survey pulses have also been shown to modify the vocal behavior of common dolphins in the open sea. Wakefield (2001) demonstrated a shift in certain whistle parameters during airgun shooting, specifically 1) there is an increase in the start, end, minimum and mean frequencies of whistles, and 2) the whistle contours become flatter. The significance of these changes is not clear, but they perhaps signify adaptation to the increased noise environment. Given the decreased airgun output at higher frequencies, there would seem a clear advantage in shifting vocalizations to higher frequencies to reduce noise masking. Whistle contours appear to become flatter as a consequence of the upper whistle frequency remaining largely unchanged.

Stone (1996, 1997a,b, 1998) reported that common dolphins, white beaked dolphins, and white sided dolphins were sighted in the vicinity of seismic surveys less often when the guns were firing than when they were not firing. These observations were statistically significant in the case of common dolphins.

Finneran et al. (2000b) discuss a behavioral response paradigm to measure masked underwater hearing thresholds in two species of odontocete – bottlenose dolphin and beluga – before and after exposure to sound signatures representing distant underwater explosions. Their experiments produced no masked TTS to simulated detonations as large as 500 kg HBX-1 charges, at ranges as close as 1.5 km. However, alterations in animals' trained behavior began to occur at levels equivalent to 5-kg charges at 9.3 km and 5-kg charges at 1.5 km (two bottlenose dolphins) and 500 kg at 1.9 km (one beluga whale).

Morton and Symonds (2002) demonstrated convincingly that displacement of odontocetes can occur in response to anthropogenic sound disturbance. During a 15-year study of killer whales in Johnstone Strait and Broughton Archipelago, a significant decrease in killer whale abundance was observed during a 7-year period where acoustic harassment devices (AHDs) were installed in the Broughton Archipelago. Morton and Symonds (2002) reported on the use of 10-kHz Airmar AHDs with source levels of 194 dB re 1 µPa at 1 m, designed to cause physical pain to seals. The authors did not report on AHD signal duration or duty cycle. Killer whale abundance returned to baseline levels after the AHDs were removed. The abundance of killer whales in the control area, Johnstone Strait, which was not disturbed, remained stable throughout the study. Airmar AHDs also have been used as a marine mammal deterrent in at least one other study. Kraus et al. (1997) employed active alarms, which emitted a broadband signal with a fundamental frequency of 10 kHz and a source level of 132 dB re 1 µPa at 1 m, well within the hearing range of harbor porpoises and harbor seals. Signal duration was ~300 ms and was repeated every 4 s. Kraus et al. (1997) utilized a lower power setting to scare porpoises away (rather than a higher setting designed to cause pain to seals).

---

[10] Use of the term "tolerable" is not necessarily equivalent to a "negligible effect." Further, a non-negligible effect on an individual must be evaluated relative to the population as a whole.

6)   *Analysis of Proposed Action*

The preceding discussion indicates that seismic surveys are unlikely to produce gross physical injury in marine mammals. Such effects, if they occur, would be limited to those animals located in close proximity to an airgun (i.e., estimated at several tens of meters) when it begins firing. The main concerns, therefore, are the potential for temporary or permanent hearing impairment and/or behavioral disturbance.

Hearing impairment. In the absence of species-specific data on auditory impacts for marine mammals, a received sound pressure level of 180 dB re 1 µPa (rms) or greater will be used as a indication of potential concern about temporary and/or permanent hearing impairment. Most of the energy emitted by seismic sources is in the 10 to 200 Hz range, but there is some energy up to 500 to 1,000 Hz and beyond (Richardson et al., 1995; Goold and Fish, 1998; also see **Appendix D, Figure D-3** and **Appendix G, Figures G-4** and **G-5**). Marine mammals with hearing in the low-frequency range and exposed to levels exceeding 180 dB re 1 µPa (rms) may be at risk for auditory impacts such as TTS. The species considered to be at most potential risk is Bryde's whale, the only mysticete species occurring regularly (although uncommon) in the Gulf. Sperm whales are also considered at risk for auditory impacts due to their relatively low frequency clicks, their documented responses to some low-frequency sounds, and their deep-diving behavior. Beaked whales may be sensitive to a wide range of sound frequencies, including those produced by seismic airgun arrays. This factor, when coupled with their similar deep-diving habits, strategic stock status (and population susceptibility to mortality losses, see **Appendix F, Section I.A - Marine Mammals**), and relatively widespread distribution in the Gulf, warrants concern for this group.

Typically, 3D seismic survey lines are about 25 m apart. Source levels as reported in **PEA Table II-3** are typically 233 to 240 dB re 1 µPa (zero-to-peak). While transmission loss varies depending on a number of factors including water depth, sound velocity profile, sediment type, etc., in general, sound pressure levels of 180 dB re 1 µPa (rms) or greater have been calculated to extend ~300 m (0.16 nmi) laterally from the survey line. Further discussion of impact zone calculations may be found in **Appendix C, Section III - Seismic Source Levels**, along with explicit assumptions (e.g., application of spherical or modified cylindrical spreading to calculate sound attenuation; non-radial nature of the isopleths) relative to G&G seismic surveys in OCS waters of the GOM. In the very shallow, cold waters of the Alaskan Beaufort Sea, a cylindrical spreading model was applied; modeling results and field verifications indicated that sound levels of 200 dB or more could occur within 100 m of the source (Richardson et al., 1995). In temperate waters offshore California, seismic operations were projected to create a 180-dB (average pulse pressure level) contour at 316 m from the source (LePage et al., 1995). Sound attenuation in the latter case was much greater, presumably a function of the spreading loss model employed and differential output and attenuation from the seismic array. Preliminary calculations of radial distance to the 160- and 180-dB isopleths (i.e., as possible impact zones) for a typical airgun array being used in the GOM are provided in **Appendix C, Section III - Seismic Source Levels**[11]. Based on

---

[11] Utilization of ~0.3 and ~3.0 km as appropriate maximum radial distances to the 180- and 160-dB (rms) isopleths is based on a typical seismic array (i.e., 4,550-in.^{3} airgun array; 240 dB re 1 µPa [zero-to-peak], 230 dB re 1 µPa [rms]) in commercial use in the GOM. Actual array output varies by operator and can be lower than the typical system employed in this analysis. Lower output systems produce smaller zones of ensonification. In addition, the frequency spectra of a seismic signal has not been accounted for in the distance calculations (i.e., radial distance to a target isopleth). It is known that a seismic signal is comprised primarily of low frequency components (i.e., significant energy at 10-1,000 Hz, peak frequency at 50-60 Hz, with lesser contributions from both mid- and high frequency components). Although some data regarding differential frequency attenuation of seismic sound have been published over the past few years, insufficient information is currently available to estimate accurately and integrate frequency components into the isopleth calculations within the framework of a PEA (e.g., due to limited field measurements, variations in seismic array composition and geometry, etc.). As a consequence, frequency spectra cannot readily be accounted for in the current impact analysis. See **Appendix C, Section III - Seismic Source Levels** for additional discussion.

these considerations, the assumption is that all points within a surveyed lease block will be ensonified to levels of at least 180 dB re 1 µPa (or even 200 dB or greater) at some time during a seismic survey (each point will actually be exposed repeatedly due to overlapping coverage of adjacent lines). The time between the first and second passage (or other passages) of a seismic array (and sound exposure for a stationary receptor) within a lease block may range from hours to days, depending upon the length of the vessel trackline and the relative position on the trackline. Consequently, seismic survey activity (number of lease blocks surveyed annually) provides a rough approximation of the total area of potential exposure, recognizing that the total number of surveys conducted and the total area covered are two of the factors considered during impact assessment.

From 1992 through 2000, annual number of lease blocks surveyed ranged from 1,825 to 6,520 (**PEA Table II-4**). It is estimated that the total number of blocks to be surveyed (using primarily 3D seismic) in the next several years will range from 1,041 to 6,520. Total numbers of lease blocks in the Central and Western Planning Areas are 9,110 and 6,516, respectively, for a total of 15,626. The percentage of the total area encompassed by the Central and Western Planning Areas to be surveyed will annually range from 7% to 42%. Lease blocks typically are 4.8 km on side, with a total area of 23.04 km^2; therefore, the total area covered by (3D) seismic surveys annually can be estimated to range from 23,985 to 150,221 km^2.

A second viable approach to describing the duration and scope of potential exposure is based on seismic activity in the vicinity of a single cetacean, the latter of which is (for the sake of this description) located 500 m from a survey trackline. As the seismic vessel moves along its survey track, the distance between the airgun array and the whale will vary between ~1,120 m and 500 m. The sound level that the whale receives will vary between approximately 174 and 168 dB re 1 µPa (rms) for those distances, with the source as described. At a maximum towing speed of 5 kn, the source moves 26 m every 10 s, with airguns firing every 10 to 15 s. If the mammal remains in place, it will experience 52 to 78 shots at distances ranging from ~1,120 m to 500 m and back to ~1,120 m. The time duration of the 2,000-m pass is approximately 13 min.

The number of streamers and the separation between the streamers will dictate how many times a survey source vessel will pass within some specified distance of a marine mammal. Survey requirements will also determine how long the ship will continue going in one direction, and will subsequently determine how soon the ship will pass back along a nearby parallel path. All of these details will determine the potential for total exposure (i.e., the total number of airgun shots for which the received level at an animal's location is above a certain pressure level in a certain period of time). For most surveys, the operative period would most likely be a day or a week, assuming the animal does not follow the source vessel around. Total potential exposure over this period is estimated to be based on several vessel passes (i.e., twice daily) by a stationary cetacean.

Abundance data from marine mammal surveys provide a basis for evaluating the magnitude of potential risk to marine mammals. It must be pointed out that survey data are known to be underestimates due to availability bias (failure to detect submerged individuals). More importantly, marine mammal abundance varies tremendously both spatially and temporally. The following calculations are intended to assess the magnitude of potential exposure, based on available population estimates and a recognition of spatial and temporal variability and survey data limitations. More precise estimates of the total numbers of individuals potentially exposed to seismic noise may become available as further marine mammal survey work in the GOM is completed. Additional species- or group-specific discussions of the potential for exposure to seismic noise and estimates of incidental take are presented in **Appendix L** (**Potential Incidental Harassment and Take).**

The minimum number of Bryde's whales in the northern GOM is estimated to be 17 (Waring et al., 1997)[12]. GulfCet II ship survey data for the oceanic northern GOM (>200 m) yield a slightly higher estimate of 22 Bryde's whales, with a confidence interval from 4 to 123 (Mullin and Hoggard, 2000). The mean density was estimated to be 0.005 individuals/100 km^2. A recent density estimate for Bryde's whales in the Eastern Gulf of Mexico Planning Area was established at 0.03 individuals/100 km^2. This species is not expected in the Western or Central Planning Areas, thus density estimates in these areas were zero (K. Mullin, NMFS, written comm., 2003). When cumulative annual seismic survey operations are considered (see **Appendix L**), perhaps one Bryde's whale may be exposed to levels of 180 dB or higher if it does not avoid exposure. Strandings and sightings data indicate that Bryde's whales are most likely to be present in the northeastern Gulf near the 100-m isobath (Davis et al., 2000; Würsig et al., 2000). Given that lower seismic survey activity levels are expected in the Eastern Gulf Planning Area (relative to the Central and Western Gulf), the probability of exposure of Bryde's whales to levels of 180 dB or higher (in the absence of avoidance) is quite low. Similarly, in the Central and Western Planning Areas, where most of the seismic surveys occur and where fewer Bryde's whales are thought to occur, the risk of auditory impairment is also quite low. If Bryde's whales are able to recognize approaching seismic operations and avoid exposure, fewer animals would be at risk.

The abundance of sperm whales in the oceanic northern GOM (including northeast and northwest slope [200 to 2,000 m] and abyssal [>2,000 m] areas, or strata) has been estimated at 1,349 individuals[12] (coefficient of variation [CV]=0.23; 95% confidence interval [CI]=869-2,093), based on estimates derived from ship-based surveys conducted in 1996 and 1997, and from 1999 to 2001 (Fulling et al., in review; K. Mullin, NMFS, written comm., 2003). Recent density estimates of sperm whales in the northern Gulf are 0.29, 0.44, 0.25 individuals/100 km^2 for the Eastern, Central, and Western Gulf of Mexico Planning Areas, respectively (K. Mullin, NMFS, written comm., 2003). When cumulative annual seismic survey operations are considered (see **Appendix L**), a total of three sperm whales may potentially be exposed to levels of 180 dB or greater if they do not avoid exposure (i.e., one sperm whale per planning area; see **Appendix L, Table L-11**). If sperm whales are able to recognize approaching seismic operations and avoid exposure, fewer animals would be at risk.

Sperm whales occur throughout northern Gulf oceanic waters (>200 m) and thus are not expected to be present in lease blocks on the continental shelf (<200 m). Further, localized aggregations of sperm whales appear to be persistent in the vicinity of the Mississippi River delta between the 500 and 2,000 m isobaths[13], and in the southeastern Gulf, west of the Dry Tortugas (approximately 25°N Lat., 84°W Long.) (Fulling et al., in review). The main risk would be in areas where these aggregations of sperm whales are commonly sighted (Davis et al., 2000; see **Figure III-2**)[14].

---

[12] NMFS has posted revised draft population estimates for select marine mammals in the GOM (NMFS, 2004). Population size and minimum population estimates for Bryde's whales are 42 and 25 individuals, respectively; for sperm whales, population size and minimum population estimates are 1,315 and 1,035 individuals, respectively.

[13] Distribution information for deep-water marine mammals is problematic, given the availability of limited survey data. Occurrence of deep-water species in water depths of 1,500-2,000 m and beyond may have been documented during offshore surveys or may represent the expert opinion of the source author(s). Statements regarding the potential distribution of particular species in deep water (e.g., to the 2,000-m isobath) may also represent an artificial endpoint.

[14] Two surveys were conducted under GulfCet I and II (August 1994; October 1996) to obtain behavioral observations and photo-identification of sperm whales. Areas surveyed centered around the Mississippi Canyon, a region previously recognized for high sperm whale densities. NMFS surveys in the GOM have been extensive (see **Figure III-2**), with sperm whale sightings throughout the region, and aggregations evident off the Mississippi River delta and near the Dry Tortugas.

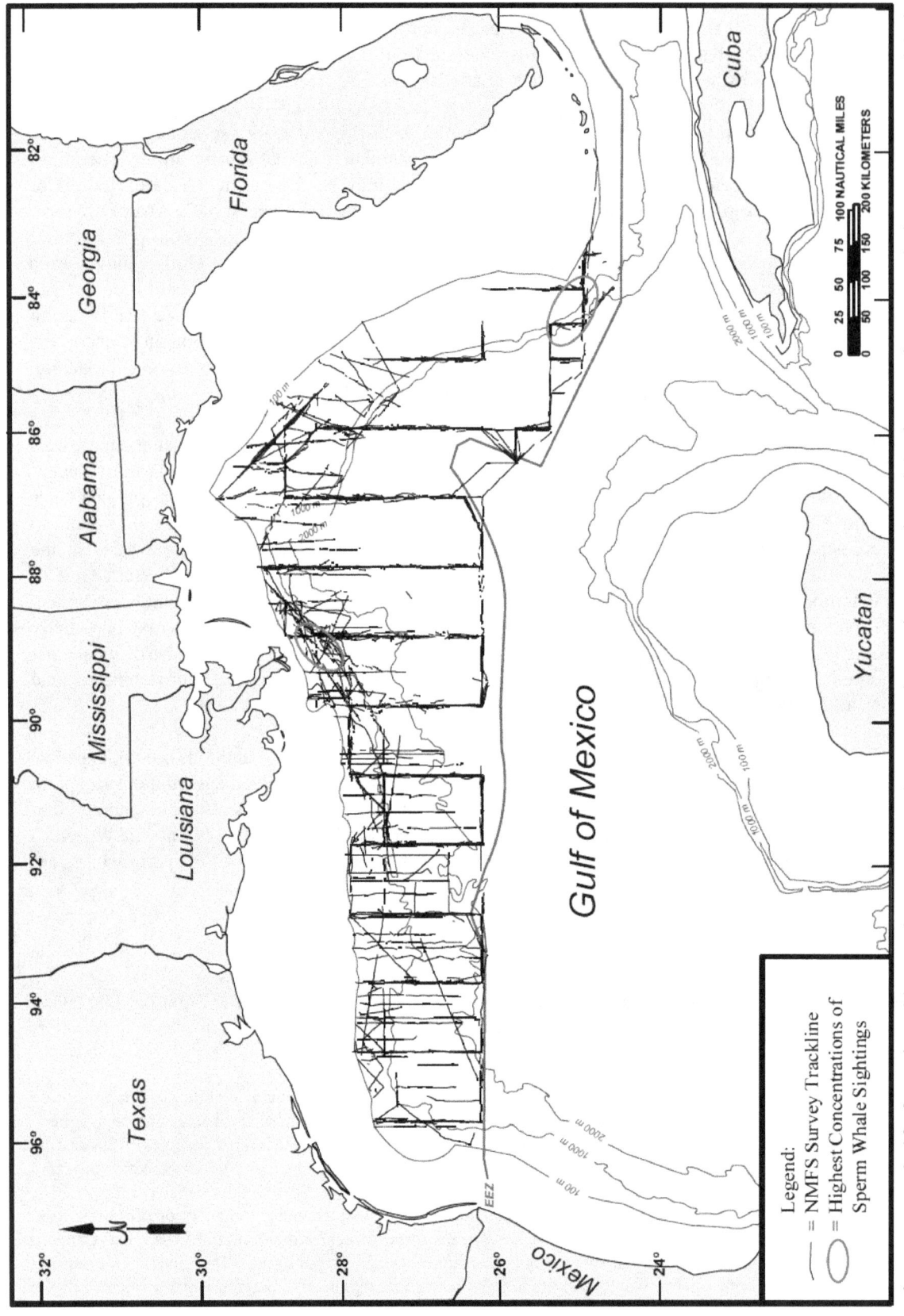

Figure III-2. Gulf of Mexico shipboard surveys conducted by the National Marine Fisheries Service (NMFS) between 1991 and 2001. Areas of highest sperm whale sightings (i.e., aggregations) denoted (Adapted from: K. Mullin, NMFS, written comm., 2003).

Waring et al. (1999), in their preparation of the marine mammal stock assessment for Gulf and Atlantic cetacean populations, calculated a minimum population estimate of Cuvier's beaked whales (*Ziphius cavirostris*) in the northern GOM at 20 individuals, as derived from surveys conducted from 1991 to 1994.[15] Minimum population estimates of Blainville's (*Mesoplodon densirostris*) and Gervais' (*M. europaeus*) beaked whales were not individually calculated in this study because of the uncertainty of species identification of sightings. However, an estimate of the average abundance of undifferentiated beaked whales was 117 individuals (Waring et al., 1997). GulfCet II ship surveys for the oceanic northern GOM yield abundance estimates of 159 Cuvier's beaked whales, 150 unidentified *Mesoplodon* spp., and 75 unidentified Ziphiids, with confidence intervals from 44 to 577, 59 to 384, and 22 to 254 for each group, respectively. Mean density estimates for each group were 0.040, 0.038, and 0.019 individuals/100 km^2, respectively. Recent density estimates of Cuvier's beaked whales in the northern Gulf are 0.03, 0.03, 0.00 individuals/100 km^2 for the Eastern, Central, and Western Gulf of Mexico Planning Areas, respectively (K. Mullin, NMFS, written comm., 2003). Recent density estimates of unidentified *Mesoplodon* spp. and unidentified ziphiids in the northern Gulf ranged from 0.02 to 0.07 and 0.01 to 0.05 individuals/100 km^2, respectively (K. Mullin, NMFS, written comm., 2003). When cumulative annual seismic survey operations are considered (see **Appendix L**), two Cuvier's beaked whales may be exposed to levels of 180 dB or greater if they do not avoid exposure (i.e., one Cuvier's beaked whale each in the Eastern and Central Planning Areas; see **Appendix L, Table L-11**). For unidentified *Mesoplodon* spp. and unidentified ziphiids, the maximum number of individuals exposed would be three and three, respectively (i.e., one unidentified *Mesoplodon* spp. and one unidentified ziphiid per planning area; see **Appendix L, Table L-11**). If beaked whales are able to recognize approaching seismic operations and avoid exposure, fewer animals would be at risk.

Behavioral responses. Behavioral responses to seismic survey noise are considered possible in any species of GOM cetacean. While the principal energy from airgun arrays is in the low frequencies where primarily mysticetes (Bryde's whale) would be likely to hear it, there is also energy at higher frequencies that can probably be detected and has caused documented responses in odontocetes. It must also be noted that seismic arrays use acoustic transponders, which emit sounds in the several kHz range (Norris et al., 2000) and obviously could be detected by odontocetes.

With a variable proportion of lease blocks in the Gulf having seismic surveys in a given year (with most activity in deeper waters of the Central and Western Gulf), it is likely that some behavioral responses are occurring. However, the biological importance of such responses is unknown (e.g., effect on the size or productivity of the species stock). At this point, there is no evidence that adverse behavioral impacts at the local population level are occurring in the GOM. Examples of adverse effects that have been attributed to acoustic sources in other settings include mass strandings of beaked whales possibly caused by high energy military sonar (Frantzis, 1998). No such impacts have been noted in the GOM, where seismic surveys have been occurring for decades. However, scientific study of possible acoustic-based impacts has only recently begun.

Individual seismic surveys can cover from a few blocks to hundreds of lease blocks and take days to months to complete (**PEA Table II-3**). As many as five regional surveys are being conducted at any one time in the GOM, with more than 30 surveys annually. On such a broad scale, if seismic surveys were resulting in avoidance behavior, one might expect abundance to differ depending on the proximity to seismic operations. Norris et al. (2000) analyzed GulfCet II acoustic survey data to determine whether abundance estimates differed depending on whether seismic signals were present. They did not detect any differences attributable to seismic survey noise. However, the seismic signals detected during this study were measured from distant vessels, and the received seismic pulses were thus relatively low in intensity and frequency.

---

[15] Draft population size and minimum population estimates for Cuvier's beaked whales in the northern GOM are 88 and 58 individuals, respectively (NMFS, 2004).

b.   *Vessel Traffic (Risk of Collisions)*

The potential effects of large vessels underway on marine mammals include direct vessel effects (direct physical contact [shipstrike] and temporary behavioral [aversive or attractive] effects), and effects associated with vessel noise. Though considered possible, especially with certain species, the possibility of a shipstrike between a relatively slow-moving seismic survey vessel at cruising speed and a marine mammal is considered low. While a seismic vessel is on survey, because of the slow towing speeds involved (e.g., 4.5 kn), the possibility of shipstrike with a marine mammal becomes negligible. Species of concern for possible shipstrike with all marine transportation while underway at speed include primarily slow-moving species (e.g., northern right whales) and deep-diving species while on the surface (e.g., sperm whales, pygmy/dwarf sperm whales, and beaked whales). Impacts between seismic vessels at speed and any of these species would possibly result in the mortality of the individual marine mammal. In the case of sperm whales, which are currently listed as endangered species, this would constitute a significant adverse impact. However, since there are no records of collisions between sperm whales and seismic vessels, it is assumed that the probability of this encounter, and thus impact, is very low. Certain species, such as bottlenose dolphins and spotted dolphins (*Stenella* spp.), actively approach any vessels while underway at speed to swim within the vessel's bow pressure wave (i.e., "bowride"). Under this circumstance, interactions between these marine mammals and vessel traffic are not considered to constitute an adverse impact. In terms of seismic vessel noise and associated impacts, International Association of Geophysical Contractors (IAGC) has noted that total vessel traffic in an area where a seismic vessel is operating typically decreases as fishing and recreational vessels move out of the survey area to avoid vessel conflicts.

c.   *Mitigation*

Under existing regulations, use of explosives is prohibited except under written authorization from the MMS (see **Appendix E, Section II.A.2 - Stipulations and Protective Measures**). This is a measure intended to protect marine mammals, sea turtles, and other marine life. (It has been assumed that no explosives would be used for G&G activities under the Proposed Action.) In addition, the Proposed Action includes several existing measures (i.e., ESA consultation) designed to protect Florida manatees from vessel strikes (see **Appendix E, Section II.A.2**). These measures are considered effective in avoiding significant impacts to manatees. Additional mitigation measures applicable to other marine mammals are included in remaining alternatives (i.e., Alternatives 2, 3, and 4).

d.   *Significance Criteria and Evaluation*

A ***significant adverse impact*** on marine mammals occurs when

- the PBR level (defined in **Appendix F, Section I.A.5 - Abundance Estimates**) is exceeded for any marine mammal stock (i.e., any mortality or serious injury would be considered an exceedance of the PBR level for any strategic stock or listed species); or
- any listed species or strategic stock is displaced from critical habitat (or key habitat if critical habitat is not formally designated) for any length of time; or
- there is long-term or permanent displacement of any species from preferred feeding, breeding, or nursery habitats (other than critical habitat); or
- there is a substantial (or chronic) disruption of behavioral patterns to an extent that may adversely affect a species or stock through effects on annual rates of recruitment or survival.

An *adverse but not significant impact* on marine mammals occurs when

- mortality or serious injury occurs to marine mammals, but not in excess of the PBR (i.e., no deaths or serious injuries of strategic stocks or listed species); or
- there is a short-term displacement of marine mammals from preferred feeding, breeding, or nursery grounds (but not critical habitat); or
- there is some disruption of behavioral patterns, but to an extent that is unlikely to affect adversely a species or stock through effects on annual rates of recruitment or survival.

A *negligible impact* on marine mammals occurs when

- no mortality or serious injury to any marine mammal results; or
- no displacement of listed species or strategic stocks from critical habitat results; or
- no displacement of any species from preferred feeding, breeding, or nursery grounds results; or
- little or no disruption of behavioral patterns or other sublethal effects are evident.

USDOI, MMS (2001b), which serves as the basis for the impact criteria noted above, also defines extent and duration of impact as follows. For marine mammal impact assessment, a "short-term" impact can be defined as infrequent and temporary and characterized by sudden onset and short duration. Short-term impacts may occur within fixed and varied geographic locations. Considering the average life spans of marine mammals, the duration of a short-term impact would be one which may last seconds, hours, or perhaps even up to several days. A "long-term" impact is an impact or series of impacts that are characterized by long duration or frequent reoccurrence, typically within a specific geographic location. Considering the average life spans of marine mammals, the duration of a long-term impact would be one that may last an appreciable fraction of an individual animal's lifetime (i.e., perhaps months to years). A "local" (or "localized") impact is one that occurs within a defined location, is not widespread or general in extent, and affects only restricted numbers of individuals of one or more species but is unlikely to affect the population status of the impacted species or stock of a species. A "regional" impact is one that may affect the status of a species or local stock of a species. The areal extent of a regional impact may vary greatly, ranging from a broad geographic area (one that encompasses one or more ecological habitats or systems) to a much smaller area, as in the case where a species, stock, or a life stage of a species is concentrated into a relatively small area (e.g., sperm whales off the Mississippi River delta). A "strategic stock" includes those stocks that are not listed under ESA but that have estimated human-caused mortality greater than PBR. The term "population stock" or "stock" means a group of marine mammals of the same species or smaller taxa in a common spatial arrangement that interbreed when mature. The term "PBR" refers to the maximum number of individuals of a particular marine mammal species, not including natural mortalities, that may be removed from a stock while allowing that stock to reach or maintain its optimum sustainable population. These significance criteria are consistent with impact definitions outlined by Swartz and Hofman (1991), particularly as they pertain to a negligible impact. Specifically, any impact that would increase the rate of decline of a declining species or population, or affect a stable species or population, such that its distribution or size would be reduced for a period greater than 1 year, generally could not be considered negligible (Swartz and Hofman, 1991).

An additional clarification of the significance criteria is warranted. How MMS defines "negligible impact" in this PEA for NEPA purposes is different from "negligible impact" that NMFS has defined in 50 CFR 216.103 for MMPA. The NMFS definition is "an impact resulting from the specified activity that cannot be reasonably expected to, and is not reasonably likely to, adversely affect the species or stock through effects on annual rates of recruitment or survival." The NMFS definition of negligible impact under MMPA overlaps the adverse but not significant and negligible impacts as defined in this EA.

From an auditory impact standpoint, one species (i.e., Bryde's whale) may potentially realize an impact based on its presumed hearing capabilities, while two other marine mammal taxa (i.e., sperm whale, beaked whales) may be at potential risk of exposure and subsequent impact on the basis of their deep-diving habits and undefined potential for low frequency sensitivity. Bryde's whales are most likely to be present in the Eastern Planning Area, where a relatively low level of future seismic survey activity (i.e., as compared to the Western and Central Gulf) is anticipated. Sperm whales are common in the northern Gulf, particularly in shelf-edge and slope waters off the Mississippi River delta. As noted by Norris et al. (2000), this is an area of intense industry activity, and it is unknown whether the site fidelity of sperm whales reflects low sensitivity to seismic noise or a high motivation to remain in the area. Historically, sightings of sperm whales have been made in the north-central GOM as far back as the mid-1800's (Townsend, 1935). Beaked whales also appear to prefer deep water, though little is known of their respective life histories. Stomach content analyses suggest that these whales feed primarily on deep water cephalopods, although they will also take fishes and some benthic invertebrates. In the Gulf, beaked whales have been sighted at the sea surface between the 700 and 2,000 m isobaths.

The Proposed Action may result in behavioral responses that could be interpreted as Level B harassment (see **Appendix B, Section I.B - Harassment Definitions**). However, due to the absence of behavioral harassment criteria, the limited data on response thresholds (particularly for odontocetes), the presumption but uncertainty of hearing sensitivities, and lack of a basis to interpret the biological importance of such behavioral responses to the marine mammal stock, it would be inappropriate to base a significance determination on behavioral responses alone. In evaluating the significance of behavioral responses, several observations are relevant. First, there presently is no evidence that significant adverse behavioral impacts attributable to seismic survey activity are occurring to marine mammals in the GOM. However, there are no historic data on marine mammal distributions or abundance estimates in the Gulf prior to the onset of seismic profiling surveys. There are no regular mass strandings or other similar events that may be attributed to impacts associated with seismic activities on the continental shelf of the northern Gulf, though seismic surveys have been occurring there with some regularity for decades. Second, there is no evidence to suggest that seismic surveys are displacing marine mammals either permanently or for significant periods of time from areas of the GOM, including those areas that may serve as important habitat (i.e., critical or preferred habitats). Areas that are currently federally-designated as critical habitat for marine mammals in the Gulf consist only of selected coastal and inshore areas in peninsular Florida (i.e., for the Florida manatee) that extend from the Crystal River area south through the Florida Keys. Preferred habitats include, but are not limited to, the productive continental slope waters off of the Mississippi River delta and in the southeastern Gulf off the Dry Tortugas for sperm whales, and a selected area on the outer shelf off northwest Florida (northeastern Gulf) for Bryde's whale (W. Hoggard, NMFS, oral comm., 2001). Other Gulf habitats that may support aggregations of marine mammals, such as cold core rings (i.e., cyclonic, mesoscale eddy features characterized by relatively colder, upwelled water at their center), are dynamic, mesoscale oceanographic features. It is very unlikely that marine mammals would be displaced entirely, or for long periods of time, from such areas. Generally, it is presumed that a spatial displacement of marine mammals from a specific seismic survey's "zone of responsiveness" would be expected to be short-term in duration. Third, unlike the Pacific coast where most observations of behavioral responses to seismic surveys have been recorded, there are no known migratory corridors of a similar nature in the GOM. Therefore, seismic surveys in the GOM are not displacing mammals from migratory corridors.

Impacts on Sperm Whales: If sperm whales are unable to avoid exposure to noise levels >180 dB re 1 µPa in their range of sensitive frequencies, then there is a reasonable potential that sperm whales may experience noise levels that could potentially cause auditory impairment and/or behavioral disturbance.

However, just entering the 180-dB zone does not necessarily mean that whales will be adversely affected by alterations in hearing ability or behavior. The areas of most concern are in the oceanic Gulf, offshore of the Mississippi River mouth (and extending east to the De Soto Canyon area in the Eastern Planning Area), where aggregations of sperm whales are sighted with regularity. Although sperm whales apparently are not being permanently displaced from this area due to seismic surveys, it is unknown whether their site fidelity reflects low sensitivity to seismic noise or a high motivation to remain in the area in spite of this noise. Because there is some evidence of sperm whale responses to low frequency noise, including possibly leaving an area where seismic surveys were occurring, it is reasonable to presume that these animals may be exposed to aversive noise levels (i.e., noise levels that would cause behavioral modification, such as avoidance or displacement) in their preferred habitat. Minor behavioral disruption typically does not adversely affect either the individual or the population. However, there is no evidence that behavioral changes prompted by seismic noise are, or are not, of sufficient magnitude to have meaningful effects on the population.

In the absence of appropriate criteria and data documenting physiological damage, a more useful approach to the determination of effects may be obtained through assessing the PBR, or the potential for displacement from preferred habitat (e.g., preferred feeding, breeding, or nursery habitat; key habitat) or substantial alteration in behavior among species of concern. Reviewing the significance criteria noted previously as they apply to sperm whales, exposure to seismic operations in the GOM is *not expected to* 1) result in any mortality or serious injury (i.e., no exceedance of the PBR level) to sperm whales; 2) displace sperm whales from key habitat; 3) result in long-term or permanent displacement of sperm whales from preferred feeding, breeding, or nursery habitats; and 4) result in a substantial or chronic disruption of behavioral patterns to an extent that may adversely affect sperm whales through effects on annual rates of recruitment or survival. There is also potential for solitary individuals to be present throughout the GOM. Therefore, impacts from seismic operations are considered potentially **adverse but not significant**. Adverse impacts, although not considered significant, could be mitigated by operational restrictions (i.e., prohibiting concurrent surveys in areas of known sperm whale aggregation), or by requiring further mitigation (i.e., any seismic surveys in this area to include passive acoustic monitoring or passive and active acoustic monitoring to ensure that sperm whales are not exposed to potentially damaging or aversive noise levels). These measures and restrictions are discussed under Alternatives 2, 3, and 4.

Impacts on Bryde's Whales: Seismic surveys, especially in the Eastern Planning Area, pose some risk of auditory impairment and/or behavioral disturbance to Bryde's whales if the whales are capable of hearing the noise and are unable to avoid exposure. Impacts in the Central and Western Planning Areas are less likely given the low frequency of sightings of this species there. However, in either case, G&G seismic activities could potentially result in auditory impairment and/or behavioral disturbance for whales unable to avoid exposure.

Reviewing the significance criteria noted previously as they apply to Bryde's whales, exposure to seismic operations in the GOM is *not expected to* 1) result in any mortality or serious injury (i.e., no exceedance of the PBR level) to Bryde's whales; 2) displace Bryde's whales from key habitat; 3) result in long-term or permanent displacement of Bryde's whales from preferred feeding, breeding, or nursery habitats; and 4) result in a substantial or chronic disruption of behavioral patterns to an extent that may adversely affect Bryde's whales through effects on annual rates of recruitment or survival. Therefore, G&G seismic operations produce a potentially **adverse but not significant** impact to Bryde's whales. Although not significant, this potential impact could be mitigated by operational restrictions (no concurrent survey activity) in areas where sightings are more common (e.g., the Eastern Planning Area), or by requiring seismic surveys to include passive acoustic monitoring or passive and active acoustic monitoring to ensure that Bryde's whales are not exposed to potentially damaging or aversive noise levels. These measures and restrictions are discussed under Alternatives 2, 3, and 4.

Impacts on Beaked Whales: Little is known of the hearing and auditory responses of beaked whales. Due to their deep diving habits, they are at a risk similar to sperm whales for auditory impairment from seismic arrays passing over them. Reviewing the significance criteria noted previously, exposure to seismic operations in the GOM is *not expected to* 1) result in any mortality or serious injury to beaked whales (i.e., no exceedance of the PBR level [PBR available only for Cuvier's beaked whale]); 2) displace beaked whales from key habitat; 3) result in long-term or permanent displacement of beaked whales from preferred feeding, breeding, or nursery habitats; and 4) result in a substantial or chronic disruption of behavioral patterns to an extent that may adversely affect beaked whales through effects on annual rates of recruitment or survival. Given the problematic nature of sightings data for these species in the Gulf, beaked whales may be more prevalent than current data sources suggest. In view of the paucity of data on these species, it is assumed that impacts of G&G seismic activities on beaked whales are, as is the case with sperm whales, potentially **adverse but not significant**.

Impacts on Other Cetaceans: Impacts of G&G seismic activities on other cetaceans are potentially **adverse but not significant**. Physical injury and auditory impairment are possible but would most likely be limited to those animals found within close proximity to a firing airgun, a zone that would presumably be avoided by marine mammals. Seismic surveys may cause behavioral responses, but there is no evidence for, nor reason to anticipate, that they are having adverse effects on cetacean populations. Mortality or serious injury are not expected, nor is substantial displacement or disruption of important behavioral patterns to other cetaceans expected.

Impacts on Manatees: Impacts of G&G seismic activities on manatees are **negligible**. Due to their coastal habitat, these animals are unlikely to come in contact with offshore G&G operations. Existing mitigation measures protect manatees against coastal vessel traffic.

3.   **Impacts of Routine Activities - Alternative 2 (Addition of Vessel-based Passive Acoustic Monitoring as a Requirement)**

Alternative 2 requires the use of passive acoustic monitoring in conjunction with existing NTL requirements (i.e., ramp-up, visual monitoring) for all seismic surveys in water depths greater than 200 m throughout the GOM and all OCS waters (regardless of water depth) in the Eastern Planning Area. This potential mitigation measure has been evaluated in **Appendix E, Section II - Mitigation Measures and Operational Restrictions**. Major summary points are noted below.

Alternative 2 – Under this alternative, acoustic monitoring is passive (i.e., no acoustic sources are used, only listening devices) and can occur either from a vessel-based system or from a hydrophone or sonobuoy array placed on the seafloor, or both. Operations conducted in the Alaskan Beaufort Sea used both approaches, while a sonobuoy array was used in the Santa Barbara Channel and hydrophones were used in Puget Sound.

There are recognizable limitations to passive acoustic monitoring. From an operational perspective, it may prove difficult to monitor for biological sound sources during a seismic survey. Determinations of range and bearing (e.g., are the sources within the exclusion zone, are the sources approaching the exclusion zone, etc.) also may be problematic. From a biological standpoint, not all marine mammal species vocalize continuously or for a high proportion of time. This measure is appropriate mitigation for sperm whales (which do vocalize for a high proportion of time) but is not expected to be useful in monitoring for Bryde's and beaked whales unless they vocalize.

Impacts on Sperm Whale: Under this alternative, impacts to sperm whales would be mitigated as follows. All seismic surveys being conducted at water depths greater than 200 m throughout the GOM and all OCS waters of the Eastern Planning Area (including areas off the Mississippi River mouth and the Eastern Planning Area where aggregations of sperm whales are sighted with regularity) would include compliance with existing NTL requirements (i.e., ramp-up, visual monitoring) to ensure that sperm whales are not present within the 500-m (radial distance) exclusion zone; and real-time passive acoustic monitoring to ensure that sperm whales are not present within the exclusion zone. As noted previously, implementation of ramp-up and visual monitoring is expected to reduce the potential for acoustic impacts to sperm whales. However, there are limitations to visual monitoring. Addition of passive acoustic monitoring will help to reduce the limitations inherent in visual monitoring. Ramp-up and visual monitoring will reduce the probability that marine mammals will move into the exclusion zone, thereby risking hearing impairment. Passive acoustic monitoring will be useful in identifying sperm whale vocalizations (clicks) and vocalizations from other marine mammals. While existing NTL-based mitigation measures offer some level of reduction in potential impact, the addition of passive acoustic monitoring as a requirement further maximizes mitigation effectiveness. In all cases, impacts are expected to remain potentially **adverse but not significant**.

Impacts on Bryde's Whale: Under this alternative, impacts to Bryde's whales would be mitigated as follows. All seismic surveys (under the water depth restrictions noted above) in the Gulf would include ramp-up procedures, and visual monitoring to ensure that Bryde's whales are not present within the exclusion zone. Implementation of passive acoustic monitoring is expected to reduce impacts to Bryde's whales in the event that whales that are present are vocalizing. Impacts are expected to remain potentially **adverse but not significant**.

Impacts on Beaked Whales: Under this alternative, impacts to beaked whales would be mitigated as follows. All seismic surveys (under the water depth restrictions noted above) in the Gulf would include ramp-up procedures and visual monitoring to ensure that beaked whales are not present within the exclusion zone. Implementation of passive acoustic monitoring is expected to reduce impacts to beaked whales in the event that whales that are present are vocalizing. Impacts are expected to remain potentially **adverse but not significant**.

Impacts on Other Cetaceans: Impacts of G&G seismic activities on other cetaceans would, in general, remain the same as compared to those projected under Alternative 1 (i.e., potentially **adverse but not significant**) because of the limitations associated with passive acoustic monitoring (i.e., lack of vocalization for some marine mammal species).

Impacts on West Indian Manatee: Impacts of G&G seismic activities on manatees would be the same as under Alternative 1, i.e., **negligible**. Due to the coastal habitat, these animals are unlikely to come in contact with offshore G&G operations. Existing mitigation measures protect manatees against coastal vessel traffic.

4.   **Impacts of Routine Activities - Alternative 3 (Addition of Both Passive and Active Acoustic Monitoring as Requirements)**

Under Alternative 3, both passive and active acoustic monitoring would be required, supplementing existing NTL-based mitigation measures (i.e., ramp-up and visual monitoring). This alternative includes the same depth restrictions as noted under Alternative 2. The same benefits noted previously (under Alternatives 1 and 2) for ramp-up and visual monitoring remain. This alternative also includes the use of passive and active acoustic monitoring further to ensure that the exclusion zone remains clear of marine mammals while the array is ramped up.

Impacts on Sperm Whale: Under this alternative, impacts to sperm whales would be mitigated as follows. All seismic surveys (i.e., in all OCS waters >200 m deep throughout the GOM and all OCS waters [regardless of water depth] in the Eastern Planning Area, including the areas off the Mississippi River mouth and the Eastern Planning Area where aggregations of sperm whales are sighted with regularity) would include NTL-based operational requirements (i.e., ramp-up procedures and visual monitoring) and the use of both passive and active acoustic monitoring to ensure that sperm whales are not present within the exclusion zone. Implementation of passive acoustic monitoring is expected to reduce the potential for noise impacts to sperm whales when whales are vocalizing. Sperm whales vocalize frequently, therefore, passive acoustic monitoring is expected to be an effective mitigation measure. Addition of passive acoustic monitoring also will help to reduce the limitations inherent in visual monitoring. Use of active acoustic monitoring may provide additional mitigation; however, there are limitations and potential impacts associated with this mitigation measure (e.g., active source may be more harmful than the seismic source being mitigated; limited detection range; inability to determine species based purely on animal size; see **Appendix E, Section II.B.4 - Acoustic Monitoring**). Passive acoustic monitoring offers an increased potential for identification of sperm whales in proximity to seismic operations; active acoustic monitoring of sperm whales is problematic. Because the potential for acoustic impact cannot be completely eliminated, impacts are expected to remain potentially **adverse but not significant**.

Impacts on Bryde's Whale: Under this alternative, impacts to Bryde's whales would be mitigated as follows. Under the same depth restrictions noted above, all seismic surveys would include NTL-based operational requirements (i.e., ramp-up and visual monitoring in water depths >200 m throughout the GOM; all OCS waters [regardless of water depth] in the Eastern Planning Area) and the use of passive and active acoustic monitoring to ensure that Bryde's whales are not present within the exclusion zone. As noted previously, implementation of passive acoustic monitoring is expected to reduce impacts to Bryde's whales when whales are vocalizing. While Bryde's whales vocalize, frequency of vocalization is unknown and, therefore, problematic. Effectiveness of passive acoustic monitoring of Bryde's whales will be limited; further, the limited distribution of Bryde's whales (i.e., Eastern Planning Area, 100-200+ m water depths) limits the utility of passive acoustic monitoring. For active acoustic monitoring, there are limitations and potential impacts associated with this mitigation, as noted above (e.g., active source may be more harmful than the seismic source being mitigated, etc.). Due to limited utility of passive acoustic monitoring and limitations associated with active acoustic monitoring, impacts to Bryde's whales under this alternative are expected to remain potentially **adverse but not significant**.

Impacts on Beaked Whales: Under this alternative, impacts to beaked whales would be mitigated as follows. Under the same depth restrictions noted above, all seismic surveys would include NTL-based operational requirements (i.e., ramp-up and visual monitoring in water depths >200 m in all Gulf waters; all OCS waters in the Eastern Planning Area) and the use of passive and active acoustic monitoring to ensure that beaked whales are not present within the exclusion zone. As noted previously, implementation of passive acoustic monitoring is expected to reduce the potential for impacts to beaked whales when whales vocalize. Beaked whales vocalize; however, the frequency of their vocalizations is unknown and, therefore, problematic. For active acoustic monitoring, there are limitations and potential impacts associated with this mitigation, as noted previously. Due to limited utility of passive acoustic monitoring and limitations associated with active acoustic monitoring, impacts to beaked whales under this alternative are expected to remain potentially **adverse but not significant**.

Impacts on Other Cetaceans: Impacts of G&G seismic activities on other cetaceans would, in general, be the same as under Alternative 1. Passive acoustic monitoring would be effective only for those cetaceans that vocalize. Active acoustic monitoring has limitations and potential impacts associated with use of this mitigation. Impacts of this alternative on other cetaceans are expected to remain the same as noted under Alternative 1 (i.e., potentially **adverse but not significant**).

Impacts on West Indian Manatee: Impacts of G&G seismic activities on manatees would be the same as under Alternative 1, i.e., **negligible**. Due to the coastal habitat, these animals are unlikely to come in contact with offshore G&G operations. Existing mitigation measures protect manatees against coastal vessel traffic.

5.   Impacts of Routine Activities - Alternative 4 (Restrict G&G Seismic Survey Activities)

Under Alternative 4, G&G seismic surveying operations would be subject to an operational restriction – prohibition of simultaneous surveys by more than one G&G survey vessel in those portions of the GOM most frequented by sperm whales and/or Bryde's whales. *Note: The second part of this alternative – cessation of all types of seismic surveys in the those portions of the GOM most frequented by sperm whales and/or Bryde's whales (i.e., spatial exclusion) – fails to fully meet the purpose and need (see **PEA Section II.E.4 - Evaluation Relative to Purpose and Need**) and was removed from further consideration. Consideration also was given to restricting all seismic surveys (in those portions of the GOM most frequented by sperm whales) on a seasonal basis (i.e., temporal exclusion). However, this restriction was removed from further consideration, as detailed in **Appendix E** (see **Section II.B.1 - Geographic and Seasonal Restrictions**).*

Alternative 4 (i.e., restrictions on concurrent seismic operations in areas most frequented by sperm and Bryde's whales) remains a viable alternative, with qualification. Current industry practice effectively eliminates concurrent seismic operations in the same general area, as discussed in **Appendix E, Section II.B.5 - Operational Restrictions**. While minimum separation distance requirements are dependent upon various factors (e.g., output of each array, physical characteristics of the environment, relative orientation of each survey vessel, etc.), it is optimal to have two seismic survey vessels far enough apart so that each vessel is hearing only ambient noise. As noted in **Appendix E, Section II.B.5**, background noise levels in the Gulf are assumed to be on the order of 140 dB re 1 μPa. Under optimal conditions, seismic survey vessels and their arrays should be separated by 3.1 to 4.0 km, the approximate dimension of a single OCS lease block. Given the impact zone calculations presented previously (i.e., ~300 m to the 180-dB isopleth), it is highly unlikely that a marine mammal would realize acoustic impact from the simultaneous firing of airguns during concurrent operations located in adjacent lease blocks (see **Appendix C, Section III - Seismic Source Levels** and **Figures C-8** and **C-9**). Details of the calculations used to derive this radial distance are provided in **Appendix C, Section III - Seismic Source Levels**.

Multiple vessels (e.g., multi-ship surveys) may also be employed, as described in **Appendix D (Section I.A - Seismic Surveys)**, to conduct ocean bottom cable surveys and vertical cable surveys. Depending upon whether 2D or 3D operations are being conducted, two ships are placed one behind the other, towing single or multiple streamer cables. Streamer length is sufficient to record seismic data at distances of 8 to 12 km. In both of these cases, only one airgun array is operational. Therefore, it would not be possible for a marine mammal to be positioned between two survey vessels, both of which would be firing their airguns. In summary, marine mammal exposure to seismic survey noise will be similar, regardless of whether the sound source originates from concurrent single or multi-ship surveys, or concurrent multiple surveys. The advantage of this mitigation rests with the elimination of continued or multiple exposure, even at very low levels, for concurrent survey activities (i.e., marine mammals hearing more than one airgun array).

Impacts on Sperm Whale: Under this alternative (Alternative 4), impacts to sperm whales would be slightly reduced from those noted under Alternative 1. Impacts are expected to remain potentially **adverse but not significant**.

Impacts on Bryde's Whale: Under this alternative (Alternative 4), impacts to Bryde's whales would be slightly reduced from those noted under Alternative 1. Impacts are expected to remain potentially **adverse but not significant**.

Impacts on Beaked Whales: Under this alternative (Alternative 4), impacts to beaked whales would be slightly reduced from those noted under Alternative 1. Impacts are expected to remain potentially **adverse but not significant**.

Impacts on Other Cetaceans: Impacts of G&G seismic activities on other cetaceans under this alternative (Alternative 4) would be slightly reduced from those noted under Alternative 1. Impacts are expected to remain potentially **adverse but not significant**.

Impacts on West Indian Manatee: Impacts of G&G seismic activities on manatees would be the same as under Alternative 1, i.e., **negligible**. Due to the coastal habitat, these animals are unlikely to come in contact with offshore G&G operations. Existing mitigation measures protect manatees against coastal vessel traffic.

6.   Impacts of Accidents

Marine mammals could be affected by accidentally spilled lubricating oil or diesel fuel from a vessel associated with G&G operations or by jet fuel leaking from a streamer array under tow by a G&G vessel. The deliberate discharge of oil or fuel in U.S. waters is prohibited under MARPOL, Annex V, Public Law 100-220 (101 Statute 1458). Effects of spilled oil on marine mammals are discussed by Geraci and St. Aubin (1987), and more recently summarized by USDOI, MMS (2002a). Oil may affect marine mammals through various pathways: surface contact with skin and mucous membranes, inhalation, ingestion (either directly or through the ingestion of oiled prey), or (in the case of mysticete whales) fouling of baleen plates (USDOI, MMS, 2001b, 2002a). The types and relative risks of G&G vessel accidents are discussed in detail within **Appendix K (Risk of Vessel Accidents)**. As discussed in **Appendix K**, the risk of G&G vessel accidents is low (e.g., three incidents reported in the GOM during 1996 and 1997 involving "research vessels," inclusive of G&G vessels). Records of the volumes of released lubricating oil or diesel fuel in documented G&G vessel accidents have generally been low (i.e., two to five minor releases per year in the GOM). Streamers containing jet fuel hold only small volumes; the unlikely release of some or all of this volatile fuel from a single streamer into the water is not expected to result in a significant impact to marine mammals. Thus, incidents involving the release of oil or fuel from vessels associated with G&G operations are not expected to result in significant impacts to marine mammals. Impacts are thus **negligible**.

7.   Cumulative Impacts

The cumulative activity scenario is presented in **Appendix J (Cumulative Activity Scenario and Associated Impacts)**. Major impact producing factors under the cumulative activity scenario include vessel traffic (i.e., cargo, tanker, military, commercial fishing, recreational boating) and its associated noise and shipstrike potential. Impacts from vessel traffic under the cumulative scenario are potentially **adverse but not significant**. In terms of vessel activity levels, seismic survey vessel activity represents a very small component of total vessel activity in Gulf waters. For example, oil and gas support vessels account for approximately one quarter of a million transits per year in Gulf waters, with commercial vessels >10,000 dead weight tons (DWT) contributing another 36,000 trips (see **Appendix F, Section II.E - Marine Transportation**). By comparison, approximately 30 seismic surveys may occur annually in the Gulf, or 0.01% of the activity from these three sources (i.e., oil and gas support operations, commercial cargo and tanker activity, seismic surveys). Commercial fishing and recreational boating, military operations, and ocean study activities also contribute to the cumulative vessel activity level, further reducing the relative contribution from seismic surveys. Therefore, G&G operations contribute an extremely minor amount of additional vessel activity in the GOM. Under the assumption that future seismic survey activity levels will generally conform to those noted in **PEA Section II.B.2 - Level of Activity** and **Appendix D (Section V - Level of Activity)**, seismic vessel activity is not expected to produce a significant incremental increase in vessel activity levels. The cumulative incremental impact attributed to G&G vessel operations is **negligible**. In summary, cumulative impacts to marine mammals from vessel traffic and noise will remain potentially **adverse but not significant**.

Measurements of ambient noise levels in the GOM are lacking. According to the predominant noise sources identified in **Appendix F (Section IV - Noise Environment)** for the GOM and their relative contributions to total noise levels, seismic surveys represent a relatively minor, intermittent, and non-stationary component of the overall noise environment. Seismic surveys produce repetitive, localized, and short-term increases in ambient noise levels, with the period between potential exposure ranging from hours to days (i.e., time between separate passes of a seismic survey vessel). In the near-field, within ~300 m or so of an array, received sound levels may reach or exceed 180 dB re 1 μPa (rms). At greater distances, sound from a seismic survey is of a similar nature to other commercial vessel activity. Given the current level of vessel activity and its associated infrastructure, future seismic survey activity is not expected to produce a significant incremental increase in ambient noise levels. Analysis of cumulative noise impacts on marine mammals in the GOM (provided in **Appendix J**) suggests that impacts are potentially **adverse but not significant**. The cumulative incremental impact attributed to G&G vessel noise is **negligible**.

C.   SEA TURTLES

1.   Affected Environment

Five species of sea turtles are known to inhabit the GOM (Pritchard, 1997). These are the loggerhead, leatherback, Kemp's ridley, green, and hawksbill turtles. All five species are listed as either endangered or threatened species under the ESA (Pritchard, 1997). Additional information on sea turtle species of the GOM is provided in **Appendix F (Section I.B - Sea Turtles)**.

The loggerhead turtle (*Caretta caretta*), a threatened species, is the most abundant sea turtle in the GOM (Dodd, 1988). Loggerhead nesting along the Gulf Coast occurs primarily along the Florida panhandle, although some nesting has been reported from Texas through Alabama as well (NMFS and USFWS, 1991a). Loggerhead turtles have been primarily sighted in waters over the continental shelf, although many surface sightings of this species also have been made over the outer slope, beyond the 1,000-m isobath. Sightings of loggerheads in waters over the continental slope suggest that they may be in transit through these waters to distant foraging sites or while seeking warmer waters during winter. Although loggerheads were widely distributed across the shelf during both summer and winter, their abundance in surface waters over the slope was greater during winter than summer (Mullin and Hoggard, 2000).

The leatherback turtle (*Dermochelys coriacea*), an endangered species, is the most abundant turtle in waters over the northern GOM continental slope (Mullin and Hoggard, 2000). It is the most pelagic and wide-ranging sea turtle, undertaking extensive migrations from the tropics to boreal waters. Leatherback nesting within the continental U.S. is limited to eastern Florida (NMFS and USFWS, 1992a; Ernst et al., 1994; Meylan et al., 1995). Leatherbacks spatially appear to use both continental shelf and slope habitats in the GOM (Fritts et al., 1983a,b; Collard, 1990; Davis and Fargion, 1996). GulfCet I and II surveys suggest that the region from Mississippi Canyon to De Soto Canyon, especially near the shelf edge, appears to be an important habitat for leatherbacks (Mullin and Hoggard, 2000). Temporal variability in leatherback distribution and abundance suggests that specific areas may be important to this species, either seasonally or for short periods of time. During the GulfCet I and II programs, leatherbacks were sighted frequently during both summer and winter (Mullin and Hoggard, 2000).

The Kemp's ridley turtle (*Lepidochelys kempi*) is an endangered species and the smallest sea turtle. Survey data from the GOM suggest that Kemp's ridley turtles occur mainly in waters over the continental shelf. Juvenile and adult Kemp's ridleys are typically found in shallow areas and especially in areas of seagrass habitat (Marquez, 1990; NMFS and USFWS, 1992b; Ernst et al., 1994). The major nesting area for this species is near Rancho Nuevo, along the northeastern coast of Mexico (Tamaulipas), although scattered nesting has also been reported in other areas of Mexico and Texas, Colombia, Florida, and South Carolina (Ernst et al., 1994). Adult Kemp's ridleys exhibit extensive interesting movements, although they appear to travel near the coast, especially within shallow waters along the Louisiana coast.

The green turtle (*Chelonia mydas*) is listed as threatened, except for the Florida breeding population, which is listed as endangered. This species is found throughout the GOM. They occur in small numbers over seagrass beds along the south Texas coast and the Florida Gulf coast. Reports of green turtle nesting along the GOM coast are infrequent, and the closest important nesting aggregations are along the Florida East coast and the Yucatan Peninsula (NMFS and USFWS, 1991b).

The hawksbill turtle (*Eretmochelys imbricata*), an endangered species, has been recorded in all the Gulf states (NMFS and USFWS, 1993). However, sightings north of Florida are rare. This is the least common sea turtle in the GOM (Marquez, 1990; Hildebrand, 1995). Hawksbill nesting within the continental United States is limited to southeastern Florida and the Florida Keys.

There are no designated critical habitats for sea turtles in the northern GOM. NMFS does recognize many coastal areas of the Gulf as preferred habitat (important, sensitive habitats that are essential for the species within a specific geographic area) – e.g., seagrass beds in Texas lagoons and other nearshore or inshore areas (including jetties) for green turtles; bays and lakes, especially in Louisiana and Texas for ridleys. Sargassum mats also are recognized as preferred habitat for hatchlings. There are no designated migratory routes for turtles in the Gulf.

2. Impacts of Routine Activities - Alternative 1 (Proposed Action)

The main concern from an impact perspective is noise from seismic surveys. In addition, most G&G seismic activities involve vessel traffic, which carries some risk of collisions with turtles.

a. Seismic Surveys

Impacts of seismic surveys on sea turtles may include auditory trauma (impact) and/or behavioral disturbance. Information on sea turtle hearing and auditory impacts is reviewed in **Appendix H (Sea Turtle Hearing and Sensitivity to Acoustic Impact)**.

1) Sea Turtle Hearing

Relatively little is documented or understood, for any sea turtle species, about their hearing ability or their dependency on sound, passive or active, for survival cues. Moreover, since sea turtles often have been reported to show a lack of response to even intense sounds (Wever, 1978), they were initially characterized as having insensitive hearing capabilities. Differences in hearing capability (frequency selectivity and threshold levels) among species and life history stages have not been documented for sea turtles in the literature. In fact only two species, loggerhead and green sea turtles, and one life history stage, juvenile, have undergone any auditory investigations.

The anatomy of the sea turtle ear does not lend itself to aerial conduction but rather is structured for sound conduction through two media, bone and water (Békésy, 1948; Lenhardt, 1982; Lenhardt and Harkins, 1983).

Auditory testing and behavioral studies show that turtles can detect low-frequency sounds such as those produced by airguns. Ridgway et al. (1969) determined that green sea turtles detected sound frequencies in the range of 200 to 700 Hz and displayed a high level of sensitivity at the low tone region of about 400 Hz. Bartol et al. (1999) studied juvenile loggerhead turtles and determined that maximum sensitivity was in the region of 250 to 1,000 Hz. The decline in sensitivity was rapid after 1,000 Hz, and the most sensitive threshold tested was found to be at 250 Hz. For various reasons noted in the following discussions, neither of these tests can be used to calculate hearing thresholds in dB re 1 μPa, which are the appropriate units for evaluating underwater sound sources. They serve mainly to indicate the frequency range that is likely to be detected by sea turtles.

2)   *Behavioral Responses to Airgun Noise*

Two studies have examined the response of juvenile loggerheads to sound in their natural environment (O'Hara and Wilcox, 1990; Moein et al., 1995). Both studies used airguns as an acoustic repelling device for sea turtles.

O'Hara and Wilcox (1990) attempted to create a sound barrier for loggerhead turtles at the end of a canal of Florida Power & Light using seismic airguns. The test results indicated that at 140 kg/cm^2, the airguns were effective as a deterrent for a distance of about 30 m. The sound output of this system was characterized as approximately 220 dB re 1 μPa at 1 m in the 25- to 1,000-Hz frequency range. However, this study did not account for the reflection of sound by the canal walls. Consequently, the stimulus frequency and intensity levels are ambiguous (O'Hara and Wilcox, 1990).

Moein et al. (1995) investigated the use of airguns to repel juvenile loggerhead sea turtles from hopper dredges. A net enclosure (approximately 18 m x 61 m x 3.6 m) was erected in the York River, Virginia to contain the turtles, and an airgun was stationed at each end of the net. Using a float attached to the posterior of the carapace, the position of the turtle was noted as the airguns fired. Sound frequencies of the airguns ranged from 100 to 1,000 Hz (Zawila, 1995). Three source levels (175, 177, and 179 dB re 1 μPa at 1 m) were used. Beginning at the 175 dB source level, one airgun fired every 5 s for 5 min, followed by a 10-min resting period. Then, the airgun at the other end fired using the same source level. This was repeated using the successively higher source levels, resulting in exposure to a total of 30 min of airgun pulses. Avoidance of the airguns was observed upon first exposure for the juvenile loggerheads. However, these animals also appeared to habituate to the sound stimuli. After three separate exposures to the airguns, the turtles no longer avoided the stimuli (Moein et al., 1995). Another possible interpretation is that the turtles gave up trying to avoid the airguns, since they could not truly escape the noise (J. Sigurdson, Department of the Navy, oral comm., 2000).

3)   *Auditory Trauma*

There are two types of potential trauma for an animal from acoustic stimuli: lethal and sublethal. Typically, trauma that produces a lethal effect, trauma so severe that an animal in close proximity to an intense sound source dies within minutes, is not limited to impacts on the auditory system (i.e., the acoustic wave damages tissues other than those associated with hearing). This effect has been documented in association with use of underwater explosives (O'Keeffe and Young, 1984; Klima et al., 1988) but not seismic surveys using airguns. Sublethal impacts of acoustic impulses on the hearing system of sea turtles have been examined in only one study.

The extent of trauma to the auditory system is usually placed into two categories, both of which represent hearing impairment: PTS and TTS (Saunders et al., 1985; Lenhardt, 1986). Threshold levels of hearing are defined as the minimum intensity required for perception of that sound by an individual. Threshold levels will differ by frequency and are usually recorded from a subject animal without the interference of background noise. Loss of sensitivity, or an increase in the threshold level due to trauma, is a threshold shift. If this loss is irreversible, then a PTS has occurred. This PTS could result from direct damage to the middle ear (i.e., the tympanum or columella) or inner ear (hair cells). If the sensitivity levels return to normal after a period of time, then the threshold shift is temporary (TTS). Many factors can determine the extent and duration of a TTS in an animal, including the frequency, intensity, and duration of the damaging sound. TTS can extend over a wide range of frequencies or be limited to a narrow band, and normal hearing can return over a period of days, or months (Melnick, 1991; Lombarte et al., 1993; Hastings et al., 1996).

Possible response to threshold shifts by an animal can vary tremendously, and behavioral responses by sea turtles to TTS are unknown. Behavioral research on sea turtles in their natural environment is difficult, and only a few studies have attempted to measure reactions of healthy animals to sound stimuli, as noted previously. However, loss of hearing, either temporary or permanent, could potentially prevent an individual animal from detecting predators or prey, from successfully navigating along migration routes or to a nesting beach, or from avoiding man-made dangers, such as boating traffic, to name a few.

One study has produced evidence of possible TTS in juvenile loggerhead sea turtles (*Caretta caretta*) (Moein et al., 1995). The turtles in this study were used to test the efficacy of seismic airguns to repel sea turtles from the path of hopper dredges. As noted previously, three source levels (175, 177, and 179 dB re 1 μPa at 1 m) were used, and the animals were exposed to airgun pulses for a total of 30 min (six 5-min intervals) (Moein et al., 1995; Zawila, 1995). Turtles were tested for stress levels and hearing thresholds before and after the airgun trials. These health tests not only ensured that all loggerhead turtles used in the net enclosure were healthy animals but also demonstrated the long-term effects of sound exposure on the turtles. A temporary alteration of blood chemistry values after exposure to the airguns indicated that these turtles might have been affected by exposure to repeated acoustic stimuli. Values indicated both an increase in the stress level of the animal as well as damage to tissues. However, the magnitude of the changes did not indicate serious injury to the turtles' organs, and levels returned to normal in approximately 2 weeks. Thus, this exposure to sound stimuli did cause minor, but reversible, changes to the turtles' tissues (Moein et al., 1995).

Hearing thresholds also were examined by Moein et al. (1995). Auditory evoked potentials were recorded from these juvenile loggerheads, and auditory thresholds were obtained before the first testing sequence in the net with the sound stimuli, within 24 h after each test, and approximately 2 weeks after the last exposure to the airgun. Half of turtles tested exhibited a shift in the latency of their auditory evoked potentials collected within 24 h of exposure (see **Appendix H, Section IV.B - Auditory Trauma**). This shift was indicative of a change in the hearing physiology. However, in all five cases, hearing capabilities of each turtle returned to normal by the end of 2 weeks, and the effect was always temporary. Turtle position was not monitored in sufficient detail to calculate the received sound pressure levels to which any of the affected turtles were exposed.

b.   *Vessel Traffic*

G&G survey vessels may collide with sea turtles, causing death or injury. All of the recovery plans for the listed sea turtles include vessel collisions as one of the factors affecting recovery, though they are not among the principal concerns for any species (NMFS and USFWS, 1991a,b, 1992a,b, 1993). Because sea turtles are submerged most of the time (Standora et al., 1984; Eckert et al., 1986, 1989;

Keinath and Musick, 1993; Keinath et al., 1996), the overall risk of such collisions is low. The slow speed of seismic survey vessels (typically 4.5 kn) and the possible avoidance of airgun noise by turtles (discussed previously) are factors minimizing the risk of such collisions.

c.   *Mitigation*

Under existing regulations, use of explosives is prohibited except under written authorization from the MMS (see **Appendix E, Section II.A.2 - Stipulations and Protective Measures**). This is a measure intended to protect sea turtles, marine mammals, and other marine life. (It has been assumed that no explosives would be used under the Proposed Action.) There are no other existing mitigation measures specifically protecting sea turtles.

d.   *Significance Criteria and Evaluation*

A *significant adverse impact* on sea turtles is one that is likely to cause

- death or life-threatening injury of one or more individuals of a sea turtle species (all are listed as endangered or threatened); and/or
- long-term or permanent displacement of any sea turtle species from critical habitat or other preferred feeding or breeding areas, nesting beaches, or migratory routes; and/or
- destruction or adverse modification of critical habitat for any listed species.

An *adverse but not significant impact* on sea turtles is one that is likely to cause

- minor injury (not life-threatening) of one or more individuals of a sea turtle species; and/or
- short-term displacement of any sea turtle species from critical habitat or other preferred feeding or breeding areas, nesting beaches, or migratory routes.

A *negligible impact* on sea turtles is one that is likely to cause

- no injury to any sea turtle; and
- no displacement of any sea turtle species from critical habitat or other preferred feeding or breeding areas, nesting beaches, or migratory routes; and
- little disruption of behavioral patterns or other non-injurious effects.

Significance criteria for sea turtles reflect their protected status under Federal law and parallel criteria established in USDOI, MMS (2001b). One of the considerations in determining significance under NEPA is "whether the action threatens a violation of Federal, State, or local law or requirements imposed for the protection of the environment" (40 CFR 1508.27).

The ESA prohibits jeopardizing endangered and threatened species or adversely modifying critical habitats essential to their survival. Section 7 of the Act requires consultation with NMFS and/or USFWS (for Federal actions) if the applicant has reason to believe that an endangered or threatened species may be present in the area affected by the project, and that implementation of the Proposed Action will likely affect such species. Such consultation normally occurs between MMS and the appropriate agencies during the EIS process for each lease sale. The Proposed Action is not likely to result in deaths or mortal injuries of any listed sea turtles and therefore would not jeopardize the existence of any endangered species. Furthermore, there are no designated critical habitats for sea turtles in the northern GOM. Therefore, seismic surveys are not displacing sea turtles from critical habitat, or adversely affecting such habitat.

III-38

Environmental Impacts

Very little is known about sea turtle hearing or the role of sound in their life cycle. Auditory testing and behavioral studies show that turtles can detect and respond to low frequency sounds (250 to 1,000 Hz) from airguns. There is some indication of avoidance, at least initially, of airgun pulses with source levels of 175 to 179 dB re 1 µPa (Moein et al., 1995). The sound pressure levels received by the affected turtles are unknown, although obviously they were less than the source levels. Such levels typically would extend several hundred meters laterally from a typical seismic array (see **PEA Section II.B.3.c - Calculated Impact Zone**). The same study produced possible evidence of a temporary effect on the auditory response of turtles exposed to a total of 30 min of airgun pulses.

All five sea turtle species are assumed to be at some risk for auditory trauma, although hearing data are available only for loggerhead and green turtles. Because sea turtles remain submerged much of the time, they may be passed over by seismic arrays and therefore exposed to the highest sound levels, which are directed downward. This may be particularly the case for leatherbacks, which are deep divers.

Hatchling sea turtles are probably at minimal risk for noise impacts. These animals inhabit sargassum mats floating on the sea surface. Due to the attenuation pattern of airgun arrays (see **Appendix D, Section II.A - Airguns**), seismic noise levels would be lowest in near-surface waters.

The Proposed Action may also result in behavioral responses such as avoidance of seismic survey arrays. There is no evidence that adverse behavioral impacts are occurring. In offshore waters, avoidance would cause only a temporary displacement from a particular geographic location during a seismic survey. Offshore seismic survey noise would not displace turtles from preferred coastal habitats such as seagrass beds in Texas lagoons, other nearshore or inshore areas (including jetties) for green turtles, or bays and lakes (especially in Louisiana and Texas) for ridleys.

Impacts on Sea Turtles: Seismic survey noise may disturb sea turtles and may produce temporary or permanent hearing impairment in some individuals but is unlikely to cause death or life-threatening injury. Neither seismic surveys nor other G&G activities are expected to cause long-term or permanent displacement from critical habitat or other preferred habitat, nor will they result in destruction or adverse modification of critical habitat. Most G&G seismic activities involve vessel traffic, which carries some risk of collisions with turtles. However, because sea turtles are submerged most of the time and may avoid seismic arrays, the risk of death or life-threatening injury is low.

Visual monitoring required under NTL No. 2004-G01 provides limited mitigation for sea turtles found within the exclusion zone. It is noteworthy that sea turtles are difficult to observe in the field, and previous mitigation monitoring efforts have underscored this limitation (e.g., Department of the Navy, 1998, 1999b, 2001b). Under optimal conditions of low sea state and proper sun angle, sea turtles may be visible within a portion of the exclusion zone. Identification of floating sargassum rafts also has been used as an indicator of sea turtle presence, particularly for juveniles (Department of the Navy, 1998).

Therefore, impacts of G&G seismic activities on sea turtles will be **negligible** most of the time, with occasional impacts being potentially **adverse but not significant** (e.g., when a sea turtle cannot avoid and is subsequently exposed to seismic survey noise).

3. Impacts of Routine Activities - Alternative 2 (Addition of Vessel-based Passive Acoustic Monitoring as a Requirement)

Passive acoustic monitoring (Alternative 2) will not reduce impacts to sea turtles, given that sea turtles do not vocalize. Lack of vocalization effectively eliminates this mitigation measure as it applies to sea turtles. Impacts to sea turtles under Alternative 2 remain the same as under Alternative 1 (i.e., impacts range from **negligible** to potentially **adverse but not significant**).

4.   Impacts of Routine Activities - Alternative 3 (Addition of Both Passive and Active Acoustic Monitoring as Requirements)

Implementation of both passive and active acoustic monitoring will not reduce impacts to sea turtles, given that this group does not vocalize. Impacts to sea turtles under Alternative 3 are expected to remain the same as under Alternative 1 (i.e., impacts range from **negligible** to potentially **adverse but not significant**).

5.   Impacts of Routine Activities - Alternative 4 (Restrict G&G Seismic Survey Activities)

Under Alternative 4, G&G seismic surveying operations would be subject to an operational restriction – prohibition of simultaneous surveys by more than one G&G survey vessel in those portions of the GOM most frequented by sperm whales and/or Bryde's whales (Alternative 4). While these operational restrictions are of primary importance to marine mammal species of concern, they are unlikely to reduce impacts to sea turtles. Only those sea turtles present within these areas would realize benefit from this mitigation. Therefore, sea turtle impacts under this alternative would be similar to those under Alternative 1 (i.e., impacts range from **negligible** to potentially **adverse but not significant**).

6.   Impacts of Accidents

Sea turtles could be affected by accidentally spilled lubricating oil or diesel fuel from a vessel associated with G&G operations or by jet fuel leaking from a streamer array under tow by a G&G vessel. The deliberate discharge of oil or fuel in U.S. waters is prohibited under MARPOL, Annex V, Public Law 100-220 (101 Statute 1458). Effects of spilled oil on sea turtles are discussed by Geraci and St. Aubin (1987) and Lutcavage et al. (1995, 1997). Oil may affect sea turtles through various pathways: surface contact with skin and mucous membranes, inhalation, or ingestion (either directly or through the ingestion of oiled prey) (USDOI, MMS, 2001b). The types and relative risks of G&G vessel accidents are discussed in detail within **Appendix K (Risk of Vessel Accidents)**. As discussed in **Appendix K**, the risk of G&G vessel accidents is low (e.g., three incidents reported in the GOM during 1996 and 1997 involving "research vessels," inclusive of G&G vessels). Records of the volumes of released lubricating oil or diesel fuel in documented G&G vessel accidents has generally been low (i.e., two to five minor releases per year in the GOM). Streamers containing jet fuel hold only small volumes; the unlikely release of some or all of this volatile fuel from a single streamer into the water is not expected to result in a significant impact to sea turtles. Thus, incidents involving the release of oil or fuel from vessels associated with G&G operations are not expected to result in significant impacts to sea turtles. Impacts are thus **negligible**.

7.   Cumulative Impacts

The cumulative activity scenario is presented in **Appendix J (Cumulative Activity Scenario and Associated Impacts)**. Major impact producing factors under the cumulative activity scenario include vessel traffic (i.e., cargo, tanker, military, commercial fishing, recreational boating) and its associated noise and shipstrike potential. Impacts from vessel traffic under the cumulative scenario are potentially **adverse but not significant**. In terms of vessel activity levels, seismic survey vessel activity represents a very small component of total vessel activity in Gulf waters. For example, oil and gas support vessels account for approximately one quarter of a million transits per year in Gulf waters, with commercial vessels >10,000 DWT contributing another 36,000 trips (see **Appendix F, Section II.E - Marine Transportation**). By comparison, approximately 30 seismic surveys may occur annually in the

Gulf, or 0.01% of the activity from these three sources (i.e., oil and gas support operations, commercial cargo and tanker activity, seismic surveys). Commercial fishing and recreational boating, military operations, and ocean study activities also contribute to the cumulative vessel activity level, further reducing the relative contribution from seismic surveys to total vessel activity levels. Therefore, G&G operations contribute an extremely minor amount of additional vessel activity in the GOM. Under the assumption that future seismic survey activity levels will generally conform to those noted in **PEA Section II.B.2 - Level of Activity** and **Appendix D (Section V - Level of Activity)**, seismic vessel activity is not expected to produce a significant incremental increase in vessel activity levels. The cumulative incremental impact attributed to G&G vessel operations is **negligible**.

Measurements of ambient noise levels in the GOM are lacking. According to the predominant noise sources identified in **Appendix F (Section IV - Noise Environment)** for the GOM and their relative contributions to total noise levels, seismic surveys represent a relatively minor, intermittent, and non-stationary component of the overall noise environment. Seismic surveys produce repetitive, localized, and short-term increases in ambient noise levels, with the period between potential exposure ranging from hours to days (i.e., time between separate passes of a seismic survey vessel). In the near-field, within ~300 m or so of an array, received sound levels may reach or exceed 180 dB re 1 μPa (rms). At greater distances, sound from a seismic survey is of a similar nature to other commercial vessel activity. Given the current level of vessel activity and its associated infrastructure, future seismic survey activity is not expected to produce a significant incremental increase in ambient noise levels. Analysis of cumulative noise impacts on sea turtles in the GOM (provided in **Appendix J**) suggests that cumulative impacts are potentially **adverse but not significant**. The cumulative incremental impact attributed to G&G vessel noise is **negligible**. In summary, cumulative impacts to sea turtles from vessel traffic and noise will remain potentially **adverse but not significant**.

D.   FISHES

1.   Affected Environment

The GOM's marine habitats, ranging from coastal marshes to the deep-sea abyssal plain, support a varied and abundant fish fauna. Distinctive fish assemblages can be recognized within broad habitat classes for the continental shelf and oceanic waters as follows: soft bottom fishes, hard bottom fishes, and coastal pelagic fishes on the continental shelf; and epipelagic, midwater fishes, and demersal fishes in oceanic waters (>200 m water depths). **Appendix F (Section I.C - Fish Resources)** presents detailed information on fish populations in the GOM.

One endangered fish species, the smalltooth sawfish (*Pristis pectinata*), occurs in the open waters of the GOM. This species normally inhabits shallow waters (≤10 m), often near river mouths or in estuarine lagoons over sandy or muddy substrates, but also may occur in deeper waters (20 m) of the continental shelf. It is considered rare in the northern GOM, with its known range presently reduced to the coastal waters of Everglades National Park in extreme southern Florida. The endangered status of the smalltooth sawfish was finalized on 1 May 2003 (50 CFR Part 224). Critical habitat has not been defined, and data are being collected on the life history and biology of this species. Information on biology and distribution is outlined in detail in NMFS (2000).

Only one threatened fish species occurs in the GOM: the Gulf sturgeon. This species occurs primarily off Florida and Alabama, where it spends winter months in estuaries and inner shelf waters (over soft bottoms). The biology and status of this species are discussed in detail by USDOI, MMS (1999, 2002a).

2.   Impacts of Routine Activities - Alternative 1 (Proposed Action)

The main concern from an impact perspective is noise from seismic surveys, as well as the high pressure pulse realized in the near-field. Other G&G activities are considered to have negligible impacts on fish populations.

a.   Seismic Surveys

The general physiology of sound detection by fishes is relatively well understood (Fay and Megela Simmons, 1999; Popper and Fay, 1999). In contrast, the usual acoustic behavior and uses of sound by fishes are less well documented. Finally, the effects of intense and potentially damaging sound on fish hearing and behavior are only poorly understood, with only a small number of studies published in the peer-reviewed literature. **Appendix I (Fish Hearing and Sensitivity to Acoustic Impact)** presents a review of literature on fish hearing and acoustic impacts.

1)   *Fish Hearing*

Two modes of hearing occur in fishes. The stimulation route common to all organs and fish species is known as the "direct" route. In this stimulation mode, acoustic particle motion or hydrodynamic motion accelerates the fish's soft tissues, including the sensory epithelium, with the surrounding water. This mode of hearing is inherently directional. In this primitive mode of hearing, sound pressure does not play a direct role, and species that apparently hear exclusively in this mode are known as "hearing generalists."

Some species have an additional mode of hearing that renders them sensitive to sound pressure. In these "hearing specialists," the swimbladder or other gas bubble is efficiently linked mechanically to the fluid systems of the ear, and thus to the otoliths. As sound is transmitted through the animal's body, sound pressure fluctuations cause the gas bladder to expand and contract, and the motions of its walls are transmitted to the sensory epithelia of, usually, one of the otolith organs. The hearing specialists are particularly sensitive to sound, with best thresholds in the region of 50 dB re 1 μPa, and hear in a relatively wide frequency range (<100 to 3,000 Hz), with best sensitivity occurring between 200 and 1,000 Hz. Sensitivity to sound additionally depends on sound duration (with approximately equal energy detected up to sound durations of 400 ms), on the level of ambient, or background noise in the frequency region of the signals (Fay, 1988), and on the physiological integrity of the auditory system.

The sounds produced by airguns are impulsive, damped sinusoids, generally with fundamental frequencies in the range between 10 and 200 Hz. All GOM fish species have best hearing frequencies within, or slightly above, this range. Thus, airguns are likely to be audible by all fishes, given sufficient received sound pressure levels.

2)   *Physiological and Anatomical Effects on Auditory Systems of Fishes*

There are only two papers in the peer-reviewed literature specifically addressing the question of impairment to the auditory system following intense sound exposure (Enger, 1981; Hastings et al., 1996). These studies are summarized below and discussed in more detail in **Appendix I**.

Hastings et al. (1996) is the more comprehensive study, focusing on the oscar (*Astronotus ocellatus*), a freshwater cichlid and hearing generalist. In this study, fish were exposed to three sound levels (100, 140, and 180 dB re 1 μPa) at two frequencies (60 and 300 Hz) in a laboratory setting. Animals were stimulated for 1 hour, some with continuous tones, and others with intermittent tone bursts of 20% duty cycle (12 s on and 48 s off per minute). The only hearing impairment observed that could be attributed to sound exposure occurred in one or both ears of four of five animals stimulated at 300 Hz (continuous tone) that had been allowed to survive the exposure for 4 days. Animals stimulated at lower sound pressure levels, using continuous tones or tone bursts, or allowed less than 4 days after exposure for examination showed no hair cell damage. These results suggest that higher frequencies are likely to be more damaging than lower frequencies (given that both are detectable), but that impairment is only slight following 180 dB continuous 300 Hz tone exposure for 1 hour. It was assumed that hearing impairment would be more likely at higher sound exposure levels, and possibly for longer exposure durations.

Enger (1981) exposed Atlantic cod (*Gadus morhua*) to pure tones of various frequencies between 50 and 400 Hz at 180 dB for from 1 to 5 hours. Hair cell damage was observed in many animals. These results show more robust damaging effects than in the Hastings et al. (1996) study on oscar. The reasons for greater damage in the cod may be due to the higher hearing sensitivity in cod compared with oscar, and thus the higher level of the stimulating sound with respect to the hearing threshold. Hastings et al. (1996) speculate that levels of 220 to 240 dB at 300 Hz would be required to produce extensive hair cell damage in hearing generalists such as the oscar, and by extension, levels possibly 40 dB less than this to cause hearing impairment in the more sensitive hearing specialists.

In studies published only as abstracts but cited in Hastings et al. (1996), Cox et al. (1986a,b, 1987) exposed goldfish (hearing specialists) to pure tones between 250 and 500 Hz at levels of 182 to 204 dB re 1 µPa for 2 hours. In these studies, 100% of the fish stimulated at 204 dB (200 Hz) had damage to hair cells, but none of the animals stimulated at any frequency at 182 dB (500 Hz) were damaged.

In summary, these data indicate that the hair cells of inner ear organs of fishes can be damaged by intense sound. The lowest levels causing documented damage are 180 dB for continuous, long-duration tones in a region of good hearing (200 to 500 Hz for many fish species). The effects of intermittent sound stimulation are not known except that a 20% duty cycle is clearly less effective in damaging hair cells than continuous sounds. It seems likely that most fishes exposed to airgun shots at a distance of a few meters could receive inner ear damage (and subsequent hearing impairment) as a result of source levels in the range between 210 and 240 dB. However, as the distance between the fish and the airgun sources increased, the probability of hearing impairment would decrease according to the nature of distance attenuation taking place in the ensonified environment.

The types of anatomical damage to the ears reviewed above are expected to cause elevated sound detection thresholds in the affected animals, and the defects would be expected to last for the period of time that the tissue damage occurred. This total or partial deafness would be termed PTS in an analogy with effects observed in terrestrial animals and humans. The effect on hearing would presumably be restricted to those stimuli that selectively activate those areas. Thus, in cod for example, increased thresholds would be expected to occur for sounds in the frequency range served by the lesioned areas, but not necessarily for sounds of other frequencies. Since the hair cells of fishes can regenerate, beginning at about 7 days post trauma (Lombarte et al., 1993), damage-induced threshold shifts may not be ultimately permanent but might last for only 1 or 2 weeks. However, there are no published functional data on PTS in fishes to evaluate these hypotheses.

More is known about transient (rapidly recovering) threshold shifts that could occur following exposures to sounds at lower sound pressure levels than those causing PTS. Studies reviewed in **Appendix I (Section II.D - Behavioral Effects of Intense Sound on Hearing in Fishes)** indicate that long duration, continuous sounds in the region of 150 to 170 dB at 800 Hz and below could cause transient hearing impairment (30 dB or so) in most fishes that could last as long as 24 hours.

3) *Auditory Masking*

Behavioral studies on several fish species show that the thresholds for tone detection can be raised by the presence of additional sounds. When the threshold for the detection of one sound (i.e., the "signal") is raised by the presentation of another sound (i.e., the "masker"), masking is said to occur. Masking is greatest when signal and masker are simultaneous (Fay and Coombs, 1988) but can also occur for brief periods of time (less than one second; typically less than 300 ms) before masker onset (backward masking) and following masker offset (forward masking) (Popper and Clarke, 1979). In general, the auditory systems of fishes are crudely frequency-selective so that masking effects are restricted to signal frequencies in the frequency region of the masker (Hawkins and Chapman, 1975; Fay et al., 1978). Thus, for example, airguns would be expected to cause masking only within or near the limits of the airgun shot's spectral profile as it impinges on the fish, and only for up to 300 ms prior to and following the shot (non-simultaneous masking). The masking effect (threshold elevation) would be proportional to the received shot level. There are no masking data in the literature for intermittent, impulsive maskers similar to airgun shots.

Recent experiments on goldfish indicate that fish are capable of what has been called "auditory scene analysis" (Bregman, 1990). This means that a sound stream of interest can be "heard out" and analyzed for its informational content independently of simultaneous, potentially interfering sounds (Fay, 1998). As demonstrated, this capacity depends on spectral and temporal pattern differences between the signal and the potentially interfering sounds. These studies were carried out using repetitive impulses or clicks as both signals and potentially interfering sounds. Thus, the presence of intermittent, audible airgun shots would not necessarily impair fishes in receiving and appropriately interpreting other, biologically relevant sounds from the environment, so long as the shot sounds were not high enough in received level to cause PTS or TTS.

4) *Effects on Behavior*

Understanding of the uses of sound by fishes in normal behavior is very poor, with very few relevant published papers in the literature. Using the existing literature, common sense, and analogy with other vertebrate species studied, it is thought that fishes use sounds to aid in predator and prey detection and in general orientation to environmental sound sources and scatterers. In addition, as summarized above, it is thought that fishes are capable of resolving the general auditory scene, made up of the collection of sound sources and scatterers that normally occupy it. Thus, impaired hearing or excessive masking could have a general effect of rendering the affected fish generally less fit for survival and reproduction. In addition, some, but not all species are known to make sounds used in communication. This topic has been reviewed recently by Zelick et al. (1999).

Sonic fishes are known to produce sounds during aggressive interactions, territorial defense, territorial advertisement, swimming, courtship, and mating itself. In general, sounds are best understood in the context of reproduction because they are associated with particular, stereotyped behaviors. Thus, it is likely that the major negative effects of impaired hearing or excessive masking would possibly be in disrupting the usual reproductive behavior, and thus the reproductive success, of some species. In addition, Myrberg (1981) has identified various categories of acoustic communication, some less obvious than others, that are used by fishes. These are startle or warning sounds that may help protect individuals and groups from predation; courting sounds used as part of the usual mating behaviors including advertisement; swimming sounds used in schooling and aggregation; aggressive sounds used when competing for mates; sounds used in other aggressive interactions (e.g., in territorial defense); sounds used by interceptor species to avoid predation or to locate prey; and sounds overheard and used to competitive advantage by competitors. This is a very wide range of sound types, some produced by specially adapted sound production organs, and some not. The uses of these sounds are very widespread, too, including listening species that produce sound using specialized mechanisms, and other species that do not produce sound, but only listen for them to some advantage.

As indicated in **Appendix I (Section II.E - Uses of Sound by Fishes)**, some species of 37 families occurring in the GOM are known to use sound in behavior, according to the definitions of communication developed by Myrberg (1981). In general, this analysis of the potential use of sound in a communication context could potentially apply to most all fish species found in the GOM.

There have been no published reports on the effects of hearing impairment or excessive masking on the acoustic communication behavior of any species. Thus we can only speculate that impairment has the potential to disrupt a wide range of behaviors in many Gulf species (up to 270 species). These behaviors include startle responses to predators, courtship and mate choice, territorial and individual advertisement, maintenance of schooling and aggregation, aggressive competition for mates and other resources, and overhearing or intercepting potential predators, prey, and competitors. In addition to these behaviors classified by Myrberg (1981) as communication, it is also likely that hearing is used to help form a general image of the auditory scene that may include both other fishes and abiotic sound sources and scatterers. Fishes unable to use the usual acoustic ambience for this scene formation would be expected to be less fit for survival, reproduction, and competition for resources. Since all fish species investigated have inner ear organs of essentially the same type, and have complex auditory brain pathways that are like most other vertebrates, it would be expected that hearing plays an important role in maintaining the fitness of all species.

5)   *Near-Field Effects*

An additional impact to fishes from seismic airgun blasts is the effects on fish eggs and larvae. Laboratory and field studies have shown that statistically significant mortality of eggs and larvae in close proximity (i.e., 2 and 3 m) to airguns can occur (Dalen and Knutsen, 1986; Holliday et al., 1987). Overall the data indicate that significant impacts on fish eggs and larvae (generalized from studies on northern anchovy) would only result from repeated exposures to full seismic arrays (Holliday et al., 1987).

6)   *Conclusions*

All fish species investigated can hear, with varying degrees of sensitivity, within the frequency range of sound produced by seismic airguns. These sounds have the potential to cause masking of the sounds normally used by fishes in their usual acoustic behaviors at levels as low as 60 to 80 dB (just above detection thresholds for many species). Levels as high as 160 dB may cause receiving fishes to change their behaviors and movements, which may temporarily affect the usual distribution of animals and commercial fishing. Continuous, long-term exposure to levels above 180 dB has been shown to cause damage to the hair cells of the ears of some fishes under some circumstances. These effects may not be permanent since damaged hair cells are repaired and/or regenerated in fishes. The effects of intermittent sound stimulation (such as airgun pulses) are not known, except that a 20% duty cycle is clearly less effective in damaging hair cells than continuous sounds. It seems likely that most fishes exposed to airgun shots at a distance of a few meters could receive inner ear damage as a result of source levels in the range between 210 and 240 dB. As the distance between the fish and the airgun sources increased, the probability of hearing impairment would decrease according to the nature of distance attenuation taking place in the ensonified environment.

Fishes of the GOM are generally representative of the species for which quantitative hearing data are available. It is thought that most, if not all, species of fish in the GOM (and elsewhere) use the sense of hearing to contribute to their general fitness for survival and reproduction. The sense of hearing is thought to be used not only in intra-specific sound communication in some species, but also in many more general contexts in which most or all fishes monitor their acoustic environment by simply listening to the ambient sounds produced by other fishes and by abiotic sources and sound scatterers. There are no experimental studies investigating the effects of excessive masking or hearing impairment on the usual behaviors of fishes. However, several studies indicate that seismic shooting can temporarily alter the behaviors and movements of several fish species when received sound pressure levels are sufficiently high. The long-term effects of these hearing-related behavioral disruptions on fish populations are unknown.

Mortality in early life history stages can greatly affect recruitment of adult fishes. Starvation, predation, pollution stress, unfavorable temperatures, disease, and other stressors acting on the larval stage can ultimately affect year class strength in adults (Houde, 1987). Understanding the contribution of seismic-induced mortality of early life stages is fraught with uncertainty due to highly variable distribution in space and time. In addition, species-specific differences in larval growth rates (and, therefore, planktonic duration) add a great deal of variability to the recruitment process. Therefore, for near-field pressure wave effects from firing airguns, the impacts of seismic operations on fish eggs and larvae in the GOM would be adverse on a very small scale and negligible when extrapolated to the population level for species affected.

b.   *Mitigation*

Under existing regulations, use of explosives is prohibited except under written authorization from the MMS (see **Appendix E, Section II.A.2 - Stipulations and Protective Measures**). This is a measure intended to protect sea turtles, marine mammals, fishes, and other marine life. (It has been assumed that no explosives would be used under the Proposed Action.) There are no other existing mitigation measures specifically protecting fish resources.

c.   *Significance Criteria and Evaluation*

A ***significant adverse impact*** on fishes is one that is likely to cause

- death or life-threatening injury of one or more individuals of a listed (endangered/threatened) species; and/or
- death or life-threatening injury of non-listed species in sufficient numbers to adversely affect species populations and the ecological functioning of the fish community; and/or
- long-term or permanent displacement of any species from preferred feeding, breeding, or nursery habitats (including critical habitat for listed species and essential fish habitat [EFH]), or migratory routes; and/or
- destruction or adverse modification of critical habitat or EFH.

An ***adverse but not significant impact*** on fishes is one that is likely to cause

- death or life-threatening injury of individuals (other than listed species) in small numbers that would not adversely affect the population or the ecological functioning of the fish community; and/or
- short-term displacement of individuals from preferred feeding, breeding, nursery grounds, or migratory routes (including critical habitat for listed species and EFH).

A ***negligible impact*** on fishes is one that is likely to cause

- no death or life-threatening injury of any individuals; and

- no displacement of any species from preferred feeding, breeding, nursery grounds, or migratory routes (including critical habitat for listed species and EFH); and
- little disruption of behavioral patterns or other sublethal effects.

Impact criteria noted above were derived from USDOI, MMS (2001b). The main concern from an impact perspective is noise from seismic surveys. Such noise may disturb fishes and may produce temporary or permanent hearing impairment in some individuals but is unlikely to cause death or life-threatening injury. Neither seismic surveys nor other G&G activities are expected to cause long-term or permanent displacement of any listed species (i.e., smalltooth sawfish and Gulf sturgeon) from critical habitat or other preferred habitat, nor to result in destruction or adverse modification of critical habitat or EFH. Therefore, potential impacts to fish resources will be **negligible** most of the time, with occasional impacts being potentially **adverse but not significant** (e.g., when fish in very close proximity to an airgun array cannot avoid exposure to seismic survey noise).

3. Impacts of Routine Activities - Alternative 2 (Addition of Vessel-based Passive Acoustic Monitoring as a Requirement)

The addition of vessel-based passive acoustic monitoring as a requirement will not reduce impacts to fishes. Therefore, impacts under this alternative would be identical to those under Alternative 1 (i.e., impacts range from **negligible** to potentially **adverse but not significant**).

4. Impacts of Routine Activities - Alternative 3 (Addition of Both Passive and Active Acoustic Monitoring as Requirements)

The addition of both passive and active acoustic monitoring as requirements will not reduce impacts to fishes. Therefore, impacts under this alternative would be identical to those under Alternative 1 (i.e., impacts range from **negligible** to potentially **adverse but not significant**).

5. Impacts of Routine Activities - Alternative 4 (Restrict G&G Seismic Survey Activities)

Under Alternative 4, G&G seismic surveying operations would be subject to an operational restriction – prohibition of simultaneous surveys by more than one G&G survey vessel in those portions of the GOM most frequented by sperm whales and/or Bryde's whales (Alternative 4). While these operational restrictions are of primary importance to marine mammal species of concern, they are unlikely to reduce impacts to fishes. Only those fish species present within these areas (and in close proximity to locations where seismic surveys would occur) would realize benefit from this mitigation. Therefore, impacts to fishes under this alternative would be identical to those under Alternative 1 (i.e., impacts range from **negligible** to potentially **adverse but not significant**).

6. Impacts of Accidents

Fishes could be affected by a spill caused by an accident involving a G&G vessel. The effects of oil spills on fishes, with an emphasis on those species present in the GOM, have recently been summarized in USDOI, MMS (2001b, 2002a). While adult fishes are at relatively low risk, pelagic fish eggs and larvae are considerably more susceptible. This susceptibility is offset by the naturally high mortality rates exhibited by marine fishes. The risk of vessel accidents is discussed in **Appendix K (Risk of Vessel Accidents)**. The probability of such incidents occurring is quite low (e.g., three incidents reported in the GOM during 1996 and 1997), with the potential for a pollution incident even lower (i.e., two to five minor releases per year in the GOM). An event involving a survey vessel could result in release of diesel fuel, but such an event has an extremely remote probability of occurring. Thus, incidents involving survey vessels are not expected to result in significant impacts on fishes. Impacts to fishes from accidents are unlikely; however, if they occur, they would be most prevalent among egg and larval forms. Impacts to fishes from accidents are considered to be **negligible**.

7. Cumulative Impacts

The cumulative activity scenario is presented in **Appendix J** (**Cumulative Activity Scenario and Associated Impacts**). The major impact producing factor under the cumulative activity scenario is noise. Measurements of ambient noise levels in the GOM are lacking. According to the predominant noise sources identified in **Appendix F** (**Section IV - Noise Environment**) for the GOM and their relative contributions to total noise levels, seismic surveys represent a relatively minor, intermittent, and non-stationary component of the overall noise environment. Seismic surveys produce repetitive, localized, and short-term increases in ambient noise levels, with the period between potential exposure ranging from hours to days (i.e., time between separate passes of a seismic survey vessel). In the near-field, within ~300 m or so of an array, received sound levels may reach or exceed 180 dB re 1 μPa (rms). At greater distances, sound from a seismic survey is of a similar nature to other commercial vessel activity. Given the current level of vessel activity and its associated infrastructure, future seismic survey activity is not expected to produce a significant incremental increase in ambient noise levels. Analysis of cumulative noise impacts on GOM fishes (provided in **Appendix J - Cumulative Activity Scenario and Associated Impacts**) suggests that cumulative impacts are potentially **adverse but not significant**. The cumulative incremental impact attributed to G&G vessel noise is **negligible**. In summary, cumulative impacts to fishes from vessel traffic and noise will remain potentially **adverse but not significant**.

E.   COMMERCIAL AND RECREATIONAL FISHERIES

1.   Affected Environment

a.   *General Description*

Commercial fisheries are very important to the economies of the Gulf coastal states (Browder et al., 1991). The GOM leads all other U.S. regions in fishery production. In 2002, commercial fishery landings in the GOM, which includes western Florida, Alabama, Mississippi, Louisiana, and Texas, were over 1.7 billion pounds worth over $700 million (NMFS, 2003). Of the individual states, Louisiana led in total landings and value in 2002 with 1.3 billion pounds landed worth $305 million (see **Appendix F, Table F-6**). Mississippi was second with landings approaching 217 million pounds worth $48 million, followed by Texas (93 million pounds, $174 million), Florida's west coast (81 million pounds, $142 million), and Alabama (23 million pounds, $34 million). As noted in **Appendix F**, pounds landed and dollar value of landings are characteristics that are both used to rank and compare landings by state. Differences in rank (i.e., pounds landed vs. dollar value) reflect differential market prices for the species targeted.

Many species are caught and landed in the GOM commercial fisheries. Browder et al. (1991) stated that the fishery includes at least 97 species from 33 families. They considered the most important species groups to be oceanic pelagic (epipelagic) fishes, reef (hard bottom) fishes, coastal pelagic species, and estuarine dependent species. Primary estuarine dependent species targeted are menhaden, penaeid shrimps (brown, white, and pink), and blue crab; oysters are important, but are not considered here because they are harvested exclusively in inshore waters. Targeted species from the other groups include yellowfin tuna and swordfish (epipelagic); king and Spanish mackerels (coastal pelagic); and spiny lobster, red snapper, red grouper, and gag (reef/hard bottom).

Each species or species group is caught using one of various methods and gear types. Shrimps are taken by bottom trawling, menhaden are caught in purse nets, yellowfin tuna are caught on surface longlines, and snapper and grouper are caught by hook and line. The main fishing practices and seasons in the GOM are summarized in **Appendix F, Table F-7**.

b.   *Fishery Management and Essential Fish Habitat*

Most fishery species in the GOM are managed by the Gulf of Mexico Fishery Management Council (GMFMC), as outlined in **Appendix F, Section II.A.4 - Essential Fish Habitat in the Gulf of Mexico**. Fishery management plans (FMPs) have been prepared, to date, for corals and coral reefs, shrimps, stone crab, spiny lobster, reef fishes, coastal pelagic fishes, and red drum. All of these FMPs were recently amended to address EFH for the managed species (GMFMC, 2003). Another group of exploited species, the highly migratory pelagic fishes, is managed by NMFS Highly Migratory Species (HMS) Management Division, Office of Sustainable Fisheries. This office recently prepared an FMP (NMFS, 1999a) for Atlantic tunas, swordfish, and sharks that inhabit a broad geographic region that encompasses the GOM. This document also addressed EFH for the managed highly migratory species.

These two documents were consulted to gather information on EFH for the Federal waters of the GOM. Tables were prepared listing those species and life stages whose EFH occurred within the Federal waters of the Gulf. For each species, the tables also indicate whether the habitat for the appropriate life stage is pelagic (oceanic or coastal) or benthic (soft bottom or hard bottom). In some cases, such as corals and some sharks and reef fishes, there was insufficient information available to accurately describe EFH.

Appendix F (Section II.A.4 - Essential Fish Habitat in the Gulf of Mexico) provides tabular information pertinent to EFH. For example, **Table F-10** presents invertebrate and reef fish species managed by the GMFMC for which EFH has been identified. Corals were not included in the table as there are many soft and hard coral species in the Gulf, but formal EFH descriptions have yet to be made by the GMFMC. **Table F-11** presents EFH information for managed coastal pelagic species and red drum. **Table F-12** gives EFH for HMS such as swordfish, tunas, and sharks managed by NMFS. Although billfish (sailfish [*Istiophorus platypterus*], blue marlin [*Makaira nigricans*], white marlin [*Tetrapterus albidus*], and longbill spearfish [*T. pfluegeri*]) are now considered under HMS, there were no EFH designations in NMFS (1999a).

Spatially limited EFH, called habitat areas of particular concern (HAPCs), have also been identified in the GOM by the GMFMC. These include Dry Tortugas (Fort Jefferson National Monument), Florida Keys National Marine Sanctuary, Florida Middle Grounds, and Flower Garden Banks National Marine Sanctuary. All of these HAPCs are important with respect to corals and coral reefs and provide habitats for reef species such as snappers, groupers, and spiny lobster. Recently, two hard bottom areas offshore of west Florida near the Florida Middle Grounds have been closed for a minimum of 5 years to fishing for reef fishes by the GMFMC and HMS of NMFS, as outlined in **Appendix F (Section II.A.4 - Essential Fish Habitat in the Gulf of Mexico)**.

EFH includes most of the substrate and water column of the GOM where the managed species commonly occur. There will be some impact by the ensonification of the water column during the various types of geophysical surveys expected for the GOM. The potential effects of sound on fishes are described in detail in **Appendix I (Section II - Effects of Sound on Fishes)**. These effects include physiological or anatomical effects on auditory systems, potential behavioral alterations, and auditory masking. Sounds produced by seismic airguns fall within the frequency range from 10 to 200 Hz (and above), which is within the audible range for GOM fishes discussed above (i.e., reef fishes, red drum, coastal pelagic fishes, and highly migratory fishes). Although rigorous supporting data are not available, several studies indicate that seismic shooting can temporarily alter the behaviors and movements of several fish species when received sound pressures are sufficiently high.

2.   Impacts of Routine Activities - Alternative 1 (Proposed Action)

a.   Seismic Surveys

There are two main ways in which geophysical surveys could affect commercial fishing: 1) seismic surveys could cause behavioral changes in target species that could make them more difficult to catch; and 2) survey vessels and towed cables could temporarily preclude fishers from productive fishing grounds.

1)   *Possible Effects of Intense Sound on Fish Distribution and Commercial Catch*

It is now well documented that intense sounds such as those produced by seismic airguns affect the spatial distribution of fishes during and following exposure, thus affecting the commercial catch by trawl or hook and line within the exposure area and for a certain period post-exposure. Klimley and Beavers (1998) have carried out the best controlled experimental study on this topic of fish movements, using intense, low frequency sounds of the type used in the Acoustic Thermometry of Ocean Climate (ATOC) study (Munc et al., 1994). The goal of this study was to observe the behaviors of penned rockfish (*Sebastes flavidus, S. ariculatus,* and *S. mystinus*) during presentations of ATOC-like sounds (narrow-band signal centered at 75 Hz at peak sound pressure levels of 145 to 153 dB re 1 μPa in the vicinity of the fish). Sounds were 25 min in duration, including a 5-min gradual ramp-up in level. Animals confined to sea pens were allowed to move up to 15 m away from the pen area closest to the source. Results showed remarkably few behaviors indicating that the fish found the intense sounds objectionable (i.e., aversive). Prior to testing, many animals were stationed in the part of the pen closest to the transducer (where sound pressure levels would be greatest). During exposure, the animals tended to remain in these high exposure areas despite the opportunity to move 15 m away from the transducers to a pen area where the sound level was more than 40 dB less. The authors concluded that the ATOC-like signals had no apparent effect on the distribution of rockfish within the pen, and thus would not be expected to cause animals to move from a location resembling a natural refuge. Of course, the major differences between the ATOC-like signal and airgun shots are the brief, impulsive nature of the airgun sounds and their rapid rise time compared to the 5-min ramp-up of the ATOC signal.

Pearson et al. (1992) investigated the effects of airgun sounds on rockfish (*S. mystinus, S. menanops, S. miniatus,* and *S. serranoides*) behavior. At levels of 180 dB re 1 μPa, animals appeared to be alarmed, and either aggregated more tightly, descended, or ascended in the water column. The lowest level causing any observed behavioral change was 161 dB.

Some species of sharks may be attracted to some sounds, and other species may move away from some sound sources. Klimley and Myrberg (1979) noted that lemon sharks (*Negaprion brevirostris*) were repulsed from a sound (noise) source at 123 dB, but only when the signal was ramped up rapidly in amplitude (96 dB/s). At slow ramp rates, no effect was observed at this pressure level. Nelson and Gruber (1963) observed that sharks were attracted by intermittent, low frequency sounds resembling those produced by struggling or injured fishes. Received levels in this experiment were not recorded.

Popper and Carlson (1998) have reviewed the literature on attempts to control and guide fish movements using sound. The major questions have concerned controlling some clupeids near water intakes at dams and power plants using very high frequency sounds, and controlling salmonid movements at dams and other manmade waterways using very low frequency sounds. Experiments using ultrasound to control clupeid behavior (e.g., Nestler et al., 1992) are not especially relevant to the question of airgun geophysical exploration since the airgun sounds are very low in frequency. It appears that ultrasound at high levels (110 to 140 kHz at 180 dB and above) can divert some herring species (e.g., American shad [*Alosa sapidissima*]) from sound sources, but this effect is restricted only to those species, like the

American shad, that have developed ultrasonic hearing (Mann et al., 1997). Apparently, not all clupeids have this ability (Higgs and Popper, 2000). Knudsen et al. (1994) were successful in causing downstream migrating salmon to reverse their direction and swim away from intense, infrasound (<20 Hz) sources. This too may not be especially relevant with respect to seismic exploration in the GOM. However, some other species (e.g., Atlantic cod [*Gadus morhua*]) are known to detect efficiently very low frequency sound (0.1 to 30 Hz) (Sand and Karlsen, 1986), and it would not be surprising to find other Gulf species with such capabilities.

There have been a number of field studies investigating the effects of seismic airgun shots on cods and herrings distributions and catch successes in marine environments (Chapman and Hawkins, 1969; Matousek et al., 1988; Løkkeberg, 1991; Skalski et al., 1992; Engås et al., 1993; Løkkeberg and Soldal, 1993). Løkkeberg (1991) and Engås et al. (1993) both reported that the cod catch (by trawl) was reduced (80% to 50% reduction) during and following seismic shooting in the North Sea off the coast of Norway. The calculated sound pressure levels received by the fish were 191 and 160 dB, respectively. Remarkably, Engås et al. (1993) found that the reduced catch lasted for at least 5 days within a 33-km radius of the shooting. In the Pacific, off the coast of California, Skalski et al. (1992) found that calculated received levels of 161 dB caused rockfish (*Sebastes* sp.) to change behavior, to show alarm reactions at 180 dB, and startle reactions at 200 to 205 dB.

Along the west coast of Scotland, Wardle et al. (2001) exposed several reef fish species in varying age classes (i.e., juvenile cod, *Gadus morhua*; adult pollack, *Pollachius pollachius*; juvenile saithe, *Pollachius virens*; and adult mackerel, *Scomber scombrus*) to airgun noise and recorded their reaction. Maximum seismic source levels from the three-airgun array were calculated at 218 dB re 1 µPa (peak-to-peak). While the fishes observed failed to move off the reef and airgun noise did not affect diurnal rhythms, involuntary reactions were elicited when firings occurred within 10 m of the fishes being observed. Wardle et al. (2001) suggest that the visual cue of a firing airgun may elicit a more pronounced response than airgun noise alone.

In summary, the consensus is that seismic airgun shooting can result in reduced trawl and longline catch of several species when the animals receive levels as low as 160 dB. These effects would tend to be less in shallower water where sound propagation at these frequencies would be less efficient, and where sound levels would be lower due to the use of smaller airgun arrays. Reduced catch by trawl probably reflects temporary movement of the target species away from the ensonified area. Reduced catch by hook-and-line could be caused by fish moving away or changing feeding behaviors. In any case, there are sufficient careful observations in the literature to conclude that airgun shooting may cause a temporary reduction in the commercial fish catch within at least several kilometers of the ensonified area.

2)   *Space-Use Conflicts*

Preclusion of fishers from productive fishing grounds constitutes a space-use conflict. The size of the area precluded to fishing will obviously depend upon the overall area of the geophysical survey. As described in **Appendix D (Section I.A - Seismic Surveys)**, seismic information is collected along predetermined tracklines that form a larger spatial grid. The size of this grid will vary with individual projects, but usually includes multiple lease blocks. The seismic vessel operates on a 24-h basis for days, weeks, or months, depending on the survey scope, towing as many as six parallel streamers that are up to 12 km long. These vessels operate under a "restricted ability to maneuver" designation, which means other vessels in the path of the survey vessel must give way. Further, because of the length of the hydrophone array, the survey vessel requires considerable turning room between tracklines. Thus, the area precluded to fishing will extend beyond the planned geophysical survey area when appropriate allowances are made for maneuvering the vessel.

Clearly, any fishing operation occurring within a geophysical survey area (and appropriate buffer zone) would be affected. The degree of impact would depend upon the relative mobility of the fishing operation (USDOI, MMS, 1995). Fixed gear such as trapping is most vulnerable, and mobile gear such as hook-and-line fishing from drifting (or trolling) boats is least vulnerable. Common fishing practices in the GOM, including bottom trawling for shrimps, purse netting for menhaden, gillnetting for sharks, bottom longlining for groupers, and surface longlining for yellowfin tuna, would also be vulnerable. These gear types are not very mobile and require considerable time to deploy and retrieve. The potential for conflicts will increase with increasing water depth along with decreasing mobility of fishing vessels. The basic characteristics of these fisheries are described in **Appendix F (Section II.A - Commercial Fisheries)**. Broadly defined commercial fishing areas were depicted for the Eastern, Central, and Western Gulf of Mexico Planning Areas by USDOI, MMS (1986a,b). These visuals do not cover fisheries occurring in the open GOM beyond the shelf break. More recent discussions of commercial fishing operations on a Planning Area basis are presented in USDOI, MMS (2001b, 2002a).

The primary fishery in the open GOM is surface longlining for tuna, sharks, and swordfish. This is an example of fishing gear that is highly susceptible to interaction with seismic survey vessels. In fact, there have been anecdotal accounts of conflicts between longliners and seismic surveyors in recent years (B. Block, Stanford University, oral comm., 2000). Surface longlining occurs in deep waters offshore of the continental shelf. The spatial distribution of longline sets made in 1998 is shown in **Figure III-3**. Effort is generally widespread throughout the GOM, particularly in the deep-water Gulf where increasing oil and gas industry activity is occurring. Surface longlines used in the GOM yellowfin tuna fishery consist of mainline averaging 30 mi but that can exceed 60 mi (NMFS, 1999b). Between 20 and 30 hooks (attached to short leaders) are attached per mile of mainline. Buoys with radar reflectors are placed regularly along the length of the passively drifting mainline, which is only attached to the vessel during deployment and retrieval.

Surface currents and wind greatly influence the movement of longlines and other drifting gear (e.g., gill nets and purse nets) and must be taken into account when assessing potential impacts. A longline deployed upstream of a geophysical survey grid could drift into the path of the survey vessel. Surface currents in the GOM vary considerably and can range from 0.5 to 3 nmi/h. Surface longlines are allowed to drift for 4 to 5 h before a 10- to 12-h retrieval period (Lopez et al., 1979; Sakagawa et al., 1987). Thus, for a worst-case 3-nmi/h current with 17-h set time, the gear should be 51 nmi upstream of the survey area to prevent a conflict. Given a 0.5-nmi/h current with a 14-h set/retrieval time, the gear should be set at least 7 nmi upstream of the survey area.

A detailed analysis of potential conflicts between deep-water fishing and oil and gas operations has recently been contracted by MMS. This project will examine potential conflicts with all types of deep-water oil and gas operations and fishing practices. The results of the study will provide additional information to evaluate impacts and potential mitigation measures.

b.   *Mitigation*

There are no existing mitigation measures specifically protecting commercial and recreational fisheries, beyond U.S. Coast Guard announcements of pending seismic surveys (e.g., location, duration) published in the *Local Notice to Mariners*. It is possible that mitigation measures will be developed in the

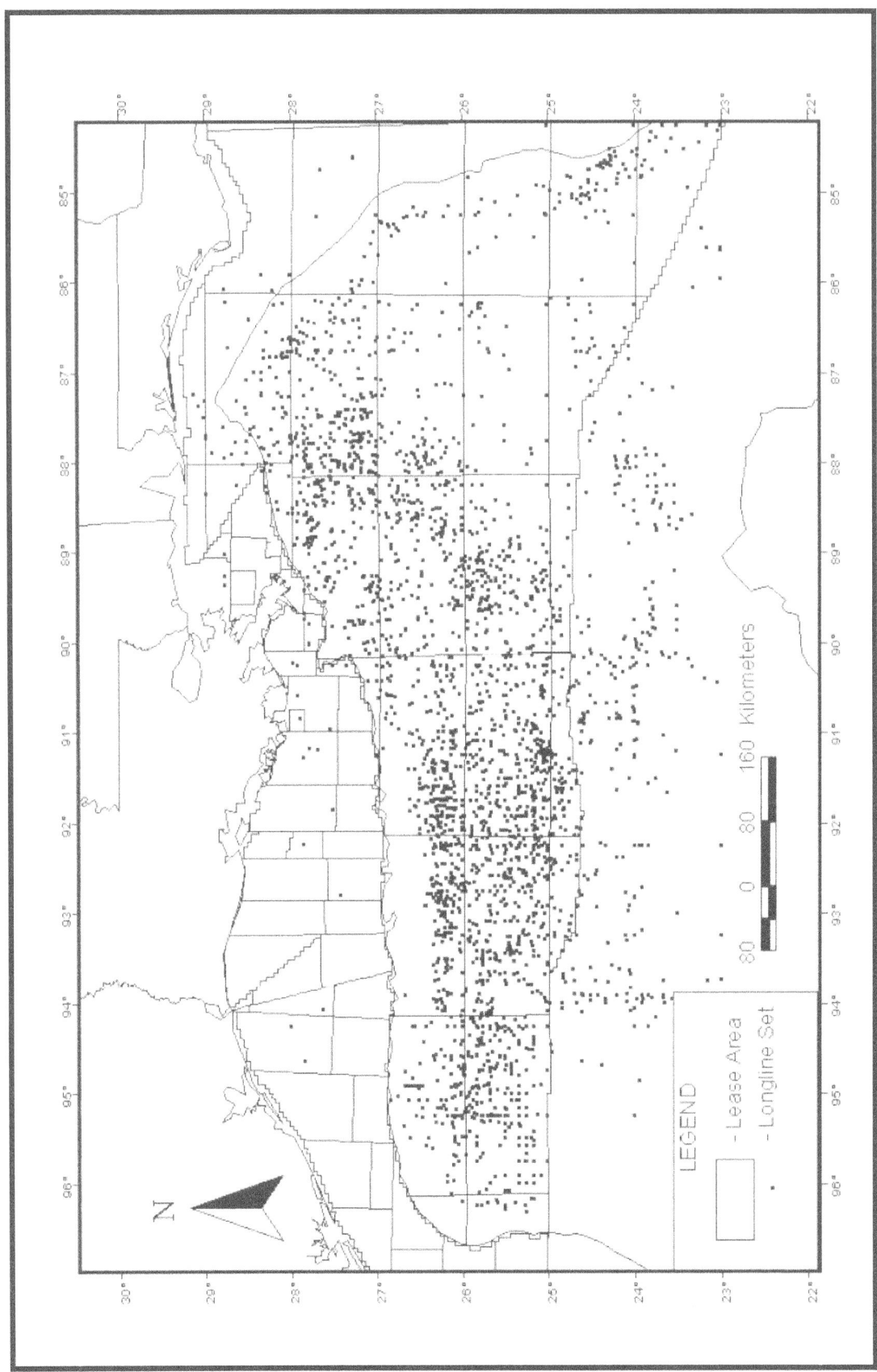

Figure III-3. Spatial distribution of surface longline sets made in the Gulf of Mexico during 1998 (Source: NMFS, 1999b).

near future to reduce the potential for space-use conflicts. Mitigation of conflicts with fishing operations such as surface longlining will require improved communication between fishers and geophysical survey personnel. Radio or cellular telephone contact is critical to allow fishers and survey personnel to communicate before, during, and after longline sets made in the vicinity of a planned survey area. Current and wind conditions could be monitored to prevent drifting gear from being set upstream of a survey area. Additional mitigation could be effected through the NMFS fishermen's contingency fund.

c. **Significance Criteria and Evaluation**

A **significant adverse impact** on commercial and recreational fisheries is one that is likely to cause

- fishers to be precluded from using 10% or more of the fishable area, for all or most of a fishing season; and/or
- 10% or more of fishers to be precluded from a fishing area (50-nmi radius) for all or most of a fishing season; and/or
- economic losses due to a decrease in catchability of target species in 10% or more of the fishable area.

An **adverse but not significant impact** on commercial and recreational fisheries is one that is likely to cause

- fishers to be precluded from using less than 10% of the fishable area for all or most of a fishing season; and/or
- less than 10% of fishers to be precluded from a fishing area (50-nmi radius) for all or most of a fishing season; and/or
- economic losses due to a decrease in catchability of target species in less than 10% of the fishable area.

A **negligible impact** on commercial and recreational fisheries is one that is likely to cause

- fishers to be occasionally precluded from fishing areas due to seismic survey activity; and/or
- possible short-term and localized decreases in catchability of target species, but not to an extent resulting in economic losses.

Space-use conflicts between seismic surveys and longline fisheries are likely until further mitigation is implemented. Due to the annual number and extent of regional seismic surveys as well as the widespread distribution of longline sets, some gear conflicts and/or exclusion of fishers from preferred fishing areas is likely. However, it is unlikely that such conflicts would preclude fishers from using 10% or more of the fishing grounds, or any specific fishing area, for all or most of a season. Seismic surveys are also likely to result in some short-term and localized decreases in catchability, but not to an extent that would be expected to result in economic losses. Overall, potential impacts on commercial and recreational fisheries are potentially **adverse but not significant**.

3. **Impacts of Routine Activities - Alternative 2 (Addition of Vessel-based Passive Acoustic Monitoring as a Requirement)**

The addition of vessel-based passive acoustic monitoring as a requirement will not reduce impacts to commercial and recreational fisheries. Therefore, impacts under this alternative would be identical to those under Alternative 1 (i.e., potentially **adverse but not significant**).

4.   Impacts of Routine Activities - Alternative 3 (Addition of Both Passive and Active Acoustic Monitoring as Requirements)

The addition of both passive and active acoustic monitoring as requirements will not reduce impacts to commercial and recreational fisheries. Therefore, impacts under this alternative would be identical to those under Alternative 1 (i.e., potentially **adverse but not significant**).

5.   Impacts of Routine Activities - Alternative 4 (Restrict G&G Seismic Survey Activities)

Under Alternative 4, G&G seismic surveying operations would be subject to an operational restriction – prohibition of simultaneous surveys by more than one G&G survey vessel in those portions of the GOM most frequented by sperm whales and/or Bryde's whales (Alternative 4). While these operational restrictions are of primary importance to marine mammal species of concern, they are unlikely to reduce impacts to commercial and recreational fisheries. Only those fish species of commercial or recreational importance that are present within these areas (and in close proximity to locations where seismic surveys would occur) would realize benefit from this mitigation. Therefore, impacts to commercial and recreational fisheries under this alternative would be identical to those under Alternative 1 (i.e., potentially **adverse but not significant**).

6.   Impacts of Accidents

Commercial and recreational fisheries could be affected by a spill caused by an accident involving a G&G vessel. The effects of oil spills on fishes, with an emphasis on those species present in the GOM, have recently been summarized in USDOI, MMS (2001b). While adult fishes are at relatively low risk, pelagic fish eggs and larvae are considerably more susceptible. This susceptibility is offset by the naturally high mortality rates exhibited by marine fishes. The risk of vessel accidents is discussed in **Appendix K (Risk of Vessel Accidents)**. The probability of such incidents occurring is quite low (e.g., three incidents reported in the GOM during 1996 and 1997), with the potential for a pollution incident even lower (i.e., two to five minor releases per year in the GOM). An event involving a survey vessel could result in release of diesel fuel, but such an event has an extremely remote probability of occurring. Thus, incidents involving survey vessels are not expected to result in significant impacts on commercial and recreational fisheries. Impacts to commercial and recreational fisheries from accidents are unlikely; however, if they occur, they would be most prevalent among egg and larval forms. Impacts to commercial and recreational fisheries from accidents are considered to be **negligible**.

7.   Cumulative Impacts

The cumulative activity scenario is presented in **Appendix J (Cumulative Activity Scenario and Associated Impacts)**. The major impact producing factors under the cumulative activity scenario are space use conflicts and noise. In terms of vessel activity levels (for space use conflicts and as a noise source), seismic survey vessel activity represents a very small component of total vessel activity in Gulf waters. For example, oil and gas support vessels account for approximately one quarter of a million transits per year in Gulf waters, with commercial vessels >10,000 DWT contributing another 36,000 trips (see **Appendix F, Section II.E - Marine Transportation**). By comparison, approximately 20 seismic surveys may occur annually in the Gulf, or 0.01% of the activity from these three sources (i.e., oil and gas support operations, commercial cargo and tanker activity, seismic surveys). Commercial fishing and recreational boating, military operations, and ocean study activities also contribute to the cumulative vessel activity level, further reducing the relative contribution from seismic surveys. Impacts from vessel operations (and associated areal preclusion) under the cumulative scenario are **negligible**. Because, G&G

operations contribute an extremely minor amount of additional vessel activity in the GOM, incremental impacts are deemed **negligible**.

Measurements of ambient noise levels in the GOM are lacking. According to the predominant noise sources identified in **Appendix F (Section IV - Noise Environment)** for the GOM and their relative contributions to total noise levels, seismic surveys represent a relatively minor, intermittent, and non-stationary component of the overall noise environment. Seismic surveys produce repetitive, localized, and short-term increases in ambient noise levels, with the period between potential exposure ranging from hours to days (i.e., time between separate passes of a seismic survey vessel). In the near-field, within ~300 m or so of an array, received sound levels may reach or exceed 180 dB re 1 μPa (rms). At greater distances, sound from a seismic survey is of a similar nature to other commercial vessel activity. Given the current level of vessel activity and its associated infrastructure, future seismic survey activity is not expected to produce a significant incremental increase in ambient noise levels. Analysis of cumulative noise impacts on GOM commercial and recreational fisheries (provided in **Appendix J**) suggests that cumulative impacts are **negligible**. The cumulative incremental impact attributed to G&G vessel noise is **negligible**. In summary, cumulative impacts to commercial and recreational fisheries from space use conflicts and noise will remain **negligible**.

F.   COASTAL AND MARINE BIRDS

1.   Affected Environment

The waters and adjacent coastal landforms of the northern GOM are inhabited by a diverse assemblage of resident and migratory birds. These include primarily seabirds, shorebirds, wetlands birds, and waterfowl. Brief descriptions of these species groups may be found in USDOI, MMS (1997a).

Birds of the GOM and adjacent landforms that are currently listed as endangered or threatened species under the ESA comprise primarily coastal or inshore species. These include the piping plover, whooping crane, eskimo curlew, brown pelican, wood stork, and bald eagle (USDOI, USFWS, 1998). Of these, the brown pelican is the only species of endangered or threatened bird that may be affected by the Proposed Action.

In the northern GOM, brown pelicans occur regularly, but in small numbers. Mississippi, Louisiana, and Caribbean populations of the brown pelican are endangered, but this designation does not apply to populations in Alabama or Florida. In addition, Louisiana has requested the removal of its brown pelican population from the endangered species list (USDOI, MMS, 1997a). Brown pelicans generally do not venture far offshore. Fritts and Reynolds (1981) observed them only within about 40 km of the shoreline, and only one was seen over the northern GOM continental slope during the GulfCet surveys (Peake, 1996).

2.   Impacts of Routine Activities - Alternative 1 (Proposed Action)

There are no data indicating that seismic surveys or other G&G activities have adverse impacts on birds. Potential impact mechanisms are noise impacts from seismic surveys, and disturbance by vessel and aircraft traffic.

a.   *Seismic Surveys*

Generally, noise produced from activities associated with seismic surveys might impact only those offshore species of birds that spend large quantities of time underwater, either swimming or plunge diving while foraging for food. Offshore GOM birds that may be classified as underwater swimmers include certain waterfowl (some diving ducks) and seabirds (loons and cormorants). Generally, these species are limited to waters of the inner continental shelf. Waterfowl and loons are both seasonal migrants (winter), whereas cormorants are resident species. GOM plunge diving birds include only certain seabirds (primarily brown pelicans, gannets, and boobies). Brown pelicans are a resident species that is also limited to waters of the inner continental shelf. Gannets and boobies, however, are seasonal migrants (gannets in winter and boobies primarily in summer) that may range throughout the GOM.

Noise from seismic surveys could adversely affect surface-feeding and diving seabirds near airgun arrays. However, there are no data indicating such impacts exist. Stemp (1985) found no effect of seismic survey activity on the distribution and abundance of seabird populations in arctic Canadian environment. Parsons (in Stemp, 1985) reported that shearwaters with their heads underwater were observed within 30 m of seismic sources (explosives) and did not respond. Because seismic pulses are directed downward and highly attenuated near the surface (see **Appendix D, Section II.A - Airguns**), birds feeding on the surface or diving just below it are unlikely to be exposed to sound levels sufficient to cause temporary or permanent hearing impairment. In any case, sound pressure levels would not be sufficient to cause death or life-threatening injury.

A possible mechanism for indirect impacts is alteration of prey concentrations. However, persistent, widespread alterations in abundance of fishes are not expected (see **PEA Section III.D - Fishes**).

b.   *Vessel and Aircraft Traffic*

The impacts of routine vessel traffic (including vessels involved in G&G surveys) on coastal and marine birds have been analyzed in recent lease sale EISs (USDOI, MMS, 1997a, 1999, 2001a, 2002a,b). Effects of vessel traffic on birds offshore are negligible. In nearshore and coastal waters, vessels must use selected nearshore and inland waterways and adhere to protocols set forth by the U.S. Coast Guard for reduced vessel speeds. Vessel traffic in coastal and inland waters may disturb coastal birds, but impacts are diminished by the routine presence and low speeds of vessels in coastal waters. Permit stipulations (see **Section III.F.2.c - Mitigation** below) avoid disturbance of islands used for nesting by the endangered brown pelican. Aircraft involved in G&G surveys must maintain altitudes of at least 2,000 ft when flying over biologically sensitive areas such as wildlife refuges and national parks.

c.   *Mitigation*

There is one existing mitigation measure that is partly or wholly designed to minimize impacts on birds. This protective measure is included as a stipulation to G&G permits. Specifically, transportation operations conducted through Aransas or Cavello Passes will avoid disturbance of the following islands used for nesting by the endangered brown pelican: Sundown Island in Matagorda Bay, Second Chain of Islands in San Antonio Bay, Long Reef in Aransas County, and Pelican Island in Nueces County, Texas.

d.   *Significance Criteria and Evaluation*

A *significant adverse impact* on coastal and marine birds is one that is likely to cause

- death or life-threatening injury of one or more individuals of a listed (endangered/threatened) species; and/or
- death or life-threatening injury of non-listed species in sufficient numbers to affect the population adversely; and/or

- long-term or permanent displacement of any species from preferred feeding, breeding, or nursery habitats (including critical habitat for listed species), or migratory routes; and/or
- destruction or adverse modification of critical habitat for any listed species.

An *adverse but not significant impact* on coastal and marine birds is one that is likely to cause

- death or life-threatening injury of individuals (other than listed species) in small numbers that would not adversely affect the population; and/or
- short-term displacement of any species from preferred feeding, breeding, nursery grounds, or migratory routes (including critical habitat for listed species).

A *negligible impact* on coastal and marine birds is one that is likely to cause

- no death or life-threatening injury of any individuals; and
- no displacement of any species from preferred feeding, breeding, nursery grounds, or migratory routes (including critical habitat for listed species); and
- little disruption of behavioral patterns or other non-injurious effects.

The two concerns from an impact perspective are noise from seismic surveys and disturbance of coastal bird populations by vessel and aircraft traffic. Although noise from seismic surveys could in theory affect surface-feeding and diving seabirds very near airgun arrays, there are no data indicating such impacts exist. Because seismic pulses are directed downward and highly attenuated near the surface, birds feeding on the surface or diving just below it are unlikely to be exposed to sound levels sufficient to cause temporary or permanent hearing impairment. Sound pressure levels would not be sufficient to cause death or life-threatening injury.

The brown pelican is the only endangered bird species likely to come into contact with offshore G&G activities. They are commonly found in the northern GOM but occur only within coastal habitats and nearshore waters (see **Appendix F, Section I.D - Coastal and Marine Birds**). Neither seismic surveys nor other G&G activities are expected to kill or injure brown pelicans, cause long-term or permanent displacement from critical habitat or other preferred habitat, or to result in destruction or adverse modification of critical habitat. Existing mitigation measures are effective in minimizing impacts of vessel and aircraft traffic on coastal bird populations. In conclusion, potential impacts to coastal and marine birds are **negligible**.

3. **Impacts of Routine Activities - Alternative 2 (Addition of Vessel-based Passive Acoustic Monitoring as a Requirement)**

The addition of vessel-based passive acoustic monitoring as a requirement will not reduce impacts to coastal and marine birds. Therefore, impacts under this alternative would be identical to those under Alternative 1 (i.e., **negligible**).

4. **Impacts of Routine Activities - Alternative 3 (Addition of Both Passive and Active Acoustic Monitoring as Requirements)**

The addition of both passive and active acoustic monitoring as requirements will not reduce impacts to coastal and marine birds. Therefore, impacts under this alternative would be identical to those under Alternative 1 (i.e., **negligible**).

5.   Impacts of Routine Activities - Alternative 4 (Restrict G&G Seismic Survey Activities)

Under Alternative 4, G&G seismic surveying operations would be subject to an operational restriction – prohibition of simultaneous surveys by more than one G&G survey vessel in those portions of the GOM most frequented by sperm whales and/or Bryde's whales (Alternative 4). While these operational restrictions are of primary importance to marine mammal species of concern, they are unlikely to reduce impacts to coastal and marine birds. Therefore, impacts to birds under this alternative would be identical to those under Alternative 1 (i.e., **negligible**).

6.   Impacts of Accidents

Coastal and marine birds could be affected by a spill caused by an accident involving a G&G vessel. The effects of oil spills on coastal and marine birds, with an emphasis on those species present in the GOM, have recently been summarized in USDOI, MMS (2001b). The risk of vessel accidents is discussed in **Appendix K (Risk of Vessel Accidents)**. The probability of such incidents occurring is quite low, with the potential for a pollution incident even lower. An event involving a survey vessel could result in release of diesel fuel, but such an event has an extremely remote probability of occurring. Thus, incidents involving survey vessels are not expected to result in significant impacts on coastal and marine birds. Impacts to birds from accidents are unlikely; however, if they occur, they would range from **negligible** to potentially **adverse but not significant**.

7.   Cumulative Impacts

The cumulative activity scenario is presented in **Appendix J (Cumulative Activity Scenario and Associated Impacts)**. The major impact producing factors under the cumulative activity scenario are coastal vessels and aircraft traffic, and associated noise. In terms of vessel activity levels, seismic survey vessel activity represents a very small component of total vessel activity in Gulf waters. For example, oil and gas support vessels account for approximately one quarter of a million transits per year in Gulf waters, with commercial vessels >10,000 DWT contributing another 36,000 trips (see **Appendix F, Section II.E - Marine Transportation**). By comparison, approximately 30 seismic surveys may occur annually in the Gulf, or 0.01% of the activity from these three sources (i.e., oil and gas support operations, commercial cargo and tanker activity, seismic surveys). Commercial fishing and recreational boating, military operations, and ocean study activities also contribute to the cumulative vessel activity level, further reducing the relative contribution from seismic surveys. Therefore, G&G operations contribute an extremely minor amount of additional vessel activity in the GOM.

Measurements of ambient noise levels in the GOM are lacking. According to the predominant noise sources identified in **Appendix F (Section IV - Noise Environment)** for the GOM and their relative contributions to total noise levels, seismic surveys represent a relatively minor, intermittent, and non-stationary component of the overall noise environment. Seismic surveys produce repetitive, localized, and short-term increases in ambient noise levels, with the period between potential exposure ranging from hours to days (i.e., time between separate passes of a seismic survey vessel). In the near-field, within ~300 m or so of an array, received sound levels may reach or exceed 180 dB re 1 µPa (rms). At greater distances, sound from a seismic survey is of a similar nature to other commercial vessel activity. Given the current level of vessel activity and its associated infrastructure, future seismic survey activity is not expected to produce a significant incremental increase in ambient noise levels. Analysis of cumulative noise impacts on the coastal and marine birds of the GOM (provided in **Appendix J**) suggests that cumulative impacts from vessel and traffic noise are **negligible**. The cumulative incremental impact attributed to G&G vessel traffic and noise is **negligible**. In summary, cumulative impacts to coastal and marine birds from vessel and aircraft traffic and noise will remain **negligible**.

G.   BENTHIC COMMUNITIES

1.   Affected Environment

Benthic communities in the GOM are described in previous EISs (USDOI, MMS, 1996, 1997a,b, 2000, 2001a, 2002a,b). Species distribution patterns are summarized in Gallaway (1988) and Pequegnat et al. (1990).

Most of the GOM has a soft, muddy bottom in which burrowing worms are the most abundant animals present. Benthic habitats of environmental concern include low relief hard bottom and seagrass/algal areas on the inner and middle continental shelves, and areas of high relief hard bottom (topographic features, pinnacles, and ridges) on the outer continental shelf and upper slope. Hard bottom and seagrass areas are described as "live bottom" habitats because they are characterized by high productivity and a diverse animal community.

Chemosynthetic communities are another benthic habitat of concern in the deep-water GOM. These communities are unique because they support dense epifaunal growth in the deep-water environment and are not directly dependent upon photosynthesis. These communities have now been found along much of the northern GOM continental slope and along the Florida Escarpment (Paul et al., 1984; MacDonald et al., 1990, 1995; MacDonald, 2000).

2.   Impacts of Routine Activities - Alternative 1 (Proposed Action)

a.   *Geological and Geochemical Sampling*

The main concern from an impact perspective is geological and geochemical sampling (bottom sampling and shallow coring). Such sampling occurs both prelease (e.g., as an exploration tool) and postlease (e.g., after a shallow hazards survey, which identifies potential benthic communities of concern such as live bottom areas and chemosynthetic communities). Existing protective measures and lease stipulations specifically protect topographic features, live bottom communities (including the pinnacle trend area), and chemosynthetic communities (see **Appendix E, Section II.A.2 - Stipulations and Protective Measures**). Therefore, no adverse impacts on sensitive benthic resources are anticipated.

Soft bottom communities are affected by bottom sampling and shallow coring, as well as deep-tow side-scan sonar surveys (which involve dragging a chain across the bottom), and placement and retrieval of bottom cables and anchors. Bottom sampling involves about 500 piston/gravity cores each year. The direct environmental consequences of piston/gravity cores (and heat flow probes) are 10-cm diameter holes in the ocean floor. Depending upon the firmness of the seafloor, the core or probe weight stand (30- to 45-cm diameter footprint) may also impact the seafloor and crush seafloor animals if the core or probe penetrates to maximum depth. However, the total area of seafloor disturbed by all bottom sampling and shallow coring activities each year is estimated to be much less than 0.01% of any given lease block. Similarly, the placement and removal of bottom cables and anchors will produce localized sediment disturbance to soft bottom communities. Remnants of prior activity may leave depressions or holes in the benthos. No overall changes in species composition, community structure, and/or ecological functioning of soft bottom communities are expected. Bottom cables or anchors deployed and retrieved as part of a prelease survey may adversely affect unidentified benthic communities of concern (i.e., live bottom areas, topographic features, pinnacles, chemosynthetic communities), should they be present when cables are positioned and removed.

b.   Mitigation

There are several existing mitigation measures that protect benthic communities (see **Appendix E, Section II.A.2**). In particular, seafloor disturbing activities are prohibited within the No-Activity Zones of certain topographic features, and anchoring is restricted within the "Pinnacle Trend" area of the northeastern GOM. Further protections apply for activities within the Flower Garden Banks National Marine Sanctuary. All seafloor disturbing activities under G&G permits require the submission of detailed plans so that the MMS can evaluate potential impacts and specify protective measures if necessary.

c.   Significance Criteria and Evaluation

A *significant adverse impact* on benthic communities is one that is likely to cause

- substantial damage to sensitive benthic resources such as chemosynthetic communities, live bottom communities (including the pinnacle trend), or hard bottom communities associated with topographic features; and/or
- changes in species composition, community structure and/or ecological functioning of soft bottom communities, with measurable change in species composition or abundance beyond that of normal variability, or ecological function within a species range, for 5 years or longer (i.e., long-term).

An *adverse but not significant impact* on benthic communities is one that is likely to cause

- changes in species composition, community structure, and/or ecological functioning of soft bottom communities beyond that of normal variability, or ecological function within a species range, that persist for less than 5 years.

A *negligible impact* on benthic communities is one that is likely to cause

- changes in abundance of individual species but no overall changes in species composition, community structure, and/or ecological functioning of soft bottom communities.

Existing protective measures and lease stipulations specifically protect topographic features, live bottom communities (including the pinnacle trend area), and chemosynthetic communities (see **Appendix E, Section II.A.2**). Therefore, no adverse impacts on known sensitive benthic resources are anticipated. However, potential damage to unidentified sensitive benthic resources could occur during prelease surveys employing bottom-founded equipment (e.g., cables); the severity of such impact will be dependant upon the extent of burial and/or scraping damage. The total area of soft bottom seafloor disturbed by all bottom sampling and shallow coring activities each year is estimated to be much less than 0.01% of any given lease block. No overall changes in species composition, community structure, and/or ecological functioning of soft bottom communities are expected. Therefore, potential impacts to soft bottom communities under this alternative are **negligible**. Impacts to unidentified sensitive benthic resources may occur due to bottom-disturbing activities. All bottom-disturbing OCS activities are reviewed by MMS for their potential for impacts to sensitive benthic biological and archaeological resources. Any OCS activities believed to present a potential for non-trivial impacts are closely reviewed and mitigations applied, as needed, to mitigate potential impacts. Therefore, impacts to unidentified sensitive benthic resources will range from **negligible** to potentially **adverse but not significant**.

3.   Impacts of Routine Activities - Alternative 2 (Addition of Vessel-based Passive Acoustic Monitoring as a Requirement)

The addition of vessel-based passive acoustic monitoring as a requirement will not reduce impacts on benthic communities. Therefore, benthic community impacts under this alternative would be identical to those under Alternative 1 (i.e., **negligible**).

4.   Impacts of Routine Activities - Alternative 3 (Addition of Both Passive and Active Acoustic Monitoring as Requirements)

The addition of both passive and active acoustic monitoring as requirements will not reduce impacts to benthic communities. Therefore, impacts under this alternative would be identical to those under Alternative 1 (i.e., **negligible**).

5.   Impacts of Routine Activities - Alternative 4 (Restrict G&G Seismic Survey Activities)

Under Alternative 4, G&G seismic surveying operations would be subject to an operational restriction – prohibition of simultaneous surveys by more than one G&G survey vessel in those portions of the GOM most frequented by sperm whales and/or Bryde's whales (Alternative 4). While these operational restrictions are of primary importance to marine mammal species of concern, they are unlikely to reduce impacts to benthic communities. Therefore, impacts to benthic communities under this alternative would be identical to those under Alternative 1 (i.e., **negligible**).

6.   Impacts of Accidents

Benthic communities are unlikely to be affected by a spill caused by an accident involving a G&G vessel. Impacts are considered to be **negligible**.

7.   Cumulative Impacts

The cumulative activity scenario is presented in **Appendix J (Cumulative Activity Scenario and Associated Impacts)**. The major impact producing factor under the cumulative activity scenario is seafloor disturbance. Cumulative impacts from other operations under the cumulative scenario were determined to be **negligible**. Cumulative incremental impacts to benthic communities from seafloor disturbance are **negligible**. In summary, cumulative impacts to benthic communities from seafloor disturbance will remain **negligible**.

H.   SUMMARY AND COMPARISON OF IMPACT DETERMINATIONS

This impact analysis represents a final determination of potential impact (or range of potential impacts) associated with G&G operations to various sensitive resources present in OCS waters of the GOM. Summary information has been compiled from the best available data sources, with appropriate qualification of recognized data gaps, limitations, or assumptions applied in the impact analysis. Detailed technical information that supports these impact determinations has been placed in a series of appendices, as appropriate. Linkage between the summary information found within this section and more comprehensive technical appendices occurs via liberal use of cross references. While this allows the reader to concentrate on the immediate issue of impact assessment without inclusion of an extensive technical discussion, it also allows for a more detailed, in-depth presentation of pertinent technical issues that support the impact determination. **Table III-4** is a summary table that concisely compares the environmental consequences of each alternative, on a resource by resource basis. Comparisons of

Table III-4

Summary and Comparison of Environmental Impacts from Each Alternative on a Resource by Resource Basis

Resource	Alternative 1 – Proposed Action[1]	Alternative 2 – Addition of Vessel-based Passive Acoustic Monitoring as a Requirement	Alternative 3 – Addition of Both Passive and Active Acoustic Monitoring as Requirements	Alternative 4 – Restrict G&G Seismic Survey Operations
Marine Mammals *Sperm whales*	Potentially *adverse but not significant* impact from potential exposure to elevated, repetitive, intermittent, and localized noise levels, resulting in possible hearing impairment No mortality or serious injury (i.e., no exceedance of the Potential Biological Removal [PBR] level); no displacement from key habitat; no long-term or permanent displacement from preferred feeding, breeding, or nursery habitats; no substantial or chronic disruption of behavioral patterns that may adversely affect sperm whales through effects on annual rates of recruitment or survival In water depths >200 m throughout the Gulf of Mexico and Federal waters <200 m deep in the Eastern Planning Area, under requirements of Notice to Lessees (NTL) No 2004-G01, visual monitoring of the 500-m exclusion zone reduces the likelihood that whales will be present in close proximity to an array When coupled with ramp-up, these measures may reduce the potential for hearing impairment or other injury to sperm whales from instantaneous start-up of an airgun array Potential for acoustic impact to sperm whales remains, as undetected individuals may enter the zone of maximum ensonification below an array during a dive Limitations: 1) visual monitoring effective during daylight, good visibility/sightability conditions; 2) ramp-up remains unproven as a mitigation measure, although recognized as a common sense measure; 3) species/group specific limitations (detectability); and 4) minor cost ramifications	Potentially *adverse but not significant.* Passive acoustic monitoring relies on passive sensing and location of whale vocalizations Potential for impact may be reduced if whales vocalize and are detected Assumed to be effective for sperm whales, as they frequently vocalize Potential for impact to sperm whales is not completely eliminated, impact level remains potentially adverse but not significant Limitations: Passive acoustic monitoring 1) only works for vocalizing whales; many animals are quiet much of the time, especially when disturbed; 2) difficult to determine the ange to the vocalizing animals when using a towed array; 3) cannot readily determine depth to vocalizing animals; 4) for fixed hydrophones (ship or bottom mounted recorders, sonobuoys, ocean bottom cables), area of coverage/detection range may be limited by noise, requiring more sensors to cover a seismic survey area; 5) requires hydrophone arrays be towed behind the survey vessel (or from an additional chase boat); 6) hydrophone performance may be affected by tow speed and the ship's acoustic characteristics, limiting detection range; and 7) cost ramifications.	Potentially *adverse but not significant.* Passive acoustic monitoring relies on passive sensing and location of whale vocalizations Active acoustic monitoring relies on an active (e.g., sonar) search for whales Potential for impact may be reduced Passive acoustic monitoring may be effective for sperm whales, as they frequently vocalize Potential for impact to sperm whales may be reduced if whales vocalize and are detected Advantages of passive acoustic monitoring relative to active acoustic monitoring include: 1) longer ranges can be achieved; 2) omnidirectional; 3) species can be potentially identified by their vocalization signature; 4) no acoustic footprint that could affect the target animals; and 5) more mature and affordable technology Advantages of active acoustic monitoring compared with passive acoustic monitoring are as follows: 1) works with non-vocalizing or cryptic whales and those species that exhibit only limited vocalization; 2) can in some cases determine 3D range and bearing, including depth of vocalizing animals; 3) avoids having to stream behind survey vessel if sound source and hydrophone/receiver are hull-mounted; and 4) may involve less bulky equipment, minimizing personnel required for handling and operation Limitations: Active acoustic monitoring has the following disadvantages: 1) active source may be more harmful than the sound source it is being used to mitigate; 2) limited detection ranges depending on power and frequency; 3) inability to identify species based purely on size; 4) limited beam width and associated problems seeing deep-diving whales at close range; 5) active systems could potentially affect the behavior of the animals themselves; 6) towfish would be required, possibly larger than passive acoustic monitoring array; and 7) current costs for development and deployment are higher	Potentially *adverse but not significant.* Restrictions on concurrent seismic operations will prevent the potential for simultaneous exposure Acoustic impacts to sperm whales would be slightly reduced; however, the potential for acoustic impacts to sperm whales remains Industry practice may already effectively implement this restriction Limitations: None

Table III-4

Summary and Comparison of Environmental Impacts from Each Alternative on a Resource by Resource Basis
(Continued)

Resource	Alternative 1 – Proposed Action [1]	Alternative 2 – Addition of Vessel-based Passive Acoustic Monitoring as a Requirement	Alternative 3 – Addition of Both Passive and Active Acoustic Monitoring as Requirements	Alternative 4 – Restrict G&G Seismic Survey Operations
Bryde's whales	Potentially *adverse but not significant* impact from potential exposure to injurious noise levels (repetitive, intermittent, and localized); same effects as sperm whales, above (e.g., no PBR exceedances, etc.) Effective visual monitoring when coupled with ramp-up may reduce the potential for hearing impairment or other injury to Bryde's whales from instantaneous start-up of an airgun array Visual monitoring of a pre-determined impact zone reduces the potential for hearing impairment The potential for acoustic impact is reduced Potential for acoustic impacts to Bryde's whales remains Limitations: Ramp-up unproven as a mitigation measure Visual monitoring effective only during daylight, during periods of good visibility; cost ramifications	Potentially *adverse but not significant*. Same as sperm whales above – potential for impact remains Bryde's whales are known to vocalize, but the frequency of their vocalizations may be problematic; passive acoustic effective only when whales vocalize; geographic and depth limits of this species may further limit mitigation effectiveness Limitations: Same as above.	Potentially *adverse but not significant*. Bryde's whales are known to vocalize; frequency of vocalization may be problematic; passive acoustic effective only when whales vocalize; geographic and depth limits of this species may further limit mitigation effectiveness Active acoustic may be problematic Potential for impact remains Limitations: Same as above	Potentially *adverse but not significant*. Same as sperm whales above – potential for impact remains the same Limitations: Same as above.
Beaked whales	Potentially *adverse but not significant* impact from potential exposure to injurious noise levels (repetitive, intermittent, and localized); same effects as sperm whales, above (e.g., no PBR exceedances, etc.)	Potentially *adverse but not significant*. Beaked whales same as Bryde's whales above – potential for impact remains Limitations: Same as above	Potentially *adverse but not significant*. Beaked whales same as Bryde's whales above – potential for impact remains Limitations: Same as above	Potentially *adverse but not significant*. Same as sperm whales above – potential for impact remains Limitations: Same as above
Other cetaceans	Potentially *adverse but not significant* impact from potential exposure to injurious noise levels (repetitive, intermittent, and localized)	Potentially *adverse but not significant*. Same as sperm whales above – potential for impact remains Limitations: Same as above	Potentially *adverse but not significant*. Same as sperm whales above – potential for impact remains Limitations: Same as above	Potentially *adverse but not significant*. Same as sperm whales above – potential for impact remains Limitations: Same as above
Manatees	*Negligible* impact due to unlikely exposure	*Negligible*. No effect of mitigation; potential for impact remains the same Limitations: Same as above	*Negligible*. No effect of mitigation; potential for impact remains the same Limitations: Same as above	*Negligible*. No effect of mitigation; potential for impact remains the same Limitations: Same as above
Sea Turtles	Impacts primarily *negligible*, but may elevate to potentially *adverse but not significant* from potential exposure to injurious noise levels (repetitive, intermittent, and localized) and vessel traffic; seismic noise may disturb sea turtles and may produce temporary or permanent hearing impairment in some individuals, but is unlikely to cause death or life-threatening injury Seismic surveys and other G&G activities are not expected to cause long-term or permanent displacement from critical habitat/preferred habitat, nor result in destruction or adverse modification of critical habitat	*Negligible* to potentially *adverse but not significant*. No effect of mitigation due to lack of vocalization; potential for impact remains the same Limitations: Same as above No effect of mitigation due to lack of vocalization	*Negligible* to potentially *adverse but not significant*. Potential for impact remains the same Limitations: Same as above	*Negligible* to potentially *adverse but not significant*. No effect of mitigation; potential for impact remains the same Limitations: Same as above

Table III-4

Summary and Comparison of Environmental Impacts from Each Alternative on a Resource by Resource Basis
(Continued)

Resource	Alternative 1 – Proposed Action[1]	Alternative 2 – Addition of Vessel-based Passive Acoustic Monitoring as a Requirement	Alternative 3 – Addition of Both Passive and Active Acoustic Monitoring as Requirements	Alternative 4 – Restrict G&G Seismic Survey Operations
Fishes	*Negligible* to potentially *adverse but not significant* impact from seismic survey noise (repetitive, intermittent, and localized); noise may disturb fish and may produce temporary or permanent hearing impairment in some individuals, but is unlikely to cause death or life-threatening injury Seismic surveys are not expected to cause long-term or permanent displacement of any listed species from critical habitat/preferred habitat, nor to result in destruction or adverse modification of critical habitat or essential fish habitat	*Negligible* to potentially *adverse but not significant.* No effect of mitigation; potential for impact remains the same Limitations: Same as above	*Negligible* to potentially *adverse but not significant.* No effect of mitigation; potential for impact remains the same Limitations: Same as above	*Negligible* to potentially *adverse but not significant.* No effect of mitigation; potential for impact remains the same Limitations: Same as above
Commercial and Recreational Fisheries	Potentially *adverse but not significant* impact from space-use conflicts (between seismic surveys and longline fisheries) and seismic survey noise (short-term and localized decreases in catchability, not to an extent that would be expected to result in economic losses)	Potentially *adverse but not significant.* No effect of mitigation; potential for impact remains the same Limitations: Same as above	Potentially *adverse but not significant.* No effect of mitigation; potential for impact remains the same Limitations: Same as above	Potentially *adverse but not significant.* No effect of mitigation; potential for impact remains the same Limitations: Same as above
Coastal and Marine Birds	*Negligible* impact from seismic surveys (repetitive, intermittent, and localized noise) and aircraft and vessel traffic (noise, disturbance)	*Negligible.* No effect of mitigation; potential for impact remains the same Limitations: Same as above	*Negligible.* No effect of mitigation; potential for impact remains the same Limitations: Same as above	*Negligible.* No effect of mitigation; potential for impact remains the same Limitations: Same as above
Benthic Communities	*Negligible* impact from geological and geochemical sampling, anchors, and bottom cables (placement, retrieval) on soft bottom communities, with *negligible* to potentially *adverse but not significant* impacts to sensitive benthic communities (if unidentified prior to bottom-related activities) No seismic related impacts[2]	*Negligible* (see footnote 2) No effect of mitigation; potential for impact remains the same Limitations: Same as above	*Negligible* (see footnote 2) No effect of mitigation; potential for impact remains the same Limitations: Same as above	*Negligible* (see footnote 2) No effect of mitigation; potential for impact remains the same Limitations: Same as above

[1] In all Gulf waters >200 m and all outer continental shelf waters (regardless of water depth) in the Eastern Planning Area; includes ramp-up and visual monitoring per NTL No 2004-G01; in water depths <200 m (except in the Eastern Planning Area), no visual monitoring or ramp-up required

[2] Negligible impacts to sensitive benthic resources are expected from bottom-related (i e, seafloor) activities due to existing protective measures and operational restrictions, coupled with proper identification of known sensitive resources; increased impact levels might be realized if sensitive resources remain unidentified (see **PEA Section III.G - Benthic Communities**)

alternatives are based on their perceived advantages relative to the Proposed Action, Alternative 1. Limitations evident in each alternative also are noted in the table.

As is evident from this table, Alternatives 2, 3, and 4 fail to offer either effective mitigation or reduction in impact (due to operational restrictions) to most of the resources listed, including fishes, commercial and recreational fisheries, coastal and marine birds, and benthic communities. Impacts to these resources remain unchanged relative to Alternative 1.

Among sea turtles, visual monitoring (Alternative 1) offers limited mitigation under those conditions where individual turtles or sargassum rafts may be sighted (perception bias). Remaining alternatives (i.e., Alternatives 2, 3, and 4) offer no reduction in impact level relative to Alternative 1.

Among the marine mammals, there are expected decreases in impacts to all vocalizing marine mammals (with the exception of manatees) under Alternative 2 (passive acoustic monitoring) and Alternative 3 (passive and active acoustic monitoring), in spite of the limitations noted for each mitigation measure. Alternative 2 (in water depths >200 m throughout the GOM and all OCS waters [regardless of water depth] in the Eastern Planning Area, where NTL No. 2004-G01 is required) offers the greatest potential for reduced impacts to vocalizing species (e.g., sperm whales); impacts to non-vocalizing marine mammals remain unchanged. A combination of existing NTL requirements and passive acoustic monitoring, while not completely eliminating the limitations inherent in each individual measure, is expected to provide the greatest degree of assurance that no marine mammals (with the exception of manatees) have ventured into the exclusion zone of an operational seismic array.

Under Alternative 3, some marine mammals (i.e., those that vocalize) may realize benefit from passive acoustic monitoring, and some species may be detectable using active acoustic monitoring techniques. However, there are limitations and potential impacts associated with active acoustic monitoring that may outweigh its potential benefits. Mitigation is not expected to be as effective under Alternative 3 as may be realized under Alternative 2.

Alternative 4 offers an indeterminate reduction in the potential for impact to sperm and Bryde's whales; however, current industry practice may already address the need to avoid concurrent seismic survey activity.

As a final note, in all cases where impacts are expected to decrease, there has been no reduction in impact designation level established under Alternative 1, due in part to the limitations inherent in each mitigation. Only the potential for impact has been reduced as a result of the mitigation measure.

I.   SUMMARY OF DATA GAPS, LIMITATIONS, AND ASSUMPTIONS

A summary of data adequacy, recognized data gaps and limitations, and assumptions used in this analysis has been compiled in **Table III-5**. Topics outlined in the table include adequacy of life history and ecological data for GOM species, adequacy of survey data, assumptions and interpolations applied to the available data, mitigation effectiveness, and significant data gaps (e.g., recognition and avoidance of seismic noise, seismic pulse frequency characteristics and the range of marine mammal hearing sensitivities, etc.). Where identified limitations or data deficiencies have been noted, a realistic approach or interpretation of available data was employed in the impact analysis.

Table III-5

Summary of Identified Data Limitations, Data Gaps, and Interpolations Utilized During Impact Determinations of Geological and Geophysical (G&G) Activities on Potentially Affected Resources

Limitation, Data Gap, or Interpolation	Comment
Adequacy of Biological Data	Life history and ecological data are available for certain species or groups found within outer continental shelf (OCS) waters of the U.S. Gulf of Mexico (GOM), while others are not as well known. Physiological data (e.g., hearing sensitivity [hearing thresholds, frequency]; sensitivity to intermittent seismic sound) are adequate for several groups (e.g., fishes), with only limited data available for marine mammals and turtles. Hearing sensitivities measured in several odontocete species, but only inferred from vocalization characteristics for mysticetes and remaining odontocete species. Similarly, only limited hearing threshold data exist for sea turtles.
Adequacy of Survey Data	Historical and current survey activity of marine mammals and sea turtles in the U.S. GOM is rather extensive, based on completion (or continuation) of nearly a dozen major survey efforts in the region since 1980 (e.g., GulfCet I and II surveys, Southeast Fisheries Science Center [SEFSC] and SEFSC surveys, SEFSC-Minerals Management Service [MMS] joint survey efforts). Line transect data are readily available for the larger marine mammals species and species/species groups that are readily visible at the sea surface (i.e., available for sighting); more cryptic species or those that are solitary or occur in small groups are not as well described. Distributional information, based on available survey data, is biased towards shallower waters (e.g., nearshore, continental shelf, upper slope), indicative of the relative survey effort. Deep water species distributional and density information should be qualified, as appropriate. Refined marine mammal density estimates became available as a result of recent survey efforts completed by National Marine Fisheries Service (NMFS). These revised density estimates, categorized by water depth (i.e., <200 m; ≥200 m) and planning area, have been integrated into this analysis.
Simplified Calculation of Radial Distance to Isopleths of Interest	A simplifying assumption in the calculation of distance to the isopleths of interest is that the ensonified zone is circular or radial. A reasonable accommodation is also made in the calculations for the array effect. As noted in **Appendices C (Section III - Seismic Source Levels)** and **D (Section II.A - Airguns)**, the seismic signal from an array is directed in a downward, vertical direction, towards the seafloor. Array geometry (e.g., airgun separation and size) and aspect (i.e., in-line versus cross-line array axes) also affect the geometry of the sound field created. As a consequence, the area of ensonification surrounding an array is elliptical when viewed from above, indicative of the fact that there is a difference in the sound attenuation rate along the water surface fore and aft of the array versus port and starboard of the array (see **Appendix C, Figure C-12; Appendix D, Figure D-4; Appendix L, Figure L-2**). Consideration of the array effect takes this fact into account. For the purpose of this analysis, the greatest radial distance to the target isopleth is calculated; this provides an estimate of the maximum distance to a particular isopleth. However, calculations that are based on this maximum radial distance produce an overestimate of total area ensonified. A review of a recent model validation study (LGL Ltd., environmental research associates, 2003) and field measurements of seismic arrays (e.g., see Goold and Fish, 1998) indicates that the actual elliptical area ensonified by either single or multiple pulses is approximately one-half of that calculated using a radial geometry. Therefore, a 50% reduction in area ensonified has been adopted in the harassment and take analysis (**Appendix L**) to account for conversion from a radial to an elliptical zone of ensonification.

Table III-5

Summary of Identified Data Limitations, Data Gaps, and Interpolations Utilized During Impact Determinations of Geological and Geophysical (G&G) Activities on Potentially Affected Resources
(Continued)

Limitation, Data Gap, or Interpolation	Comment
Recognition and Avoidance of Seismic Noise	National Oceanic and Atmospheric Administration Fisheries (NOAA-F) has established 160 and 180 dB re 1 µPa (rms) as possible exposure levels to which marine mammals may exhibit behavioral response or physiological injury. These levels are subject to change as more data regarding sound exposure levels and marine mammal sensitivity and response become available. Specific information about the reactions of some baleen and odontocete whales to low frequency noise pulses has come from observing responses of bowhead whales in the Beaufort Sea, migrating gray whales off California, and migrating or lingering humpback whales off western Australia to pulses from airguns and other non-explosives (Richardson et al. 1986; Malme et al., 1983; 1984; McCauley et al. 1998, 2000). Avoidance reactions to seismic sounds at received levels of about 160 to 170 dB re 1 µPa (rms) and, in some cases, somewhat lower levels, have been documented. Humpback whales showed avoidance at a mean received sound level of 140 dB re 1 µPa (rms). About 50% of feeding gray whales will cease feeding at an average peak pressure level of 173 dB (Malme et al., 1988), and about 50% of migrating gray whales will avoid a seismic source at about the same average received sound levels (Malme and Miles, 1985). Similar observations for GOM marine mammals exposed to seismic are generally lacking. Further, the question of acclimation to seismic noise (i.e., use of seismic arrays in a mature oil producing region such as the GOM) remains unanswered. In addition, marine mammal researchers have suggested that the orientation of the marine mammal at risk relative to the sound source also may affect how much an animal may be affected (e.g., whether temporary threshold shift or permanent threshold shift results from the exposure). Currently, there is no provision in the incidental take calculation methodology to accommodate the potential for recognition and avoidance of the seismic source by marine mammals in the general area (e.g., beyond the 160-dB [rms] isopleth).
Seismic Pulses and the Range of Marine Mammal Hearing Sensitivities	There is only limited accommodation within the impact analysis to account for the predominant frequencies that comprise seismic signals and the potential for overlap (or, more importantly, the lack thereof) between seismic signals and the vocalization/estimated hearing frequencies of species of concern. As outlined in **Appendix G (Section I.E.1.b - Biological Context)**, hearing threshold data for marine mammals are very limited, particularly for the larger whale species. Hearing thresholds have been inferred from vocalization data. In general, most toothed cetaceans are good mid- to high-frequency vocalizers (i.e., sonic to ultrasonic specialists, 100 Hz to 100 kHz), while baleen whales generally vocalize in the lower frequency bands (i.e., sensitive to infrasonic sounds, 10 Hz and upwards; see **Appendix G, Section I.A – Marine Mammal Hearing**). More specifically, mysticetes exhibit inferred hearing thresholds of 10 to 31,000 Hz, with dominant frequencies of 16 to 25,000 Hz; odontocete hearing thresholds are highly species-specific, with dominant frequencies of 120 to 60,000 Hz in delphinids (**Appendix G, Tables G-1 and G-2**; see Ketten, 1998). For this G&G Programmatic Environmental Assessment, impact determinations (and take estimations) considered the possibility of frequency overlap between the predominant components of a seismic pulse and the inferred hearing capabilities of GOM marine mammal species. One major consideration in an assessment of impact should be the output frequencies of the seismic source and the hearing/frequency sensitivities of the species of concern.
Intermittent vs. Continuous Noise Exposure	Sensitive resources, particularly those that are mobile, are not continuously exposed to seismic noise (e.g., along the entire transect or series of transects surveyed during a seismic survey). Airgun arrays are fired intermittently, every 12 to 16 s, and in some cases slightly longer. The duration of each pulse is measured in terms of milliseconds, as are the reflections of the pulse off the seafloor or sea surface. In addition, both the sound source and a majority of biological resources (e.g., marine mammals, sea turtles, fish) are moving. Relative movement of vessel and the resource (e.g., whales) either away from or towards one another over a period of minutes may be significant. Both the type of exposure (i.e., intermittent, pulsed from G&G seismic sources) and the duration of the exposure (i.e., due to resource proximity and location in the water column [surface vs. submerged], ability to move away from a potentially adverse sound source) are important elements in estimating exposure and potential for impact.

Table III-5

Summary of Identified Data Limitations, Data Gaps, and Interpolations Utilized During Impact Determinations of Geological and Geophysical (G&G) Activities on Potentially Affected Resources

(Continued)

Limitation, Data Gap, or Interpolation	Comment
Effect of Biotic and Abiotic Factors in Marine Mammal Distribution	Oceanographic and other conditions may strongly influence the distribution and numbers of marine mammals present in an area. Therefore, for some species, available density estimates may not be fully representative of the densities that may be encountered during future seismic operations (e.g., a species distribution within a particular planning area). In addition, limited life history and ecological data are available for certain species, while others are well known. The results of this assumption may either underestimate or overestimate impact. This is particularly problematic for species that often congregate (i.e., show clumped distribution patterns). To represent the potential presence of a particular species in shallow or deep water more accurately, within one of three GOM Planning Areas, refined density estimates became available as a result of recent survey efforts completed by NMFS. These revised density estimates, categorized by water depth (i.e., <200 m; $\geq$200 m) and planning area, have been integrated into this analysis. The problem persists, however, for smaller scale oceanographic features (e.g., rings), which may attract select species, or cross depth or planning area boundaries.
Effectiveness of Existing Mitigation Measures	MMS Notice to Lessees No. 2004-G01 was implemented to provide assurance that marine mammals and turtles (i.e., at water depths $\geq$200 m throughout the GOM; all OCS waters in the Eastern Planning Area) will not be subjected to relatively high sound levels in close proximity to an array. The use of ramp-up procedures, coupled with protected species observation (visual monitoring) and reporting and establishment of a 500-m (radial) zone (exclusion zone) around the center of the towed airgun array and the area within the immediate vicinity of the survey vessel, are intended to 1) provide confidence to the G&G operator that the initiation of ramp-up will occur without whales, other marine mammals, and sea turtles in close proximity, and 2) provide an audible warning to marine mammals in the general area that seismic operations are being initiated and to allow sufficient time for those animals to leave the immediate vicinity. Visual monitoring is not expected to be completely effective – e.g., marine mammals and sea turtles may remain submerged (and unsighted) during visual monitoring, cryptic species may only surface for a very short time, weather and sea surface conditions may limit an observer's ability to sight animals at the surface.
Other Assumptions	• Assumption that 3D seismic survey activity levels realized in 2002 will, in general, remain relatively constant and distributed between water depths and planning areas in similar fashion. Limited accommodation is made in this analysis for increased activity in select planning areas where no activity occurred in 2002 (i.e., Eastern Gulf Planning Area). • Utilization of 0.3 and 3.0 km as appropriate maximum radial distances to the 180- and 160-dB (rms) isopleths is based on a "typical" seismic array (i.e., 4,550-in.^{3} airgun array; 240 dB re 1 μPa [zero-to-peak], 230 dB re 1 μPa [rms]) in use in the GOM (see **Appendix C, Section III – Seismic Source Levels**). Actual array output varies by operator and can be lower than the typical system employed in this analysis. Lower output systems produce smaller zones of ensonification. For example, for an array with a specified source level 10 dB less than the typical array, the range to the 180- and 160-dB isopleths would be on the order of 100 and 1,000 m, respectively.

IV. CONSULTATION, COORDINATION, AND DOCUMENT REVIEW

In addition to the MMS personnel contacted during the project, the following Federal agency personnel were consulted during the preparation of the PEA:

NOAA - F (formerly NMFS), Silver Spring, Maryland:
- Therese Conant
- Roger Gentry
- Ken Hollingshead
- Terry Rowles
- Barbara Schroeder
- Kathy Wang

NMFS, Pascagoula, Mississippi:
- Keith Mullin

NMFS, Miami, Florida:
- Blair Mase
- Nancy Thompson

NMFS, La Jolla, California:
- Jay Barlow
- Shannon Rankin

National Oceanic and Atmospheric Administration, Pacific Marine Environmental Laboratory, Newport, Oregon:
- Chris Fox

Office of Naval Research, Arlington, Virginia:
- Robert Gisiner

U.S. Fish and Wildlife Service, Panama City, Florida:
- Lorna Patrick

U.S. Fish and Wildlife Service, Jacksonville, Florida:
- Linda Walker

U.S. Fish and Wildlife Service, Washington, DC:
- Ronald Britton

The following personnel from the private sector and academia were consulted during the preparation of the PEA:

- Roy Bampton, Veritas Marine Acquisition
- William Kerry Behrens, Fugro GeoServices
- Gary Bernhardt, Benthos
- Doug Bremner, WesternGeco
- Mike Brunham, Petroleum Geo-Services
- Jack Caldwell, WesternGeco
- Paul Chelminski, Bolt Technology Corporation
- David Crockett, WesternGeco
- Stephen Cryer, WesternGeco
- Stuart Denny, Veritas Marine Acquisition

- Bill Dragoset, WesternGeco
- William Evans, Texas A&M University, Galveston
- Philip M. Fontana, Veritas Marine Acquisition
- Kurt Fristrup, Cornell University, Cornell Laboratory of Ornithology
- Elvind Fromyr, Petroleum Geo-Services
- G.C. Gill, International Association of Geophysical Contractors
- Earnest Hall, Oyo Corporation
- Scott Hammond, Bell GeoSpace
- Ted Hampton, Fugro GeoServices
- Richard Henman, WesternGeco
- Alf Hesthag, WesternGeco
- Todd Jones, GXT
- Jeff Mayville, WesternGeco
- Frank Morrison, University of California, Berkeley
- Michael W. Norris, WesternGeco
- Tom Parker, Petroleum Geo-Services
- Thomas O. Pickens, Petroleum Geo-Services
- Bill Pramik, Petroleum Geo-Services
- David Pryer, Petroleum Geo-Services
- Dan Quinn, Baker Hughes
- W. John Richardson, LGL Ltd., Environmental Research Associates
- Peter Seidel, Petroleum Geo-Services
- Peter Stewart, GXT
- Bob Tatham, University of Texas, Austin
- Aaron Thode, Massachusetts Institute of Technology
- Peter Tyack, Woods Hole Oceanographic Institution

MMS, responsible agency, industry, and public review of the Draft PEA resulted in valuable contributions, recommendations, and additional technical information pertinent to G&G activities and projected impacts. A broad representation of interested parties contributed review comments to finalize the PEA.

MMS and other responsible agencies or entities reviewed the preliminary and draft versions of the PEA. The following individuals have provided comments and recommendations regarding this assessment, listed alphabetically by office or agency:

MMS Gulf of Mexico OCS Region, Office of Leasing and Environment:
- Warren Barton
- Tom Bjerstedt
- Dennis Chew
- Jeff Childs
- Richard Defenbaugh (Contracting Officer's Technical Representative [COTR])
- Stephanie Gambino
- Bill Lang
- Debbie Miller (LE Technical Editor)
- Michelle Morin
- Terry Scholten
- Sarah Tsoflias

MMS Headquarters Office, Environmental Division:
- Judy Wilson

MMS Headquarters Office, Resource Evaluation Division:
- Dave Zinzer

MMC:
- MMC staff

NOAA-F (previously NMFS), Headquarters Office:
- Ken Hollingshead
- Simona Roberts

The Draft PEA also was released for public review and comment. An abbreviated chronology and summary of public comments follows. In mid-August 2002, the Draft PEA was made available for review and comment to interested non-Federal parties. The opportunity for comment expired on 19 September 2002. Comments were received from several parties. These parties are listed below in the sequence that comments were received, and with the comments synopsized.

1) E-mail message, R. Dykstra (Orchids & Egrets, Inc.) to MMS dated 5 September 2002; subject "Programmatic Environmental Assessment (PEA) of G&G on the Gulf of Mexico's OCS."

   Mr. Dykstra's message asks MMS to consider the potential negative environmental impacts of G&G exploration on marine life of the Gulf OCS that outweigh the "short term and supposed" strategic need for domestic fuel sources. He comments that the PEA is needed and should not be rushed. He encourages conservation and accelerated development of sustainable energy sources. Mr. Dykstra's general comments did not result in any specific guidance by MMS for technical revisions of the PEA.

2) E-mailed letter, G.C. Gill (International Association of Geophysical Contractors) to MMS dated 19 September 2002; captioned "IAGC comments on a Draft Programmatic Environmental Assessment on Geological and Geophysical Exploration in the Gulf of Mexico (EA)."

   Mr. Gill's letter notes the role of the IAGC as an international trade association representing the geophysical services industry, which acquires, interprets, and provides data and/or interpretations to the oil and gas industry. The letter notes that IAGC reviewers found "numerous technical errors"; they reviewed some sections of the PEA, but not others; and the IAGC requests that the PEA be made internally consistent, so that each time a matter is addressed, corrections be made, even if the IAGC review did not address that place in the document. The letter comments that the PEA recognizes that seismic activities pose few risks to the marine environment, and essentially none that have been documented by scientific evidence. The letter notes that IAGC and MMS are research partners in the Sperm Whale Seismic Study, a multi-year research program intended to provide precise information on the effects of seismic acoustic emissions on marine mammals, especially sperm whales. In addition to these general comments, the IAGC provided several pages of additional general comments and more specific technical comments. The IAGC endorses [Draft PEA] Alternative 1[1] and believes the other mitigation measures are speculative, disadvantageous, and should not be required. Some

---

[1] Alternatives evaluated in the Draft PEA have been revised to reflect current MMS requirements on G&G operations; see **Appendix A**.

of the IAGC's specific technical comments were incorporated into the MMS guidance for revision of the Draft PEA.

3) E-mailed letter, D. Quinn (Baker Atlas) to MMS dated 19 September 2002; responding to the G&G Draft Programmatic Environmental Assessment with comments on the regulations with respect to borehole seismic surveys.

Mr. Quinn's letter comments that the Draft PEA discusses seismic operations in the Gulf solely in the context of "surface seismic surveys" without adequate discussion of borehole seismic surveys, which use airguns as a sound source and so should be included in the scope of the PEA. The letter provides a brief overview of borehole seismic operations and suggests revisions to the suite of mitigations that comprise Alternative 3 to reduce environmental impact while still allowing for borehole seismic data acquisition. Mr. Quinn's comments resulted in guidance by MMS to expand the scope of the PEA to better address borehole seismic survey operations.

4) Letter, K. Harb (National Ocean Industries Association [NOIA]) to MMS dated 19 September 2002; captioned "Draft Programmatic Environmental Assessment on Geological and Geophysical Exploration in the Gulf of Mexico Outer Continental Shelf."

Ms. Harb notes NOIA's role as a national trade association representing all segments of the offshore energy industry. The NOIA letter comments on the importance of seismic surveys to the offshore oil and gas industry and expresses strong support for Alternative 1, the "Proposed Action." The letter lists the mitigation measures associated with each of the four alternatives discussed in the Draft PEA and expresses concern that the additional measures and restrictions proposed in [Draft PEA] Alternatives 2, 3, and 4 would slow OCS exploration and development. NOIA's general comments did not result in any specific guidance by MMS for technical revisions of the PEA.

5) E-mailed comments, C. Sarthou (Gulf Restoration Network) to MMS dated 19 September 2002; headed "Comments submitted by the Gulf Restoration Network on the MMS' G&G … Programmatic Draft Environmental Assessment."

Ms. Sarthou's comments briefly describe the scope and interests of the Gulf Restoration Network (GRN), then express several specific concerns regarding the Draft PEA. The GRN does not believe the finding of no significant adverse impact to marine mammals in Gulf of Mexico is justified by the analyses. The GRN believes the Proposed Action may have a significant effect and so should be addressed by preparation of an EIS, the Proposed Action violates the mandates of the OCS Lands Act, and authorization for "taking" marine mammals is needed to avoid violation of the Endangered Species Act and the Marine Mammal Protection Act. The GRN provides detailed arguments to support these assertions, generally citing the analyses or scientific information presented in the PEA.

The GRN general comments did not result in any specific guidance for technical revisions of the Draft PEA. MMS has taken several other actions that address the GRN concerns:
- MMS has issued NTLs that require mitigations similar to those described in the Draft PEA under Alternative 3.
- MMS has submitted a petition to NOAA, requesting rulemaking to authorize small "takes" of sperm whales incidental to OCS seismic operations in the Gulf of Mexico.

- Prior to public release of the Draft PEA, MMS determined that Alternative 3 was preferred over Alternative 1. This was communicated to Federal reviewers, but not to early non-Federal recipients of the Draft PEA, including GRN.

6) Letter, L.F. Griffin (Florida Department of Environmental Protection) to R. Defenbaugh (MMS) dated 2 December 2002; captioned "Draft Programmatic Environmental Assessment – geological & geophysical exploration for mineral resources on the Gulf of Mexico outer continental shelf."

   Ms. Griffin briefly comments on the State's review of the PEA, the reasons for preparation of the PEA, previous pertinent documents, and the PEA alternatives. Florida concurs that [Draft PEA] Alternative 3 is the preferable alternative but does not believe the information and analysis in the PEA supports a conclusive finding of no significant impact. Florida recommends that MMS prepare a draft EIS on G&G activities, with [Draft PEA] Alternative 3 as the Proposed Action. Ms. Griffin notes that the Federal actions described in the PEA (i.e., permitting G&G operations) are subject to State CZM consistency review. Florida provided several detailed technical comments, some of which have been incorporated into the MMS guidance for revision of the Draft PEA. Florida suggests that additional studies be conducted, including a study of the effectiveness of ramp-up as a mitigation measure.

V. LITERATURE CITED

Andre, M., M. Terada, and Y. Watanabe. 1997. Sperm whale (*Physeter macrocephalus*) behavioral response after the playback of artificial sounds. Reports of the International Whaling Commission 47, SC/48/NA13:499-504.

Au, W.W.L., M.O. Lammers, and R. Aubauer. 1999. A portable broadband data acquisition system for field studies in bioacoustics. Mar. Mammal Sci. 15(2):526-531.

Balcomb, K.C., and D.E. Claridge. 2001. A mass stranding of cetaceans caused by naval sonar in the Bahamas. Bahamas J. Sci. 5:2-12.

Bartol, S.M., J.A. Musick, and M. Lenhardt. 1999. Auditory evoked potentials of the loggerhead sea turtle (*Caretta caretta*). Copeia 99(3):836-840.

Baud, R., R.H. Peterson, G.E. Richardson, L.S. French, J. Regg, T. Montgomery, T.S. Williams, C. Doyle, and M. Dorner. 2002. Deepwater Gulf of Mexico: America's Expanding Frontier. U.S. Department of the Interior, Minerals Management Service, Gulf of Mexico OCS Region, New Orleans, LA. OCS Report MMS 2002-021. 133 pp.

Békésy, G. 1948. Vibration of the head in a sound field, and its role in hearing by bone conduction. J. Acoust. Soc. Am. 20:749-760.

Bowles, A.E., B. Smultea, B. Würsig, D.P. DeMaster, and D. Palka. 1994. Relative abundance and behavior of marine mammals exposed to transmissions from the Heard Island Feasibility Test. J. Acoust. Soc. Am. 96(4):2469-2484.

Bregman, A.S. 1990. Auditory Scene Analysis: The Perceptual Organization of Sound. MIT Press, Cambridge, MA.

Brooks, J.M., M.C. Kennicutt II, and B.D. Carey, Jr. 1986. Offshore surface geochemical exploration. Oil and Gas Journal, 20 October, p. 6.

Brooks, J.M., B.B. Bernard, J.D. Stonebraker, C.F. Schiefelbein, K.A. Allen, and T.J. McDonald. 1997. Design of multidisciplinary surface geochemical exploration surveys to identify active petroleum systems offshore West Africa, p. 3. *In:* Hedberg Research Symposium, "Petroleum Systems of the South Atlantic Margin," Rio de Janeiro, 16-19 November 1997 (extended abstract).

Browder, J., B. Brown, W. Nelson, and A. Bane. 1991. Multispecies fisheries in the Gulf of Mexico. ICES Mar. Sci. Symp.:194-197.

Caldwell, D.K., and M.C. Caldwell. 1987. Underwater echolocation type clicks by captive stranded pygmy sperm whales, *Kogia breviceps*, p. 8. *In:* Abstracts, Seventh Biennial Conference on the Biology of Marine Mammals, Miami, Florida, December 5-9, 1987.

Carder, D.A., and S. Ridgway. 1990. Auditory brainstem response in a neonatal sperm whale. J. Acous. Soc. Am. 88, Suppl. 1:S4.

Chapman, C.J., and A.D. Hawkins. 1969. The importance of sound in fish behaviour in relation to capture by trawls. FAO Fish. Rept. 62:717-729.

Collard, S. 1990. Leatherback turtles feeding near a water mass boundary in the eastern Gulf of Mexico. Marine Turtle Newsletter 50:12-14.

Cox, M., P. Rogers, A. Popper, W. Saidel, and R. Fay. 1986a. Frequency regionalization in the fish ear. J. Acoust. Soc. Am., Suppl. 1, 79:S80 (abstract).

Cox, M., P. Rogers, A. Popper, and W. Saidel. 1986b. Anatomical effects of intense tone stimulation in the ear of bony fishes. J. Acoust. Soc. Am., Suppl. 1, 80:S75 (abstract).

Cox, M., P. Rogers, A. Popper, W. Saidel, R. Fay, and S. Coombs. 1987. Anatomical effects of intense tone stimulation in the goldfish ear. J. Acoust. Soc. Am., Suppl. 1, 89:S7 (abstract).

Crane, N.L., and K. Lashkari. 1996. Sound production of gray whales along their migration route, a new approach to signal analysis. J. Acoust. Soc. Am. 100(3):1878-1886.

Cranford, T.W., M. Amundin, and K.S. Norris. 1996. Functional morphology and homology in the odontocete nasal complex: implications for sound generation. J. Morphology 228:223-285.

Curry, B.E. 1999. Stress in mammals: the potential influence of fishery-induced stress on dolphins in the Eastern Tropical Pacific Ocean. U.S. Department of Commerce, La Jolla, CA. NOAA Tech. Memo. NMFS-SWFSC-260. 121 pp.

Dalen, J., and G.M. Knutsen. 1986. Scaring effects of fish and harmful effects of eggs, larvae and fry from offshore seismic explorations. 12th ICA Associated Symposium on Underwater Acoustics, 16-18 July 1986, Halifax, Nova Scotia, Canada. 10 pp.

Davis, R.W., and G.S. Fargion (eds.). 1996. Distribution and abundance of cetaceans in the north-central and western Gulf of Mexico: Final Report. Volume II: Technical Report. OCS Study MMS 96-0027. Prepared by the Texas Institute of Oceanography and the National Marine Fisheries Service. U.S. Dept. of the Interior, Minerals Management Service, Gulf of Mexico OCS Region, New Orleans, LA. 357 pp.

Davis, R.A., D.H. Thompson, and C.I. Malme. 1998. Environmental assessment of seismic exploration on the Scotian Shelf. Prepared by LGL Limited Environmental Research Associates, King City, Ontario for Canada/Nova Scotia Offshore Petroleum Board, Halifax, Nova Scotia. 181 pp. + app.

Davis, R.W., W.E. Evans, and B. Würsig (eds.). 2000. Cetaceans, Sea Turtles and Seabirds in the Northern Gulf of Mexico: Distribution, Abundance and Habitat Associations. Volume II: Technical Report. Prepared by Texas A&M University at Galveston and the National Marine Fisheries Service. U.S. Department of the Interior, U.S. Geological Survey, Biological Resources Division, USGS/BRD/CR-1999-005 and Minerals Management Service, Gulf of Mexico OCS Region, New Orleans, LA. OCS Study MMS 2000-003.

Dawson, S., J. Barlow, and D. Ljungblad. 1998. Sounds recorded from Baird's beaked whale, *Berardius bairdii*. Mar. Mammal Sci.(2):335-344.

Dellagiarino, G., P. Fulton, and D. Zinzer. 1997. Geological & geophysical data acquisition, a twenty-year retrospective, 1976-1996. Resource Evaluation Program Report, U.S. Department of the Interior, Minerals Management Service, Resource Evaluation Division, Herndon, VA. OCS Report MMS 97-0035. 39 pp.

Dellagiarino, G., P. Fulton, K. Meekins, and D. Zinzer. 1998. Geological & geophysical data acquisition, outer continental shelf through 1997. Resource Evaluation Program Report, U.S. Department of the Interior, Minerals Management Service, Resource Evaluation Division, Herndon, VA. OCS Report MMS 98-0027. 32 pp.

Dellagiarino, G., P. Fulton, K. Meekins, and D. Zinzer. 2000. Geological & geophysical data acquisition, outer continental shelf through 1999. Resource Evaluation Program Report, U.S. Department of the Interior, Minerals Management Service, Resource Evaluation Division, Herndon, VA. OCS Report MMS 2000-071. 32 pp.

Department of the Navy. 1998. Final environmental impact statement: Shock testing the SEAWOLF submarine. Southern Division, Naval Facilities Engineering Command, North Charleston, SC.

Department of the Navy. 1999a. Draft overseas environmental impact statement and environmental impact statement for surveillance towed array sensor system, low frequency active (SURTASS LFA) sonar. Chief of Naval Operations, Washington, D.C.

Department of the Navy. 1999b. Draft environmental impact statement: Shock trial of the WINSTON S. CHURCHILL (DDG 81). Southern Division, Naval Facilities Engineering Command, North Charleston, SC.

Department of the Navy. 2001a. Final overseas environmental impact statement and environmental impact statement for surveillance towed array sensor system low frequency active (SURTASS LFA) sonar. Prepared for the Department of the Navy, Chief of Naval Operations, Washington, D.C. January 2001. 2 vols.

Department of the Navy. 2001b. Final environmental impact statement: Shock trial of the WINSTON S. CHURCHILL (DDG 81). Southern Division, Naval Facilities Engineering Command, North Charleston, SC. February 2001. 229 pp. + apps.

dilithium Press, Ltd. 1989. Webster's Encyclopedic Unabridged Dictionary of the English Language. Random House, New York. 2078 pp.

Dodd, Jr., C.K. 1988. Synopsis of the Biological Data on the Loggerhead Turtle, *Caretta caretta* (Linnaeus, 1758). U.S. Fish and Wildlife Service Biol. Rep. 88. 100 pp.

Ebrom, D. 2000. Strongest signals sought in Teal South multi-component seismic coupling. Offshore, PennWell Publishing Co., Houston, TX. March 2000. 4 pp.

Eckert, S.A., D.W. Nellis, K.L. Eckert, and G.L. Kooyman. 1986. Diving patterns of two leatherback sea turtles (*Dermochelys coriacea*) during internesting intervals at Sandy Point, St. Croix, U.S. Virgin Islands. Herpetologica 42(3):381-388.

Eckert, S.A., K.L. Eckert, P. Ponganis, and G.L. Kooyman. 1989. Diving and foraging behavior of leatherback sea turtles (*Dermochelys coriacea*). Can. J. Zool. 67:2834-2840.

Engås, A., S. Løkkeborg, E. Ona, and A.V. Soldal. 1993. Effects of seismic movements on catches and availability of cod and haddock. Fisken Og Havet 3 (March 1993):1-111.

Enger, P.S. 1981. Frequency discrimination in teleosts - central or peripheral?, pp. 243-255. *In:* W.N. Tavolga, A.N. Popper, and R.R. Fay (eds.), Hearing and Sound Communication in Fishes. Springer-Verlag, New York.

Ernst, C.H., R.W. Barbour, and J.E. Lovich. 1994. Turtles of the United States and Canada. Smithsonian Institute Press, Washington, D.C. 578 pp.

Fay, R.R. 1988. Hearing in Vertebrates: A Psychophysics Databook. Hill-Fay Associates, Winnetka, IL.

Fay, R.R. 1998. Auditory stream segregation in goldfish (*Carassius auratus*). Hear. Res. 120:69-76.

Fay, R.R., and S.L. Coombs. 1988. Psychophysics and neurophysiology of frequency selectivity and masking in the goldfish, pp. 169-176. *In:* H. Duifhuis, W. Horst, and H. Wit (eds.), Basic Issues in Hearing. Groningen University Press, Groningen, The Netherlands.

Fay, R., and A. Megela Simmons. 1999. The sense of hearing in fishes and amphibians, pp. 269-318. *In:* R. Fay and A. Popper (eds.), Comparative Hearing: Fish and Amphibians. Springer Handbook of Auditory Research, Vol. 11. Springer-Verlag, New York.

Fay, R.R., W.A. Ahroon, and A.A. Orawski. 1978. Auditory masking patterns in the goldfish (*Carassius auratus*): Psychophysical tuning curves. J. Exp. Biol. 74:83-100.

Finneran, J.J., C.E. Schlundt, R. Dear, D.A. Carder, and S.H. Ridgway. 2000a. Masked temporary threshold shift (MTTS) in odontocetes after exposure to single underwater impulses from a seismic watergun. J. Acoust. Soc. Am. 108(5), part 2 (November 2000):2515.

Finneran, J.J., C.E. Schlundt, D.A. Carder, J.A. Clark, J.A. Young, J.B. Gaspin, and S.H. Ridgway. 2000b. Auditory and behavioral responses of bottlenose dolphins (*Tursiops truncatus*) and a beluga whale (*Delphinapterus leucas*) to impulsive sounds resembling distant signatures of underwater explosions. J. Acoust. Soc. Am. 108 (1):417-431.

Frantzis, A. 1998. Does acoustic testing strand whales? Nature 392:29.

Fritts, T.H., and R.P. Reynolds. 1981. Pilot study of the marine mammals, birds, and turtles in the OCS areas of the Gulf of Mexico. Rept. FWS/OBS-81/36. U.S. Fish and Wildlife Service, Washington, D.C. 139 pp.

Fritts, T.H., W. Hoffman, and M.A. McGehee. 1983a. The distribution and abundance of marine turtles in the Gulf of Mexico and nearby Atlantic waters. J. Herpetol. 17:327-344.

Fritts, T.H., A.B. Irvine, R.D. Jennings, L.A. Collum, W. Hoffman, and M.A. McGehee. 1983b. Turtles, birds, and mammals in the northern Gulf of Mexico and nearby Atlantic waters. Final report for the U.S. Department of the Interior, U.S. Fish and Wildlife Service, Division of Biological Services, Washington, D.C. FWS/OBS-82/65. 455 pp.

Fulling, G.L., K.D. Mullin, and C.W. Hubard. In review. Abundance and distribution of cetaceans in outer continental shelf waters of the U.S. Gulf of Mexico. Submitted to Mar. Mammal Sci.

Fulton, P. 1998. Prelease permits for oil, gas, and sulphur on the outer continental shelf. U.S. Department of the Interior, Minerals Management Service, Resource Evaluation Division, Herndon, VA. OCS Study MMS 98-0009. 119 pp.

Gallardo, V.A., D. Arcos, M. Salamanca, and L. Pastene. 1983. On the occurrence of Bryde's whale in an upwelling area off central Chile. Reports of the International Whaling Commission 33:481-488.

Gallaway, B.J. 1988. Northern Gulf of Mexico Continental Slope Study, Final Report, Year 4. Volume II: Synthesis Report. U.S. Department of the Interior, Minerals Management Service, Gulf of Mexico OCS Region, New Orleans, LA. OCS Study MMS 88-0053.

Geraci, J.R., and D.J. St. Aubin. 1987. Effects of offshore oil and gas development on marine mammals and turtles, pp. 587-617. *In:* D.F. Boesch and N.N. Rabalais (eds.), Long Term Environmental Effects of Offshore Oil and Gas Development. Elsevier Applied Science Publ. Ltd., London and New York.

Goold, J.C. 1996. Acoustic assessment of populations of common dolphin, *Delphinus delphis*, in conjunction with seismic surveying. J. Mar. Biol. Assoc. U.K. 76(3):811-820.

Goold, J.C. 1999. Behavioral and acoustic observations of sperm whales in Scapa Flow, Orkney Islands. J. Mar. Biol. Assoc. U.K. 79:541-550.

Goold, J.C., and M.R. Clarke. 2000. Sound velocity in the head of the dwarf sperm whale, *Kogia simus*, with anatomical and functional discussion. J. Mar. Biol. Assoc. U.K. 80(3):535-542.

Goold, J.C., and P.J. Fish. 1998. Broadband spectra of seismic survey airgun emissions, with reference to dolphin auditory thresholds. J. Acoust. Soc. Am. 103(4):2177-2184.

Goold, J.C., and S.E. Jones. 1995. Time and frequency domain characteristics of sperm whale clicks. J. Acoust. Soc. Am. 98(3):1279-1281.

Gordon, J.C.D., D. Gillespie, J. Potter, A. Frantzis, M. Simmonds, and R. Swift. 1998. The effects of seismic surveys on marine mammals. *In:* M.L. Tasker and C. Weir (eds.), Proceedings of the Seismic and Marine Mammals Workshop, London, 23-25 June 1998.

Gulf of Mexico Fishery Management Council. 2003. Draft environmental impact statement for the generic essential fish habitat amendment to the following fishery management plans of the Gulf of Mexico: shrimp fishery of the Gulf of Mexico; red drum fishery of the Gulf of Mexico; reef fish fishery of the Gulf of Mexico; stone crab fishery of the Gulf of Mexico; coral and coral reef fishery of the Gulf of Mexico; spiny lobster fishery of the Gulf of Mexico and South Atlantic; and coastal migratory pelagic resources of the Gulf of Mexico and South Atlantic, Gulf of Mexico Fishery Management Council, Tampa FL.

Hastings, M.C., A.N. Popper, J.J. Finneran, and P.J. Lanford. 1996. Effect of low frequency underwater sound on hair cells of the inner ear and lateral line of the teleost fish *Astronotus ocellatus*. J. Acoust. Soc. Am. 99:1759-1766.

Hawkins, A.D., and C.J. Chapman. 1975. Masked auditory thresholds in the cod, *Gadus morhua* L. J. Comp. Physiol. A 103:209-226.

Higgs, D., and A. Popper. 2000. Possible morphological correlates to ultrasound detection by the American shad (*Alosa sapidissima*), p. 280. *In:* Abstracts of the Association for Research in Otolaryngology, Mid-Winter meeting, #971.

Hildebrand, H.H. 1995. A historical review of the status of sea turtle populations in the western Gulf of Mexico, pp. 447-453. *In:* K.A. Bjorndal (ed.), Biology and Conservation of Sea Turtles, Second Edition, Smithsonian Institution Press, Washington, D.C.

Holliday, D.V., R.E. Pieper, M.E. Clarke, and C.F. Greenlaw. 1987. The effects of airgun energy releases on the eggs, larvae, and adults of the northern anchovy (*Engraulis mordax*). A report prepared for the American Petroleum Institute, Washington, D.C., by Tracor Applied Sciences. Tracor Document No. T-86-06-7001-U. 98 pp.

Houde, E.D. 1987. Fish early life dynamics and recruitment variability. American Fisheries Society Symposium 2:17-27.

Jefferson, T.A. 1995. Distribution, abundance, and some aspects of the biology of cetaceans in the offshore Gulf of Mexico. Ph.D. dissertation, Texas A&M University, College Station, TX. 107 pp.

Jefferson, T.A., and A.J. Schiro. 1997. Distributions of cetaceans in the offshore Gulf of Mexico. Mammal Rev. 27(1):27-50.

Johnson, B.L., O.E. DeWald, and B.W. Smith. 1984. Environmental assessment for Gulf of Mexico outer continental shelf geological and geophysical exploration activities. Prepared by the U.S. Department of the Interior, Gulf of Mexico OCS Regional Office, New Orleans, LA. September 1984. 66 pp. + apps.

Keinath, J.A., and J.A. Musick. 1993. Movements and diving behavior of a leatherback turtle, *Dermochelys coriacea*. Copeia 1993(4):1010-1017.

Keinath, J.A., J.A. Musick, and D.E. Barnard. 1996. Abundance and distribution of sea turtles off North Carolina. U.S. Department of the Interior, Minerals Management Service, Gulf of Mexico OCS Region, New Orleans, LA. OCS Study MMS 95-0024. 156 pp.

Ketten, D.R. 1995. Estimates of blast injury and acoustic trauma zones for marine mammals from underwater explosions, pp. 391-407. *In:* R.A. Kastelein, J.A. Thomas, and P.E. Nachtigall (eds.), Sensory Systems of Aquatic Mammals. De Spil Publishers, Woerden, The Netherlands.

Ketten, D.R. 1998. Marine Mammal Auditory Systems: A summary of audiometric and anatomical data and its implications for underwater acoustic impacts. NOAA Technical Memorandum NMFS-SWFSC-256.

Klima, E.F., G.R. Gitchlag, and M.L. Renaud. 1988. Impacts of the explosive removal of offshore petroleum on sea turtles and dolphins. Mar. Fish. Rev. 50:33-42.

Klimley, A., and S. Beavers. 1998. Playback of acoustic thermometry of ocean climate (ATOC) – like signal to bony fishes to evaluate phonotaxis. J. Acoust. Soc. Am. 104:2506-2511.

Klimley, A., and A. Myrberg. 1979. Acoustic stimuli underlying withdrawl from a sound source by adult lemon sharks, *Negaprion brevirostris* (Poey). Bull. Mar. Sci. 29:447-458.

Knudsen, F., P. Enger, and O. Sand. 1994. Avoidance responses to low-frequency sound in downstream migrating Atlantic salmon smolt, *Salmo salar*. J. Fish. Biol. 45:227-233.

Kraus, S.D., A.J. Read, A. Solow, K. Baldwin, T. Spradlin, E. Anderson, and J. Williamson. 1997. Acoustic alarms reduce porpoise mortality. Nature 388:525.

Lenhardt, E. 1986. Clinical Aspects of Inner Ear Deafness. Springer-Verlag, New York.

Lenhardt, M.L. 1982. Bone conduction hearing in turtles. J. Aud. Res. 22:153-160.

Lenhardt, M.L., and S.W. Harkins. 1983. Turtle shell as an auditory receptor. J. Aud. Res. 23:251-260.

LePage, K., C.I. Malme, R. Mlawski, and P. Krumhansl. 1995. Exxon SYU Sound Propagation Study, November 1995. Submitted to Exxon Exploration Company, Houston, TX, by BBN Corporation, Acoustic Technologies Division, Cambridge, MA. BBN Rep. 8120. 32 pp.

LGL, Ltd., environmental research associates. 2003. Marine mammal monitoring during Lamont-Doherty Earth Observatory's Acoustic Calibration Study in the northern Gulf of Mexico, 2003. Prepared for Lamont-Doherty Earth Observatory and National Marine Fisheries Service by LGL Ltd., environmental research associates, King City, Ontario. September 2003. LGL Report TA2822-12. var pp.

Ljungblad, D.K., B. Würsig., S.L. Swartz, and J.M. Keene. 1988. Observations of the behavioral responses of bowhead whales to active geophysical vessels in the Alaskan Beaufort Sea. Arctic 41(3):183-194.

Løkkeberg, S. 1991. Effects of geophysical survey on catching success in longline fishing. ICES CM 40:1-9.

Løkkeberg, S., and A. Soldal. 1993. The influence of seismic exploration with airguns on cod (*Gadus morhua*). ICES Mar. Sci. Symp. 196:62-67.

Lombarte, A., H. Yan, A. Popper, J. Chang, and C. Platt. 1993. Damage and regeneration of hair cell ciliary bundles in a fish ear following treatment with gentamycin. Hear. Res. 66:166-174.

Lopez, A.M., D.B. McClellan, A.R. Bertolino, and M.D. Lange. 1979. The Japanese longline fishery in the Gulf of Mexico. Mar. Fish. Rev.:23-28.

Lutcavage, M.E., P.L. Lutz, G.D. Bossart, and D.M. Hudson. 1995. Physiologic and clinicopathologic effects of crude oil on loggerhead turtles. Arch. Environ. Contam. Toxicol. 28:417-422.

Lutcavage, M.E., P. Plotkin, B. Witherington, and P.L. Lutz. 1997. Human Impacts on Sea Turtle Survival. pp. 387-410. *In:* P.L. Lutz and J.A. Musick (eds.), The Biology of Sea Turtles. CRC Press, Boca Raton, FL. 432 pp.

Lynn, S.K., and D.L. Reiss. 1992. Pulse sequence and whistle production by two captive beaked whales, *Mesoplodon* species. Mar. Mammal Sci. 8(3):299-305.

MacDonald, I.R. 2000. Seep communities, Chapter 8. *In:* Continental Shelf Associates, Inc., Deepwater Program: Gulf of Mexico Deepwater Information Resources Data Search and Literature Synthesis. Volume I: Narrative Report. U.S. Department of the Interior, Minerals Management Service, Gulf of Mexico OCS Region, New Orleans, LA. OCS Study MMS 2000-049. 340 pp.

MacDonald, I.R., N.L. Guinasso, J.F. Reilly, J.M. Brooks, W.R. Callender, and S.G. Gabrielle. 1990. Gulf of Mexico hydrocarbon seep communities: Patterns in community structure and habitat. Geo-Marine Letters 10(4):244-252.

MacDonald, I.R., W.W. Schroeder, and J.M. Brooks. 1995. Chemosynthetic Ecosystems Studies Final Report. Volume II, Technical Report. U.S. Department of the Interior, Minerals Management Service, Gulf of Mexico OCS Region, New Orleans, LA. OCS Study MMS 95-0022.

Madsen, P.T., and B. Mohl. 2000. Sperm whales (*Physeter catodon* L.) do not react to sounds from detonators. J. Acoust. Soc. Am. 107(1):668-671.

Malme, C.I., and P.R. Miles. 1985. Behavioral responses of marine mammals (gray whales) to seismic discharges, pp. 253-280. *In:* Proc. Workshop on Effects of Explosive Use in the Marine Environment, January 1985, Halifax, Nova Scotia. Technical Report No. 5, Canadian Oil and Gas Lands Administration, Environmental Protection Branch, Ottawa, Ontario. 398 pp.

Malme, C.I., P.R. Miles, C.W. Clarke, P. Tyack, and J.E. Bird. 1983. Investigations of the potential effects of underwater noise from petroleum industry activities on migrating gray whale behavior. BBN Rep. 5366. Report from Bolt Beranek & Newman Inc., Cambridge MA, for U.S. Department of the Interior, Minerals Management Service, Anchorage, AK. NTIS PB86-174174.

Malme, C. I., P.R. Miles, C.W. Clarke, P. Tyack, and J.E. Bird. 1984. Investigations of the potential effects of underwater noise from petroleum industry activities on migrating gray whale behavior/Phase II: January 1984 migration. BBN Rep. 5586. Report by Bolt Beranek & Newman Inc., Cambridge, MA, for the U.S. Department of the Interior, Minerals Management Service, Anchorage, AK. NTIS PB86-218377.

Malme, C.I., P.R. Miles, C.W. Clarke, P. Tyack, and J.E. Bird. 1985. Investigations of the potential effects of underwater noise from petroleum industry activities on feeding humpback whale behavior. BBN Rep. 5851. OCS Study MMS 85-0019. Report from BBN Labs Inc., Cambridge MA, for the U.S. Department of the Interior, Minerals Management Service, Anchorage. NTIS PB86-218385.

Malme, C.I., B. Würsig, J.E. Bird, and P. Tyack. 1986. Behavioral responses of gray whales to industrial noise: Feeding observation and predictive modelling. BBN Rep. 6265. OCS Env. Assess. Progr. Final Rep. Princ. Invest., NOAA, Anchorage, AK 56:393-600. OCS Study MMS 88-0048. NTIS PB88-249008.

Malme, C.I., B. Würsig, J.E. Bird, and P. Tyack. 1988. Observations of feeding gray whale responses to controlled industrial noise exposure, pp. 55-73. *In:* W.M. Sackinger et al., Port and ocean engineering under arctic conditions, volume II. Geophys. Inst., Univ. Alaska, Fairbanks. 111 pp.

Mann, D., Z. Lu, and A. Popper. 1997. Ultrasound detection by a teleost fish. Nature (London) 389:341.

Marquez, M.R. 1990. FAO species catalogue. Vol. 11: Sea Turtles of the World. An Annotated and Illustrated Catalogue of Sea Turtle Species Known to Date. FAO Fisheries Synopsis. No. 125, Vol. 11. FAO, Rome. 81 pp.

Mate, B.R., K.M. Stafford, and D.K. Ljungblad. 1994. A change in sperm whale (*Physeter macrocephalus*) distribution correlated to seismic surveys in the Gulf of Mexico. J. Acoust. Soc. Am. 96(5):3268-3269.

Matousek, J., A. Wells, and P. McGroddy. 1988. Field testing of behavioral barriers for fish exclusion at cooling-water intake systems. Electric Power Research Institute, Project # 221406. Report No. EPRI CS-5995.

McCauley, R.D., M.N. Jenner, C. Jenner, K.A. McCabe, and J. Murdoch. 1998. The response of humpback whales (*Megaptera novaengliae*) to offshore seismic survey noise: preliminary results of observations about a working seismic vessel and experimental exposures. APPEA Journal 1998:692-707.

McCauley, R.D., J. Fewtrell, A.J. Duncan, C. Jenner, M.N. Jenner, J. Penrose, R.I.T. Prince, A. Adhitya, J. Murdoch, and K. McCabe. 2000. Marine seismic surveys - A study of environmental implications. APPEA Journal 40:692-708.

McDonald, M.A., J.A. Hildebrand, S. Webb, L. Dorman, and C.G. Fox. 1993. Vocalizations of blue and fin whales during a mid-ocean ridge airgun experiment. J. Acoust. Soc. Am. 94(3):1849.

Melnick, W. 1991. Human temporary threshold shifts (TTS) and damage risk. J. Acoust. Soc. Am. 90:147-155.

Meylan, A., B. Schroeder, and A. Mosier. 1995. Sea turtle nesting activity in the State of Florida 1979-1992. Florida Marine Research Publications, State of Florida, Dept. of Environmental Protection, No. 52.

Moein, S.E., J.A. Musick, J.A. Keinath, D.E. Barnard, M.L. Lenhardt, and R. George. 1995. Evaluation of seismic sources for repelling sea turtles from hopper dredges, pp. 90-93. *In:* L.Z. Hales (compiler), Sea Turtle Research Program: Summary Report. Prepared for U.S. Army Engineer Division, South Atlantic, Atlanta, GA, and U.S. Naval Submarine Base, Kings Bay, GA. Technical Report CERC-95-. 145 pp.

Morton, A.B., and H.K. Symonds. 2002. Displacement of *Orcinus orca* (L.) by high amplitude sound in British Columbia, Canada. ICES J. of Mar. Sci. 59(1):71-80.

Mullin, K.D., and W. Hoggard. 2000. Visual surveys of cetaceans and sea turtles from aircraft and ships, Chapter 4. *In:* R.W. Davis, W.E. Evans, and B. Würsig (eds.), Cetaceans, Sea Turtles and Seabirds in the Northern Gulf of Mexico: Distribution, Abundance and Habitat Associations. Volume II: Technical Report. Prepared by Texas A&M University at Galveston and the National Marine Fisheries Service. U.S. Department of the Interior, U.S. Geological Survey, Biological Resources Division, USGS/BRD/CR-1999-005 and Minerals Management Service, Gulf of Mexico OCS Region, New Orleans, LA. OCS Study MMS 2000-003.

Munc, W., R. Spindel, A. Baggeroer, and T. Birdsall. 1994. The Heard Island feasibility test. J. Acoust. Soc. Am. 96:2330-2342.

Myrberg, A. 1981. Sound communication and interceptions by fishes, pp. 395-426. *In:* W. Tavolga, R. Fay, and A. Popper (eds.), Hearing and Sound Communication in Fishes. Springer-Verlag, New York.

National Marine Fisheries Service. 1999a. Fishery Management Plan for Atlantic tunas, swordfish, and sharks, Volume II. National Marine Fisheries Service, Division of Highly Migratory Species, Office of Sustainable Fisheries, Silver Spring, MD. 302 pp.

National Marine Fisheries Service. 1999b. Pelagic longline logbook data. Southeast Fisheries Science Center, Miami, FL.

National Marine Fisheries Service. 2000. Fisheries of the U.S., 1999. Prepared by M.C. Holliday and B.K. O'Bannon. NOAA, NMFS, Silver Spring, MD.

National Marine Fisheries Service. 2003. Landing Statistics web page, URL: www nmfs noaa.gov.

National Marine Fisheries Service. 2004. Draft 2003 marine mammal stock assessment report. www nmfs noaa.gov/prot_res/readingrm/draft_2003_sars/sefsc2003sar_revision_whales_atl.pdf. 35 pp.

National Marine Fisheries Service and U.S. Fish and Wildlife Service. 1991a. Recovery plan for U.S. populations of loggerhead turtle. National Marine Fisheries Service, Washington, D.C. 64 pp.

National Marine Fisheries Service and U.S. Fish and Wildlife Service. 1991b. Recovery plan for U.S. population of Atlantic green turtle. National Marine Fisheries Service, Washington, D.C. 52 pp.

National Marine Fisheries Service and U.S. Fish and Wildlife Service. 1992a. Recovery plan for leatherback turtles in the U.S. Caribbean, Atlantic, and Gulf of Mexico. National Marine Fisheries Service, Washington, D.C. 65 pp.

National Marine Fisheries Service and U.S. Fish and Wildlife Service. 1992b. Recovery plan for the Kemp's ridley sea turtle (*Lepidochelys kempii*). National Marine Fisheries Service, St. Petersburg, FL. 40 pp.

National Marine Fisheries Service and U.S. Fish and Wildlife Service. 1993. Recovery plan for hawksbill turtles in the U.S. Caribbean Sea, Atlantic Ocean, and Gulf of Mexico. National Marine Fisheries Service, St. Petersburg, FL. 52 pp.

National Research Council. 2003. Ocean noise and marine mammals. The National Academies Press, Washington, D.C. ISBN 0-309-08563-5. 192 pp. + plates.

Nelson, D., and S. Gruber. 1963. Sharks: Attraction by low-frequency sounds. Science 142:975-977.

Nestler, J., G. Ploskey, J. Pickens, J. Menezes, and C. Schildt. 1992. Responses of blueback herring to high-frequency sound and implications for reducing entrainment at hydropower dams. N. Amer. J. Fish. Man. 12:667-683.

Norris, J.C., W.E. Evans, and S. Rankin. 2000. An acoustic survey of cetaceans in the northern Gulf of Mexico, Chapter 5. *In:* R.W. Davis, W.E. Evans, and B. Würsig (eds.), Cetaceans, Sea Turtles and Seabirds in the Northern Gulf of Mexico: Distribution, Abundance and Habitat Associations. Volume II: Technical Report. Prepared by Texas A&M University at Galveston and the National Marine Fisheries Service. U.S. Department of the Interior, U.S. Geological Survey, Biological Resources Division, USGS/BRD/CR-1999-005 and Minerals Management Service, Gulf of Mexico OCS Region, New Orleans, LA. OCS Study MMS 2000-003.

Occupational Employment Statistics. 1999. National Occupational Employment and Wage Estimates, 49-9092 Commercial Divers. http://www.bls.gov/OES/1999/oes499092.htm

O'Hara, J., and J.R. Wilcox. 1990. Avoidance responses of loggerhead turtles, *Caretta caretta*, to low frequency sound. Copeia 1990(2):564-567.

O'Keeffe, D.J., and G.A. Young. 1984. Handbook on the environmental effects of underwater explosions. Naval Surface Weapons Center. NWC TR 83-240.

Paul, C.K., B. Hecker, R. Commeau, R.P. Freeman-Lynde, C. Neummann, W.P. Corse, S. Golubic, J.E. Hook, E. Sikes, and J. Curray. 1984. Biological communities at the Florida escarpment resemble hydrothermal vent taxa. Science 226:965-967.

Peake, D.E. 1996. Bird surveys, pp. 271-304. *In:* R.W. Davis and G.S. Fargion (eds.), Distribution and abundance of cetaceans in the north-central and western Gulf of Mexico, Final Report. Volume II: Technical Report. U.S. Department of the Interior, Minerals Management Service, Gulf of Mexico OCS Region, New Orleans, LA. OCS Study MMS 96-0027. 357 pp.

Pearson, W., J. Skalski, and C. Malme. 1992. Effects of sounds from a geophysical survey device on behavior of captive rockfish. Can. J. Fish. Aquatic Sci. 49:1343-1356.

Pequegnat, W.E., B.J. Gallaway, and L.H. Pequegnat. 1990. Aspects of the ecology of the deep-water fauna of the Gulf of Mexico. Am. Zool. 30:45-64.

Pierson, M.O., J.P. Wagner, V. Langford, P. Birnie, and M.L. Tasker. 1998. Protection from, and mitigation of, the potential effects of seismic exploration on marine mammals. *In:* M.L. Tasker and C. Weir (eds.), Proceedings of the Seismic and Marine Mammals Workshop, London, 23-25 June 1998. http://www.smru.st-and.ac.uk/seismic/index.html

Popper, A., and T. Carlson. 1998. Application of sound and other stimuli to control fish behavior. Trans. Amer. Fish. Soc. 127:673-707.

Popper, A.N., and N.L. Clarke. 1979. Non-simultaneous auditory masking in the goldfish *C. auratus*. J. Exp. Biol. 83:145-158.

Popper, A., and R. Fay. 1999. The auditory periphery in fishes, pp. 43-100. *In:* R. Fay and A. Popper (eds.), Comparative Hearing: Fish and Amphibians. Springer Handbook of Auditory Research, Vol. 11. Springer-Verlag, New York.

Pritchard, P.C.H. 1997. Evolution, Phylogeny, and Current Status, pp. 1-28. *In:* P.L. Lutz and J.A. Musick (eds.), The Biology of Sea Turtles. CRC Press, Boca Raton, FL. 432 pp.

Reeves, R.R., D.K. Ljungbald, and J.T. Clarke. 1984. Bowhead whales and acoustic seismic surveys in the Beaufort Sea. Polar Research 22(138):271-280.

Richardson, W.J. 1998. Reactions of bowhead whales and ringed seals to an open water seismic program in the Alaskan Beaufort Sea. *In:* M.L. Tasker and C. Weir (eds.), Proceedings of the Seismic and Marine Mammals Workshop, London, 23-25 June 1998.

Richardson, W.J., and C.I. Malme. 1993. Man-made noise and behavioral responses, pp. 631-700. *In:* The Bowhead Whale. Spec. Publ. 2. Society for Marine Mammalogy, Lawrence KS. 787 pp.

Richardson, W.J., B. Würsig, and C.R. Greene. 1986. Reaction of bowhead whales to seismic exploration in the Canadian Beaufort Sea. J. Acoust. Soc. Am. 79(4):1117-1128.

Richardson, W.J., C.R. Greene, Jr., C.I. Malme, and D.H. Thomson. 1995. Marine Mammals and Noise. Academic Press, San Diego. 576 pp.

Ridgway, S.H., E.G. Wever, J.G. McCormick, J. Palin, and J.H. Anderson. 1969. Hearing in the giant sea turtle, *Chelonia mydas*. Proc. Nat. Acad. Sci. 64:884-890.

Ridgway, S.H., D.A. Carder, R.R. Smith, T. Kamolnick, C.E. Schlundt, and W.R. Elsberry. 1997. Behavioral responses and temporary shift in masked hearing threshold of bottlenose dolphins, *Tursiops truncatus*, to 1-second tones of 141 to 201 dB re 1 μPa. Technical Report 1751, July 1997. Naval Command, Control and Ocean Surveillance Center, RDT&E Division, San Diego, CA. 16 pp. + app.

Rivers, J.A. 1997. Blue whale, *Balaenoptera musculus*, vocalizations from the waters off central California. Mar. Mammal Sci. 13(2):186-195.

Rogers, T.L., and S.M. Brown. 1999. Acoustic observations of Arnoux's beaked whale (*Berardius arnuxii*) off Kemp Land, Antarctica. Mar. Mammal Sci. 15(1):192-198.

Ross, D. 1976. Mechanics of underwater noise. Pergamon Press, New York. 375 pp.

Sakagawa, G.T., A.L. Coan, and N.W. Bartoo. 1987. Patterns in longline fishery data and catches of bigeye tuna, *Thunnus obesus*. Mar. Fish. Rev. 49(4):57-66.

Sand, O., and H. Karlsen. 1986. Detection of infrasound by the Atlantic cod. J. Exp. Biol. 125:197-204.

Sassen, R., J.M. Brooks, I.R. MacDonald, M.C. Kennicutt II, N.L. Guinasso, Jr., and A.G. Requejo. 1993. Association of oil seeps and chemosynthetic communities with oil discoveries, upper continental slope, Gulf of Mexico. Trans. Gulf Coast Assoc. Geol. Soc. 43:349-355.

Saunders, J.C., S.P. Dear, and M.E. Schneider. 1985. The anatomical consequences of acoustical injury: A review and tutorial. J. Acoust. Soc. Am. 78(3):833-860.

Schlundt, C.E., J.J. Finneran, D.A. Carder, and S.H. Ridgway. 2000. Temporary shift in masked hearing thresholds of bottlenose dolphins, *Tursiops truncatus*, and white whales, *Delphinapterus leucas*, after exposure to intense tones. J. Acoust. Soc. Am. 107(6):3496-3508.

Skalski, J., W. Pearson, and C. Malme. 1992. Effects of sounds from a geophysical survey device on catch per unit effort on a hook-and-line fishery for rockfish (*Sebastes spp.*). Can. J. Fish. Aquatic Sci. 49:1357-1365.

Sodal, A. 1999. Measured underwater acoustic wave propagation from a seismic source. *In:* Proceedings of the Airgun Environmental Workshop, London, July 6, 1999.

Stafford, K.M., C.G. Fox, and D.S. Clark. 1998. Long range acoustic detection and localization of blue whale calls in the northeast Pacific Ocean. J. Acoust. Soc. Am. 104(6):3616-3625.

Stafford, K.M., S.L. Nieukirk, and C.G. Fox. 1999. Low frequency whale sounds recorded on hydrophones moored in the eastern tropical pacific. J. Acoust. Soc. Am. 106(6):3687-3698.

Standora, E.A., J.R. Spotila, J.A. Keinath, and C.R. Shoop. 1984. Body temperatures, diving cycles, and movement of a subadult leatherback turtle, *Dermochelys coriacea*. Herpetologica 40(2):169-176.

Steevens, C.C., R. Sylvester, and J. Clark. 1997. Effects of low-frequency water-borne sound on divers: Open water trial. Naval Submarine Medical Research Laboratory Report VI (24). 44 pp.

Stemp, R. 1985. Observations on the effects of seismic exploration on seabirds, pp. 217-233. *In:* G.D. Greene, F.R. Engelhardt, and R.J. Paterson (eds.), Proceedings of the Workshop on Effects of Explosives Use in the Marine Environment, January 29-31, 1985, Halifax. Canada Oil and Gas Lands Administration, Environmental Protection Branch, Technical Report No. 5.

Stone, C.J. 1996. Cetacean observations during a seismic survey of the *M.V. Mintrop*, west of Shetland. Report to Conoco (UK) Limited.

Stone, C.J. 1997a. Cetacean and seabird observations in Tranche 52 during 1997. Report to Conoco (UK) Limited.

Stone, C.J. 1997b. Cetacean observations during seismic surveys in 1996. JNCC Reports, No. 228. Joint Nature Conservation Committee, Aberdeen, Scotland.

Stone, C.J. 1998. Cetacean observations during seismic surveys in 1997. JNCC Reports, No. 278. Joint Nature Conservation Committee, Aberdeen, Scotland.

Swartz, S.L., and R. Hofman. 1991. Marine mammal and habitat monitoring: Requirements; principles; needs; and approaches. Marine Mammal Commission, Washington, D.C. NTIS No. PB91-215046. 18 pp.

Thompson, P.O., L.T. Findley, and O. Vidal. 1990. 20-Hz pulses and other vocalizations of fin whales, *Balaenoptera physalus*, in the Gulf of California, Mexico. J. Acoust. Soc. Am. Suppl. 1, Vol. 88, S5.

Townsend, C.H. 1935. The distribution of certain whales as shown by logbook records of American whale ships. Zoologica 19:3-50.

United Nations Educational, Scientific, and Cultural Organization. 1996. Scientific diving: a general code of practice. Best Publishing Company – UNESCO Publishing, Flagstaff, AZ. 278 pp.

U.S. Department of the Interior, U.S. Fish and Wildlife Service. 1996. Florida manatee recovery plan (*Trichechus manatus latirostris*), second revision. Prepared by the Florida Manatee Recovery Team for the Southeast Region, U.S. Fish and Wildlife Service, Atlanta, GA. Approved January 29, 1996.

U.S. Department of the Interior, U.S. Fish and Wildlife Service. 1998. Endangered and Threatened Wildlife and Plants. 50 CFR 17.11 and 17.12. December 31, 1998. 56 pp.

U.S. Department of the Interior, Geological Survey. 1976. Final environmental impact statement. Regulations pursuant to geological and geophysical explorations of the outer continental shelf. FES 76-23.

U.S. Department of the Interior, Minerals Management Service. 1986a. Visual 2E, Commercial fisheries and endangered and threatened species. U.S. Department of the Interior, Minerals Management Service, Gulf of Mexico OCS Region, New Orleans, LA.

U.S. Department of the Interior, Minerals Management Service. 1986b. Visual 2, Commercial fisheries and endangered and threatened species. U.S. Department of the Interior, Minerals Management Service, Gulf of Mexico OCS Region, New Orleans, LA.

U.S. Department of the Interior, Minerals Management Service. 1995. OCS environmental assessment of a proposed 3-dimensional seismic survey, Santa Ynez Unit. U.S. Department of the Interior, Minerals Management Service. Pacific OCS Region, Camarillo, CA.

U.S. Department of the Interior, Minerals Management Service. 1996. Final Environmental Impact Statement, Outer Continental Shelf Oil & Gas Leasing Program 1997-2002. U.S. Department of the Interior, Minerals Management Service, Herndon, VA.

U.S. Department of the Interior, Minerals Management Service. 1997a. Gulf of Mexico OCS Oil and Gas Lease Sales 169, 172, 175, 178, and 182, Central Planning Area. Final Environmental Impact Statement. U.S. Department of the Interior, Minerals Management Service, Gulf of Mexico OCS Region, New Orleans, LA.

U.S. Department of the Interior, Minerals Management Service. 1997b. Gulf of Mexico OCS Oil and Gas Lease Sales 171, 174, 177, and 180, Western Planning Area. Final Environmental Impact Statement. U.S. Department of the Interior, Minerals Management Service, Gulf of Mexico OCS Region, New Orleans, LA.

U.S. Department of the Interior, Minerals Management Service. 1999. Destin Dome 56 unit development and production plan and right-of-way-pipeline application. Draft Environmental Impact Statement. Volume I: Sections I-X. U.S. Department of the Interior, Minerals Management Service, Gulf of Mexico OCS Region, New Orleans, LA. OCS EIS/EA MMS 99-0040.

U.S. Department of the Interior, Minerals Management Service. 2000. Gulf of Mexico Deepwater Operations and Activities - Environmental Assessment. U.S. Department of the Interior, Minerals Management Service, Gulf of Mexico OCS Region, New Orleans, LA. OCS EIS/EA MMS 2000-001. 267 pp.

U.S. Department of the Interior, Minerals Management Service. 2001a. Gulf of Mexico OCS Oil and Gas Lease Sale 181, Eastern Planning Area. Final Environmental Impact Statement. U.S. Department of the Interior, Minerals Management Service, Gulf of Mexico OCS Region, New Orleans, LA. OCS EIS/EA MMS 2001-055. 2 vols.

U.S. Department of the Interior, Minerals Management Service. 2001b. Proposed Use of Floating Production, Storage, and Offloading Systems on the Gulf of Mexico Outer Continental Shelf – Western and Central Gulf Planning Areas. Final Environmental Impact Statement. U.S. Department of the Interior, Minerals Management Service, Gulf of Mexico OCS Region, New Orleans, LA. OCS EIS/EA MMS 2000-090.

U.S. Department of the Interior, Minerals Management Service. 2002a. Gulf of Mexico OCS Oil and Gas Lease Sales: 2003-2007. Central Planning Area Sales 185, 190, 194, 198, and 201. Western Planning Area Sales 187, 192, 196, and 200. Final Environmental Impact Statement. U.S. Department of the Interior, Minerals Management Service, Gulf of Mexico OCS Region, New Orleans, LA. OCS EIS/EA MMS 2002-052. 2 vols.

U.S. Department of the Interior, Minerals Management Service. 2002b. Gulf of Mexico OCS Oil and Gas Lease Sales 189 and 197. Eastern Planning Area. Draft Environmental Impact Statement. U.S. Department of the Interior, Minerals Management Service, Gulf of Mexico OCS Region, New Orleans, LA. OCS EIS/EA MMS 2002-056. 2 vols.

Van Parijs, S.M., and P.J. Corkeron. 2001. Boat traffic affects the acoustic behaviour of Pacific humpback dolphins, *Sousa chinensis*. J. Mar. Biol. Assoc. U.K. 81:533-538.

Wakefield, E.D. 2001. The distribution and vocal behaviour of common dolphin *(Delphinus delphis* L.) in the Celtic sea and adjacent waters, with particular reference to the effects of seismic surveying. M.Sc. Thesis, University of Wales, Bangor.

Wardle, C.S., T.J. Carter, G.G. Urquhart, A.D.F. Johnstone, A.M. Ziolkowski, G. Hampson, and D. Mackie. 2001. Effects of seismic air guns on marine fish. Cont. Shelf Res. 21(2001):1005-1027.

Waring, G.T., D.L. Palka, K.D. Mullin, J.H.W. Hain, L.J. Hansen, and K.D. Bisack. 1997. U.S. Atlantic and Gulf of Mexico Marine Mammal Stock Assessments - 1996. NOAA Tech. Mem. NMFS-NE-114. 250 pp.

Waring, G.T., D.L. Palka, P.J. Clapham, S. Swartz, M. Rossman, T. Cole, L.J. Hansen, K.D. Bisack, K. Mullin, R.S. Wells, D.K. Odell, and N.B. Barros. 1999. U.S. Atlantic and Gulf of Mexico marine mammal stock assessments - 1999. NOAA Tech. Mem. NMFS-NE-153. 196 pp.

Wartzok, D., and D.R. Ketten. 1999. Marine mammal sensory systems, pp. 117-175. *In:* J.E. Reynolds III and S.A. Rommel (eds.), Biology of marine mammals. Smithsonian Institution Press, Washington, D.C. 590 pp.

Watkins, W.A. 1981. Reaction of three species of whales to implanted radio tags. Deep-Sea Res. 28A(6):589-599.

Watkins, W.A., and W.E. Scheville. 1975. Sperm whales react to pingers. Deep-Sea Res. 22:123-129.

Watkins, W.A., M.A. Daher., K.M. Fristrup, T.J. Howald, and G.N. Disciara. 1993. Sperm whales tagged with transponders and tracked underwater by sonar. Mar. Mammal Sci. 9:55-67.

Wever, E.G. 1978. The Reptile Ear: Its Structure and Function. Princeton University Press, Princeton.

Willis, P.M., and R.W. Baird. 1998. Status of the dwarf sperm whale, *Kogia simus*, with special reference to Canada. Can. Field Nat. 112:114-125.

Würsig, B., S.K. Lynn, and K.D. Mullin. 1996. Behavior of cetaceans relative to survey vessels. *In:* R.W. Davis and G.S. Fargion (eds.), Distribution and abundance of cetaceans in the north-central and western Gulf of Mexico, final report. Volume II: Technical report. Prepared by the Texas Institute of Oceanography and the National Marine Fisheries Service. U.S. Department of the Interior, Minerals Management Service, Gulf of Mexico OCS Region Office, New Orleans, LA. OCS Study MMS 96-0027. 357 pp.

Würsig, B., T.A. Jefferson, and D.J. Schmidley. 2000. The marine mammals of the Gulf of Mexico. Texas A&M University Press, College Station, TX. 232 pp.

Yost, W.A. 1994. Fundamentals of hearing: An introduction. Academic Press, New York.

Zawila, J.S. 1995. Characterization of a seismic air gun acoustic dispersal technique at the Virginia Institute of Marine Science sea turtle test site, pp. 88-89. *In*: L.Z. Hales (compiler), Sea Turtle Research Program: Summary Report. Prepared for U.S. Army Engineer Division, South Atlantic, Atlanta, GA, and U.S. Naval Submarine Base, Kings Bay, GA. Technical Report CERC-95-. 145 pp.

Zelick, R., D. Mann, and A. Popper. 1999. Acoustic communication in fishes and frogs, pp. 363-412. *In:* R. Fay and A. Popper (eds.), Comparative Hearing: Fish and Amphibians. Springer Handbook of Auditory Research, Vol. 11. Springer-Verlag, New York.

VI. LIST OF PREPARERS AND CONTRIBUTORS

This PEA was prepared by Continental Shelf Associates, Inc. (CSA) under Contract No. 1432-01-99-CT-30987 for the MMS Gulf of Mexico OCS Region, New Orleans, Louisiana. The following individuals contributed to the technical content of the PEA:

- Brian J. Balcom (CSA) – Responsibilities: project manager; co-author, Sections I, II, and III and Appendices A, B, C, D, E, I, K, and L. Degrees: M.S. Marine Biology; B.S. Biology. 23 years of experience.

- Robert B. Cady (CSA) – Responsibilities: co-author, Appendix B. Degrees: M.S. Oceanography; B.S. Marine Biology. 9 years of experience.

- Luis M. Lagera, Jr. (CSA) – Responsibilities: co-author, Appendices F and J. Degrees: Ph.D. Environmental Sciences (Ecology); M.S. Biological Sciences; B.S. Zoology. 14 years of experience.

- Neal W. Phillips (CSA) – Responsibilities: co-author, Sections I, II, and III and Appendix F. Degrees: Ph.D. Ecology; M.S. Marine Studies; B.S. Biological Sciences. 20 years of experience.

- David B. Snyder (CSA) – Responsibilities: co-author, Section III and Appendix F (fish and fisheries). Degrees: M.S. Marine Biology and Ichthyology; B.S. Zoology. 21 years of experience.

- M. John Thompson (CSA) – Responsibilities: co-author, Section III and Appendix F (benthic communities, cultural resources, and military, recreational, and other commercial uses). Degrees: M.S. Marine Biology; B.S. Biology. 29 years of experience.

- Stephen T. Viada (CSA) – Responsibilities: co-author, Section III and Appendix F (marine mammals; sea turtles; coastal and marine birds). Degrees: M.S. Biological Oceanography; B.S. Zoology. 21 years of experience.

- Soraya Moein Bartol (Woods Hole Oceanographic Institution) – Responsibilities: co-author, Section III and Appendix H. Degrees: Ph.D. Marine Science; M.S. Marine Science; B.A. Biology. 11 years of experience.

- Douglas C. Biggs (Texas A&M University) – Responsibilities: co-author, Section III and Appendix F. Degrees: Ph.D. Biological Oceanography; B.A. Biology. 32 years of experience.

- James M. Brooks (TDI-Brooks International, Inc.) – Responsibilities: co-author, Section II and Appendix D. Degrees: Ph.D. Oceanography; M.S. Oceanography; B.S. Chemistry. 27 years of experience.

- Tim Chambers (Reese-Chambers Systems Consultants, Inc.) – Responsibilities: author, Appendix K. Degrees: M.S. Mathematics; B.S. Mathematics. 31 years of experience.

- Richard R. Fay (Parmly Hearing Institute, Loyola University Chicago) – Responsibilities: co-author, Section III and Appendix I. Degrees: Ph.D. Experimental Psychology and Sensory Physiology; M.S. Experimental Psychology; B.A. Psychology. 32 years of experience.

- John C. Goold (Institute of Environmental Science, University of Wales, Bangor) – Responsibilities: co-author, Section III and Appendices C and G. Degrees: Ph.D. Oceanography; B.Sc. Marine Biology and Zoology. 12 years of experience.

- Gordon M. Greve (Orion Consultants) – Responsibilities: co-author, Section II; author, Appendix D. Degrees: Ph.D. Geophysics; M.S. Geophysics; B.S. Electrical Engineering. 42 years of experience.

The PEA represents the culmination of a coordinated effort among scientists from government, academia, and the private sector. MMS, as well as other interested agencies (e.g., NMFS, Office of Naval Research) and the G&G industry, have provided valuable and constructive input during the review process leading to this final PEA. A complete listing of individuals consulted during the project is provided in **PEA Section IV - Consultation, Coordination, and Document Review**. MMS contributors have included

- Richard Defenbaugh (Deputy Regional Supervisor for Leasing and Environment, MMS Gulf of Mexico OCS Region) – Project Officer (Contracting Officer's Technical Representative) for the completion phase of the project (September 2000 to present). Responsible for review of the entire document, evaluation of the scope and content of the PEA, and acceptance by the MMS.

- Bill Lang (Oceanographer, MMS Gulf of Mexico OCS Region) – Project Officer (Contracting Officer's Technical Representative) for the initial phase of the project. Responsible for project planning, procurement, and project guidance leading to preparation of the June 2000 preliminary draft of the PEA. Senior technical expert on marine mammals and key technical reviewer of the entire document.

- Dennis Chew (Chief, NEPA and Coastal Zone Management Coordination Unit, MMS Gulf of Mexico OCS Region) – Responsible for review of the PEA for NEPA procedures and compliance, and consistency with other MMS NEPA documents.

- Hammond Eve (Regional Supervisor for Leasing and Environment, MMS Gulf of Mexico OCS Region) – Responsible for overall project approach and outcome, for management of internal (MMS) relationships, and responsibilities of key MMS personnel until his retirement in January 2004. Reviewed and critiqued the PEA for consistency with MMS NEPA policy.

- Ron Brinkman (Geophysicist, Data Acquisition and Special Projects Unit, Resource Evaluation, MMS Gulf of Mexico OCS Region) – Technical expert on equipment and techniques used to acquire data under Pre-lease Geological and Geophysical Permits and provides perspective into historical levels and locations of data gathering activities. Reviewed and critiqued the PEA for technical accuracy in area of expertise.

- Gary Lore (Regional Supervisor for Resource Evaluation, MMS Gulf of Mexico OCS Region) – Responsible for the program under which Pre-lease Geological and Geophysical Permits are issued to collect data for exploration of oil, gas, and other minerals on the Gulf of Mexico OCS and for assignment of key MMS personnel. Reviewed and critiqued the PEA.

- Dave Cooke (Chief, Resource Studies Section, Resource Evaluation, MMS Gulf of Mexico OCS Region) – Responsible for the day-to-day management of the Pre-lease Geological and Geophysical Permit program on the Gulf of Mexico OCS. Responsible for review of the PEA regarding data acquisition procedures and historical information.

- David Zinzer (Geophysicist, Resource Evaluation Division, MMS Headquarters) – Reviewed and contributed to the PEA.

Review comments on technical and regulatory issues pertinent to G&G activities also were provided by MMS senior staff at the Alaska OCS Region Office (Anchorage, Alaska), Pacific OCS Region Office (Camarillo, California), and at MMS Headquarters (Herndon, Virginia). The recommendations and constructive criticisms offered by all MMS staff are recognized and greatly appreciated.

VII. GLOSSARY[1]

Acoustic: Pertaining to the sense or organs of hearing, to sound, or to the scientific study of sound, especially of its generation, transmission, and reception.

Active acoustic monitoring: A sound-producing and listening/recording system designed to detect the presence of an inanimate or animate object (or objects) in the water.

Adverse impact: Any impact arising from an activity (e.g., a Proposed Action) that is non-beneficial. An adverse impact can manifest itself as a direct, indirect, and/or cumulative impact, and may affect one or more resources in the environment. Adverse impacts may be further divided into negligible, adverse but not significant, or significant, based on appropriate resource-specific impact evaluation criteria.

Allision: The collision between one ship in motion with another stationary vessel, structure, or facility.

Ambient noise: The typical or persistent environmental background noise present in the ocean.

Anthropogenic noise: Noise related to or produced by human activities.

Attenuation: Energy losses in a pressure wave, or sound, which occur as the wave travels through a medium. In seawater, the associated wavefront diminishes due to the spreading of the sound over an increasingly larger volume and the absorption of some of the energy; also known as transmission loss.

Baleen: Filtering plates that hang from the upper jaw of baleen whales.

Baleen whales: Filter-feeding whales, also known as mysticetes.

Bin or bin-size: One of a set of discrete areas into which a survey area is divided. A bin refers to an area with in-line and cross-line dimension used for 3D seismic data processing. Bin-size, numbers of bins, and number of hard copy records vary depending upon the type and objectives of the specific G&G survey being conducted.

Block (= lease block): A geographical area portrayed on official MMS protraction diagrams or leasing maps. MMS OCS blocks are typically square, measure 3 statute miles along each side, and comprise approximately 2,331 hectares. Lease blocks are those that have been offered for sale by the MMS and have been subsequently leased by a designated lease holder (e.g., operator or consortium) for mineral exploration and development.

Cetacean: Of, or belonging to, the order Cetacea. Aquatic placental mammals with anterior flippers, no hind limbs, a blow hole, and a horizontal tail fluke. Commonly known as whales, dolphins, and porpoises.

Conservative: As used in the context of this PEA, an estimate or judgment that errs on the side of precaution and protection of sensitive environmental resources.

---

[1] Glossary definitions derived or adapted from dilithium Press, Ltd. (1989), Richardson et al. (1995), USDOI, MMS (2001b, 2002a), and/or Department of the Navy (2001b).

Critical habitat: Designated geographic areas where one or more species of concern reside, characterized by the presence of physical and/or biological attributes (characteristics) deemed essential to the survival of the species, or the area surrounding such habitats, which are essential to the survival of the species.

Decibel: A dimensionless unit used to express the ratio of two quantities on a logarithmic scale. In underwater acoustics, sound pressure is measured in decibels relative to a reference pressure of 1 μPa, as 20 times the logarithm of a measured level divided by the reference level. Acoustic intensity is measured as 10 times the logarithm of a measured intensity divided by the intensity of a plane wave of rms pressure equal to 1 μPa.

de minimus: An entity (or attribute) that is smallest or least significant.

Endangered species: Defined in 16 U.S.C. 1532 as any species that is in danger of extinction throughout all or a significant portion of its range (other than a species of Class Insecta designated as a pest). Federally endangered species are listed in 50 CFR 17.11 and 17.12.

Ensonification: The introduction of noise into the environment.

Frequency: The number of times that a periodic function or vibration occurs within 1 second. The number of cycles per second is expressed in Hertz (1 Hz = 1 cycle per second; 10 Hz = 10 cycles per second, etc.).

Geological and geophysical (G&G) activities: Activities designed to collect data pertinent to subsurface geologic structure, including seismic surveys (e.g., high-resolution site surveys and various types of seismic exploration and development surveys), deep-tow side-scan sonar surveys, electromagnetic surveys, geological and geochemical sampling, and remote sensing.

Habitat: Environment where an animal or plant normally lives, often characterized by a dominant plant form or physical characteristic.

Harassment: Under the MMPA, any act of pursuit, torment, or annoyance that has the potential to a) injure a marine mammal or marine mammal stock in the wild; or b) disturb a marine mammal or marine mammal stock in the wild by causing disruption of behavioral patterns, including, but not limited to, migration, breathing, nursing, breeding, feeding, or sheltering.

Hertz: The unit of frequency, defined as the number of cycles per second.

KiloHertz (kHz): One thousand (1,000) Hz, or 1,000 cycles per second.

Masking: The obscuring of sounds of interest by interfering sounds, generally at similar frequencies (see Richardson et al., 1995).

Mysticete: Any of several whale species of the suborder Mysticeti characterized by the presence of symmetrical skulls, paired blow holes, and plates of whale bone (baleen plates) instead of teeth.

Negligible impact: An adverse impact to a resource (or resources) that is extremely minor in both context and intensity.

Odontocete: Cetaceans of the suborder Odontoceti, characterized by a single blow hole, an asymmetric skull, and teeth (e.g., sperm whales, orcas, dolphins).

Passive acoustic monitoring: Listening/recording systems that detect sounds created by an object (source) in the water (i.e., a one-way transmission of sound waves traveling through the water from the source to the receiver).

Pelagic: Living in the water column, including marine plants and animals that are free-floating and drift passively, or animals that are strong swimmers.

Permanent threshold shift (PTS): The permanent deterioration of hearing due to prolonged or repeated exposure to high level sounds, and/or permanent hearing damage from exposure to rapid high pressure events (e.g., explosive discharges).

Pinniped: Of, or belonging to the Pinnipedia, aquatic mammals such as seals, sea lions and walruses. Pinnipeds are carnivorous, have paddle shaped hind limbs for locomotion, and must "haul out" on land or ice to pup.

Ramp-up: The progressive build up of power output from a seismic airgun array, achieved by gradually increasing the number of guns fired simultaneously within an array, until full power is achieved with all guns firing simultaneously.

Received level (RL): The sound level, either pressure or intensity, received at a given point at a certain distance from a sound source.

Root mean squared (rms): The square root of the arithmetic mean of the squares of a set of numbers.

Sirenian: An herbivorous aquatic mammal of the order Sirenia, which include the manatee and dugong.

Sound Navigation and Ranging System (sonar): A method for detecting and locating objects submerged in water by means of the sound waves they reflect or produce.

Sound pressure level (SPL): The pressure of a sound wave, described logarithmically as 20 times the logarithm of the measured pressure divided by a reference pressure of 1 μPa.

Sound speed: The velocity that sound waves travel through a medium (e.g., water, air). Sound speed through seawater is ~1,500 m/s (4,920 ft/s), with variability due to differences in water temperature, salinity, and depth (pressure). Sound speed increases with increases in temperature and pressure (depth) and to a lesser extent with increase in salinity.

Source level (SL): The sound pressure or intensity level at 1-m range from the center of an acoustic source. With large devices such as airgun arrays, sound level is measured at ranges greater than 1 m, and the source level back calculated to 1-m range.

SURTASS LFA sonar: Long-range, all-weather low frequency (between 100 and 500 Hz) sonar system composed of both active and passive components. SURTASS (Surveillance Towed Array Sensor System) is the passive component. LFA (Low Frequency Active) is the active component.

Take: Under the Marine Mammal Protection Act, to harass, harm, pursue, hunt, shoot, wound, kill, trap, capture, or collect, or to attempt any of these activities.

Temporary threshold shift (TTS): Similar to PTS; however, deterioration in hearing ability is temporary. Repeated occurrence of TTS is thought to lead to PTS.

Threatened species: Any species that is likely to become an endangered species within the foreseeable future throughout all or a significant portion of its range. Threatened species are listed in 50 CFR 17.12.

Time-lapse surveys: Seismic surveys that are repeated over the same area after sufficient time has passed; also known as four dimensional (or 4D) surveys, where the fourth dimension is time.

Tract (= lease tract): An OCS block, or portion of an OCS block, or group of OCS blocks offered for lease at an OCS lease sale.

Transmission loss: Energy losses in a pressure wave, or sound, which occur as the wave travels through a medium. In seawater, the associated wavefront diminishes due to the spreading of the sound over an increasingly larger volume and the absorption of some of the energy. Also known as attenuation.

Wavelength: The distance between corresponding points of two successive waves.

VIII. KEY WORD INDEX

Appendices

Appendix A
Environmental Assessment Document Development and Revisions
to Project Alternatives

Environmental Assessment Document
Development and Revisions to Project Alternatives

The Final Programmatic Environmental Assessment (PEA) dealing with geological and geophysical (G&G) operations in the Gulf of Mexico has undergone a series of revisions and updates since the Preliminary Draft PEA was initially reviewed by the Minerals Management Service (MMS) in 2001. The following summary was provided by MMS to document the evolution of the PEA, as reflected in the comment and review process of earlier draft versions of the PEA and changes in G&G operational requirements.

Early phases of the PEA process included the following:

- MMS solicitation and award of a competitive contract to Continental Shelf Associates, Inc., Jupiter, FL, to prepare a PEA – June 1999;
- Preparation of the first preliminary draft PEA and review by MMS, other Federal agencies, and pertinent experts - August 2000; and
- Preparation of the second draft PEA document and review by MMS only - June 2001.

Following several revisions and amendments to the draft working document, a formal Draft PEA was submitted to the MMS for review in mid-February 2002 and subsequently released for formal review in late February 2002. Formal review and comments were requested from the National Marine Fisheries Service (NMFS; now called NOAA Fisheries; NOAA-F) Headquarters and Southeast Region offices; the Marine Mammal Commission (MMC); and MMS Gulf of Mexico outer continental shelf (OCS) Region and Headquarters offices.

Timely comments were received from MMC[1, 2] and MMS reviewers, and informally from NOAA-F reviewers. Formal comments were received from NOAA-F in June 2002[3]. The Draft PEA was released for private sector and public review in mid-August 2002. Consolidation of reviewers' comments was started in March 2002 and continued intermittently through 2002 as comments were received from Federal reviewers, then from private sector and public commenters, and from State agency reviewers.

During the extended period of time that this Draft G&G PEA was under review, a number of events have occurred, and documents were prepared that affected the MMS position on seismic surveys. These include

- The St. Petersburg office of NOAA-F prepared and issued the Biological Opinion (BO) for the Endangered Species Act (ESA) Section 7 Consultation on OCS Sale 184[4]. This BO drew heavily on the Draft G&G PEA and imposed non-discretionary terms and conditions on MMS (generally reflecting implementation of the ramp-up and visual monitoring alternative of the Draft G&G PEA) for seismic surveys related to leases issued under Sale 184.
- The MMS mentioned the NOAA-F non-discretionary terms and requirements in the Final Sale Notice Package for Sale 184[5] and prepared a Notice to Lessees (NTL No. 2002-G07)[6] to implement these requirements.
- Representatives of the seismic and oil and gas industries objected to new requirements at the Final Notice of Sale stage of the lease sale process, in the absence of prior notification or opportunity for public comment. The International Association of Geophysical Contractors (IAGC) sent letters to MMS Director Burton[7] and NOAA-F Assistant Administrator Hogarth[8] commenting on the BO[9] and providing information on seismic technologies[10].

During discussions among MMS, NOAA-F, and representatives of the seismic industry regarding the NTL requirements and implementation plans, these representatives requested an opportunity to review and comment on the Draft G&G PEA. The Draft G&G PEA was released for public review and comment in mid-August 2002. Comments were received from the seismic industry[11, 12], the oil & gas industry[13], an environmental group[14], a State agency[15], and one interested citizen[16]. During the same general period, the NOAA-F prepared the Draft BO for ESA Section 7 Consultation[17] on the MMS proposal for several OCS Sales in the Western and Central Gulf of Mexico over a 5-year period (commonly called the "Multi-sale EIS"). A copy of the Draft BO was provided to MMS and was released to the private sector. The final BO was issued by NOAA-F in November 2002[18].

During summer 2002, a major multi-year study of Gulf sperm whales and other cetaceans sponsored by MMS, with cooperation and support from the seismic industry (i.e., the Sperm Whale Seismic Study, SWSS) was successfully initiated. Major accomplishments included tagging numerous sperm whales with satellite-monitored radio tags, and field testing a passive acoustic listening system for its ability to detect and locate sperm whales, relative to the effectiveness of visual observers. SWSS (2002) summarized the results of the survey effort – the passive acoustic monitoring system was far superior to visual observers, as it could detect cetaceans underwater and at distances or in sea states where visual observations are not reliable.

During discussions among MMS, NOAA-F, and representatives of the seismic and oil and gas industries, the MMS position on requirements to mitigate impacts of seismic surveys to cetaceans continued to evolve, resulting in issuance of Addendum 1[19] to NTL No. 2002-G07 in October 2002. Subsequently, NTL No. 2003-G08 was issued effective June 2003; this NTL further refined the requirements of G&G operators working in water depths ≥200 m throughout the U.S. Gulf of Mexico, and superceded the mitigation measures outlined in NTL No. 2002-G07 and its amendment. In March 2004, MMS implemented its latest regulations pertinent to Gulf of Mexico G&G seismic operations – NTL No. 2004-G01[20]. This latest NTL does not introduce any new mitigation measures; however, the previous restrictions adopted in NTL No. 2003-G08 specifically to protect sperm whales are now applicable to all whales. Performance of these mitigation measures is a condition of the approval of applications for geophysical permits. These mitigation measures apply to geophysical activities conducted under lease terms, for all seismic survey operations conducted in waters >200 m (656 ft) throughout the Gulf of Mexico and, in the Eastern Gulf of Mexico Planning Area, for all seismic survey operations conducted regardless of water depth. All marine mammals, including dolphins, and sea turtles continue to be protected effectively by implementing existing seismic survey mitigation measures requiring ramp-up, protected species observer training, visual monitoring, and reporting.

In December 2002, MMS submitted a request to NOAA-F for a Notice of Proposed Rulemaking under Section 101(a)(5) of the Marine Mammal Protection Act. This request was pursued to authorize the incidental taking of small numbers of sperm whales in the Gulf of Mexico as a result of seismic survey activity. NOAA-F formally announced the request and solicited comments in March 2003[21]. In April 2003, MMS submitted a letter to NOAA-F updating its earlier petition, and requesting inclusion of new information in the review, comment, and rulemaking approval process[22].

All of these events have caused a continuing evolution of the MMS position on measures that must or should be taken to prevent potential harm to sperm whales, and have rendered certain aspects of the February 2002 Draft G&G PEA obsolete. Specifically, Alternative 1 as described in the Draft G&G PEA is no longer the status quo. While the status quo continues to evolve, the format of the Final PEA has been modified to support that evolution. Specifically, the alternatives outlined in this Final PEA have been revised to reflect the current status quo (as of May 2004). Old and new alternative definitions are outlined in **Table A-1**.

Table A-1
Comparison of Alternatives – Final versus Draft Programmatic Environmental Assessment (PEA)

Scope of Current Alternative (Final PEA)	Scope of Old Alternative (Draft PEA)
Alternative 1 – Continuation of the Status Quo. Specifically, geological & geophysical (G&G) operators must adhere to the requirements of Notice to Lessees (NTL) No. 2004-G01 by utilizing ramp-up, employing visual monitoring using trained observers, completing reporting requirements, and having the option of initiating seismic operations during nighttime and periods of limited visibility using passive acoustic monitoring techniques. The purpose of these measures is to establish and maintain a 500-m exclusion (or impact) zone around the seismic array; this zone is to remain free of whales. When a whale is located within (or approaching) the exclusion zone, the array is to be shut down until the animal has cleared the zone. Resumption of survey activities requires verification (via visual observers during daylight and good visibility; or optionally during nighttime and/or poor visibility using passive acoustic monitoring) that the exclusion zone is clear of whales.	Alternative 1 – Continuation of the Status Quo. Under this alternative, G&G operators may continue to operate in Gulf waters in the absence of any mitigation measures to minimize or eliminate the potential for acoustic impacts to marine mammals.
Alternative 2 – Addition of Vessel-based Passive Acoustic Monitoring as a Requirement. This alternative includes the same requirements as Alternative 1. In addition, all seismic operations must utilize passive acoustic monitoring techniques prior to ramp-up in conjunction with visual monitoring. The purpose of these measures is to establish and maintain a 500-m exclusion (or impact) zone around the seismic array; this zone is to remain free of marine mammals. When a whale is located within (or approaching) the exclusion zone, the array is to be shut down until the animal has cleared the zone. Resumption of survey activities requires verification (via visual observers and passive acoustic monitoring results) that the exclusion zone is clear of whales.	Alternative 2 – Implement Additional Mitigation Measures. Under this alternative, all seismic surveys conducted 1) west of the 88th meridian (88° W Long) at water depths >200 m would include ramp-up procedures (i.e., where seismic sources are started at low levels and increased in intensity until full strength is achieved), visual monitoring, and/or passive acoustic monitoring; and 2) east of the 88th meridian within Federal outer continental shelf (OCS) waters of the Gulf of Mexico would include ramp-up procedures, visual monitoring, and/or passive acoustic monitoring.

Table A-1
Comparison of Alternatives – Final versus Draft Programmatic Environmental Assessment (PEA)
(Continued)

Scope of Current Alternative (Final PEA)	Scope of Old Alternative (Draft PEA)
Alternative 3 – Addition of Both Passive and Active Acoustic Monitoring as a Requirement. This alternative includes the same requirements as Alternative 1. In addition, all seismic operations must utilize both passive acoustic monitoring and active acoustic monitoring techniques prior to ramp-up in conjunction with visual monitoring. The purpose of these measures is to establish and maintain a 500-m exclusion (or impact) zone around the seismic array; this zone is to remain free of marine mammals. When a whale is located within (or approaching) the exclusion zone, the array is to be shut down until the animal has cleared the zone. Resumption of survey activities requires verification (via visual observers and passive acoustic monitoring results) that the exclusion zone is clear of whales.	Alternative 3 – Implement a Suite of Mitigation Measures. Under this alternative, the same depth and geographic restrictions noted under Alternative 2 would apply – ramp-up procedures (only during daylight hours), visual monitoring using trained marine mammal observers, and allowance of continuous (day and night) seismic operations. However, if a marine mammal is spotted within or immediately ahead of the predetermined impact zone (i.e., isopleth of 180 dB re 1 μPa, rms) surrounding the array, an immediate shut down of the array is required. Subsequent restart of the array, using ramp-up and visual observers, will only be allowed during daylight hours and following clearance of the impact zone. Similar restrictions apply to airgun arrays shut down for maintenance.
Alternative 4 – Restrict G&G Activities. This alternative includes the same requirements as Alternative 1. In addition, G&G operators would be precluded from conducting simultaneous seismic operations in those portions of the Gulf of Mexico most frequented by sperm whales and Bryde's whales (i.e., water depths >200 m). The purpose of this measure is to remove the potential for simultaneous exposure to seismic noise from concurrent surveys in the same general area. Minimum distance for concurrent seismic operations is 4.0 km, the approximate dimension of an OCS lease block (i.e., concurrent seismic operations cannot occur in adjacent blocks).	Alternative 4 – Restrict G&G Activities. Existing mitigation measures (protective measures routinely specified in permit requirements and lease stipulations) described previously for the Proposed Action would also be included in this alternative; however, simultaneous survey activity is prohibited within those portions of the Gulf of Mexico most frequented by sperm whales and Bryde's whales.

[1] Letter, R. H. Mattlin (MMC) to R. E. Defenbaugh (MMS), dated 26 April 2002, formally providing comments on the Draft PEA; with attached paper by Swartz & Hofman.

[2] Attachment to R. H. Mattlin letter dated 26 April 2002; copy of paper: Swartz, S. L. and R. J. Hofman. 1991. Marine mammal and habitat monitoring: requirements, principles, needs, and approaches. Marine Mammal Commission, Washington, D.C. 18 pp.

[3] Letter, D. R. Knowles (NMFS) to R. E. Defenbaugh (MMS), dated June 21, 2002, formally transmitting Office of Protected Resources review of the final draft PEA regarding G&G exploration operations in the Gulf of Mexico.

[4] National Marine Fisheries Service (NOAA Fisheries) Southeast Regional Office. July 11, 2002. Endangered Species Act – Section 7 Consultation Biological Opinion for Gulf of Mexico Outer Continental Shelf Lease Sale 184. 80 pp.

[5] MMS Final Sale Notice Package, OCS oil and gas lease sale 184, Western Gulf of Mexico, August 21, 2002, New Orleans, LA. See "Information to Lessees" section; page 11, "Lease stipulation for protected species..." (see: http://www.gomr mms.gov/homepg/lsesale/184stipf.pdf)

[6] MMS Notice to Lessees and Operators (NTL) No. 2002-G07, dated August 22, 2002; titled "Implementation of seismic survey mitigation measures." (see: http://www.gomr.mms.gov/homepg/regulate/regs/ntls/ntl02-g07 html)

[7] Letter, G. C. Gill (IAGC) to J. Burton, dated August 2, 2002, captioned "Regulation of geophysical operations for the protection of marine mammals."

[8] Letter, G. C. Gill (IAGC) to W. S. Hogarth, dated August 2, 2002, captioned "NOAA Fisheries Biological Opinion/ MMS OCS Lease Sale 184."

[9] Attachment to G. C. Gill (IAGC) letters dated August 2, 2002; titled "IAGC comments on the NOAA Fisheries Biological Opinion for OCS Lease Sale 184."

[10] Attachment to G. C. Gill (IAGC) letters dated August 2, 2002; titled "Airgun arrays and marine mammals."

[11] E-mailed letter, G. C. Gill (IAGC) to MMS dated September 19, 2002; captioned "IAGC comments on a Draft Programmatic EA on G&G Exploration in the Gulf of Mexico."

[12] E-mailed letter, D. Quinn (Baker Atlas) to MMS dated September 19, 2002; responding to the draft G&G PEA with comments on the regulations with respect to borehole seismic surveys.

[13] Letter, K. Harb (National Ocean Industries Association) to MMS dated September 19, 2002; captioned "Draft programmatic environmental assessment on geological and geophysical exploration in the Gulf of Mexico outer continental shelf."

[14] E-mailed comments, C. Sarthou (Gulf Restoration Network) to MMS dated September 19, 2002, headed "Comments submitted by the Gulf Restoration Network on the MMS' G&G ... PDEA."

[15] Letter, L. F. Griffin (Florida Department of Environmental Protection) to R. Defenbaugh (MMS) dated December 2, 2002, captioned "Draft programmatic environmental assessment – geological & geophysical exploration for mineral resources on the Gulf of Mexico outer continental shelf."

[16] E-mail message, R. Dykstra to MMS dated September 5, 2002, subject "Programmatic EA of G&G on the Gulf of Mexico's OCS."

[17] NOAA Fisheries, Southeast Regional Office. Undated draft (ca. September 2002). Endangered Species Act – Section 7 Consultation Biological Opinion for Gulf of Mexico Outer Continental Shelf Lease Multi-sale (185, 187, 190, 192, 194, 196, 198, 200, 201). 107 pp.

[18] NOAA Fisheries, Southeast Regional Office. November 29, 2002. Endangered Species Act – Section 7 Consultation Biological Opinion for Gulf of Mexico Outer Continental Shelf Multi-Lease Sale (185, 187, 190, 192, 194, 196, 198, 200, 201). Consultation No. F/SER/2002/00718. 146 pp.

[19] MMS Addendum 1 to Notice to Lessees and Operators (NTL) No. 2002-G07, dated October 15, 2002; titled "Implementation of seismic survey mitigation measures." (see: http:www.gomr.mms.gov/homepg/regulate/regs/ntls/ntl02-g07add1 html)

[20] A copy of NTL No. 2004-G01 follows as an attachment.

[21] NOAA-F Federal Register notice – Taking and Importing Marine Mammals; Taking Marine Mammals Incidental to Conducting Oil and Gas Exploration Activities in the Gulf of Mexico. Federal Register, Vol. 68, No. 41, Monday, March 3, 2003, pp. 9991-9996.

[22] Letter, Thomas A. Readinger, Associate Director of Offshore Minerals Management, MMS, Washington, D.C. to Chief, Marine Mammal Conservation Division, NOAA-F, dated 15 April 2003, with enclosure.

**Attachment
NTL No. 2004-G01**

UNITED STATES DEPARTMENT OF THE INTERIOR
MINERALS MANAGEMENT SERVICE

NTL No. 2004-G01
Effective Date: March 1, 2004

NOTICE TO LESSEES AND OPERATORS OF FEDERAL OIL, GAS, AND
SULPHUR LEASES IN THE OUTER CONTINENTAL SHELF,
GULF OF MEXICO OCS REGION
Implementation of Seismic Survey Mitigation Measures and
Protected Species Observer Program

The Minerals Management Service (MMS) issued NTL No. 2003-G08, effective June 5, 2003, pursuant to 30 CFR 250.103, to explain how you implement seismic survey mitigation measures, including special ramp-up procedures and protected species observation and reporting. That NTL applied to all seismic operations throughout the Gulf of Mexico OCS in waters greater than 200 m (656 ft) in depth. The measures contained therein apply to all on-lease seismic surveys you conduct under 30 CFR 250.201 and all off-lease seismic surveys you conduct under 30 CFR 251.

This NTL supersedes and replaces NTL No. 2003-G08. It does not introduce any new types of mitigation measures; however, the previous restrictions adopted in NTL No. 2003-G08 specifically to protect sperm whales now will be applied to all whales (as defined herein), some of which may be encountered in water depths less than 200 m in certain areas of the Gulf of Mexico. All marine mammals, including dolphins (as defined herein), and sea turtles continue to be protected effectively by implementing existing seismic survey mitigation measures requiring ramp-up, protected species observer training, visual monitoring and reporting. This NTL also updates the Paperwork Reduction Act language to reflect the Office of Management and Budget's approval of the information collection hourly burdens and assignment of a new approval expiration date.

Background

The use of an airgun or airgun arrays while conducting seismic operations may have an impact on marine wildlife, including marine mammals and sea turtles. Some marine mammals, such as the sperm whale (*Physeter macrocephalus*), and all sea turtles that inhabit the Gulf of Mexico are protected under the Endangered Species Act (ESA). All marine mammals are protected under the Marine Mammal Protection Act (MMPA).

In July 2002, the MMS completed a formal Section 7 consultation under the ESA with the National Oceanic and Atmospheric Administration Fisheries Service (NOAA Fisheries) on Lease Sale 184 activities. The NOAA Fisheries issued a Biological Opinion (BiO) for Gulf of Mexico (GOM) Outer Continental Shelf (OCS) Lease Sale 184 requiring seismic operators to use ramp-up and visual observation procedures for seismic surveys in water depths equal to or greater than 200 meters (656 feet). Ramp-up procedures are already in use on seismic vessels in the GOM. The MMS also conducted a formal Section 7 consultation for lease sales in the Central and Western Gulf of Mexico Planning Areas included in the Five-Year Oil and Gas Leasing Program for 2002-2007. NOAA Fisheries issued a Multi-Lease Sale BiO with Conservation Recommendations to protect marine mammals and sea turtles during seismic operations. In response, the MMS issued NTL No. 2003-G08, which emphasized protection of sperm whales. However, additional whale species potentially may be impacted by seismic operations. Therefore, the MMS is expanding application of the seismic survey mitigation measures to include water depths less than 200 m in the Eastern Planning Area of the Gulf of

Mexico and to prohibit ramp-up and to require shut-down when any whale is observed within the exclusion zone.

Procedures for ramp-up, protected species observer training, visual monitoring and reporting are described in detail in this NTL. Performance of these mitigation measures is a condition of the approval of applications for geophysical permits. These mitigation measures apply to geophysical activities conducted under lease terms, for all seismic survey operations conducted in waters deeper than 200 meters (656 feet) throughout the GOM and, in the Eastern Planning Area of the GOM, for all seismic survey operations conducted regardless of water depth. You must demonstrate your compliance with these mitigation measures by submitting to MMS certain reports detailed in this NTL.

Definitions

Terms used in this NTL have the following meanings:
1. Airgun means a device that releases compressed air into the water column, creating an acoustical energy pulse with the purpose of penetrating the seafloor.
2. Ramp-up means the gradual increase in emitted sound levels from an airgun array by systematically turning on the full complement of an array's airguns over a period of time.
3. Visual monitoring means the use of trained observers to scan the ocean surface visually for the presence of marine mammals and sea turtles. These observers must have successfully completed a visual observer training program as described below. The area to be scanned visually includes, but is not limited to, the exclusion zone. Visual monitoring of an exclusion zone and adjacent waters is intended to establish and, when visual conditions allow, maintain a zone around the sound source and seismic vessel that is clear of marine mammals and sea turtles, thereby reducing or eliminating the potential for injury.
4. Exclusion zone means the area at and below the sea surface within a radius of 500 meters surrounding the center of an airgun array and the area within the immediate vicinity of the survey vessel.
5. Whales means all marine mammals in the Gulf of Mexico except dolphins (see definition below) and manatees. This includes all species of baleen whales (Suborder *Mysticeti*), all species of beaked whales (*Ziphius cavirostris* and *Mesoplodon sp.*), sperm whales (*Physeter macrocepahalus*), and pygmy and dwarf sperm whales (*Kogia sp.*). Of the baleen whales, only the Bryde's whale (*Balaenoptera edeni*) is expected to be present in the northern Gulf of Mexico and is considered uncommon. This species has primarily been sighted in water depths less than 200 m in the eastern Gulf of Mexico. Sightings of other baleen whale species are highly unlikely.
6. Dolphins means all marine mammal species in the Family *Delphinidae*. In the GOM, this includes, among others, killer whales, pilot whales, and all of the "dolphin" species.

Ramp-up Procedures

The intent of ramp-up is to warn marine mammals and sea turtles of pending seismic operations and to allow sufficient time for those animals to leave the immediate vicinity. Under normal conditions, animals sensitive to these activities are expected to move out of the area. For all seismic surveys, use the ramp-up procedures described below to allow whales, other marine mammals, and sea turtles to depart the exclusion zone before seismic surveying begins. Measures to conduct ramp-up procedures during all seismic survey operations are as follows:
1. Visually monitor the exclusion zone and adjacent waters for the absence of whales for at least 30 minutes before initiating ramp-up procedures. If no whales are detected, you

may initiate ramp-up procedures. <u>Do not initiate</u> ramp-up procedures at night or when you cannot visually monitor the exclusion zone for whales if your minimum source level drops below 160 dB re 1 μPa-m (rms) (see measure 5).

2. Initiate ramp-up procedures by firing a single airgun. The preferred airgun to begin with should be the smallest airgun, in terms of energy output (dB) and volume (in^3).

3. Continue ramp-up by gradually activating additional airguns over a period of at least 20 minutes, but no longer than 40 minutes, until the desired operating level of the airgun array is obtained.

4. Immediately shut down all airguns ceasing seismic operations at any time a whale is detected entering or within the exclusion zone. You may recommence seismic operations and ramp-up of airguns only when the exclusion zone has been visually inspected for at least 30 minutes to ensure the absence of whales.

5. You may reduce the source level of the airgun array to maintain a minimum source level of 160 dB re 1 μPa-m (rms) for routine activities, such as making a turn between line transects, or for maintenance needs. This procedure may be conducted during periods of impaired visibility (e.g., darkness, fog, high sea states) and does not require a 30-minute visual clearance of the exclusion zone before the airgun array is again ramped up to full output.

Protected Species Observer Program

Visual Observers

Visual observers who have completed a protected species observer training program as described below are required on all seismic vessels conducting operations in water depths greater than 200 meters (656 ft) throughout the Gulf of Mexico as of August 31, 2003. Visual observers are required on all seismic vessels conducting operations in OCS water depths less than 200 meters (656 ft.) in the Eastern Planning Area of the Gulf of Mexico as of the effective date of this NTL. At least two protected species visual observers will be required on watch aboard seismic vessels at all times during daylight hours (dawn to dusk) when seismic operations are being conducted, unless conditions (fog, rain, darkness) make sea surface observations impossible. If conditions deteriorate during daylight hours such that the sea surface observations are halted, visual observations must resume as soon as conditions permit. Operators may engage trained third party observers, may utilize crew members after training as observers, or may use a combination of both third party and crew observers. During these observations, the following guidelines shall be followed: (1) other than brief alerts to bridge personnel of maritime hazards, no additional duties may be assigned to the observer during his/her visual observation watch (if conditions warrant more vigilant look-outs when navigating around or near maritime hazards, additional personnel must be used to ensure that watching for protected species remains the primary focus of the on-watch observers), (2) no observer will be allowed more than 4 consecutive hours on watch as a visual observer, (3) a "break" time of no less than 2 hours must be allowed before an observer begins another visual monitoring watch rotation (break time means no assigned observational duties), and (4) no person (crew or third party) on watch as a visual observer will be assigned a combined watch schedule of more than 12 hours in a 24-hour period. Due to the concentration and diligence required during visual observation watches, operators who choose to use trained crew members in these positions are encouraged to select only those crew members who demonstrate willingness as well as ability to perform these duties.

Training

All visual observers must have completed a protected species observer training course. The MMS will not sanction particular trainers or training programs. However, basic training criteria have been established and must be adhered to by any entity that offers observer training. Operators may utilize observers trained by third parties, may send crew for training conducted by third parties, or may develop their own training program. All training programs offering to fulfill the observer training requirement must (1) furnish to the MMS, at the address listed in this NTL, a course information packet that includes the name and qualifications (i.e., experience, training completed, or educational background) of the instructor(s), the course outline or syllabus, and course reference material; (2) furnish each trainee with a document stating successful completion of the course; and (3) provide the MMS with names, affiliations, and dates of course completion of trainees.

The training course must include the following elements:
  I. Brief overview of the MMPA and the ESA as they relate to seismic acquisition and protection of marine mammals and sea turtles in the Gulf of Mexico
  II. Brief overview of seismic acquisition operations in the Gulf of Mexico
  III. Overview of seismic mitigation measures (NTLs) and the protected species observer program in the Gulf of Mexico
  IV. Discussion of the role and responsibilities of the protected species observer in the Gulf of Mexico, including:
      a. Legal requirements (why you are here and what you do)
      b. Professional behavior (code of conduct)
      c. Integrity
      d. Authority of protected species observer to call for shut-down of seismic acquisition operations
      e. Assigned duties
          1. What can be asked of the observer
          2. What cannot be asked of the observer
      f. Reporting of violations and coercion
  V. Identification of Gulf of Mexico marine mammals and sea turtles, with emphasis on whales
  VI. Cues and search methods for locating marine mammals, especially whales, and sea turtles.
  VII. Data collection and reporting requirements:
      a. Forms and reports to MMS via email protectedspecies@mms.gov on the 1st and 15th of each month
      b. Whale in exclusion zone/shut-down report within 24 hours

Visual Monitoring Methods

The observers on duty will look for whales, other marine mammals, and sea turtles using the naked eye and hand-held binoculars provided by the seismic vessel operator. The observers will stand watch in a suitable location that will not interfere with navigation or operation of the vessel and that affords the observers an optimal view of the sea surface. The observers will provide 360° coverage surrounding the seismic vessel and will adjust their positions appropriately to ensure adequate coverage of the entire area. These observations must be consistent, diligent, and free of distractions for the duration of the watch.

Visual monitoring will begin no less than 30 minutes prior to the beginning of ramp-up and continue until seismic operations cease or sighting conditions do not allow observation of the sea surface (e.g., fog, rain, darkness). If a marine mammal (whale or dolphin) or sea turtle is observed, the observer should note and monitor the position (including lat./long. of vessel and relative bearing and estimated distance to the animal) until the animal dives or moves out of visual range of the observer. Make sure you continue to observe for additional animals that may surface in the area, as often there are numerous animals that may surface at varying time intervals. At <u>any</u> time a whale is observed within an estimated 500 meters (1,640 feet) of the sound source array ("exclusion zone"), whether due to the whale's movement, the vessel's movement, or because the whale surfaced inside the exclusion zone, the observer will call for the immediate shut-down of the seismic operation and airgun firing (the vessel may continue on its course but all airgun discharges must cease). The vessel operator must comply immediately with such a call by an on-watch visual observer. Any disagreement or discussion should occur only after shut-down. When no whales are sighted for at least a 30-minute period, ramp-up of the source array may begin. Ramp-up cannot begin unless conditions allow the sea surface to be visually inspected for whales for 30 minutes prior to commencement of ramp-up (unless the method described in the section entitled "Experimental Passive Acoustic Monitoring" is used). Thus, ramp-up cannot begin after dark or in conditions that prohibit visual inspection (fog, rain, etc.) of the exclusion zone. Any shut-down due to a whale(s) sighting within the exclusion zone must be followed by a 30-minute all-clear period and then a standard, full ramp-up. Any shut-down for other reasons, including, but not limited to, mechanical or electronic failure, resulting in the cessation of the sound source for a period greater than 20 minutes, must also be followed by full ramp-up procedures. In recognition of occasional, short periods of the cessation of airgun firing for a variety of reasons, periods of airgun silence **not exceeding 20 minutes** in duration will not require ramp-up for the resumption of seismic operations if: (1) visual surveys are continued diligently throughout the silent period (requiring daylight and reasonable sighting conditions), and (2) no whales, other marine mammals, or sea turtles are observed in the exclusion zone. If whales, other marine mammals, or sea turtles are observed in the exclusion zone during the short silent period, resumption of seismic survey operations must be preceded by ramp-up.

Reporting

The importance of accurate and complete reporting of the results of the mitigation measures cannot be overstated. Only through diligent and careful reporting can the MMS, and subsequently NOAA Fisheries, determine the need for and effectiveness of mitigation measures. Information on observer effort and seismic operations are as important as animal sighting and behavior data. In order to accommodate various vessels' bridge practices and preferences, vessel operators and observers may design data reporting forms in whatever format they deem convenient and appropriate. Alternatively, observers or vessel operators may adopt the United Kingdom's Joint Nature Conservation Committee forms (available at their website http://www.jncc.gov.uk/). At a minimum, the following items should be recorded and included in reports to the MMS:

Observer Effort Report: Prepared for each day during which seismic acquisition operations are conducted. Furnish an observer effort report to MMS on the 1st and the 15th of each month that includes:
- Vessel name
- Observers' names and affiliations
- Survey type (e.g., site, 3D, 4D)

- MMS Permit Number (for "off-lease seismic surveys") or OCS Lease Number (for "on-lease seismic surveys")
- Date
- Time and lat./long. when daily visual survey began
- Time and lat./long. when daily visual survey ended
- Average environmental conditions while on visual survey, including

  - Wind speed and direction
  - Sea state (glassy, slight, choppy, rough or Beaufort scale)
  - Swell (low, medium, high or swell height in meters)
  - Overall visibility (poor, moderate, good)

Survey Report: Prepared for each day during which seismic acquisition operations are conducted and the airguns are being discharged. Furnish a survey report to MMS on the 1st and the 15th of each month during which operations are being conducted that includes

- Vessel name
- Survey type (e.g., site, 3D, 4D)
- MMS Permit Number (for "off-lease seismic surveys") or OCS Lease Number (for "on-lease seismic surveys")
- Date
- Time pre-ramp-up survey begins
- What marine mammals and sea turtles were seen during pre-ramp-up survey?
- Time ramp-up begins
- Were whales seen during ramp-up?
- Time airgun array is operating at the desired intensity
- What marine mammals and sea turtles were seen during survey?
- If whales were seen, was any action taken (i.e., survey delayed, guns shut down)?
- Reason that whales might not have been seen (e.g., swell, glare, fog)
- Time airgun array stops firing

Sighting Report: Prepared for each sighting of a marine mammal (whale or dolphin) or sea turtle made during seismic acquisition operations. Furnish a sighting report to MMS on the 1st and the 15th of each month during which operations are being conducted that includes

- Vessel name
- Survey type (e.g., site, 3D, 4D)
- MMS Permit Number (for "off-lease seismic surveys") or OCS Lease Number (for "on-lease seismic surveys")
- Date
- Time
- Watch status (Were you on watch or was this sighting made opportunistically by you or someone else?)
- Observer or person who made the sighting
- Lat./long. of vessel
- Bearing of vessel
- Bearing and estimated range to animal(s) at first sighting
- Water depth (meters)
- Species (or identification to lowest possible taxonomic level)
- Certainty of identification (sure, most likely, best guess)
- Total number of animals

- Number of juveniles
- Description (as many distinguishing features as possible of each individual seen, including length, shape, color and pattern, scars or marks, shape and size of dorsal fin, shape of head, and blow characteristics)
- Direction of animal's travel – compass direction
- Direction of animal's travel – related to the vessel (drawing preferably)
- Behavior (as explicit and detailed as possible; note any observed changes in behavior)
- Activity of vessel
- Airguns firing? (yes or no)
- Closest distance (meters) to animals from center of airgun or airgun array (whether firing or not)

Note: If this sighting was of a whale(s) within the exclusion zone that resulted in a shut-down of the airguns, include in the sighting report the observed behavior of the whale(s) before shut-down, the observed behavior following shut-down (specifically noting any change in behavior), and the length of time between shut-down and subsequent ramp-up to resume the seismic survey (note if seismic survey was not resumed as soon as possible following shut-down). Send this report to MMS **within 24 hours of the shut-down**. These sightings should also be included in the first regular semi-monthly report following the incident.

Additional information, important points, and comments are encouraged. All reports will be submitted to MMS on the 1st and the 15th of each month (with one exception noted above). Forms should be scanned (or data typed) and sent via email to protectedspecies@mms.gov. Please note that these marine mammal and sea turtle reports are in addition to any reports you submit under NTL No. 98-20, dated September 15, 1998, and NTL No. 2002-G01, effective March 15, 2002, and reports required as a condition of your geophysical permit.

Borehole Seismic Surveys

Borehole seismic surveys differ from surface seismic surveys in a number of ways, including the use of much smaller airgun arrays, having an average survey time of 12-24 hours, utilizing a sound source that is not usually moving at 4-5 knots, and requiring the capability of moving the receiver in the borehole between shots. Due to these differences, the following altered mitigations apply only to borehole seismic surveys:

- During daylight hours, when visual observations of the exclusion zone are being performed as required in this NTL, borehole seismic operations will not be required to ramp-up for shutdowns of 30 minutes or less in duration, as long as no whales, other marine mammals, or sea turtles are observed in the exclusion zone during the shutdown. If a whale, other marine mammal, or sea turtle is sighted in the exclusion zone, ramp-up is required and may begin only after visual surveys confirm that the exclusion zone has been clear for 30 minutes.
- During nighttime or when conditions prohibit visual observation of the exclusion zone, ramp-up will not be required for shutdowns of 20 minutes or less in duration. For borehole seismic surveys that utilize passive acoustics during nighttime and periods of poor visibility, ramp-up is not required for shutdowns of 30 minutes or less.
- Nighttime or poor visibility ramp-up is allowed only when passive acoustics are used to ensure that no whales are present in the exclusion zone (as for all other seismic surveys). Operators are strongly encouraged to acquire the survey in daylight hours when possible.

- Protected species observers must be used during daylight hours, as required in this NTL, and may be stationed either on the source boat or on the associated drilling rig or platform if a clear view of the sea surface in the exclusion zone and adjacent waters is available.
- All other mitigations and provisions for seismic surveys as set forth in this NTL will apply to borehole seismic surveys.
- Reports should reference OCS Lease Number, Area/Block and Borehole Number

Experimental Passive Acoustic Monitoring

Whales, especially sperm whales, are very vocal marine mammals, and periods of silence are usually short and most often occur when these animals are at the surface and may be detected using visual observers. However, sperm whales are at the greatest risk of potential injury from seismic airguns when they are submerged and under the airgun array. Passive acoustic monitoring appears to be very effective at detecting submerged and diving sperm whales, and some other marine mammal species, when they are not detectable by visual observation. The MMS strongly encourages operators to participate in an experimental program by including passive acoustic monitoring as part of the protected species observer program. Inclusion of passive acoustic monitoring does **not** relieve an operator of any of the mitigations (including visual observations) in this NTL **with the following exception**: Monitoring for whales with a passive acoustic array by an observer proficient in its use will allow ramp-up and the subsequent start of a seismic survey during times of reduced visibility (darkness, fog, rain, etc.) when such ramp-up otherwise would not be permitted using only visual observers. If you use passive acoustic monitoring, include an assessment of the usefulness, effectiveness, and problems encountered with the use of that method of marine mammal detection in the reports described in this NTL. A description of the passive acoustic system, the software used, and the monitoring plan should also be reported to MMS at the beginning of its use.

Paperwork Reduction Act of 1995 (PRA) Statement

The PRA (44 U.S.C. Chapter 35) requires us to inform you that we collect the information described in this NTL to ensure that you conduct operations in a manner that will not jeopardize threatened or endangered species or destroy or adversely modify critical habitat that has been designated for those species. We protect all proprietary information submitted according to the Freedom of Information Act and 30 CFR 250.196. An agency may not conduct or sponsor a collection of information unless it displays a currently valid Office of Management and Budget (OMB) control number. You are not obligated to respond until the OMB has approved this collection of information. We estimate the hour burden to be 24 hours per day for visual marine mammal observation, 8 hours for training each observer, 1 hour per marine mammal observation report, ½ hour for each submittal of training materials and certifications, and 1 hour for each voluntary submittal of experimental passive acoustic monitoring information. Direct comments regarding the burden or any other aspect of this information collection to the Information Collection Clearance Officer, Mail Stop 4230, Minerals Management Service, 1849 C Street, N.W., Washington, DC 20240.

In addition, this NTL refers to information collection requirements under 30 CFR 250, subpart B. The OMB has approved all of the information collection requirements in these regulations and assigned OMB control number 1010-0049.

Contact

Any questions regarding this NTL should be submitted in writing to:
protectedspecies@mms.gov.

Submittals by mail may be directed to:

Minerals Management Service, Gulf of Mexico OCS Region, Attention: Environmental Sciences
Unit (MS 5430), 1201 Elmwood Park Blvd., New Orleans, LA 70123-2394

Chris C. Oynes
Regional Director

Appendix B
**Implications of the Marine Mammal Protection Act on Geological
and Geophysical Survey Activities in the Gulf of Mexico**

Implications of the Marine Mammal Protection Act on Geological and Geophysical Survey Activities in the Gulf of Mexico

The purpose of this appendix is to outline applicable regulations (i.e., those parts of the U.S. Code [31 USC 1361 et seq.] and Code of Federal Regulations [50 CFR 216 et seq.]) that pertain to protection of marine mammals, and to discuss the implications of these regulations to geological and geophysical (G&G) activities in the Gulf of Mexico. Appropriate statutory or regulatory sections that are relevant to the following topics are identified, followed by a concise "plain language" summary.

I.   REVIEW OF PERTINENT REGULATIONS

A.   PROHIBITIONS ON THE "TAKING" OF MARINE MAMMALS

The term "take" means to harass, hunt, capture, or kill, or attempt to harass, hunt, capture, or kill any marine mammals. The Marine Mammal Protection Act (MMPA) (50 CFR 216.11) states that it is unlawful for

1) Any person, vessel, or other conveyance under the jurisdiction of the United States to "take" marine mammals on the high seas.

2) Any person, vessel, or other conveyance to "take" any marine mammal in water or on lands under the jurisdiction of the United States, unless otherwise expressly provided for by an international treaty, convention, or agreement to which the United States is a part before the date of the MMPA.

3) Any person to use any port, harbor, or other place under the jurisdiction of the United States to "take" or import marine mammals or marine mammal products or for any person to possess a mammal or a product from a marine mammal that was "taken" in violation of the MMPA.

Interested parties should also review 16 USC 1372 for regulations pertinent to a moratorium on the taking of marine mammals and 16 USC 1373 for regulations on taking of marine mammals.

B.   HARASSMENT DEFINITIONS

Under the 1994 amendments to the MMPA, Congress statutorily defined and divided the term "harassment" into two levels. Levels of harassment include Level A and Level B (16 USC 1362, 50 CFR 216.3); establishment of two harassment levels was designed to enable the U.S. Fish and Wildlife Service (USFWS) and the National Marine Fisheries Service (NMFS; also referred to as National Oceanic and Atmospheric Administration, Fisheries [NOAA-F]) to implement a general authorization for marine mammal research expected to result in nothing more than Level B harassment. Level A harassment pertains to any act of pursuit, torment, or annoyance, which has the potential to injure a marine mammal and/or group of marine mammals in the wild. Level B harassment has the potential to disturb (and not injure) a marine mammal and/or group of marine mammals in the wild by causing disruption of behavioral patterns, including, but not limited to, migration, breathing, nursing, breeding, feeding, or sheltering. To date, NMFS is still in the developmental process of determining the criteria to clearly define acoustic "take."[1]   Points of discussion pertinent to auditory thresholds center on the nature of sound and sound exposure (i.e., continuous, single pulse, single ping equivalent, repetitive), as well as species-specific sensitivities and the relative importance of avoidance behavior (e.g., to the sustainability of a marine mammal population).

C.   AUTHORIZATIONS OF SMALL, INCIDENTAL TAKES, AND THE GENERAL TERMS AND CONDITIONS OF SUCH AUTHORIZATIONS

Small take authorizations are issued for a specified activity (other than commercial fishing) within a specific geographical region only in situations where the "takings" are of small numbers, the "takings" have a negligible impact on the stocks of the marine mammals involved (including species that are endangered or threatened; see **Section I.D** of this appendix for further details), and there is no unmitigable impact on the availability of the affected species or stock for subsistence taking by Alaskan Natives. There are two types of "small take" authorizations:

1) A Letter of Authorization, or LOA (16 USC 1371[a][5][A] and 50 CFR 216.106); and

2) An Incidental Harassment Authorization, or IHA (16 USC 1371[a][5][D] and 50 CFR 216.107).

An LOA involves the incidental, but not intentional taking of a small number of marine mammals during a maximum period of 5 years, while an IHA involves the incidental, but not intentional taking by harassment of a small number of marine mammals during a maximum period of 1 year.

If the applicant can show that there is no potential for serious injury (i.e., any injury that will likely result in mortality), mortality, or behavioral disruptions that have non-negligible effects on growth, longevity, or reproduction, or the potential for serious injury or mortality can be negated through mitigation requirements that could be required under the authorization, an IHA is required. However, if these two stipulations cannot be met, the applicant requires an LOA.

D.   APPLICATION AND APPROVAL PROCESSES FOR AUTHORIZATIONS

Upon determining the type of authorization required, the applicant must submit a written request to the NMFS Office of Protected Resources and the appropriate NMFS Regional Office (i.e., the appropriate region where the specified activity is planned). The request must include items 1) through 14) (see the following discussion in **Section I.I** of this appendix for details) before being considered by NMFS. However, incomplete applications and applications that do not contain or refer to the necessary National Environmental Policy Act (NEPA) documentation (if applicable) will be returned to the applicant with an explanation.

When an application for a small take authorization is received, a summary of the application is published in the Federal Register, newspapers of general circulation, and appropriate electronic media in the coastal areas that may be affected by the specified activity. NMFS then invites any interested party (within 30 days following the publication of the summary) to submit their written data or views on the

---

[1] Recent Final Rules that address acoustic criteria include take of marine mammals incidental to 1) naval activities (i.e., shock testing; see NMFS, 2001b and Department of the Navy, 2001); 2) operation of a low frequency sound source by the North Pacific Acoustic Laboratory (acoustic thermometry of ocean climate [ATOC]; see NMFS, 2001c and associated documents); and operation of seismic sources in the northern Gulf of Mexico for research purposes (see NOAA-F, 2003 and associated documents).

"taking" proposed in the application. All information and suggestions will be considered by NMFS in developing the most effective regulations governing the issuance of an LOA or conditions governing the issuance of an IHA.

The Assistant Administrator shall then evaluate each request to determine, based on the best scientific data available, whether the taking by the specified activity within the specified geographic region will have a negligible impact on the species or stock and will not have an unmitigable adverse impact on the availability of such species or stock for subsistence uses by Alaskan Natives. If the Assistant Administrator finds that the mitigating measures would successfully negate the impacts of the specified activity, a finding of negligible impact will be made. Any such preliminary findings shall be proposed for public comment along with either the proposed IHA or LOA or the proposed regulations for the specified activity.

If, following the public review period, the Assistant Administrator finds that the taking by the specified activity would have more than a negligible impact on the species or stock or would have an unmitigable adverse impact on the availability of such species or stock for subsistence uses, the Assistant Administrator will publish the negative finding in the Federal Register along with the basis for denying the request.

For all petitions for regulations governing the taking of marine mammals, applicants must provide the information requested in 50 CFR 216.104 on their activity as a whole, which includes, but is not necessarily limited to, an assessment of total impacts by all persons conducting the activity. For allowed activities that may result in incidental takings of small numbers of marine mammals by harassment, serious injury, death, or a combination thereof, specific regulations will be established for each allowed activity that set forth the following:

1) The permissible methods of taking.

2) The means of effecting the least practicable adverse impact on the species and its habitat and on the availability of the species for subsistence uses.

3) The requirements for monitoring and reporting, including requirements for the independent peer-review of proposed monitoring plans where the proposed activity may affect the availability of a species or stock for taking for subsistence uses.

Regulations will be established based on the best available information. As new information is developed, through monitoring, reporting, or research, the regulations may be modified, in whole or in part, after notice and opportunity for public review.

E.   CONDITIONS GOVERNING AUTHORIZATIONS

As stated in **Section I.C** of this appendix, LOAs may be issued for periods up to 5 years, while IHAs may be issued for no more than one year; both an LOA and an IHA are renewable. An LOA can be withdrawn or suspended, either on an individual or class basis, as appropriate, if, after notice and opportunity for public comment, the Assistant Administrator determines (50 CFR 216.106) that

1) The regulations prescribed are not being substantially complied with; or

2) The taking allowed is having, or may have, more than a negligible impact on the species or stock; or

3) The taking allowed is having, or may have, an unmitigable adverse impact on the availability of the species or stock for subsistence uses.

An IHA can be modified, withdrawn, or suspended if, after notice and opportunity for public comment, the Assistant Administrator determines (50 CFR 216.107) that

1) The conditions and regulations prescribed in an authorization are not being substantially complied with; or

2) The authorized taking, either individually or in combination with other authorizations, is having, or may have, more than a negligible impact on the species or stock; or

3) The authorized taking, either individually or in combination with other authorizations, is having, or may have, an unmitigable adverse impact on the availability of the species or stock for subsistence uses.

If the "takings" have more than a negligible impact on the species or stocks of the marine mammals involved, then a waiver of the MMPA's moratorium on the taking of marine mammals may be issued, per 16 USC 1371(a)(3) and 16 USC 1373. The Secretary, on the basis of the best scientific evidence available and in consultation with the Marine Mammal Commission, is authorized and directed to waive the requirements of the MMPA provided that

1) The Secretary is assured that the taking is in accord with sound principles of resource protection and conservation as provided in the purposes and policies of the MMPA.

2) No marine mammal or no marine mammal product may be imported into the United States unless the Secretary certifies that the program for taking in the country of origin is consistent with the provisions and policies of the MMPA.

A waiver will not be issued for the taking of any marine mammal that has been designated by the Secretary as depleted, and no importation may be made of any such mammal, except for scientific research purposes, photography for educational or commercial purposes, or enhancing the survival or recovery of a species or stock as provided for in an LOA or IHA.

All of the requirements for a waiver are found in Section 16 USC 1371(a)(3) and 16 USC 1373. In order to waive the moratorium, the Secretary must prescribe limiting regulations with respect to the taking and importing of marine mammals as he deems necessary and appropriate to ensure that such taking will not be to the disadvantage of those species and population stocks, and will be consistent with the purposes and policies set forth in the MMPA. In prescribing such regulations, the Secretary shall give full consideration to all factors that may affect the extent to which marine mammals may be taken or imported, including but not limited to the effect of such regulations on

1) Existing and future levels of marine mammal species and population stock.

2) Existing international treaty and agreement obligations of the United States.

3) The marine ecosystem and related environmental considerations.

4) The conservation, development, and utilization of fishery resources.

5) The economic and technological feasibility of implementation.

The prescribed regulations may include, but are not limited to, restrictions with respect to

1) The number of marine mammals that may be taken or imported in any calendar year according to permits that authorize the taking or importation of marine mammals.

2) The age, size, or sex (or any combination thereof) of marine mammals that may be taken or imported.

3) The season or other period of time within which marine mammals may be taken or imported.

4) The manner and locations in which marine mammals may be taken or imported.

5) Fishing techniques that have been found to cause undue fatalities to any species of marine mammal in a fishery.

Regulations prescribed with respect to any species or stock of marine mammals must be made on the record after opportunity for an agency hearing on both the Secretary's determination to waive the moratorium pursuant to Section 103 of 16 USC 1373 and on any prescribed regulations. The Secretary shall also publish and make available to the public either before or in conjunction with the publication of notice in the Federal Register of his intention to prescribe regulations under Section 103 of 16 USC 1373 the following information:

1) A statement of the estimated existing levels of the species and population stocks of the marine mammal concerned.

2) A statement of the expected impact of the proposed regulations on the optimum sustainable population of such species or population stock.

3) A statement describing the evidence before the Secretary upon which he proposes to base such regulations.

4) Any studies made by or for the Secretary or any recommendations made by or for the Secretary or the Marine Mammal Commission that relate to the establishment of such regulations.

Any regulation prescribed according to Section 103 of 16 USC 1373 will be periodically reviewed, and may be modified from time to time in such manner as the Secretary deems consistent with and necessary to carry out the purposes of the MMPA. Every 12 months after the effective date of the MMPA (21 June 1973), the Secretary shall report to the public through publication in the Federal Register and to the Congress on the current status of all marine mammal species and population stocks subject to the provisions of the MMPA. The report will describe any actions taken and any measures believed necessary, including where appropriate, the issuance of permits pursuant to 16 USC 1373 to assure the well-being of the marine mammals subject to the provisions of the MMPA.

F.   OTHER CONSIDERATIONS

Generally, LOAs apply to a specific activity and can be issued to different holders, giving them authorization to take marine mammals incidental to the activity. IHAs are issued to a specific entity (usually the principal participant in the specific activity).

G.   PENALTIES

Violation of the MMPA, or any permit or regulation issued under it, is subject to a civil penalty of up to $10,000 per violation. Each unlawful "taking" or importation is a separate offense. Knowing violation of the MMPA, or any permit or regulation issued under it, is subject to a criminal penalty of up to $20,000 per violation and/or imprisonment for a maximum of 1 year (16 USC 1375, 50 CFR 216.40). Any vessel subject to U.S. jurisdiction used in the unlawful taking of a marine mammal is subject to seizure and forfeiture of its cargo (or the monetary value thereof) and liable for a civil penalty of up to $25,000 (16 USC 1376).

H.   TIMELINES

Simplified flow charts, with approximate timelines, are provided in **Figures B-1** (50 CFR 216.106) and **B-2** (50 CFR 216.107). These two figures reflect the various processes and typical time requirements involved in securing an LOA or IHA, respectively. Interested readers are also directed to 50 CFR 216.104.

I.   OVERVIEW

Each request for an IHA or an LOA must include the following 14 items before being considered by NMFS (16 USC 1374[b], 50 CFR 216.104):

1) A detailed description of the specific activity or class of activities that can be expected to result in incidental "taking" of marine mammals.

2) The date(s) and duration of such activity and the specific geographical region where it will occur.

3) The species and numbers of marine mammals likely to be found within the specific geographical region.

4) A description of the status, distribution, and seasonal distribution of the affected species or stock of marine mammals likely to be affected by the specified activity.

5) The type of incidental "taking" authorization that is being requested (e.g., "takes" by harassment only; "takes" by harassment, injury and/or death) and the method of incidental "taking."

6) A list, categorized by species, that includes the number, age, sex, and reproductive condition of marine mammals that may be "taken" by each type of "taking" identified in 5), above, and the number of times such "takings" by each type are likely to occur.

7) The anticipated impact of the activity upon the species or stock of marine mammals.

8) The anticipated impact of the activity on the availability of the species or stocks of marine mammals for subsistence uses (e.g., "taken" by Alaskan Natives for food, clothing, etc.).

9) The anticipated impact of the activity upon the habitat of the marine mammal populations, and the likelihood of restoration of the affected habitat.

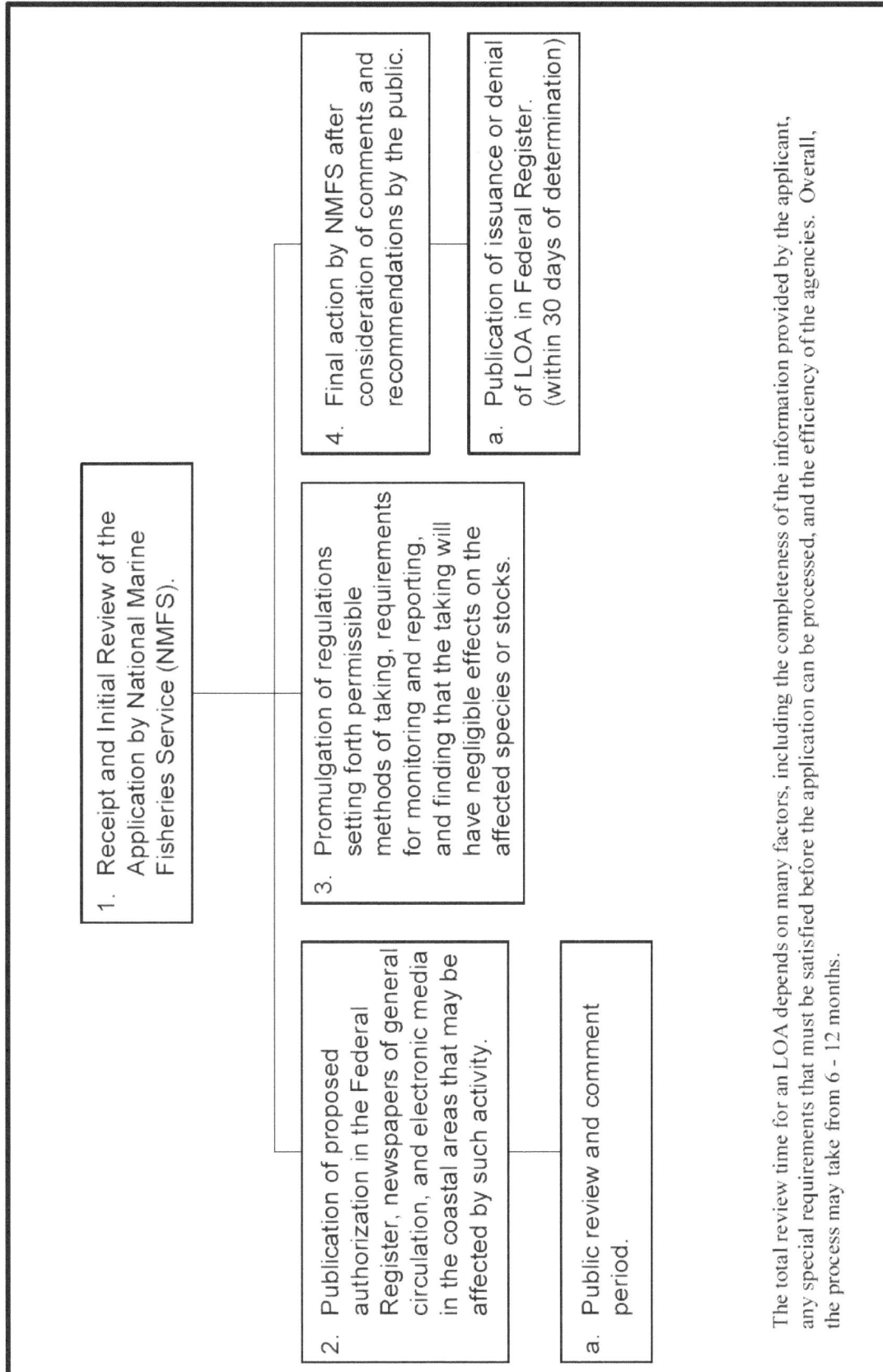

Figure B-1. Permit application review timeline for a Letter of Authorization (LOA) (Adapted from: Marine Mammal Commission, 2000).

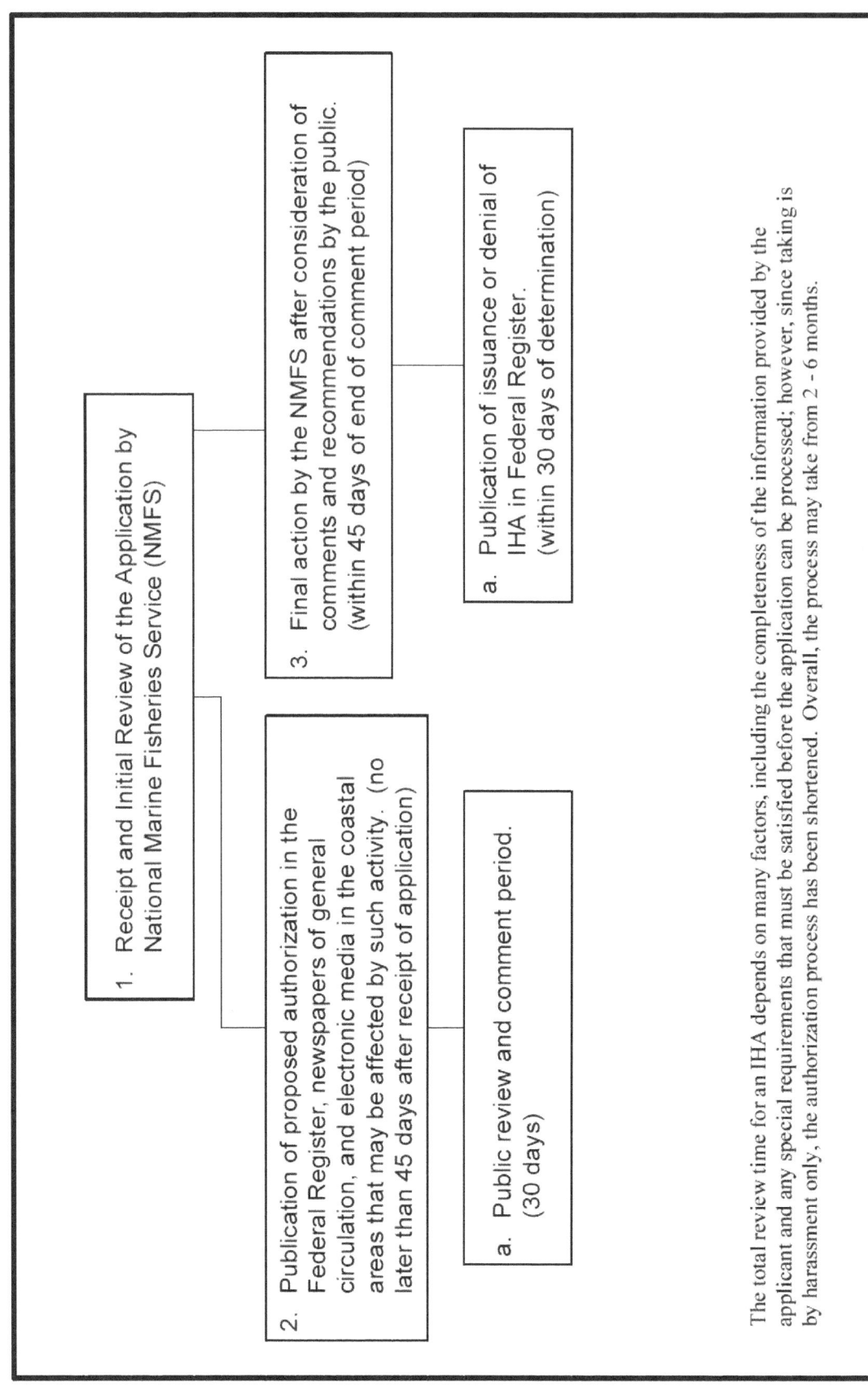

Figure B-2.   Permit application review timeline for an Incidental Harassment Authorization (IHA) (Adapted from: Marine Mammal Commission, 2000).

10) The anticipated impact of the loss or modification of the habitat on the marine mammal populations involved.

11) The economic and technological availability and feasibility of equipment, methods, and manner of conducting such activity or other means of effecting the least practicable adverse impact upon the affected species or stocks, their habitat, and on their availability for subsistence uses, paying particular attention to rookeries, mating grounds, and areas of similar significance.

12) Where the proposed activity would take place in or near a traditional arctic subsistence hunting area and/or may affect the availability of a species or stock of marine mammal for arctic subsistence uses, the applicant must submit either a plan of cooperation or information that identifies what measures have been taken and/or will be taken to minimize any adverse effects on the availability of marine mammals for subsistence uses. A plan of cooperation includes

   a) A statement that the applicant has notified and provided the affected subsistence community with a draft plan of cooperation.

   b) A schedule for meeting with the affected subsistence communities to discuss proposed activities and to resolve potential conflicts regarding any aspects of either the operation or the plan of cooperation.

   c) A description of what measures the applicant has taken and/or will take to ensure that proposed activities will not interfere with the affected communities, both prior to and while conducting the activity, to resolve conflicts and to notify the communities of any changes in the operation.

13) The suggested means of accomplishing the necessary monitoring and reporting that will result in increased knowledge of the species, the level of taking or impacts on populations of marine mammals that are expected to be present while conducting activities, and suggested means of minimizing burdens by coordinating such reporting requirements with other schemes already applicable to persons conducting such activity. Monitoring plans should include a description of the survey techniques that would be used to determine the movement and activity of marine mammals near the activity site(s), including migration and other habitat uses, such as feeding. Guidelines for developing a site-specific monitoring plan may be obtained by writing to the Director, Office of Protected Resources.

14) Suggested means of learning of, encouraging, and coordinating research opportunities, plans, and activities relating to reducing such incidental "taking" and evaluating its effects.

II. ENDANGERED SPECIES ACT CONSTRAINTS

Any taking of marine mammals listed as threatened or endangered under the Endangered Species Act (ESA) must be authorized under both the ESA and the MMPA. The ESA takes are authorized by either an Incidental Take Statement (ITS) under Section 7 (for Federal agency actions) or a Section 10 permit (for private citizens).

Section 7 of the ESA requires a Federal authorizing or action agency to consult with NMFS on any actions that might affect endangered or threatened species. If the agency or NMFS determines an action is likely to adversely affect a species, formal consultation is required. In such a situation, NMFS prepares a Biological Opinion (BO), which assesses whether the action is likely to jeopardize the existence of the species. The BO may include binding and/or discretionary recommendations to reduce impact. An ITS is attached to the BO as an appendix, and it is this statement that allows the incidental "take." An ITS cannot be authorized for a listed marine mammal until the MMPA authorization is completed.

A private citizen or State action not involving a Federal agency that would result in the incidental "taking" of an endangered or threatened species must be authorized under Section 10 of the ESA. In this situation, the applicant is required to develop a Conservation Plan that describes the action, evaluates the effect of the "take," and establishes the level of "take." The Federal action of NMFS issuing a permit under Section 10 is considered an action that itself requires Section 7 consultation. Also, an MMPA authorization is a Federal action requiring Section 7 consultation. As a result, normally NMFS waives a requirement for Section 10 if an authorization for a marine mammal is being requested, and that request includes marine mammal species listed under the ESA.

III.   NMFS AND ACOUSTIC TAKE

In 1995, NMFS formed an agency acoustic team as a direct result of the increasing numbers of requests for authorizations to "take" marine mammals from activities that produced sound. By 1998, the NMFS Office of Protected Resources added a person to the acoustics team to attend to pragmatic matters and to convene a workshop to gather information for new acoustic criteria in order to define "takes" from acoustic sources.

The Acoustics Program workshop was organized in order to complete the following tasks:

1) Draft acoustic criteria to clearly define acoustic "takes" under the MMPA.

2) Construct a network for monitoring ocean noise on a global basis.

3) Provide contact with other agencies, industry, professional societies, environmental non-governmental organizations (NGOs), and news media on acoustic matters.

4) Outline research that is needed to improve guidelines or regulations on acoustics.

5) Obtain additional funding for all aspects of the acoustics program, including research.

The workshop consisted of 10 acoustic experts who met for two and a half days to answer specific questions posed to them by the NMFS acoustics team. In addition, the general public was invited to attend and to also address the panel. The acoustic criteria are presently being drafted based on the workshop results and other information from the field of animal acoustics. At present, 16 USC 1371 Sections 101 (a)(5)(A) and (a)(5)(D) involving small take authorizations in a specified activity are the only citations that currently deal with this topic. Interested parties are also directed to the NMFS website (i.e., see http://www.nmfs.noaa.gov/prot_res/PR2/ Acoustics_Program/acoustics.html) for additional information.

A G&G operator would require an authorization for a "take" in any situation that involves harassing, hunting, capturing, or killing, or attempting to harass, hunt, capture, or kill any marine mammals. An authorization is also needed if any of the following occurs:

1) The collection of dead animals or their parts.

2) The restraint or retention of a marine mammal, no matter how temporary.

3) The tagging of a marine mammal.

4) The negligent or intentional operation of an aircraft or vessel, or of the doing of any other negligent or intentional act that results in the disturbing or molesting of a marine mammal.

Additional situations include any act of pursuit, torment, or annoyance that has the potential to injure a marine mammal and/or group of marine mammals in the wild, as well as harassment that has the potential to disturb (and not injure) a marine mammal and/or group of marine mammals in the wild by causing disruption of behavioral patterns, including, but not limited to, migration, breathing, nursing, breeding, feeding, or sheltering. Activities with potential to harass by noise include seismic airguns, ship and aircraft noise, high energy sonars, and explosives detonations.

IV.   PRELIMINARY FINDINGS RELATIVE TO MMPA VIOLATIONS

Significance criteria for marine mammals reflect their protected status under Federal law. One of the considerations in determining significance under NEPA is "whether the action threatens a violation of Federal, State, or local law or requirements imposed for the protection of the environment" (40 CFR 1508.27).

The ESA, as amended, prohibits jeopardizing endangered and threatened species or adversely modifying critical habitats essential to their survival. Section 7 of the Act requires consultation with NMFS and/or USFWS if the applicant has reason to believe that an endangered or threatened species may be present in the area affected by a proposed Federal action, and that implementation of the Proposed Action will likely affect such species.

As noted in **Section III.B.2 - Impacts of Routine Activities - Alternative 1 (Proposed Action)** of the Programmatic Environmental Assessment (PEA), the Proposed Action (Alternative 1) is not likely to result in deaths or mortal injuries of any listed marine mammals and therefore would not jeopardize the existence of any endangered species. However, it "may affect" a listed species (i.e., the sperm whale) through potential auditory impairment and/or behavioral disturbance as discussed below.

While the Proposed Action (Alternative 1) is not likely to result in deaths or mortal injuries of marine mammals in violation of the MMPA, the situation regarding harassment is not as clear. In practice, the NMFS has limited consideration of harassment to injury or behavioral changes of sufficient magnitude to have meaningful biological effects on the population, such as disrupting migrations. Examples are provided by several incidental take permits issued for seismic surveys in arctic waters, most recently for small take of marine mammals incidental to shallow water hazard activities in the Alaskan Beaufort Sea and seismic research operations in the U.S. Gulf of Mexico (NMFS, 2001a; NOAA-F, 2003). NMFS has also cited temporary threshold shift (TTS) as an example of an impact that could be considered harassment (60 FR 28379, May 31, 1995) and accepted the use of TTS as a harassment criterion in its Final Rule for the *Seawolf* (63 FR 66069, 1 December 1998) and *Winston S. Churchill* shock tests (66 FR 22450, 4 May 2001). A summary of the 160- and 180-dB "take" criteria is outlined in **Table B-1**.

While the use of a 180-dB (re 1 μPa-m [rms]) criterion for TTS for an impulse noise is the currently accepted level, it should be emphasized that there is a potential for behavioral effects from seismic noise at levels below 180 dB. Per the NMFS comments offered during Final G&G PEA

Table B-1
Summary of Take and Non-Take Criteria Relative to the 160- and 180-dB re 1 µPa Sound Exposure Levels

Level and Criterion	Source(s)
Level A Take - 180 dB re 1 µPa intermittent/continuous underwater sound	March 2001: Federal Register 66:15375-15394. National Marine Fisheries Service (NMFS) Proposed Rule (Subpart Q) to allow small takes of marine mammals incidental to Surveillance Towed-Array Sensor System/Low Frequency Active Sonar (SURTASS/LFAS). Establishes 180 dB re 1 µPa as the take threshold for this activity.
	August 2001: Federal Register 66:43442-43459. NMFS Final Rule (Subpart P) to allow small takes of marine mammals incidental to the acoustic thermometry of ocean climate (ATOC) project. Addresses acoustic energy impacts to marine mammals (see page 43447 for discussion of the 180-dB re 1 µPa criterion). Also, confirms the concept of some behavioral disruptions as being below a level that constitutes an incidental take (p. 43444).
	July 2002: NMFS published the SURTASS/LFAS Final Rule in Federal Register 67:46712-46789; 16 July 2002.
	May 2003: Federal Register 68:32460-32464. NMFS Draft Rule to allow small takes of marine mammals incidental to the seismic research activities of the *R/V Maurice Ewing*, operated by Lamont-Doherty Earth Observatory (LDEO); proposed operations slated for May-June 2003 in the northern Gulf of Mexico. Addresses acoustic energy impacts to marine mammals. LDEO proposed several airgun arrays (2-, 6-, 10-, 12-, and 20-airgun arrays). Approach: 1.5 times the 180-dB re 1 µPa radii predicted by modeling as the safety radii for cetaceans, or 75 and 1,425 m, respectively, for the 2 generator-injector (GI) guns and 20-airgun arrays.
Level B Take - 182 dB re 1 µPa and 12 psi - impulse sounds (underwater explosions)	February 2001: Final EIS for Shock Trial of the *Winston S. Churchill*. Navy EIS (Navy = lead agency; NMFS = cooperating agency). Provides a good discussion of acoustics issues, and establishes the dual criteria of 182 dB and/or 12 psi. See FEIS Appendix E for the detailed technical discussion.
	May 2001: Federal Register 66:22450-22467. NMFS Final Rule to allow small takes incidental to the *Winston S. Churchill* ship shock trials. These criteria are discussed throughout the comments/responses portion of the notice, but are formally established on p. 22464. A limitation of the 182-dB re 1 µPa criterion relative to seismic airguns is noted (see Federal Register 66:22450-22467, NMFS response to comment 13).

Table B-1
Summary of Take and Non-Take Criteria Relative to the 160- and 180-dB re 1 µPa Sound Exposure Levels
(Continued)

Level and Criterion	Source(s)
Level B Take - 160 dB re 1 µPa – intermittent/continuous underwater sound	July 2002: The Surveillance Towed-Array Sensor System Low Frequency Active (SURTASS LFA) Biological Opinion, and EIS (and incorporated in the rule making; see Federal Register 67:46712-46789; 16 July 2002) go into some detail about behavioral response at 160 dB re 1 µPa as well as the National Research Council's Low Frequency Sound and Marine Mammals. SURTASS LFA "Low Frequency Sound Scientific Research Program" controlled exposure experiments on blue, fin, gray, and humpback whales detected behavioral responses at estimated received levels between 120 to 155 dB re 1 µPa. Their acoustic modeling risk determination indicated that there was a 50% risk of behavioral response between 155 and 165 dB re 1 µPa. Prior to that, essentially research by Malme et al., 1983, 1984, 1988; Richardson et al., 1986, 1999; Richardson and Malme 1993; Dalheim and Ljungblad, 1990, indicates that the broad band levels of low-frequency sounds causing avoidance by about 50% of gray and bowhead whales in the areas of observation occurred when the received levels were around 115 to 120 dB re 1 µPa (water standard) for continuous sounds and about 160 to 170 dB re 1 µPa for the pulsed sounds. These received levels are overall levels, and the sources produced energy over a few octaves. Finally, the expert panel at the High Energy Seismic Survey (HESS) Workshop concluded that behavioral responses by marine mammals to seismic sounds would most likely occur at received levels above 140 dB re 1 re 1 µPa (rms). Throughout the discussions, there was continual reference to 190, 180, and 160 dB re 1 µPa, which represent Level A (pinnipeds), Level A (cetaceans), and Level B (cetaceans), respectively.
Non-Take Levels	August 2001: Federal Register 66:43442-43459. NMFS Final Rule (Subpart P) to allow small takes of marine mammals incidental to the ATOC project. Confirms the concept of some behavioral disruptions (such as a "momentary reaction") as being below a level that constitutes an incidental take (p. 43444).

development, the 180-dB criterion is a guidepost to ensure that a marine mammal is protected from unnecessary injury. For impulse noise, a guideline is established for a sound pressure level (SPL) of 160 dB (re 1 μPa-m [rms]) and for intermittent noise (e.g., offshore oil and gas drilling activities), an SPL of 120 dB (re 1 μPa-m [rms]) has been suggested for determining when behavioral impacts may result. While it has been recently shown that SPLs causing behavioral effects for marine mammals may be related to the location relative to the marine mammal activity, they provide preliminary guidance for when small take authorization requests should be considered.

It is noteworthy that incidental take permits issued by NMFS for Alaska and California offshore operations have been restrictive (see **Appendix E Section II.B - Possible Mitigation Measures and Operational Restrictions**), based on the presence of sensitive species and the potential for disruption of certain activities (i.e., migration, feeding). For the Gulf of Mexico, separate determinations are warranted (i.e., what activities may be affected by seismic operations, what is the importance of those affected activities relative to the individual and the population?).

Marine mammal life history and distribution summaries for Gulf of Mexico species have been provided in **Appendix F Section I.A - Marine Mammals**. No marine mammal species present in the Gulf of Mexico have documented or suggested migratory patterns, nor are there any designated critical habitats for marine mammals in Federal waters of this region. However, the presence of persistent sperm whale aggregations off the Mississippi River delta and their importance to this listed species remain to be fully studied. Similarly, the perceived hearing sensitivities of Bryde's and beaked whales (strategic stock) are not fully understood, suggesting further study is required for these species or species groups.

V.   OTHER FACTORS AND CONSIDERATIONS

Literal interpretation of the MMPA and the definitions of harassment suggest that there may be a technical violation of the law if sperm whales (a listed species) realize injurious auditory effects (e.g., TTS) or changes in behavior (e.g., avoidance behavior, moving away from a seismic noise source) from exposure to G&G surveys. However, it may also be argued that behavioral reactions to seismic noise are pragmatically below the level B "take" threshold of MMPA. The point of this discussion is to explore the ramifications of a lower threshold of the Level B "take"- one that considers the *de minimis* significance of any impact, and the apparent short duration of the impact.

Specifically, several factors must be considered in this analysis regarding the short term exposure of a marine mammal to seismic survey noise. These factors include

- Exposure duration, short term – both the seismic vessel/seismic sound source and the individual are moving; characteristics of the sound and sound exposure in the short term (i.e., on the order of minutes) must be considered;
- Exposure duration, long term – seismic survey activity is variable, with annual completion of up to 30+ surveys, Gulf-wide; on occasion, multiple vessel surveys may be undertaken, with up to five boats working concurrently; the ambient and anthropogenic noise characteristics of the broader environment (e.g., protraction areas, such as Mississippi Canyon; Planning Areas, such as the Central Planning Area) should be considered;
- Characteristics of the sound – for individuals at distance (i.e., not within several hundred meters of an array as the system is activated), there will be a gradual increase in received sound level by the individual as the seismic vessel approaches. At the point where the individual and the seismic array are closest, maximum received sound levels will be realized. Further, there will be a gradual decrease in received sound level as the vessel moves away;

- Ambient sound environment – comprehensive data specific to the Gulf of Mexico are lacking (see **Appendix F Section IV - Noise Environment**); sound sources present are represented by high volumes of commercial vessel traffic (tankers, cargo vessels, >10,000 dead weight tonnage), outer continental shelf oil and gas support operations, and commercial and recreational fishing activity (mobile sound sources), as well as several thousand fixed sound sources (i.e., platforms). Ambient and anthropogenic noise is quite variable throughout the Gulf;
- Ability to move – life history information suggests that marine mammals' response to noise is extremely variable, with anecdotal observations and scientific findings ranging from inquisitive investigation to avoidance. To date, there are no data that associate either decreases in abundance or strandings with seismic survey activity levels;
- Adequate warning – ramp-up, as a required operational measure, is expected to serve as a reliable mechanism to alert those marine mammals potentially at auditory risk of pending seismic operations; and
- Mitigation or operational restrictions – visual and/or passive acoustic monitoring offer additional mechanisms for monitoring of a predetermined impact zone, with inherent limitations evident in each; little benefit was noted among the operational restrictions considered in this analysis.

Based on discussions with MMS and NMFS, there are several aspects to consider in addressing this issue, including (but not necessarily limited to) MMPA regulatory compliance, the Minerals Management Service (MMS) NEPA process (i.e., categorical exclusion [CATEX] or EA, and determination of environmental impact:

- Compliance with MMPA – means either the action has no potential for a "take" to occur, or the applicant has authorization from NMFS (e.g., IHA or LOA) for incidental but unintentional take as a consequence of the action. Compliance with MMPA does not necessarily equate to "no take." Take can still occur with proper authorization (IHA, LOA) and the potential to mitigate the take still exists.
- The MMS NEPA process relative to most MMS-permitted G&G activities generally proceeds as a CATEX. However, MMS notes that the possibility of a violation under MMPA (or other laws or regulations) requires preparation of an EA.
- Environmental impact, while based on threshold levels and species- or resource-specific significance criteria (see **PEA Section III.A.1 - Significance Criteria**), must also consider the issue of "take." Even if an applicant has an LOA or IHA, and is authorized, a "take" could still harm a cetacean.

The potential of such a take is the core of the question. Further, one of the goals of this analysis should be a discussion of how G&G surveys should not violate the MMPA by causing impacts that result in a "take." Any activity that violates the MMPA will require MMS to review the G&G permit application via an EA, rather than as a CATEX.

To characterize those situations that can cause a violation of MMPA (i.e., "take"), this PEA has compiled the following elements for consideration:

- the spectrum of noise intensity and frequencies from G&G activities (i.e., typical source levels of 240 dB re 1 μPa [zero-to-peak], or 210 dB re 1 μPa [rms], depending upon the array effect and zero-to-peak to rms conversion; predominantly low frequency energy, with diminished energies in mid- to high frequencies; see **Appendices C [Sections II and III], D [Section II], and G**);

- identification of cetacean species for which hearing in each frequency range is especially important; in this analysis, three species or species groups have been identified as being of concern – sperm whales (listed species), Bryde's whales, and beaked whales (strategic stock);
- identification of areas of concentration of species or species groups of concern (i.e., sperm whale congregations over slope environments off the Mississippi River delta and in similar depths approximately 300 km east and offshore of the Texas-Mexico border; Bryde's whale occurrence in the Eastern Gulf Planning Area; no site preference noted for beaked whales); based on available data, no migratory routes are known or suggested, while areas of apparent congregation have been noted (i.e., based on current level of knowledge) only for the sperm whale;
- delineation of the area(s) affected by low frequency seismic survey noises, considering both the survey area, and the distances that these noises are broadcast at a level sufficient to cause a Level B harassment ("take"); calculations presented in **PEA Section II.C.e - Predetermined Impact Zone** and **Appendix C, Section III.A - Horizontal Distance of the 180-dB Isopleth** and **Appendix C, Section III.B - Horizontal Distance of the 160 - dB Isopleth** suggest that the 160- and 180-dB re 1 μPa (rms) levels conservatively extend into surface and near surface waters approximately 3,000 and 300 m from the airgun array, respectively. Simplifying assumptions regarding these calculations are also outlined in **Section III** of **Appendix C**.

NMFS has been working with the marine mammal research community and other agencies for several years in an attempt to compile appropriate acoustic criteria. New draft acoustic criteria (i.e., noise exposure criteria) continue to be developed by NOAA-F's expert panel (i.e., noise exposure criteria group); the panel is expected to release its recommended noise exposure criteria in the 2004-2005 timeframe. In the absence of NMFS guidelines on acoustic take, the permit history (i.e., for LOAs or IHAs) has been reviewed (see **Section IV** of this appendix). While there are slight variations in these criteria from one location to another, in general, NMFS has approved the use of impact zones (termed "safety zones"), which correspond to the estimated 190-dB and 180-dB re 1 μPa (rms) isopleths for pinnipeds and cetaceans, respectively. There are also differences in the duration of sound exposure (i.e., single impulsive vs. continuous; low duty cycle vs. single ping equivalent; see NMFS, 2001b,c for additional discussion). In the present analysis, both 160- and 180-dB re 1 μPa (rms) isopleths have been used in calculations of a cetacean impact zone, as noted previously. These isopleths are based on a review of recent take authorizations, conforming to accepted acoustic criteria.

There is also considerable variability regarding perceived hearing capabilities among marine mammals (see **PEA Figure III-1**). Those species or species groups perceived to be at risk were subsequently identified, with proper qualification of 1) assumptions inherent in equating vocalizations to hearing capability, and 2) missing data for a measurable proportion of the world's marine mammals.

The characteristics of seismic survey sound were also described (see **Appendices C [Sections II and III], D [Section II], and G**). Given the propagation characteristics of low and high frequency sound (i.e., attenuation of high frequency sound is greater; low frequencies travel farther than their high frequency counterpart) and the relative energy levels of these frequencies (i.e., high energy levels for low frequencies, <1,000 Hz; considerably lower energy levels for high frequencies), the low frequencies are of greatest concern. This concern has been expressed by various entities, including MMS, NMFS, and the National Research Council (1994, 2000, 2003).

Cetacean behavior in the presence of anthropogenic noise and vessel activity is variable. Scientific data and anecdotal observations are such that some species avoid seismic operations (e.g., gray whales, bowhead whales), while others (e.g., delphinids) may approach a seismic vessel during a survey to ride the bow wave. It remains unclear whether harm may occur to individuals (from exposure to seismic survey noise) under these circumstances.

Finally, consideration has also been given to the cumulative effects of seismic survey noise on marine mammals of the Gulf. The noise environment of the Gulf has been summarized in **Appendix F Section IV - Noise Environment**, based on available data. Comprehensive data on the Gulf noise environment are lacking. As an alternative approach, the relative level of activity and noise characteristics of major vessel operations in the Gulf were compiled. In a cumulative sense, seismic survey activities represent a potentially significant yet transient component of the overall noise environment. While seismic survey activities constitute a very minor portion of total vessel activity in the Gulf, such surveys do produce repetitive, mobile, and short-term increases in ambient noise levels. The period between potential exposure ranges from hours to days (i.e., time between separate passes of a seismic survey vessel). In the near-field, within several hundred meters or so of an array, received sound levels may reach or exceed 180 dB re 1 µPa (rms). At greater distances, sound from a seismic survey is of a similar nature to other commercial vessel activity.

VI. PROMULGATION OF REGULATIONS REGARDING ACOUSTIC TAKE

Resolution of the "take" question vis-à-vis future G&G operations in the Gulf of Mexico lies with the promulgation of regulations, or in a more informal mode, development of guidelines. MMS has requested preliminary rulemaking (i.e., NMFS promulgation of regulations) for the incidental take of marine mammals resulting from G&G activities in the Gulf of Mexico. The MMS request for rulemaking forges a new subpart of the MMPA regulations and is expected to result in the issuance of an LOA (R. Defenbaugh, MMS, oral comm., 2003). MMS has used the draft version of the G&G PEA and its accompanying finding of no significant impact (FONSI) as supporting documentation.

A fundamental need in considering this option lies with development of acoustic take criteria. At present, NMFS continues with its efforts to establish species- or group-specific take criteria for impulsive sound (R. Gentry, NOAA-F, oral comm., 2003). Such criteria, termed guidelines by NMFS, are still being developed, and the current timeline for completion remains unclear.[2] Once NMFS has reached a point where draft acoustic take criteria have been finalized, the agency will solicit comment from its group of independent acoustic experts. Revisions to those guidelines may be expected. Once this process is complete, NMFS intends to utilize those guidelines in its review of project-specific applications (as it has done in recent IHA and LOA reviews). Whether or not such guidelines (not regulations) go through the normal public notification process (i.e., publication in the Federal Register) remains to be determined; however, NMFS does expect that the new guidelines for acoustic take will realize a broad distribution. According to NMFS, new acoustic take guidelines may become formal regulations in the future, depending upon the results of current and future research in this area.

---

[2] NMFS has noted that acoustic take criteria are nearing completion. A major question remaining to be addressed is whether single impulse criteria (e.g., 180 dB re 1µPa [rms] for cetaceans) is applicable to repetitive impulses (i.e., as might be experienced with exposure to seismic noise or low frequency active sonar). The U.S. Navy (Office of Naval Research) is also summarizing for NMFS the current state of knowledge regarding experimental work with impulsive sound and resulting tissue damage attributed to resonance phenomena (see **Appendix G**); a workshop to discuss this issue was held in October 2001, following a planning session on this issue in July 2001.

VII. LITERATURE CITED

Dahlheim, M.E., and D.K. Ljungblad. 1990. Preliminary hearing study on gray whales, (*Eschrichtius robustus*), in the field, pp. 335-346. *In*: J. Thomas and R. Kastelein (eds.), Sensory abilities of cetaceans: laboratory and field evidence. Plenum Press, NY.

Department of the Navy. 2001. Final environmental impact statement: Shock trial of the WINSTON S. CHURCHILL (DDG 81). Southern Division, Naval Facilities Engineering Command, North Charleston, SC. February 2001. 229 pp. + apps.

Malme, C.I., P.R. Miles, C.W. Clarke, P. Tyack, and J.E. Bird. 1983. Investigations of the potential effects of underwater noise from petroleum industry activities on migrating gray whale behavior. BBN Report 5366. Report by Bolt, Beranek, and Newman Inc., Cambridge, MA, for the U.S. Department of the Interior, Minerals Management Service, Anchorage, AK. NTIS PB86-174174.

Malme, C.I., P.R. Miles, C.W. Clark, P. Tyack, and J.E. Bird. 1984. Investigations of the potential effects of underwater noise from petroleum industry activities on migrating gray whale behavior/Phase II: January 1984 migration. BBN Report 5586. Report by Bolt, Beranek, and Newman Inc., Cambridge, MA, for the U.S. Department of the Interior, Minerals Management Service, Anchorage, AK. NTIS PB86-218377.

Malme, C.I., B. Würsig, J.E. Bird, and P. Tyack. 1988. Observations of feeding gray whale responses to controlled industrial noise exposure, pp. 55-73. *In*: W.M. Sackinger et al. (eds.), Port and ocean engineering under arctic conditions, vol II. Geophys. Inst., Univ Alaska, Fairbanks. 111 pp.

Marine Mammal Commission. 2000. Annual Report to Congress, 2000. Marine Mammal Commission, Washington, D.C.

National Marine Fisheries Service (NMFS). 2001a. Small take of marine mammals incidental to specified activities; shallow water hazard activities in the Beaufort Sea. Request for comments. Federal Register 66(104):29287-29293.

National Marine Fisheries Service (NMFS). 2001b. Taking Marine Mammals Incidental to Naval Activities. Federal Register 66(87):22450-22467.

National Marine Fisheries Service (NMFS). 2001c. Taking Marine Mammals Incidental to Operation of a Low Frequency Sound Source by the North Pacific Acoustic Laboratory. Federal Register 66(160):43442-43459.

National Oceanic and Atmospheric Administration Fisheries (NOAA-F). 2003. Small Takes of Marine Mammals Incidental to Specified Activities; Marine Seismic Testing in the Northern Gulf of Mexico. Federal Register 68(104):32460-32464.

National Research Council. 1994. Low-frequency sound and marine mammals: Current knowledge and research needs. National Academy Press, Washington, D.C. 75 pp.

National Research Council. 2000. Marine mammals and low-frequency sound: Progress since 1994. Prepublication copy. National Academy Press, Washington, D.C. 134 pp.

National Research Council. 2003. Ocean Noise and Marine Mammals. National Academy Press, Washington, D.C. 204 pp.

Richardson, W.J., and C.I. Malme. 1993. Man made noise and behavioral responses, pp. 631-700. *In*: The Bowhead Whale. Spec Publ. 2. Society for Marine Mammology, Lawrence KS, 787 pp.

Richardson, W.J., B. Würsig, and C.R. Greene. 1986. Reaction of bowhead whales to seismic exploration in the Canadian Beaufort Sea. J. Acoust. Soc. Am. 79(4):1117-1128.

Richardson, W.J., G.W. Miller, and C.R. Greene, Jr. 1999. Displacement of migrating bowhead whales by sounds from seismic surveys in shallow waters of the Beaufort Sea. J. Acoust. Soc. Am. 106(4,Pt.2):2281.

Appendix C
Introduction to Acoustics and Seismic Surveys

Introduction to Acoustics and Seismic Surveys

This appendix describes, in lay terms, the mechanisms by which sound is produced during geological and geophysical (G&G) operations. Particular attention is paid to defining proper terminology and units.

I.   SEISMIC SURVEYS

Seismic surveys are the primary means by which G&G contractors explore the seabed for oil and gas reserves. The fundamentals of the process are basically those of a sonar (i.e., a projector sends out sound waves and receivers "listen" for returning echoes of those sound waves). In the case of a simple submarine sonar, a sonar ping may be projected horizontally into the water column, which will reflect off objects (such as the hulls of other submarines), and can be received by sensitive listening equipment (hydrophones) aboard the first submarine. This is the basic "source" and "receiver" system.

Seismic surveys also use this principle of a source and receiver system. The source is typically an array of airguns, which are towed behind a survey vessel, usually at distances in the order of 100 m astern (see **Appendix D**, **Figure D-2**). Airguns are essentially small compressed air chambers that, upon command, can be made to vent their compressed air underwater in a single, rapid event. This pressure release event creates an acoustic pressure pulse, and a bubble of air that rapidly expands and contracts. The bubble will oscillate in volume, creating additional pressure events, and hence a single gun does not normally produce the desired level or characteristic of sound pulse. Multiple guns of varying size, mounted in an array, are used to offset this effect. Guns within the array may be arranged in clusters, which has the effect of simulating a single gun with larger chamber volume than any individual gun. Clusters of airguns are suspended from a float system, typically referred to as a sub-array. A number of laterally spaced floats forms the entire array. Firing all guns in the array boosts the sound levels, and creates a primary low frequency sound pulse directed towards the seabed (i.e., see **Appendix D**, **Figure D-4**).

These sound pulses penetrate the seabed and will reflect strongly off major discontinuities in the rock strata, such as oil and gas pockets. These reflected sounds are received by arrays of hydrophones towed behind the survey vessel in long streamers. Complex computer processing is then undertaken to interpret these "echo-returns" and construct a sub-surface map of the seabed. A seismic survey operation is illustrated in **Figure C-1**. It is worth noting that, although the primary low frequency pulse is directed towards the seabed, both this and higher frequency sounds generated by the guns also transmit into the wider water column.

II.   UNDERWATER ACOUSTIC MEASUREMENTS IN BRIEF

A.   A FEW FUNDAMENTALS

Underwater sound is essentially the transmission of energy via compression and rarefaction of particles in the conducting medium (i.e., in this case, seawater). The pressure pulse from a sound source (such as an airgun) propagates outwards in an expanding spherical shell at approximately 1,500 m/sec (in seawater). As the shell expands, the energy contained within it is dispersed across an ever increasing surface area, and the energy per unit area decreases in proportion to the square of the distance travelled from source. This is known as the inverse square law, or more generally as spherical spreading. In addition, there may be losses due to molecular absorption of energy as the particles in the medium

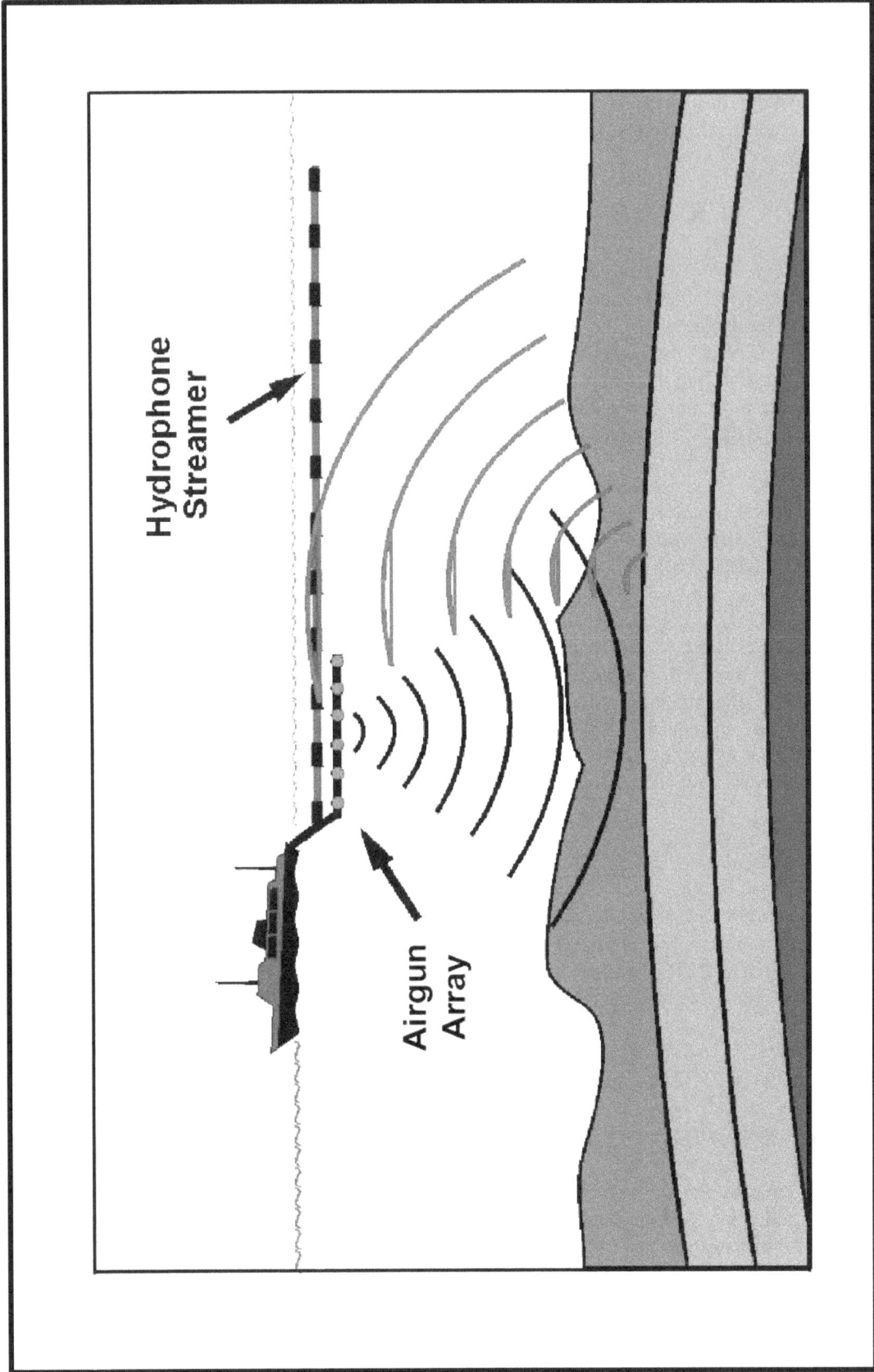

Figure C-1. Illustration of a seismic survey operation (figure courtesy of John C. Goold).

oscillate. These effects are small at low frequencies (<1 kHz), but become quite significant at high frequencies (>1 kHz).

Measurement of underwater sound levels has historically been complicated by a system of inconsistent and confusing units. It is not uncommon for technical or scientific literature to complicate the problem, using sound units that are poorly defined, misquoted, or simply not defined at all.

Sound waves are normally multiple, repeating oscillations of the compressive and rarefactive wave passing through the medium. A simple sound wave has a clearly defined amplitude and a frequency. The term "wave" itself gives an indication as to the form in which the signal is represented. It is worth illustrating this in terms of voltage on an oscilloscope, especially as sound waves are received and transduced by hydrophones as electrical signals.

B. PEAK LEVELS AND PEAK-TO-PEAK LEVELS

Figure C-2(a) shows a simple sine wave. The peak amplitude of the signal is defined as the amplitude of the wave peak above the zero crossing point (i.e., zero-to-peak), and is 1 volt in this case. The peak-to-peak amplitude of the signal is defined as the amplitude difference between the positive peak and the negative peak, which is 2 volts in this case. There are three complete cycles of this oscillating signal in a period of 1 second, and as such, this signal is defined to have a frequency of 3 Hz (3 cycles per second). A signal with the same amplitude, but at a frequency of 10 Hz, is illustrated in **Figure C-2(b)**, to clarify the concept of frequency.

Acoustic amplitude is expressed on a pressure, rather than voltage scale, but the principle and the graphical representation are identical to the electrical analogy (i.e., voltage amplitude is replaced by pressure amplitude on the vertical axis). Further, hydrophones convert pressure to voltage, so it is actually an electrical signal that is examined. Acoustic amplitudes of interest, both in air and underwater, cover such a large range that it is common to express them on a logarithmic scale, known as the decibel scale. A dB is a dimensionless unit that simply expresses the ratio of one quantity to a reference quantity on a logarithmic scale. In acoustic terminology, the dB may be the ratio of a received pressure level to a reference pressure level and is commonly used as a measurement of sound pressure levels (SPLs), thus

- $SPL = 20\log_{10}(\text{received pressure level/reference pressure level})$

As with the representation in **Figure C-2**, the received peak pressure level of a simple sound wave would be the peak amplitude of the sine wave. In underwater acoustics, the reference pressure has been standardized as 1 μPa. For example, a received sound pressure level of 1,000 μPa would be expressed as 60 dB re 1 μPa [20log(1000/1)]. The peak-to-peak level would be 66 dB re 1 μPa [20log(2000/1)]. For simple signals, peak-to-peak levels compute as 6 dB greater than zero-to-peak levels.

C. rms LEVELS

In descriptions of seismic survey noises and noise impacts, it is common (although far from uniform) to see the term "rms" quoted. This notation stands for root-mean-square, and its purpose is linked to the derivation of power measurements from oscillating signals. This can best be illustrated with the electrical analogy. For a simple sine wave signal, the rms amplitude is the peak amplitude divided by the square root of 2. In the case of a simple sine wave of 1-volt peak amplitude, the rms voltage is 0.707 volts, as illustrated in **Figure C-3**.

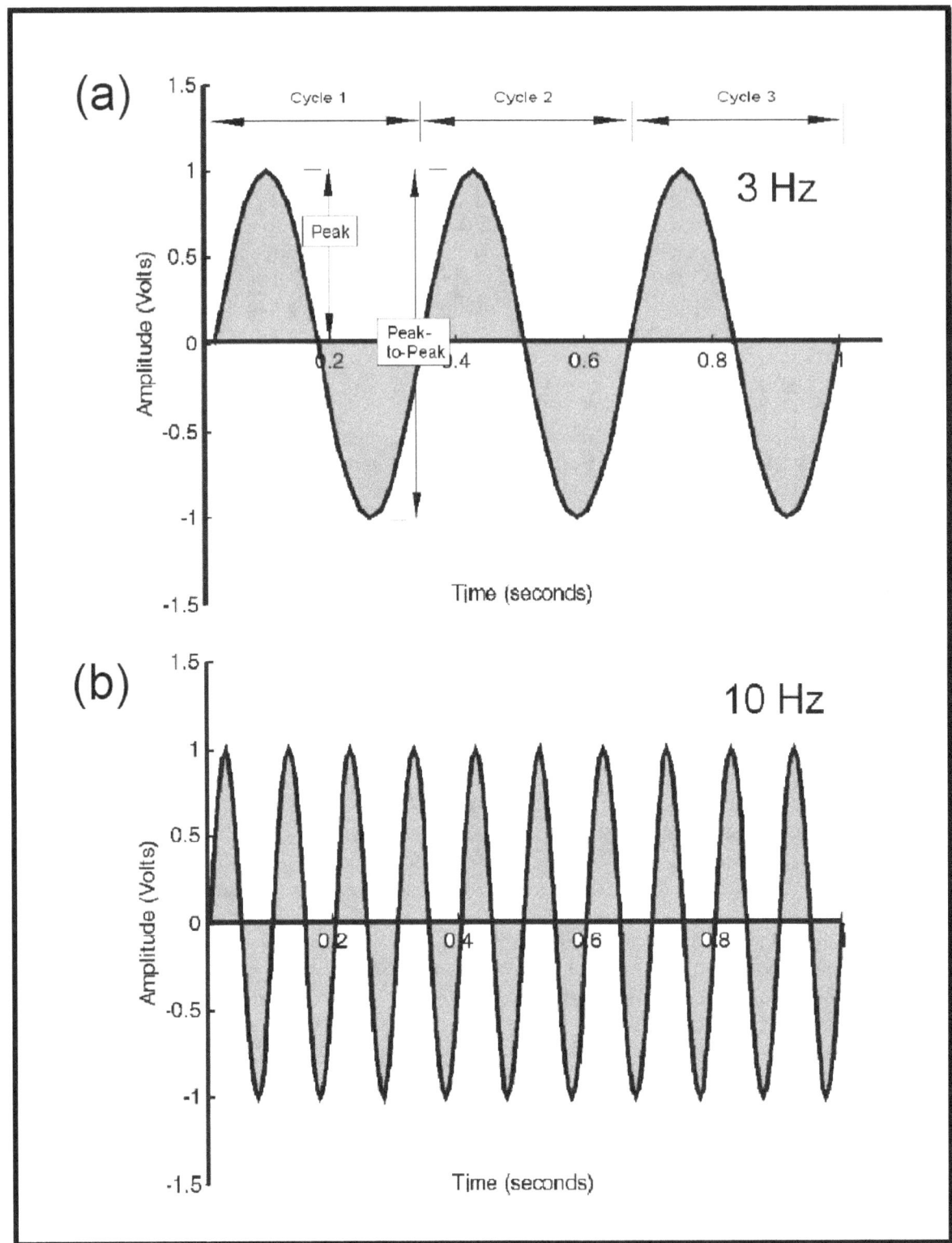

Figure C-2. Sine wave with (a) peak amplitude of 1 v, peak-to-peak amplitude of 2 v, and frequency of 3 Hz, and (b) peak amplitude of 1 v, peak-to-peak amplitude of 2 v, and frequency of 10 Hz (figure courtesy of John C. Goold).

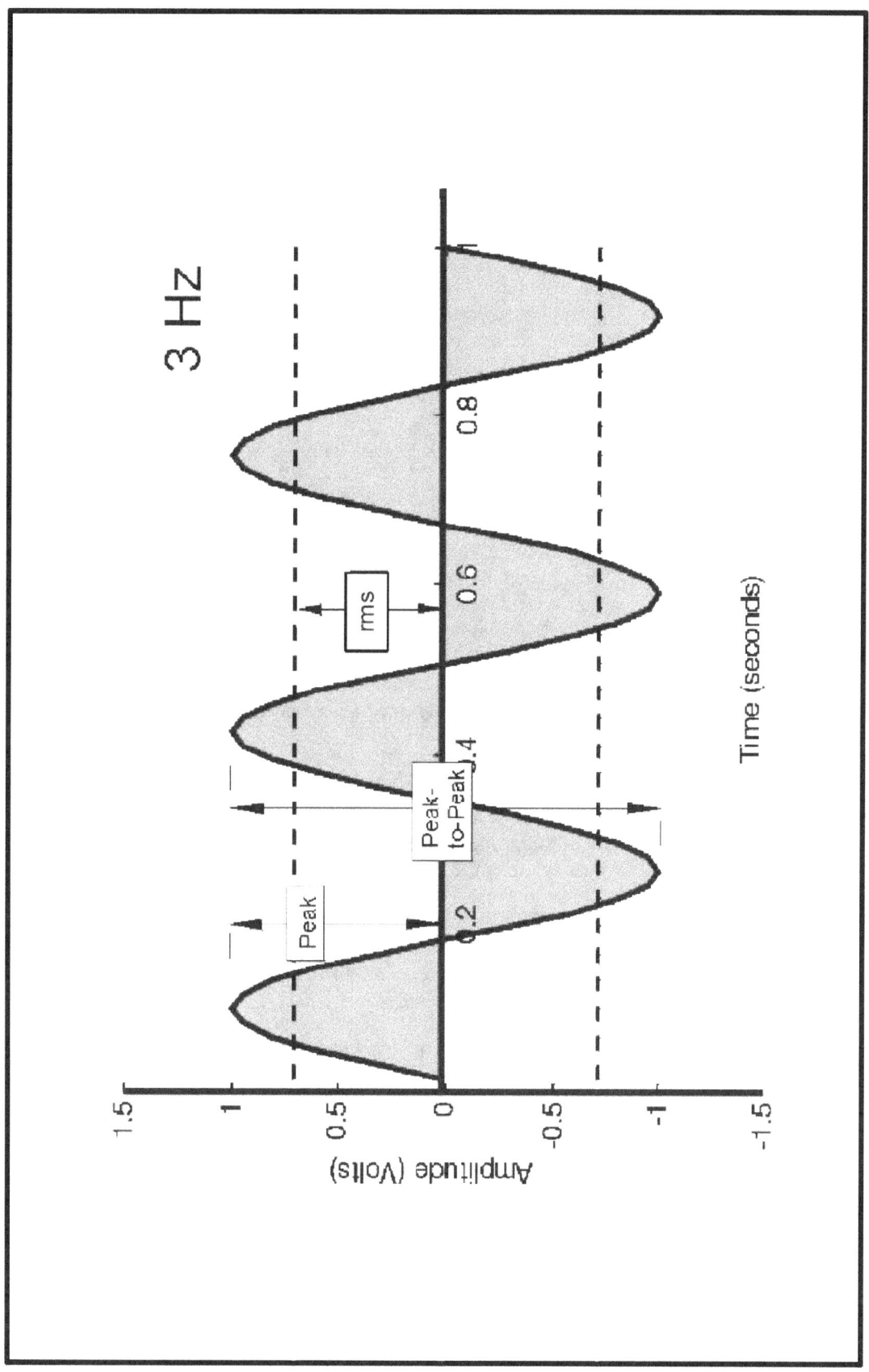

Figure C-3. Sine wave with peak amplitude of 1 v, peak-to-peak amplitude of 2 v, and frequency of 3 Hz. An rms level of 0.707 v is indicated (figure courtesy of John C. Goold).

In terms of domestic electricity generation, the product of rms voltage and rms current is used to measure the power delivery (e.g., watts). In the United Kingdom, mains electricity supply is 240 volts at 50 Hz; in the United States it is 110 volts at 60 Hz. These voltages are in fact rms values, and the true peak voltages are 339 volts and 155 volts, respectively, and would appear something like the representations in **Figure C-4** if compared on an oscilloscope. Peak voltages are recovered by multiplying the rms value by the square root of 2.

The concept is similar, but somewhat more complex, in terms of measuring rms values of airgun output from seismic surveys. The actual definition of rms is the square root of the average of the squares of a set of numbers or quantities, and it is in this form that it is more useful in the derivation of power levels from complex signals such as seismic survey pulses. Seismic survey pulses are short, generally non-periodic waveforms that do not lend themselves to the simple rms calculation. Although low frequencies dominate, seismic pulses are a mix of sounds at different frequencies and may be characterized through spectral density analysis. If the correct routines are chosen, this can be used to decompose the pulse into the energy levels of its constituent spectral components, and thereby some measure of rms values obtained. However, given the dominance of the low frequency pressure spike, there is a tendency to read peak pressures directly in terms of signal amplitude.

Calculations of energy and rms levels should also take into account pulse durations, which are typically much less than 1 second. Power is defined as energy per second. Strictly speaking, for complex waveforms, the rms levels will depend upon the individual waveform characteristics as devolved through fourier analysis, and cannot be easily generalized.

By way of crude example, **Figure C-5** illustrates a signal that is a composite of three pure sine waves, each of 1 volt peak amplitude, at different frequencies, and the spectral decomposition of the signal. Three distinct peaks are evident, which describe the three components of the original signal. There is a small amount of spectral "smearing," but when the values under the peaks are summed, they yield the rms values of the original sine waves.

If the signal in **Figure C-5** is truncated and enveloped into a short pulse, it results in a signal as shown **Figure C-6**. This signal is not dissimilar to a seismic survey pulse. The energy within the pulse is much reduced. As a result, the spectral decomposition yields much lower rms values - i.e. a longer set of sine waves of much lower amplitude would deliver the same energy as the short, high amplitude pulse. In this case, the major energy within the pulse occurs in about 1/100th of the time occupied by non-pulsed signal in **Figure C-5**. As a result, the spectral estimates are about 1/100th the magnitude, or 20 dB less. There is also considerably more spectral smearing due to the discontinuous, pulsed nature of the signal. In marine mammal sound exposure and impact studies, a quantity known as Sound Exposure Level may be used, which is analogous to this example of a short pulsed signal.

It is worth noting at this point that dBs can also be used to express ratios of energy or power in acoustics. In this form it is typical to see representations of a measured intensity to a reference intensity. The intensity of an acoustic wave is the amount of energy passing, per second, through a cross sectional area of 1 m squared, and is proportional to the square of the signal amplitude. Due to the amplitude squared relationship, the multiplier in the dB equation is 10, as opposed to 20 in the case of sound pressure level. Sound intensity in dB is expressed thus:

- $10\log_{10}$(measured intensity/reference intensity)

The reference intensity is the intensity of a plane wave of rms pressure equal to 1 μPa. Strictly speaking, power or intensity measurements in dBs should carry the dB re 1 μPa^2 notation, but this is often omitted.

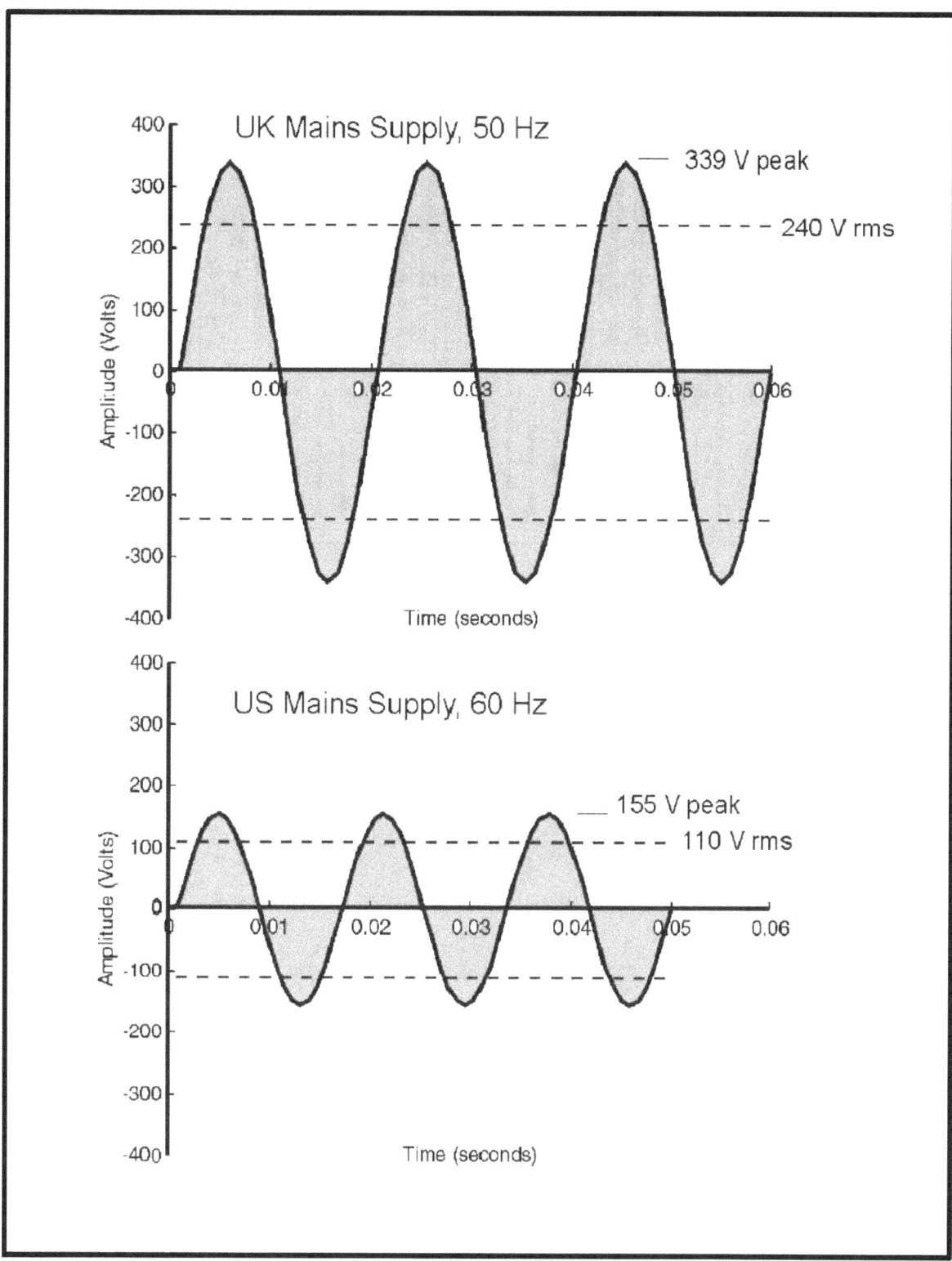

Figure C-4. Comparison of UK and US mains voltage supplies. The commonly accepted values of 240 and 110 v are in fact rms values. The peaks occur at 339 and 155 v, respectively (figure courtesy of John C. Goold).

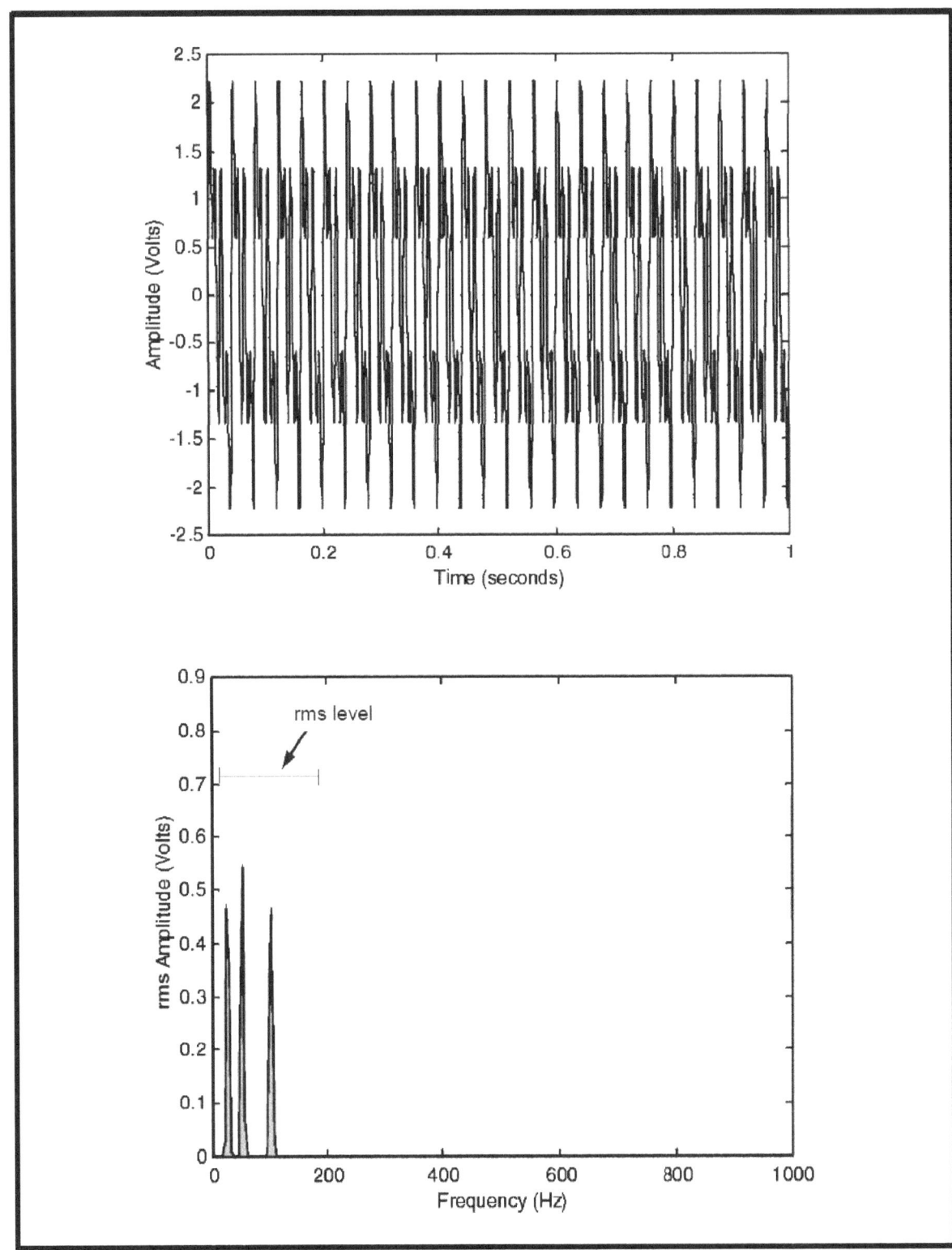

Figure C-5. Composite sine wave signal containing three distinct frequencies, and the spectral decomposition of the signal, with rms level indicated (figure courtesy of John C. Goold).

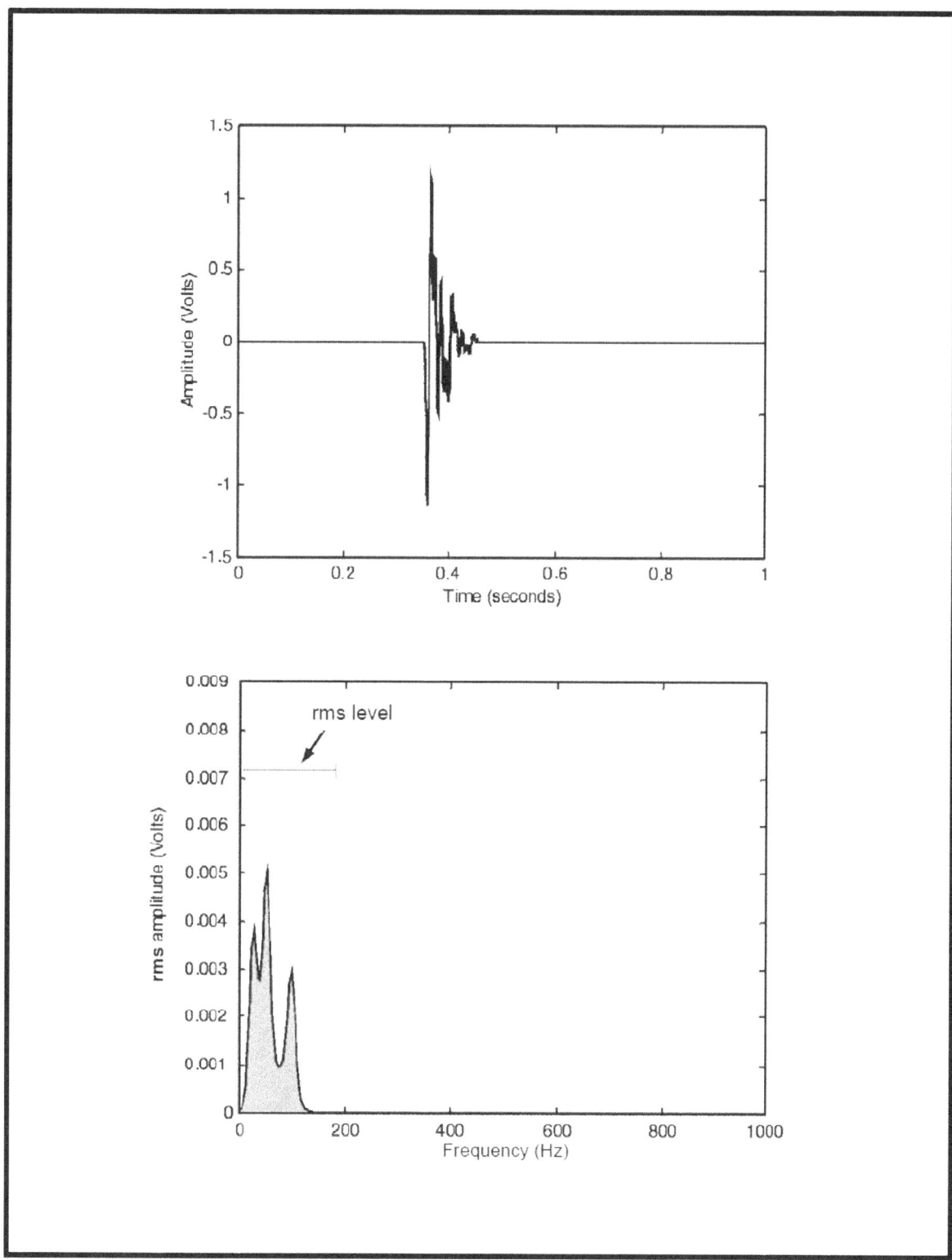

Figure C-6.   Pulsed version of composite sine wave signal, and the spectral decomposition. The rms levels are about 1/100th the magnitude of a 1-s signal (i.e., 20 dB quieter) due to the short pulse duration (figure courtesy of John C. Goold).

Quotation of seismic noise levels throughout the literature is inconsistent. Some state that values are peak-to-peak, others as zero-to-peak, others as rms, and many are not defined. This makes standardization of units within this document an impossible task, and in general, we have abstracted quantities at face value from the literature. We may suggest that unless stated otherwise, all dB values should be considered as rms. However, on making such an assumption, it should be acknowledged that many of the sound level values quoted, and calculations based upon them, could have plus or minus errors of several dB.

D.   AIRGUN PULSE rms

A typical far-field airgun array signature is illustrated in **Figure C-7(a)** for a 4,800 cubic inch array recorded with 2 ms sampling (data courtesy GXT; G. Greve, oral comm., 2003). In this figure, the signature initially shows a positive peak followed by a somewhat higher negative trough and, subsequently, a series of low amplitude peaks and troughs associated with air bubble oscillations. The sound pressure level pulse maximum is ~-60 Bar-m. The rms values are calculated by squaring each value, adding this value to the sum of the squares of the previous values, dividing by the number of previous values plus one, and then taking the square root of the sum. Values are converted to dB down from peak by multiplying 20 times the log of the ratio of this value to the peak value. The rms values drop from 50 dB down to a low of 4 dB down, then rise gradually to 16 dB down from the peak value (trough) of the airgun array signature. As is evident from **Figure C-7(b)**, the determination of rms values depends upon where in the pulse the measurement is taken and illustrates why rms measurements are used primarily for continuous signals and not for impulsive waveforms. While these data indicate that rms conversion is variable (i.e., 4 to 16 dB) and, for impulsive signals, highly dependent upon where in the pulse the value is determined, a conservative estimate for rms conversion is 10 dB. Alternatively, adoption of 12 dB might be argued because it would include measurement of the pulse plus the first bubble. Similar calculations for rms levels for other airgun array signatures should be very close to this example (i.e., 10 to 12 dB) (G. Greve, oral comm., 2003). A conservative estimate of 10 dB has been adopted in this analysis (i.e., zero-to-peak to rms).

III.   SEISMIC SOURCE LEVELS

Airgun output is usually specified by the G&G industry in terms of zero-to-peak or peak-to-peak levels, and the size of airguns is of concern operationally. Airgun sizes are quoted as chamber volumes in cubic inches, and individual guns may vary in size from a few tens to a few hundreds of cubic inches (see **Appendix D, Section II.A - Seismic Sources, Airguns**). Airgun array sizes are quoted as the sum of their individual airgun volumes, and again can vary greatly. 3D seismic survey array sizes may be on the order of 3,000 to 4,000 cubic inches, or slightly larger.

Sound producing devices, such as airguns, are usually ascribed to have a particular source level. A source level is typically defined as the sound pressure level at 1 m range from the source of the sound, and it is common to see the notation of dB re 1 μPa-m used to represent it. In the case of small transducers, such as certain piezo-ceramic elements, sound pressure at 1 m range can be measured directly. In the case of relatively large devices, it becomes unrealistic to measure sound levels at 1 m range, because

- it is not always possible to define where the center of the sound source is, and
- large devices are greater than 1 m in dimension themselves.

Further, for powerful sources such as airguns, close range measurements suffer from a so called "near-field" effect, which is the complex interaction of direct pressure waves from the compressed air

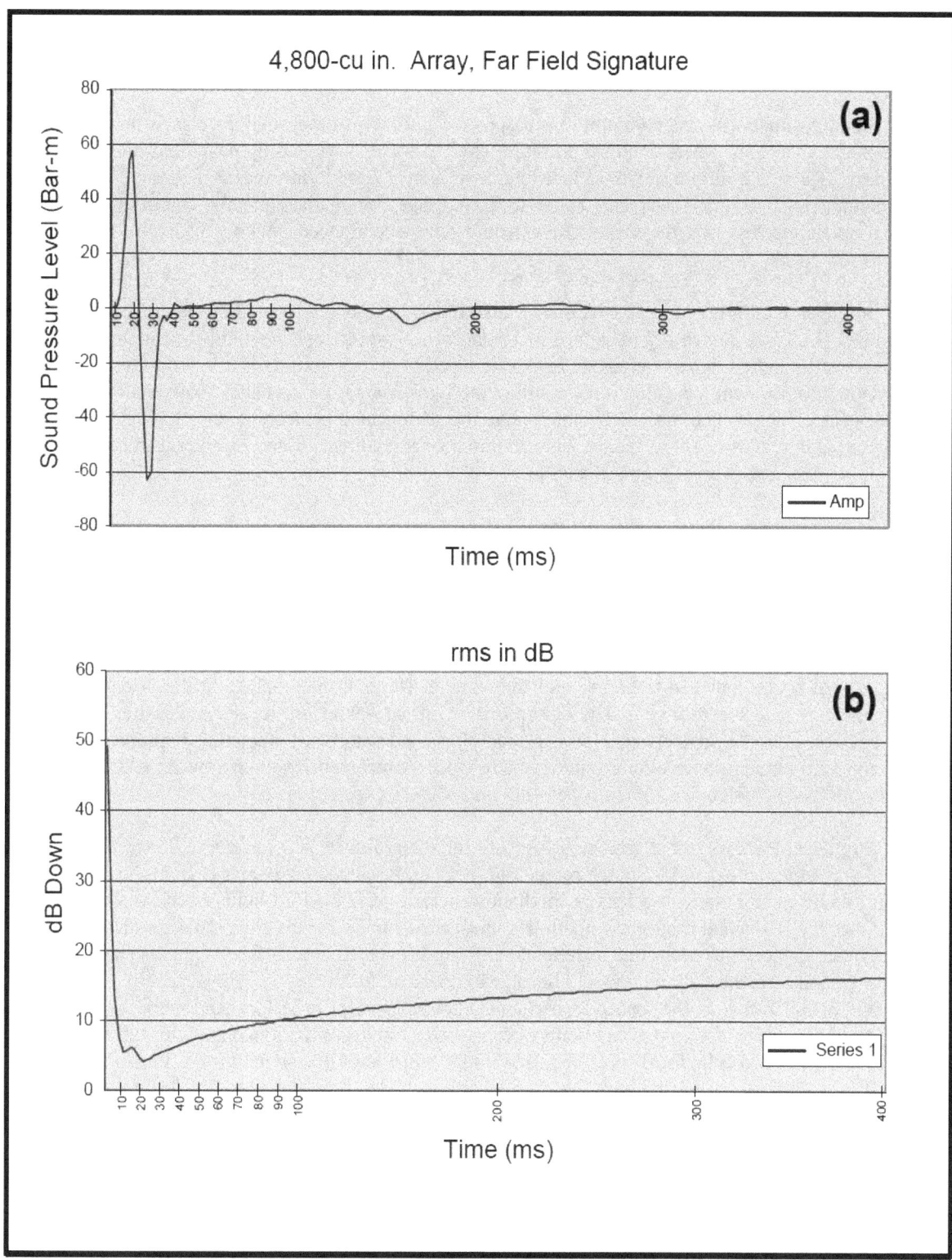

Figure C-7. Single airgun pulse signature from a vertical seismic profile survey, reflecting both
(a) sound pressure level, and (b) dB down from peak. The pulse is approximately
55-60 Bar-m above background (255 dB re 1 μPa-m). The rms level is dependent
upon where in the time series the measurement is taken. Measuring from the peak
of the first bubble pulse (time = 92 ms) gives a value of 10 dB (figure courtesy of
Gordon M. Greve).

release, and the particle compressions that compose the sound waves. These interactions at close range to powerful sources make measurements in the near-field unreliable.

In such situations, it is common to measure sound levels in the far-field, that is, at considerably greater ranges than 1 m, where acoustic pressure waves are not corrupted by near-field pressure effects. In the case of a single airgun, far-field measurements may be made at ranges of tens of meters. The airgun source level is then back calculated to 1 m range, by mathematically compensating for the spherical spreading loss that the sound wave would have experienced on its way to the measurement point.

In the case of airgun arrays, which may have individual guns separated by several meters, a point source level is clearly something of an abstract concept. Nevertheless, far-field measurements may be made at several hundred meters range and back calculated to a level notionally 1 m from the center of the array. This measurement has value in calculating the likely ranges of acoustic "isopleths" from seismic operations (i.e., rings around the vessel that receive the same sound pressure levels or acoustic intensities on their circumferences). These ranges have importance in setting mitigation criteria in relation to the effects of seismic survey noise on marine biota.

Determination of the sound level received by a marine mammal or other organism is a central issue. Determination of what constitutes a safe exposure level is equally important, and the two factors come together in estimations of safe ranges and isopleths (contours of equal loudness).

Unfortunately, neither the determination of received level, nor the determination of safe exposure levels, is easily resolved. Received level depends upon the source level of the seismic array, the specific propagation of sound between the array and the receiver (marine mammal), and the orientation of the array with respect to the receiver. The determination of safe exposure levels is a matter of scientific debate and research. Levels between 180 dB and 160 dB are commonly quoted as appropriate exposure levels, although certain observations suggest disturbance at received levels as low as 120 dB. For the purposes of this discussion, we will consider 180- and 160-dB exposure levels.

Figure C-8 shows sound spreading loss as a function of distance. The curve illustrating free field spherical spreading (20log[R]) is shown by the red line. In unbounded seawater (i.e., in the deep oceanic locations, or at close ranges to a source in shallower shelf waters), free field spherical spreading will occur. Once the horizontal propagation path becomes substantially greater than the water depth, a ducted form of spreading tends to occur due to reflections from the seabed and surface. In a duct with perfectly reflective boundaries, the spreading would become cylindrical, following a 10log[R] decay. However, in reality, the boundaries (i.e., the seabed in particular) are not perfect reflectors, and there is some loss of energy from the water column as the sound propagates. In this situation, the propagation is usually referred to as modified cylindrical spreading; and initial estimate of this attenuation is to adopt a 15log[R] decay. The precise rate at which loss will occur is variable and will be site-specific, depending upon such factors as seabed type. Modified cylindrical spreading is a phenomenon that tends to occur more with quasi-continuous sounds than with short impulse sounds. Short pulses tend to form a series of short reverberations rather than a sustained interference pattern, hence 20log[R] is more appropriate for the initial pulse of an airgun array. However, multiple reverberation of the pulse may occur in a ducted situation, thereby lengthening the sound exposure and contributing more energy overall than would be predicted by simplistic application of spherical spreading to the initial pulse only.

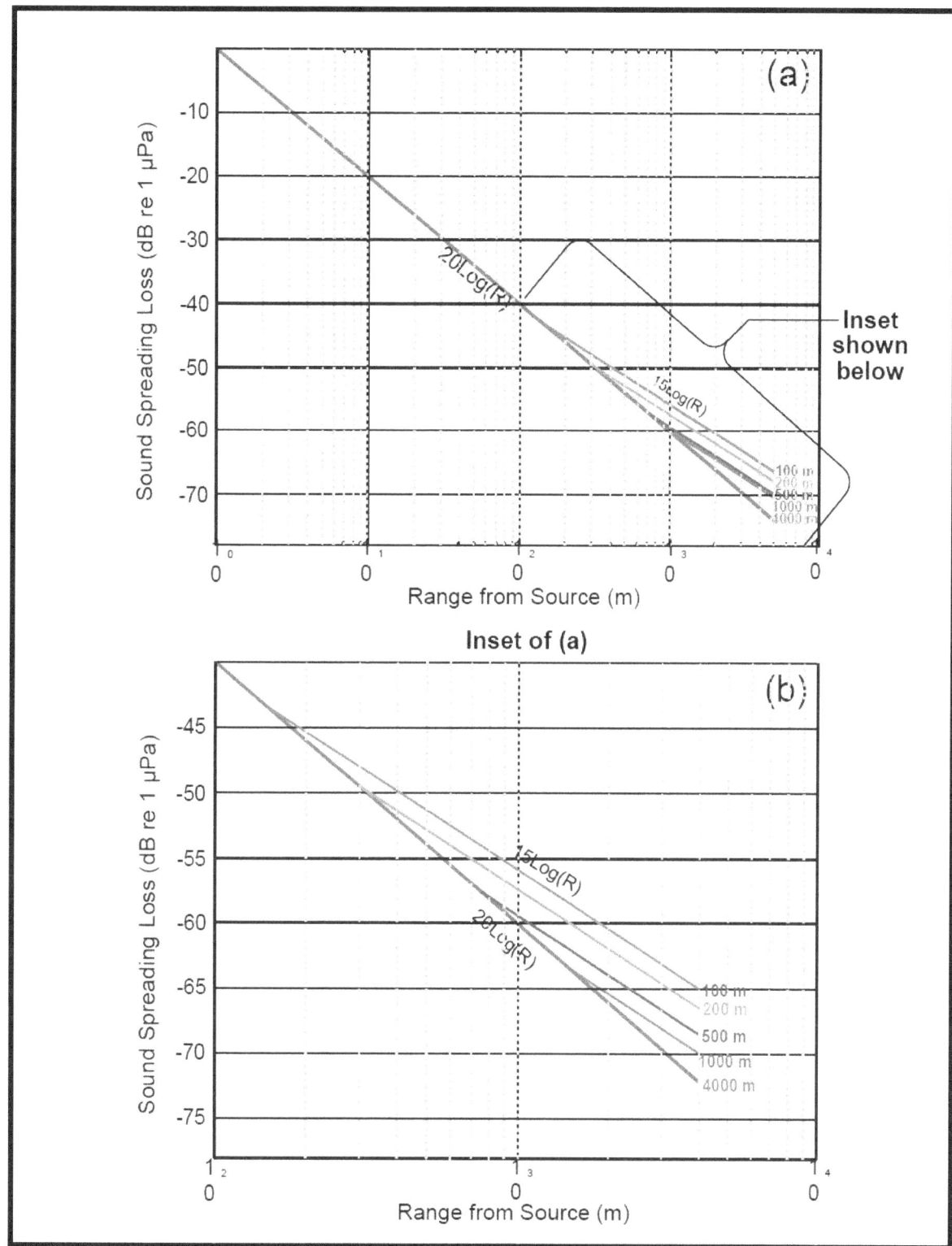

Figure C-8. Sound spreading loss versus range from source showing (a) free field spherical spreading (20log[R]: red line) and (b) detail of the transition to modified cylindrical spreading (15log[R]), the latter illustrated as several diverging branches from the free field curve at specific water depths. Transition to 15log[R] is considered to occur when the horizontal range is 1.5 times the water depth, and is more pertinent to quasi-continuous sounds than to short pulsed sounds (figure courtesy of John C. Goold).

Figure C-8 illustrates a number of 15log[R] curves branching from the free field spherical spreading (20log[R]) curve, which can be used as rough guides to read off spreading loss with distance from source. The branches relate to different water depths, arbitrarily chosen as 100, 200, 500, 1,000, and 4,000 m. To estimate sound spreading loss for quasi-continuous sounds in water 100 m deep, for instance, one should follow the free field curve and then branch along the 100 m, 15log[R] line. Each branch is positioned to take the curve into modified cylindrical spreading conditions at horizontal ranges 1.5 times their respective water depth. Therefore, the 100-m curve diverges to 15log[R] at 150-m range; the 200-m curve diverges to 15log[R] at 300-m range; and so on. It is considered that at horizontal ranges in excess of one and a half times the water depth, there will be reasonable interaction between the direct path signal and bottom reflections of quasi-continuous sounds, leading to ducted spreading. If one takes only the initial pulse in a reverberant seismic signal, then 20log[R] is likely to apply. However, if one wishes to account for total energy in the water column, including reverberations of an initial pulse, then adoption of a modified cylindrical spreading loss may be more appropriate. In the calculations to isopleths of interest (i.e., 180- and 160-dB) provided below, a simple spherical spreading model has been adopted for calculation of isopleth ranges. This model could easily be adapted to modified cylindrical spreading, if desired.

Airgun array source levels are typically expressed as zero-to-peak or peak-to-peak values. Protocol for the conversion from zero-to-peak to rms levels seems to vary, but the general consensus is that a conversion factor of -10 dB is appropriate albeit conservative, as discussed in the previous section (see **Figure C-7** and associated text). Further, the phenomenon known as the "horizontal array effect" reduces effective source levels in the horizontal plane relative to the vertical plane. Again, there is some variation of opinion regarding the array effect, but the general consensus is that the array effect causes at least a 20-dB reduction in the horizontal plane and is highly frequency dependent (i.e., higher frequencies attenuate to a greater degree than lower frequencies; see Coates, 1990; higher frequencies from airgun arrays are also produced at a lower level; directivity in the horizontal plane is likely to vary due to differential interference patterns with frequency). Finally, the output of the array is unlikely to be uniform in all directions. Unfortunately, only limited field measurements of the polar properties of a seismic array are currently available; as a result, there are insufficient data available at this time to make appropriate generalizations or informed estimates of the polar properties of airgun arrays (see **Section IV** of this appendix).

Given these considerations, preliminary calculations of radial distance to specific isopleths (i.e., contours of identical sound levels) have been developed concurrent with an explanation of the rationale employed in this approach. These basic calculations outline output at the source of a typical GOM seismic array (see **Appendix D**, **Section II.A - Seismic Sources, Airguns**) and account for rms conversion and horizontal array effects. Because of the programmatic nature of this assessment, however, precise calculations associated with specific airgun arrays or specific survey locations cannot be developed.

The following subsections calculate estimated maximum radial distances to the 180- and 160-dB (rms) isopleths based on a typical seismic array (i.e., 4,550-in^3 airgun array; 240 dB re 1 μPa [zero-to-peak], 230 dB re 1 μPa [rms]) in use in the Gulf of Mexico. Actual array output varies by operator and can be lower than the typical system employed in this analysis.

Note: The following calculations detail the methodology used to determine, in a simplistic approach, the distances to target isopleths of interest. For the 4,550-in.^{3} array considered in this analysis, the 240-dB (zero-to-peak) source level represents an estimated vertical source level, but not a real point source measurement. Real point source measurements are problematic and not easily measured in the near-field, as noted above. For the purposes of these calculations, a hypothetical source level of 260 dB (zero-to-peak) has been employed based on an extrapolation (back to the source) from spherical *spreading in the far-field. To avoid confusion, the extrapolated 260 dB (zero-to-peak) level is used only in the calculations presented herein; 240 dB (zero-to-peak) is cited throughout the remainder of this document as the real vertical source level of a 4,550-in.[3] array.*

A. HORIZONTAL DISTANCE OF THE 180-dB ISOPLETH

Taking a characteristic source level for an airgun array of 240 dB re 1 µPa produces a flawed estimation of the isopleth distance. This is because of the behavior of sound in the near-field, and the transition from near-field to far-field. It is possible to make a *flawed calculation* as follows:

240 dB	(vertical sound source level of array, zero-to-peak)
-10 dB	(rms conversion)
-20 dB	(horizontal array effect)
= 210 dB	(effective source level)
-180 dB	(target isopleth)
= 30 dB	(level in excess of 180-dB isopleth)

Using 20log [R] attenuation, a sound level drop of 30 dB will occur at approximately 30 m from source:

Range = antilog (30/20)

Erroneous range to target isopleth = ~30 m

However, this range is calculated from a notional point source (i.e., assuming that the 240-dB output of the array was actually coming from a point), which is not the case.

It should be recognized that a range of 30 m from the center point of an airgun array will be in the near-field (**Figure C-9**). This plot indicates that the near-field around a 4,550-cubic inch airgun array extends to ~100 m from the center point of the array, and that sound levels remain relatively high in this zone. However, the sound decay has a shallower gradient in the near-field than it does in the far-field. The actual source level of the array in **Figure C-9** is close to 240 dB re 1 µPa, but the extrapolation to a notional point source (under which assumption the isopleth calculation above was made) is closer to 260 dB re 1 µPa. Therefore, for the notional point source calculation above, a source level of 260 dB should have been used as a starting point. Recognizing this flaw in the calculation, one can start again with a higher extrapolated point source level.

Given that the plot in **Figure C-9** illustrates the zero-to-peak levels in the vertical plane below the array, we again take -20 dB for the array effect and -10 dB for rms conversion.

260 dB	(extrapolated vertical sound source level of the array, zero-to-peak)
-10 dB	(rms conversion)
-20 dB	(horizontal array effect)
= 230 dB	(effective horizontal sound source level)
-180 dB	(target isopleth)
= 50 dB	(level in excess of 180-dB isopleth)

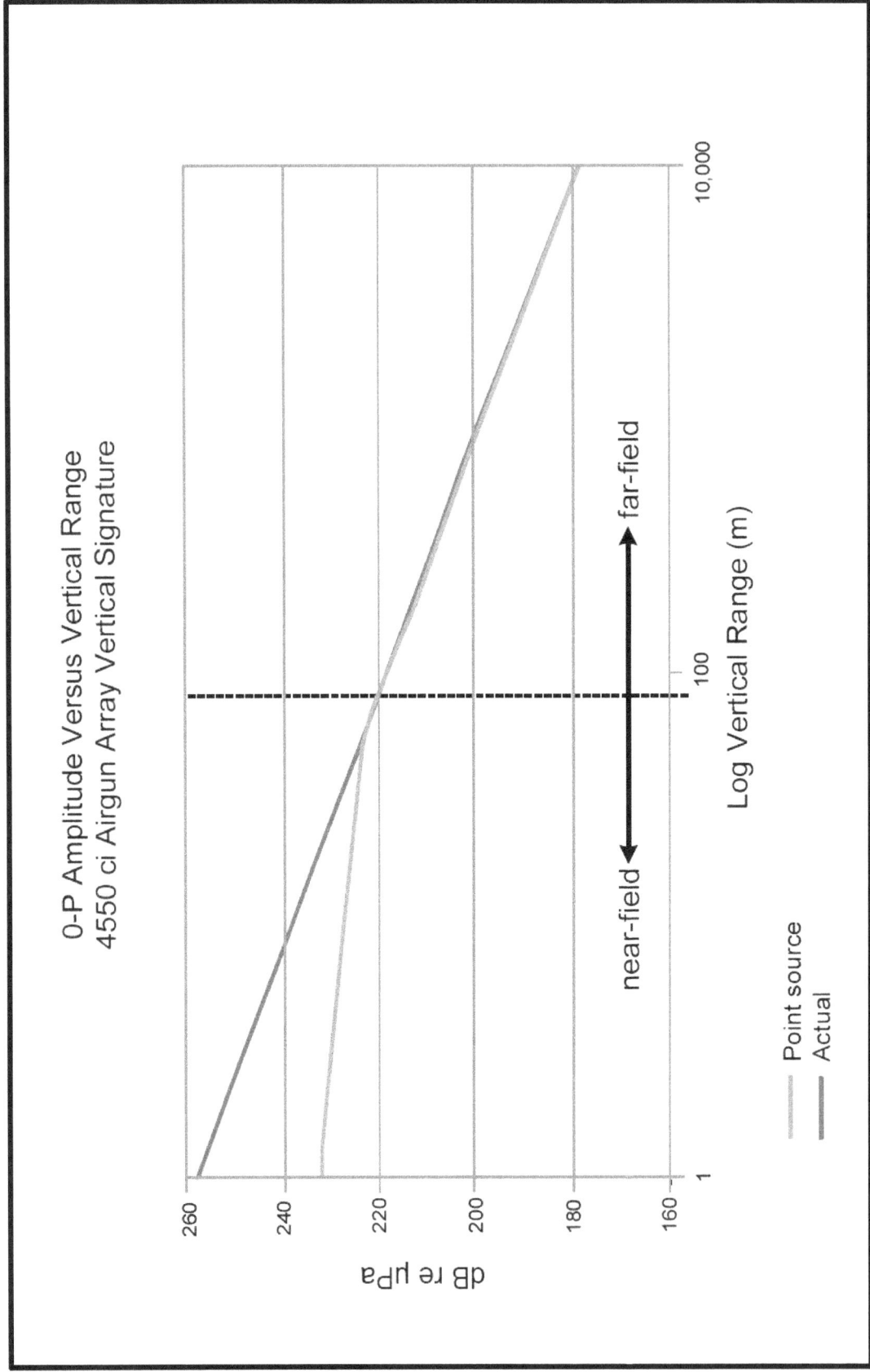

Figure C-9. Plot of sound level transition from near-field to far-field for a 4,550-cubic inch airgun array (figure adapted from: Phil Fontana, Veritas Marine Acquisition).

Using 20log [R] attenuation, a sound level drop of 50 dB will occur at approximately 300 m from source:

Range = antilog (50/20)

Actual range to target isopleth   =   ~300 m

B.   HORIZONTAL DISTANCE OF THE 160-dB ISOPLETH

Similarly, a calculation can be made to extrapolate the range of the 160-dB isopleth. Referring to **Figure C-9**, we again consider sound propagation from an extrapolated point source level:

260 dB   (extrapolated vertical sound source level of the array, zero-to-peak)
-10 dB   (rms conversion)
-20 dB   (horizontal array effect)

= 230 dB   (effective horizontal sound source level)
-160 dB   (target isopleth)

= 70 dB   (level in excess of 160-dB isopleth)

Using 20log[R] attenuation, a sound level drop of 70 dB will occur at approximately 3,000 m from source:

Range   = antilog (70/20)

Actual range to target isopleth   =   ~3,000 m

C.   FURTHER CONSIDERATIONS IN THE CALCULATION OF HORIZONTAL DISTANCE OF THE 180- AND 160-dB ISOPLETHS

The calculations of maximum radial distances to the 180- and 160-dB (rms) isopleths presented previously have been based on a typical seismic array (i.e., 4,550-in.[3] airgun array; 240 dB re 1 µPa [zero-to-peak], 230 dB re 1 µPa [rms]) in commercial use in the Gulf of Mexico. As noted in **Appendix D**, the actual array output varies by G&G operator and may be lower than the typical system evaluated in this analysis. In non-commercial applications, array size can be considerably larger with commensurate increases in array output (e.g., Lamont-Doherty Earth Observatory 20-airgun system deployed from the R/V *Maurice Ewing*).

Lower output systems produce smaller zones of ensonification relative to the typical seismic array characterized previously. For example, for an array with a specified source level 10 dB less than the typical array, the range to the 180- and 160-dB isopleths would be on the order of 100 and 1,000 m, respectively. Assuming a simplified radial geometry, the total area ensonified by this smaller array would be one-ninth that of a typical seismic array. While it is beyond the scope of this programmatic environmental assessment (PEA) to evaluate all potential array configurations, it is noteworthy that the actual ranges to the 180- and 160-dB isopleths will vary depending upon array geometry (i.e., airgun separation and size) and aspect (i.e., in-line versus cross-line array axes). Further, a smaller array would have a smaller zone of ensonification, the latter of which influences calculations of potential incidental harassment and take (see **Appendix L**).

The frequency spectra of a seismic signal has not been accounted for in the distance calculations (i.e., radial distance to a target isopleth) noted previously. It is known that a seismic signal is comprised primarily of low frequency components (i.e., significant energy at 10-1,000 Hz, peak frequency at 50-60 Hz, with contributions from both mid- and high frequency components). Insufficient information is available to accurately estimate and integrate frequency components into the isopleth calculations. As a consequence, frequency spectra cannot readily be accounted for in the current analysis.

IV.   POLAR PROPERTIES OF AIRGUN ARRAYS IN THE ENVIRONMENT

As stated previously in **Section III** of this appendix, there currently are insufficient experimental data on the directional properties of airgun arrays to construct detailed isopleths around airgun arrays. For this reason, estimates of the horizontal array effect have remained conservative, a natural prerequisite of the precautionary principle.

As discussed previously, airgun array output realizes spreading loss with increases in distance from the source (e.g., free field spherical spreading, 20log[R], and modified cylindrical spreading, 15log[R]), and is further influenced by water depth, water column structure, and the proximity and characteristics of the sea surface and seafloor. Output from a seismic array characteristically exhibits a "horizontal array effect" where differential sound attenuation occurs in the vertical versus horizontal direction. Due to array configuration, there is also variability in isopleth geometry, specifically, a near-surface directional variability.

While very little data showing isopleth geometry are present in the peer-reviewed literature, Goold (unpublished) has measured output from an airgun array used in the North Sea. **Figures C-10** and **C-11** reflect the near-surface directional variability in sound level around a 2,940-cubic inch airgun array, measured as the spectrum level in an octave band centered at 200 and 1,000 Hz, respectively, and extrapolated to 500-m range from the array. Data from port, bow, and starboard passes of the airgun array are shown; data for the stern are not available. The plot, constructed from an analysis of at-sea measurements made around an operating seismic vessel, shows a complex situation highlighted by the presence of major and minor transmission lobes around the array. Differential attenuation between 200 Hz and 1 kHz is evident. The plot shows the number of dBs down from the loudest point - it does not show actual sound pressure levels. The loudest point is closest to the outer reference ring at 0 dB; levels are plotted relative to (i.e., down from) the 0-dB ring at the angles illustrated.

Recently, modeling of airgun output and the attenuation of the seismic pulse was discussed as part of an environmental assessment of the proposed seismic research to be undertaken by Lamont-Doherty Earth Observatory (*R/V Maurice Ewing*) in the northern Gulf of Mexico (LGL Ltd., environmental research associates, 2003). Employing model results, **Figure C-12** depicts projected received sound levels from a 20-gun array and clearly shows the elliptical nature of the dB isopleths. Not only is the array directive, but the directivity varies with frequency. **Appendix D, Section II.A - Seismic Sources - Airguns** also discusses the polar or elliptical nature of airgun array output.

Data indicate that the horizontal transmission from a typical seismic array is not perfectly circular, but rather is more elliptical and irregular in nature. In the absence of modeling results that are directly applicable to this PEA (i.e., specific to a typical seismic array in use in the Gulf of Mexico), a circular geometry is applied in establishing zones of ensonification and potential impact. This approach quantifies the loudest point in a polar field and ascribes a circular zone at that radius. Directivity and the elliptical nature of seismic pulses are discussed further in **Appendix D, Section II.A - Seismic Sources, Airguns** and **Appendix L**.

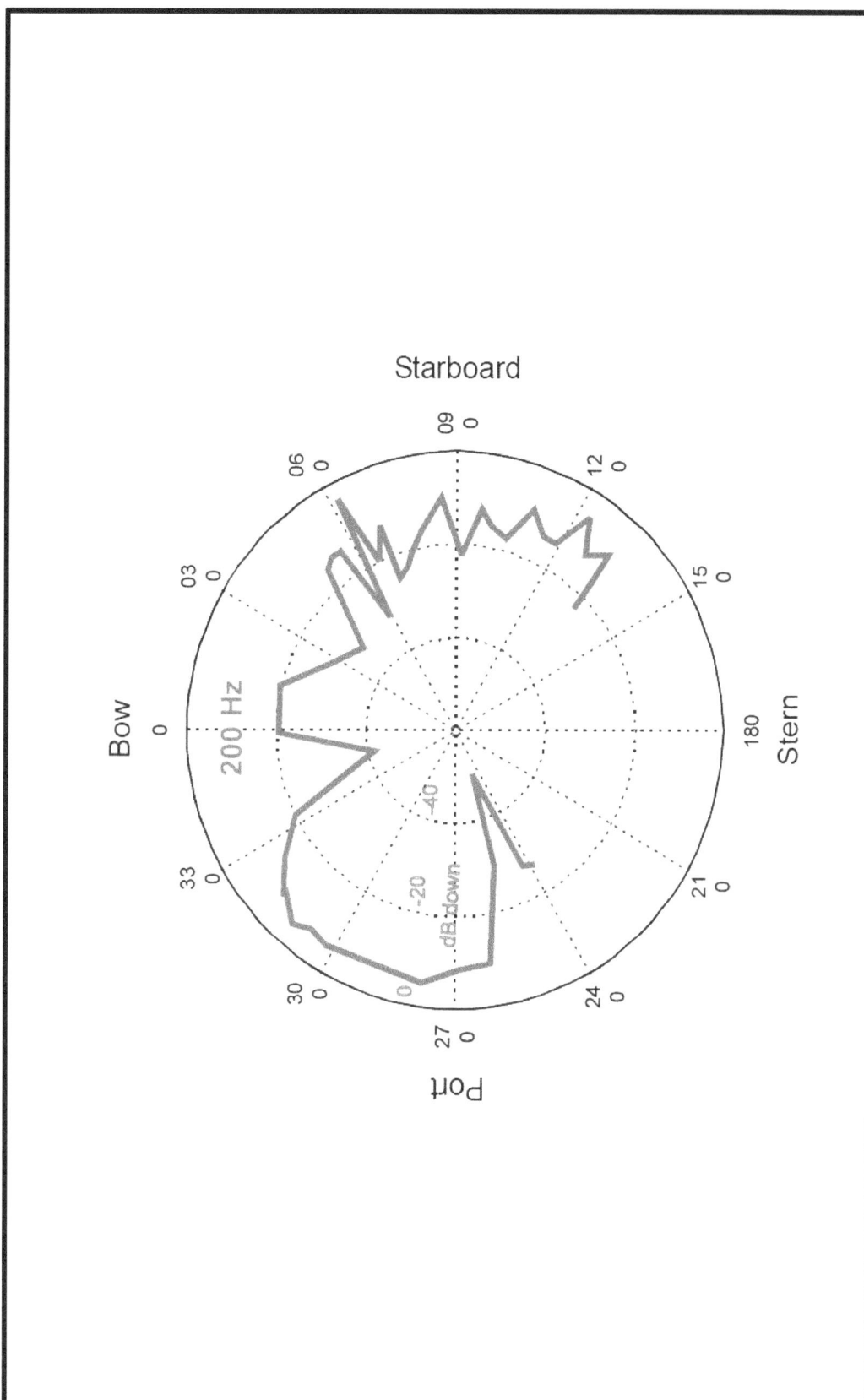

Figure C-10. Near-surface directional variability in sound level around a 2,940-cubic inch airgun array, measured as the spectrum level in an octave band centered at 200 Hz, and extrapolated to 500-m range from the array. The plot shows the number of dBs down from the loudest point – it does not show actual sound pressure levels. The loudest point is closest to the outer reference ring at 0-dB; levels are plotted relative to (i.e., down from) the 0-dB ring at the angles illustrated. The plot is a preliminary analysis derived from measurements at sea, with port, bow, and starboard passes of the gun array; data for the stern are not available. This is a specific example and should not be taken as a generalization for all gun arrays and configurations (figure courtesy of John C. Goold).

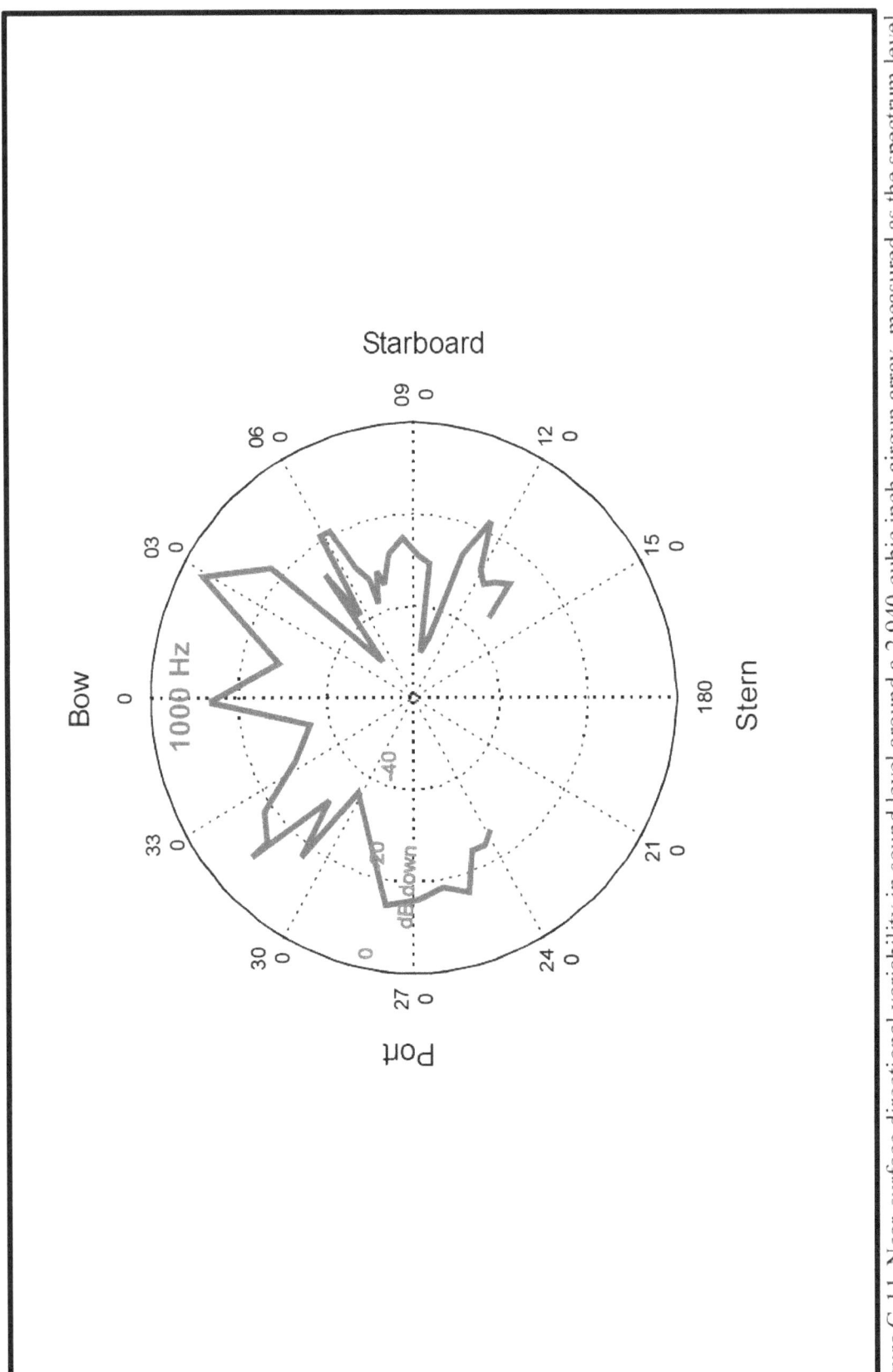

Figure C-11. Near-surface directional variability in sound level around a 2,940-cubic inch airgun array, measured as the spectrum level in an octave band centered at 1,000 Hz, and extrapolated to 500-m range from the array. The plot shows the number of dBs down from the loudest point - it does not show actual sound pressure levels. The loudest point is closest to the outer reference ring at 0-dB; levels are plotted relative to (i.e., down from) the 0-dB ring at the angles illustrated. The plot is a preliminary analysis derived from measurements at sea, with port, bow, and starboard passes of the gun array; data for the stern are not available. This is a specific example and should not be taken as a generalization for all gun arrays and configurations (figure courtesy of John C. Goold).

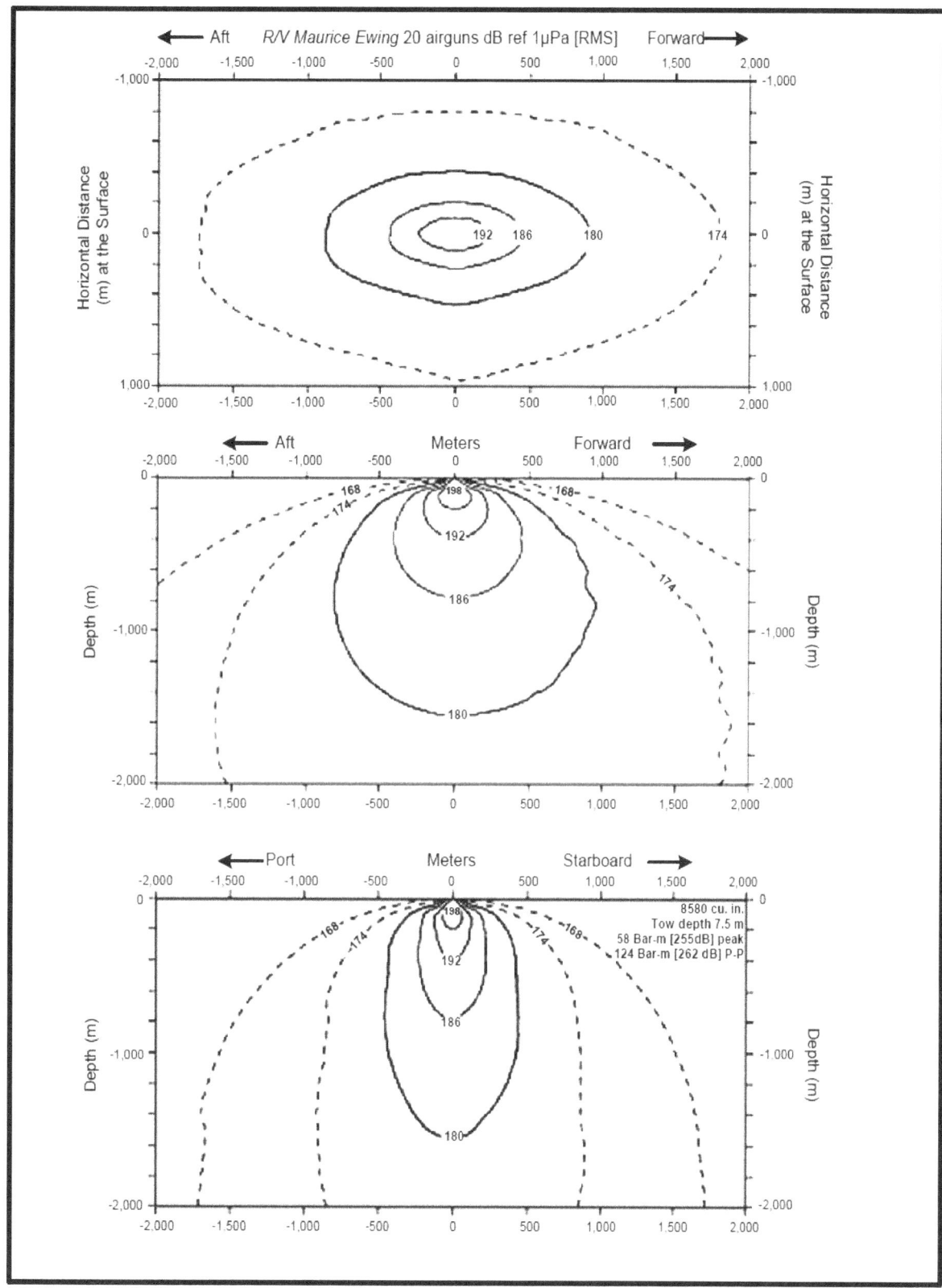

Figure C-12. Modeled received sound levels from a 20-gun array used during seismic surveys conducted by the *R/V Maurice Ewing* in the northern Gulf of Mexico, May-June 2003 (Adapted from: LGL Ltd., environmental research associates, 2003).

V. LITERATURE CITED

Coates, R.F.W. 1990. Underwater Acoustic Systems. Macmillan Publishers, London.

Goold, J.C. Unpublished. Near-surface directional variability in sound level around a 2,940-cubic inch airgun array. Provided with permission by J.C. Goold, University of Wales, Bangor, Gwynedd, Wales, UK.

LGL, Ltd., environmental research associates. 2003. Environmental assessment of marine seismic testing conducted by the *R/V Maurice Ewing* in the northern Gulf of Mexico, May – June 2003. Prepared for Lamont-Doherty Earth Observatory and the National Science Foundation by LGL Ltd., environmental research associates, King City, Ontario. 7 March 2003. LGL Report TA2822-3. 99 pp.

**Appendix D
Characterization of Geological and Geophysical
Activities and Equipment**

Characterization of Geological and Geophysical Activities and Equipment

This appendix describes the various geological and geophysical (G&G) exploration techniques currently being used (and expected to be employed by operators in the near future) in the Gulf of Mexico. Particular attention is paid to seismic techniques and especially the role of seismic sources (e.g., airguns). While this appendix is not intended to be a comprehensive analysis of all G&G activities, it is designed to provide fundamental details of various G&G techniques and methods used in the Gulf of Mexico. Such fundamental details serve as a basis for assessing the environmental impact of these operations (i.e., identification of impact producing factors or agents, determination of impact level; see **Section III - Environmental Impacts** of the Programmatic Environmental Assessment [PEA]).

I.   DESCRIPTION OF ACTIVITIES

A.   SEISMIC SURVEYS

1.   High-Resolution Site Surveys

High-resolution site surveys are conducted to investigate the shallow subsurface for geohazards and soil conditions, as well as to identify potential benthic biological communities (or habitats) and archaeological resources in support of review and mitigation measures for outer continental shelf exploration and development plans. Information also can be recovered at much greater depths, so that some surveys are used for exploration purposes. A typical operation consists of a ship towing an airgun (about 25 m behind the ship) and a 600-m streamer cable with a tail buoy (about 700 m behind the ship). The ship travels at 3 to 3.5 kn (5.6 to 6.5 km/hour), and the airgun is fired every 12.5 m (or about every 7 to 8 s). Typical surveys cover one lease block, which is 4.8 km on a side. Typically, the ship steams in one direction for about an hour, then turns around (about 20 to 30 minutes) and surveys the next track. Minerals Management Service (MMS) regulations require information be gathered on a 300- by 900-m grid, which amounts to about 129 line km of data per lease block. If the MMS has identified a block as having a high probability for the presence of historic archaeological resources (i.e., shipwrecks), grid points must be on a 50-m spacing (i.e., pursuant to Notice to Lessees No. 2002-G01). Including line turns, the time to survey one block is about 36 hours; however, before surveying the block, the streamer cable has to be balanced to enable it to stream at the proper depth, and the streamer and airgun must be deployed. These ancillary operations add to the total survey time.

Recently, 3D high-resolution site surveys using ships towing multiple streamer cables have become available. Since multiple streamers are towed, the ships tend to be slightly larger (47 vs. 37 m). Up to six streamers 100 to 200 m long are used with a tri-cluster of 8- to 10-cubic inch generator-injector (GI) airguns. With this system, 66 sail lines are necessary per block, which take about 5 days to collect. The final bin size after processing is 6.25 by 12.5 m.

2.   2D Surveys

Seismic 2D surveys are conducted in the Gulf of Mexico by geophysical contractors mainly for sale by them. These surveys are known in the industry as "spec" (speculative) surveys. Other surveys are shot for the exclusive use of a client and are known as proprietary surveys. "Spec" surveys are conducted over large multi-block areas, whereas proprietary surveys cover only a few blocks. Although the number of 2D surveys is small, compared to 3D surveys, they are important as the survey can cover a larger area in less detail resulting in a lower cost per area covered.

The ships conducting these surveys (see **Section IV - Ships** of this appendix for details) are generally 60 to 90 m long and tow a single source array 100 to 200 m behind the ship. Each source array (see **Section II.A - Seismic Sources, Airguns** of this appendix) is about 20 m long and 24 m wide. Following behind the source array another 100 to 200 m is a single streamer on the order of 8 to 12 km long. Attached to the end of each streamer cable is a tail buoy with radar reflectors.

The seismic survey ship tows the above apparatus at a speed of 4.5 kn (8.3 km/hour). About every 16 s (37.5 m), the airgun array is fired. The time between airgun firings is the desired spacing between firings, so the firings vary depending upon the actual speed of the ship.

To complete a survey, the ship will sail down a track from 12 to 20 hours (100 to 166 km), depending upon the size of the survey area. Reaching the end of the track, the ship will take 2 to 3 hours to turn around and start down another track. The spacing between tracks is usually on the order of 2 km. This procedure takes place day and night and may continue for days, weeks, or months, depending upon the size of the survey.

3.   3D Surveys

As with 2D surveys, almost all 3D seismic surveys in the Gulf of Mexico are conducted by geophysical contractors as "spec" surveys, conducted over large, multi-block areas. Proprietary surveys are usually conducted over only a few blocks.

The ships conducting these surveys are generally 80 to 90 m long, or slightly larger than those used in 2D surveys since they are towing more equipment. These ships typically tow two source arrays, at equal distances, 100 to 200 m behind the ship. The two source arrays are identical and are the same as used in the 2D surveys described previously. Following another 100 to 200 m behind the dual source arrays are anywhere from 6 to 12 streamer cables (see **Section III.A - High Resolution Site Surveys** of this appendix for details) 3 to 8 km long and spread out over a breadth of 600 to 1,500 m.

The seismic survey ship tows the above apparatus at a speed of 4.5 kn (8.3 km/hr). About every 16 s (i.e., a distance of 37.0 m for a vessel travelling at 4.5 kn), one of the dual airgun arrays is fired. Sixteen seconds later the other array is fired. The timing between firings is governed by the desired spacing between firings so the firings vary depending upon the actual speed of the ship.

Keeping track of where the airguns are fired, the position of the streamer cables, and the depth of the streamer cables is controlled by an integrated navigational system. Streamer depth is regulated by automated depth controllers called "birds." The streamer cable lateral position is calculated from a network of active acoustic devices. The end of the cable is tracked using global positioning system (GPS) satellites. Radar reflectors are routinely placed on tail buoys for detection by other vessels.

To complete a survey, the ship will continue down a track from 12 to 20 hours (100 to 166 km), depending upon the size of the survey area. Reaching the end of the track, the ship will take 2 to 3 hours to turn around and start down another track. This procedure takes place day and night and may continue for days, weeks, or months, depending upon the size of the survey. The surface grid size is normally 24 m by 48 m. Regardless of the size of the survey, subsurface information is usually obtained in bins measuring 12 m by 24 m on a side.

4.   Ocean Bottom Cable Surveys

Ocean bottom cable surveys were originally designed to enable seismic surveys in congested areas, such as producing fields, with their many platforms and producing facilities. Recently, these surveys have been found to be useful for obtaining multi-component (i.e., seismic pressure, vertical, and the two horizontal motions of the water bottom, or seafloor) information. This multi-component information allows more information to be extracted from the seismic data and hence greater information about the fluids and rock characteristics in the subsurface. These surveys have the additional advantage of lower noise levels in the data because the cables are stationary rather than moving through the water, as is the case with streamer cables. In addition, if gas effects resulting from leaky reservoir seals obscure the seismic information, these multi-component surveys can result in better structural definition of the hydrocarbon trap. Standard hydrophones are limited to water depths not exceeding 183 m. However, recent advances in hydrophone and geophone technology have enabled these devices to be used up to depths of 2,500 m or more. Of course, as these systems are deployed at deeper and deeper depths, both the physical demands upon the system and the concern for the integrity of the system increase. According to the International Association of Geophysical Contractors (IAGC), depth limitations are imposed due to the construction of the carrier systems (e.g., pressure seals, strength of cable). Whether the carrier system is comprised of streamer-type or logging-type cables, one fundamental question remains as to whether they are capable of sustaining their own weight over the vertical distance between the sea surface and the seafloor, as well as stresses placed on these systems during deployment and retrieval.

Ocean bottom cable surveys require the use of multiple ships (i.e., usually two ships for cable layout/pickup, one ship for recording, one ship for shooting, and two utility boats). These ships are generally smaller than those used in streamer operations, and the utility boats can be very small (see **Section IV - Ships** of this appendix for more information). Operations are conducted "around the clock" and begin by dropping the cables off the back of the layout boat. Length of the cable depends upon the survey demands; it is typically 4.2 km but can be up to 12 km. Groups of seismic detectors, usually hydrophones and vertical motion geophones, are attached to the cable in intervals of 25 to 50 m. Multiple cables are laid parallel to each other using this layout method with a 50 m-interval between cables. Dual airgun arrays are used. When the cable is in place, a ship towing an airgun array (which is the same airgun array used for streamer work) passes between the cables, firing every 25 m. Sometimes a faster source ship speed of 6 kn, instead of the normal 4.5 kn speed, is used with a decrease in time between gun firings. After a source line is shot, the source ship take about 10 to 15 minutes to turn around and pass down between the next two cables. This shooting and recording system results in a survey with subsurface information collected in bins 25 by 25 m. Some surveys modify the shooting and cable geometry to result in a bin size of 12.5 by 25 m, but the same approach to collecting the data is used. When a cable is no longer needed to record seismic data, it is picked up by the cable pickup ship and is moved over to the next position where it is needed. A particular cable can lay on the bottom anywhere from 2 hours to several days, depending upon operation conditions. Normally a cable will be left in place about 24 hours.

Location of the cables on the bottom is done by acoustic pingers located at the detector groups and by using the time of first arrival of the seismic pulse at the detector group. To obtain more accurate first arrival times, the seismic data are recorded with less electronic filtering than is normally used. This detailed location is combined with normal GPS navigational data collected on the source ship. In deep water, the process of accurately locating bottom cables is more difficult because of the effects of irregular water bottoms and of the thermal layers, which affect travel times and travel paths, thus causing positioning errors.

5.   Time-Lapse (4D) Surveys

The purpose of time-lapse surveys is to monitor the depletion of the reservoir and to locate zones of by-passed production in an already discovered oil or gas field. Not all fields are candidates for time-lapse surveys, and careful analysis must be done on each field to determine its feasibility for a time-lapse survey. A time-lapse survey requires repeat surveys with highly accurate navigation to ensure the same subsurface points are measured on each repeat survey. Time-lapse surveys are usually repeated every 6 months to a year, but occasionally the repeat interval can be as short as 4 months.

Time-lapse surveys can use either seismic streamer cables or, occasionally, ocean bottom cables to house the seismic detectors, depending upon which system was used to discover the field. This results in cheaper acquisition costs. Although a field could be discovered with a system using streamer cables and the subsequent time-lapse surveys done with ocean bottom cables, this would increase the cost. Whether the time-lapse surveys use streamer cables or seafloor cables to record the seismic signals, the procedure closely resembles the ones previously described in the 3D seismic survey section or the ocean bottom cable section. The main difference is in the size of the survey. Since the oil or gas field has already been located, the survey is much smaller and the time spent conducting the survey much shorter than an exploration survey. An average survey takes 2 to 4 weeks and can cover 20 square kilometers.

Although the technique began using streamer cables, the difficulty in locating the sensors with suitable precision led to the use of bottom cables, then to fixed bottom cables. When fixed bottom cables are used, the survey time after the first survey is much shorter since all that has to be done is connect the fixed bottom cable to the recording instruments and start shooting. In shallow waters, the fixed cable is easily located by attaching an acoustic transponder that responds when activated to the fixed cable. In deeper waters, locating bottom cables becomes more difficult due to irregular seafloor topography, the presence of thermal layers, and other effects that may limit position accuracy to ~20 m in 2,000-m water depths.

6.   Vertical Cable Surveys

Vertical cable surveys, although uncommon, are similar to ocean bottom surveys in that the receivers are deployed and then shot into by a source boat. However, they are substantially different from ocean bottom surveys in that the receivers are located in vertical cables anchored to the ocean bottom. Two identically configured boats are used during the survey. At the beginning of the survey, both boats are used to place the vertical cables. During the survey, one boat is used as a source boat and the other to recover and re-deploy the vertical cables.

The vertical cables are deployed on two overlapping grids. On each grid, vertical cables are deployed every 2 km. One grid is staggered 1 km to the other such that any one vertical cable is no more than 1.4 km from its closest neighbor. Normally 28 or 32 vertical cables are deployed at any one time (see **Figure D-1** for details on this geometry). Placing the cables in a known fixed position is a very critical part of the placement process. To aid in the positioning, an acoustic transponder is attached to the bottom of each vertical cable. Each vertical cable consists of an active section and lead-in section. At the bottom of the active section is not only the transponder but also an anchor composed of 680 kg of steel. In the active section are 16 specially constructed hydrophones spaced 25 m apart, which makes the active section 375 m long. At the top of the active section are placed buoyant floats to keep the cable as vertical as possible. Also attached to the top of the active section is the lead-in, which leads to the surface where the buoyant floats and a 16 channel 24-bit recorder are located.

Once the cables are in place, the source boat begins shooting in such a way that each vertical cable receives shots at a distance of 5 km in all directions. This is accomplished by sailing down lines parallel

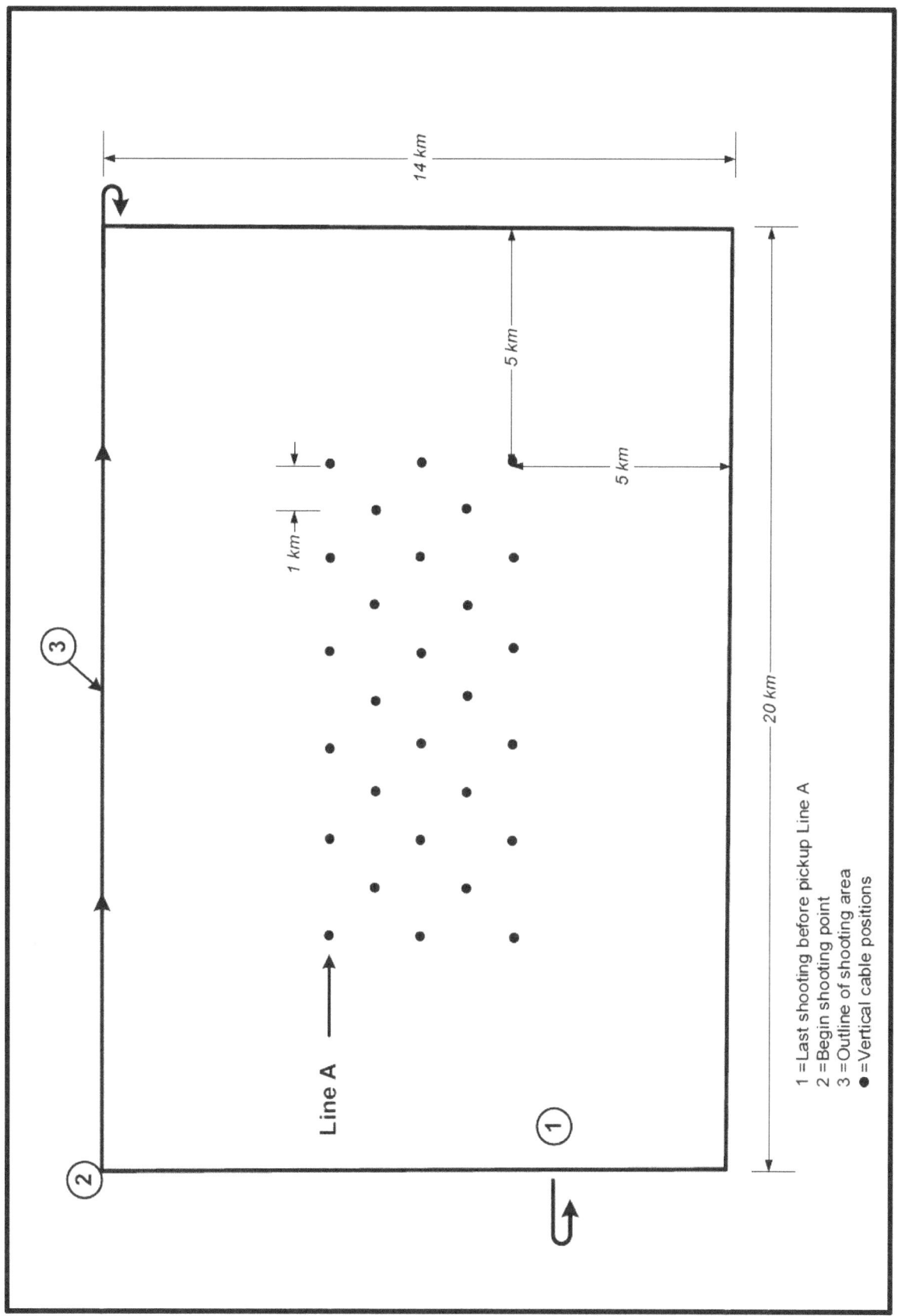

1 = Last shooting before pickup Line A
2 = Begin shooting point
3 = Outline of shooting area
● = Vertical cable positions

Figure D-1. Vertical cable acquisition geometry.

to the grid of vertical cables. (Turn around from line to line is only 15 minutes as only the source boat with its attached array has to make the turn.) This forms a shooting box around the grid of vertical cables, which extends 5 km outside the vertical cable grid. This makes the operational grid 14 by 20 km. However, once the shooting boat shoots a line 1 km beyond the first row of vertical cables, the first row of vertical cables is recovered and re-deployed (see **Figure D-1** for details). Cables may be left in place for hours or days depending upon the size of the survey and operating conditions. Shots are taken every 50 m, and the shot lines are 80 m apart. The source boat uses the same source array as is normally used in 3D streamer surveys. This array is composed of three six-gun sub-arrays with a total volume of 3,090 cubic inches. The largest airgun in the array is a Bolt Long Life 250-cubic inch airgun. Vessel speed is normally 4.5 kn, and operations are conducted "around the clock."

Surveys are normally conducted at water depths up to 1,700 m; however, since specially constructed hydrophones are used, surveys can be conducted in water depths up to 2,500 m.

7.   Vertical Seismic Profile (VSP) Surveys

VSP surveys are surveys where seismic data are recorded from sensors placed in a borehole (i.e., a hole vertical to the ocean surface or seafloor) with seismic sources deployed in various geometries around the vertical array of sensors. When the seismic source is placed very near to this vertical array of sensors, the survey is called a zero offset VSP, or check shot survey. These surveys are commonly used to correlate geologic data to seismic data. When the seismic source is placed in a series of positions along a radial line from the vertical array of sensors, the survey is called a walk-away survey. These surveys are used to obtain information about the nature of the seismic signal as well as more information about the geology surrounding the vertical array of sensors. Zero offset and walk-away VSP surveys are by far and away the most common VSP surveys conducted in the Gulf of Mexico. Less common are 3D VSP surveys, where the source is deployed in an area surrounding the vertical array of sensors. 3D VSP surveys provide more detailed information in the area surrounding the vertical array of sensors and are especially beneficial where salt layers stand between the geologic objective and the surface. In some cases, where salt is present, 3D VSP surveys are the only way to obtain geologic information below the salt layers.

In all VSP surveys, sensors are lowered down a borehole before production tubing is placed in the well bore or the well is abandoned. This means time is of the essence since drilling rig time can be very expensive (as much as $400,000 to $500,000 per day for deep-water drilling rigs). To reduce VSP survey costs, gun arrays are sometimes run at deeper depths to avoid the effects of weather, and when an airgun in the array malfunctions, the survey may continue rather than stop to replace the faulty airgun.

The sensors lowered down the borehole can be connected together in strings of 16 to 36 receivers spaced from 15 m to 150 m apart depending upon the objective of the survey, the contractor conducting the survey, and the manufacturer of the equipment used for the survey. After lowering the sensor string to the lowest portion of the borehole to be surveyed, the sensor clamps (lever arms) are activated, which push the sensors against one side of the well bore to ensure good signal quality. Once all the seismic signals are recorded at that level, the sensor clamps are retracted, and the sensor string is raised to a new level where the receiver clamps are reactivated.

The seismic sources used to generate the seismic waves detected by the sensors are the same as those used in conventional seismic surveys. Zero offset surveys are conducted using a small volume single airgun suspended by a crane located on the deck of the drilling rig. Walk-away surveys utilize a work boat with only four to eight airguns. 3D VSP surveys use the same airgun arrays as used for conventional 2D and 3D surveys. These airgun arrays can vary from 1,000 cubic inches to 5,000 cubic inches, depending upon the depth of the objective. Two or three arrays are towed behind the source vessel. When two arrays are used, the centers of the arrays are from 60 to 80 m perpendicular to the centerline of the vessel axis. When a third array is added, it is placed between the outside two arrays (along the centerline of the vessel). Typical airgun array depths are 7 to 10 m below the surface.

One method used to provide 3D coverage is for the source vessel to travel in a spiral track. The source vessel begins the spiral track at a distance of 200 m from the borehole and keeps the distance between spirals equal to the number of arrays times the array separation (e.g., two arrays spaced 120 m apart equals a 240-m track spacing). First, one airgun array will fire, then 12 to 14 seconds later the other airgun array will fire. At a typical vessel speed of 4.5 to 5 kn, the distance between firings is between 28 and 36 m. The source vessel continues on the spiral out to a distance of up to 9 km. If the borehole sensor string needs to be raised to another level, the whole procedure is repeated.

Total time spent on VSP surveys depends upon the type of survey, the objectives of the survey, the cost of the drilling rig, and the equipment used. For a zero offset survey or a walk-away survey, the survey can take less than a day, in the absence of any serious equipment failures. For a zero offset survey, the airguns are fired four to eight times for 20 seconds, followed by a 5- to 20-minute quiet time during which the sensor string is raised; the airguns are fired again for four to eight times for 20 seconds and so on until the survey is completed. A 3D survey may require up to 10 days to complete; however, 30% of that time may be with the airguns in standby mode.

Recording of VSP data and control of the operation is done on the drilling rig floor. The control center issues the fire command to the source vessel, and the vessel transmits back the actual firing time as well as the position (using differential GPS) of the source array.

8.   Multi-Ship Surveys

Multi-ship surveys are an integral part of ocean bottom surveys and vertical cable surveys. Moreover, in the quest for seismic data recorded greater distances from the source, multi-ship surveys are becoming more prevalent. Generally, this technique is used to either obtain converted wave data (shear wave) or to penetrate hard seafloor layers. Two sorts of operations are in use: one 2D and the other 3D.

The 2D operation places two ships, one behind the other, towing single streamer cables at a distance apart and a streamer length sufficient to record seismic data at distances of 8 to 12 km from the source. The 3D operation places two ships, one behind the other, each towing multiple (up to 12) streamers 2,500 m long a sufficient distance apart to record seismic data at distances of 9 km from the source. In both 2D and 3D surveys, only the lead ship is used as a seismic source boat.

9.   Undershooting

Undershooting, where an obstacle stands between the seismic source and the receiver, is used most commonly in conjunction with ocean bottom surveys. This is because the surveys typically are located in areas of congestion and the cost is low. Operationally the technique is identical to normal ocean bottom operations, except the obstructing structure lies between the source boat and the ocean bottom cable.

Undershooting using 2D or 3D streamer cable ships is rare and is required when an obstacle, most often a production platform, interferes with the routine collection of data in the immediate vicinity of the obstacle. The usual technique is for one vessel towing a seismic source, and perhaps a streamer cable, to pass on one side of the obstacle while another vessel towing a streamer cable, and perhaps a seismic source, passes on the other side of the same obstacle. The energy from the seismic source passes through the subsurface beneath the obstacle and is recorded by the vessel on the other side of the obstacle, hence the name "undershooting." Undershooting using ocean bottom cables as receivers is sometimes necessary when the geometric disposition of multiple obstacles is complex. Ocean bottom cable is generally more expensive than using surface streamers as receivers.

B.   DEEP-TOW SIDE-SCAN SONAR SURVEYS

Deep-tow side-scan sonar surveys are conducted in the Gulf of Mexico primarily for engineering studies involving the placement of production facilities and pipelines. These surveys provide information on the presence of sand flows, hydrates, and seeps as well as bottom topography (e.g., hard bottom).

Operations are conducted from ships towing cables up to 7 km long, which enables operations in water depths up to 3,000 m deep. Close to the end of the cable is a 30- to 45-m long section of chain to keep the sensor package (fish) tracking at approximately 25 to 30 m above the bottom. To do this requires the chain to drag along the seafloor, causing an approximately 10 cm wide by 15 cm deep (4 inch wide by 6 inch deep) trench to be cut in the seafloor. In situations where the chain can become entangled in shipwrecks, well heads, or other obstructions or where reef colonies live, the chain is removed, and the sensor package is kept above the seafloor by adjusting the length of the tow cable. Maintaining a constant elevation above the seafloor by adjusting the cable length is very difficult, and the elevation above the seafloor is somewhat greater in this case.

The sensor package is housed in a fish about 3.6 to 4.3 m long and 1 m in diameter. The sensor package consists of two sonar devices – one mounted on the port side and the other on the starboard side of the fish. The sonar devices emit, perpendicular to the fish, 100-kHz pulses with an intensity of 10 kw. The beam width of each of the sonar sources is 1° to 2°, which combined with the height above the seafloor of 20 m, yields continuous seafloor coverage. Also included in the sensor package is a 3.5-kHz (10-kw) pinger for subbottom profiling.

During operations, the ship pulls the cable, chain, and fish along a track, then circles and sails along a parallel track 300 m from the first. This gives about a 100-m overlap in coverage at the far ranges, which compensates for fish position errors.

C.   ELECTROMAGNETIC SURVEYS

There are two practical electromagnetic techniques applicable to marine surveys, and one technique that could be applied but has not yet been deployed. The two practical techniques, the magneto-telluric (MT) and the bi-pole, have been primarily applied in a marine environment in a research mode. However, the MT technique has seen limited use in the Gulf of Mexico where about 400 stations have been occupied. There is talk of turning the bi-pole device into a commercial device, but there has been no action as of this date. The third technique, developed by scientists in Russia, can be applied in a marine environment but has not yet been tried.

The MT technique was developed by scientists at Scripps Institute of Oceanography (SIO) and a consortium of major oil companies. In this technique, no electrical currents are induced into the earth, but the receiver device detects the natural electrical and magnetic fields present in the earth. Ships about 43 m long are used to deploy and retrieve the recording devices. These recording devices are about 1.5 m high by 1 m on a side and are attached to a concrete anchor about 60 cm on a side, 15 cm high, and weighing about 136 kg. Also attached to the recording device are four arms sticking out from each side of the box with an electrode on each end. These arms are about 20 m long and made of 5-cm plastic PVC pipe. Inside the recording box is a magnetometer and a long term recording device, which allows the box to remain on the seafloor for days at a time. The recording box is retrieved by using an acoustic pinger that releases the anchor from the recording box, which then floats to the surface.

The bi-pole technique was also developed by scientists at SIO. In this technique, two cables, joined together, with the second cable a few hundred feet longer than the first, are towed around by a ship. Attached to the end of each cable is a metal cylinder about 3 m long and 0.3 m in diameter. At regular intervals the ship stops, the cables sink to the bottom, and an electrical signal of about 50 volts and 100 amps is input through the cables and into the seafloor. These electrical signals are detected by previously deployed receivers 2 to 10 km away from the source and arranged in a linear line or profile. The receiver boxes are attached to concrete blocks like those used in the MT technique. Inside the receiver boxes are recording devices, which allow for recording for a few days. When the recording is finished, an acoustic pinger releases the recording box from the anchor, and the recording box floats to the surface for retrieval.

The third technique, developed by Russian scientists, relies upon detecting signals from iron pyrite, produced by chemical changes caused by hydrocarbon gases, percolating from leaky hydrocarbon reservoirs. The technique has not been applied in a marine environment.

D. GEOLOGICAL AND GEOCHEMICAL SAMPLING

Geological and geochemical sampling is conducted to obtain samples of the seafloor for physical and/or chemical analyses. Physical analyses are used in engineering studies for placement of structures such as platforms and pipelines. Chemical analyses (surface geochemical prospecting) are based on the premise that upward migrated petroleum from deep source rocks and reservoirs can be detected in near-surface sediments and are used to evaluate exploration potential.

Bottom sampling involves devices that penetrate only a few centimeters to several meters below the seafloor. Samples of surficial sediments are typically obtained by dropping a piston core or gravity core ("dart"), essentially a weighted tube, to the ocean floor and recovering it with an attached wire line. Samples can also be obtained using a grab, which is a device with a jaw-like mechanism, or with a dredge, which is a wire cage dragged along the seafloor. Shallow coring is done by conventional rotary drilling equipment from a drilling barge or boat. Penetration is usually limited to the recovery of several feet of consolidated rock. Usually, a program of bottom sampling and shallow coring is conducted simultaneously using a small marine drilling vessel (U.S. Department of the Interior, Geological Survey, 1976).

1. Piston and Gravity Coring

Surface geochemical prospecting is a petroleum exploration technique based on the premise that upward migrated petroleum from deep source rocks and reservoirs can be detected in near-surface sediments and used to evaluate exploration potential. In deep-water exploration, a crucial and arguably the most critical single issue is whether oil has been generated, and if it has, what can be determined, before making costly decisions, regarding the quality, maturity, and age of the source succession. Fortunately, since oil and gas accumulations are invariably leaky, surface geochemical exploration (SGE) can be used to recover and type migrant hydrocarbons in the initial exploration phases.

For the last 15 years, almost all the SGE techniques in the northern Gulf of Mexico have involved piston coring to obtain sediment samples for the analysis of upward migrated hydrocarbons (Brooks et al., 1986, 1997; Sassen et al., 1993). The typical piston core is a 6-m long, 7.5-cm diameter pipe with a 910-kg core weight. In some early studies, lighter and shorter gravity cores were used. Over the last 20 years, it is estimated that 10,000 to 15,000 piston and gravity cores have been taken in the northern Gulf for these SGE studies, primarily in continental slope water depths.

In gravity coring, wire is paid out from the coring winch at a fairly fast speed, allowing the corer to hit the bottom with a force proportional to the weight of the corer and the speed at which it is deployed (the payout speed of the winch). Penetration into the bottom is limited by the sediment type, friction of the sediment on the outside and inside walls of the core barrel, and the resistance of the water exiting the top of the core barrel. In contrast, a piston corer uses a "free fall" of the coring rig (independent of payout speed, and not restricted by trailing wire drag through the water column) to achieve a greater initial force on impact, and a sliding piston inside the core barrel to reduce inside wall friction with the sediment and to assist in the evacuation of displaced water from the top of the corer. The core barrel dimensions are generally 6-m long by 7.6-cm internal diameter by 9-cm outer diameter. Coring at lengths greater than 6 m is possible but not generally conducted for SGE studies.

2.   Heat Flow Measurements

Another tool in common use in deep-water exploration is a heat flow probe. Heat flow measurements serve critical purposes in oil exploration and production. The measured background or equilibrium heat flow, and measured sediment thermal conductivity provide strict constraints to geochemical models that determine regional scale maturation of basins with respect to oil and gas. In addition, area-wide heat flow surveys provide significant geological information on fluid flow from faults, lineaments, and around structures. Heat flow measurements, in conjunction with seismic and seafloor geochemical studies, provide a mechanism to assess fault and structural seals and contribute to a better understanding of regional hydrodynamics and hydrocarbon occurrence.

Heat flow measurements are conducted with a device that looks much like a piston corer. The device measures both temperature gradient and thermal conductivity in situ over subbottom depth intervals of up to 6 m. The footprint and impact on the sediments are almost identical to that of piston or gravity coring.

3.   Hydrocarbon "Sniffers"

In the 1970's and early 1980's, hydrocarbon "sniffers" were commonly used in SGE exploration to measure dissolved gases in near bottom waters. These "sniffers" consisted of a deep-water pumping system that was towed behind the vessel that pumped water to the surface from depths as deep as 150 to 200 m. However, little water column hydrocarbon "sniffing" has been undertaken in the 1980's and later. It is an almost obsolete technique because the amount of geochemical information available from the sediments is much greater than can be found in the water column.

4.   Shallow Coring

Shallow coring is done by conventional rotary drilling equipment from a drilling barge or boat. Penetration is usually limited to the recovery of several feet of consolidated rock.

E.   REMOTE SENSING

1.   Radar Imaging

Radar imaging is currently used to detect oil slicks on the sea surface. This is possible because when the oil molecules reach the sea surface, they form a thin layer that dampens the ocean surface capillary waves. The detection of oil slicks requires quiet water conditions and consequently is limited by sea state as well as by satellite position and frequency of coverage. The resolution of the radar images ranges from 8 to 100 m with a swath width range of 50 to 500 km. The radar satellite orbit is a near polar orbit and orbits at an altitude of 798 km. The cycle time for a duplicate orbit is 24 days, but a common spot on the earth can be revisited every 5 days and surveyed with different viewing parameters. The radar frequency is 5.3 GHz (5.6-cm wavelength) with a peak power output of 5 kw and an average power output of 300 watts. The pulse length is 42.0 μs. The MMS does not permit nor approve radar imaging surveys.

2.   Aeromagnetic Surveys

Aeromagnetic surveys are conducted in the Gulf of Mexico to look for deep crustal structure, salt related structure, and intrasedimentary anomalies. The surveys are flown by twin engine fixed wing aircraft, typically Cessna 404 or 208, or Piper Aerostar or Navajos. The flight lines are on the order of 400 km long, are at a height of 75 to 150 m above the surface, and are flown at speeds of about 220 km/hr. Flight line spacing ranges from 500 to 800 m apart with cross lines every 2,000 to 3,000 m. Acquisition rates are on the order of 1,000 to 2,000 km of data per day. The earth's magnetic field is measured by either a proton precision or cesium vapor magnetometer mounted in a "stinger" projection from the tail of the aircraft.

On occasion two magnetometers are used to measure not only the total magnetic field but also the vertical gradient of the field. This configuration requires the use of two stingers with about a 2-m vertical separation between them.

Magnetometers can also be towed behind a ship. This usually is in conjunction with a seismic survey but can be run as a separate survey. The magnetometers towed behind the ship are enclosed in a "bird" about 1 m long with a diameter of about 15 cm and weight of about 16 kg.

3.   Gravity Surveys

Marine gravity data can be collected with instruments on the seafloor, in boreholes, in ships, or in helicopters. Originally data were collected on the seafloor, but modern technology has moved the collection point to ships. Marine gravity meters have, in some cases, been housed in a ship while it is conducting a seismic survey. However, the preferred method has been to use dedicated ships (about 50 m long) in order to acquire more precise data. With the advent of GPS navigation systems and larger, more stable seismic ships, it is now possible to achieve the same order of accuracy with meters placed in seismic ships as in dedicated ships. Data grids for gravity surveys range from 1.6 km x 8 km to 9.7 km x 32 km.

Helicopter surveys are rare in the Gulf of Mexico because of the logistics required to keep the craft in the air for extended periods far from shore.

4.   Gravity Gradiometry

Measuring the earth's gravity gradient is now possible with the release of Defense Department technology. The instrument is housed in a 1-cubic meter box and located in the center of a 60-m survey ship. In shallow water the ship sails a 0.25-km by 1-km grid, and in deep water, a 1-km by 2-km grid. Typically, a 20-block area is selected for survey with the ship traveling down a line 12 blocks long, turning, and traveling back 12 blocks with a spacing between the lines double the final grid spacing. At the end of the final block, the ship returns to fill in the lines to achieve the final grid spacing. Cross lines are then sailed in a similar fashion. Using the system and sailing at 11 kn, a ship can survey about 10 blocks per day.

5.   Marine Magnetic Surveys

Marine magnetic surveys measure the earth's magnetic field for the purpose of determining structure and sedimentary properties of subsurface horizons. These surveys are usually conducted in conjunction with a seismic survey, allowing the navigation information to be used for both surveys. The development of low power digital sensors has allowed the sensor package to be towed behind the seismic source array, which has greatly improved operational efficiency of magnetic surveys. The sensor is housed in a cylindrical package measuring approximately 1 m long and 15 to 20 cm in diameter and weighing about 14 kg. Sensors are typically one of three types – an Overhauser effect sensor, a cesium vapor sensor, or a proton precession sensor. Presently, the Overhauser effect sensor is the most popular, while the proton precession sensor is gradually being phased out (Weber, 2001). The electronics package inside the case contains about 1 L of chemically inert fluid that is non-toxic. The sensor is towed behind one of the sub-arrays of the seismic source array at distances of 50, 100, or 150 m (behind the array), although 100 m is the most common. The sensor is towed at a depth of 3 m and makes use of depth devices mounted on the cable to maintain a constant depth.

F.   OTHER TECHNOLOGY

1.   Pingers/Transponders

Pingers and transponders are used in three different ways in G&G operations in the Gulf of Mexico. First, the devices are used to triangulate and provide precise location of ocean bottom cables. Second, the devices are placed between an anchor and instrument package to enable recovery of instrument packages such as vertical cables, and electromagnetic detector packages. Third, the devices are used as an altimeter to measure height above bottom in deep-tow side-scan surveys. Although three applications are identified, only two different devices are employed.

The first device, which not only releases instrument packages from a bottom anchor, but also is used to precisely locate bottom cables, is a recoverable acoustic transponder. These transponders are designed to work in different depths and hold differing weight instrument packages. When used to recover instrument packages, an acoustic signal is sent from the recovery vessel to the transponder to "wake it up." The transponder signals back with an acoustic pulse signifying it is awake and allowing the recovery vessel to determine more precisely the location of the transponder. A second signal is sent from the recovery vessel telling the transponder to release the anchor, thereby allowing the instrument package to float to the surface for recovery. This entire operation takes less than 30 minutes. When used to determine the position of a bottom cable, the device is attached to the cable. A survey vessel with differential GPS equipment wakes up the transponder with a ping, and the device sends back a ping that is used to echo range to the transponder. The survey vessel activates the transponder from several locations and is able to triangulate on the transponder and fix its position. Numerous transponders can be used with each transponder responding to a special "wake up code." The characteristics of the signal are a frequency of 7 to 40 kHz (although 10 to 15 kHz is most often used), a pulse length of about 10 ms, and a source strength on the order of 192 dB re 1 µPa at 1 m.

The second device is used to determine height from the bottom for deep-tow side-scan surveys. This device is mounted in the side-scan instrument package. Pings are transmitted from the device every second and are received by the tow vessel. The difference between the direct arrival and the ping reflected from the bottom gives the two-way time between the device and the bottom. Dividing this time by two and multiplying by the water velocity gives the height above bottom of the device. The characteristics of signals from this device are a pulse length of 2 ms (normally), a frequency of 12 kHz, and a signal strength of approximately 90 dB re 1 µPa at 1 m.

II. SEISMIC SOURCES

A. AIRGUNS

Airguns manufactured by either Bolt, Sodera, or Input/Output are the primary (almost exclusive) sources used in marine seismic surveys. The guns store compressed air in their chambers at 2,000 psi. The air is quickly discharged through a gun port, creating a pressure pulse and air bubble in the water. As the bubble travels to the surface of the water, it oscillates in volume, creating other pressure events. *(Note: **Appendix C - Introduction to Acoustics and Seismic Surveys** provides a further explanation of sound generated during seismic surveys.)* To release more energy into the pressure pulse and offset the deleterious effects of bubble oscillations on the pressure pulse, multiple airguns with various chamber sizes are used. These airgun chamber sizes vary from 20 to 380 cubic inches. In some cases, two or three airguns are placed in a cluster to increase the effective chamber size. Another technique to reduce the effect of air bubble oscillations is the GI airgun. The GI gun injects air into the bubble after it has formed to slow its collapse. The individual guns are suspended in the water from a float system referred to as a sub-array.

Each sub-array contains six or seven individual airguns spaced from 2.5 to 3 m apart, making the total sub-array length 14 to 17 m long. Typically three (sometimes four) sub-arrays are combined to form an array. When three sub-array elements are used, the spacing is 8 m between sub-arrays, and when four sub-arrays are used, the spacing is 12 m. Thus, the overall width of the array is generally 16 to 36 m (see **Figure D-2** for a typical array configuration). The array is towed at a depth of 5 to 7 m. For high-resolution work, the array is towed at a 3 m depth or shallower to improve the high-frequency characteristics of the pressure pulse. For VSP work, the array may be towed at 10 m to avoid downtime caused by sea state.

The sound pressure level (SPL) produced by the array is normally calculated at a distance at which the pressure pulse can be considered to be in the far-field (see **Appendix C, Section II.B - Peak Levels and Peak-to-Peak Levels**). The far-field begins at a distance below the center of the array where the seismic energy from all the sources in the array arrives at close to the same time. This depends on both frequency and array geometry. The formula for calculating the near-field/far-field distance is: distance equals frequency times the square of the longest dimension of the array divided by the velocity of sound in water. For the previously mentioned array, the near-field/far-field boundary is 69 m. The calculated SPL is normalized by assuming spherical spreading loss back to a point 1 m below the center of the array. Although this procedure allows comparison of relative energy outputs from different array designs, it does not represent the actual energy present at the center of the array. In fact, inside the array, the amplitude will depend upon which airgun is closest to the observation point. In the near-field, the SPL will be equal to the source strength of the largest airgun in the array minus some spreading loss. See **Table D-1** for airgun source levels. Note in **Table D-1** that airgun output data have been processed through low and high filters (i.e., there is a difference between filtered and unfiltered outputs). When airguns have significant energy above the filtered frequencies, the measured outputs are significantly different.

When the SPL is measured, a calibrated hydrophone is deployed (Johnston et al., 1988) approximately 300 m below the array, in the far-field, and sufficiently above the seafloor to prevent interference from seismic events reflected from the seafloor. Comparisons between measured and calculated results are very close (within 5%). Source levels measured and calculated this way are the source levels published by the various seismic contractors. Generally, the published source levels for single airguns are no more than 240 dB re 1 µPa at 1 m (zero-to-peak, "0-Pk" in **Table D-1**). The rms

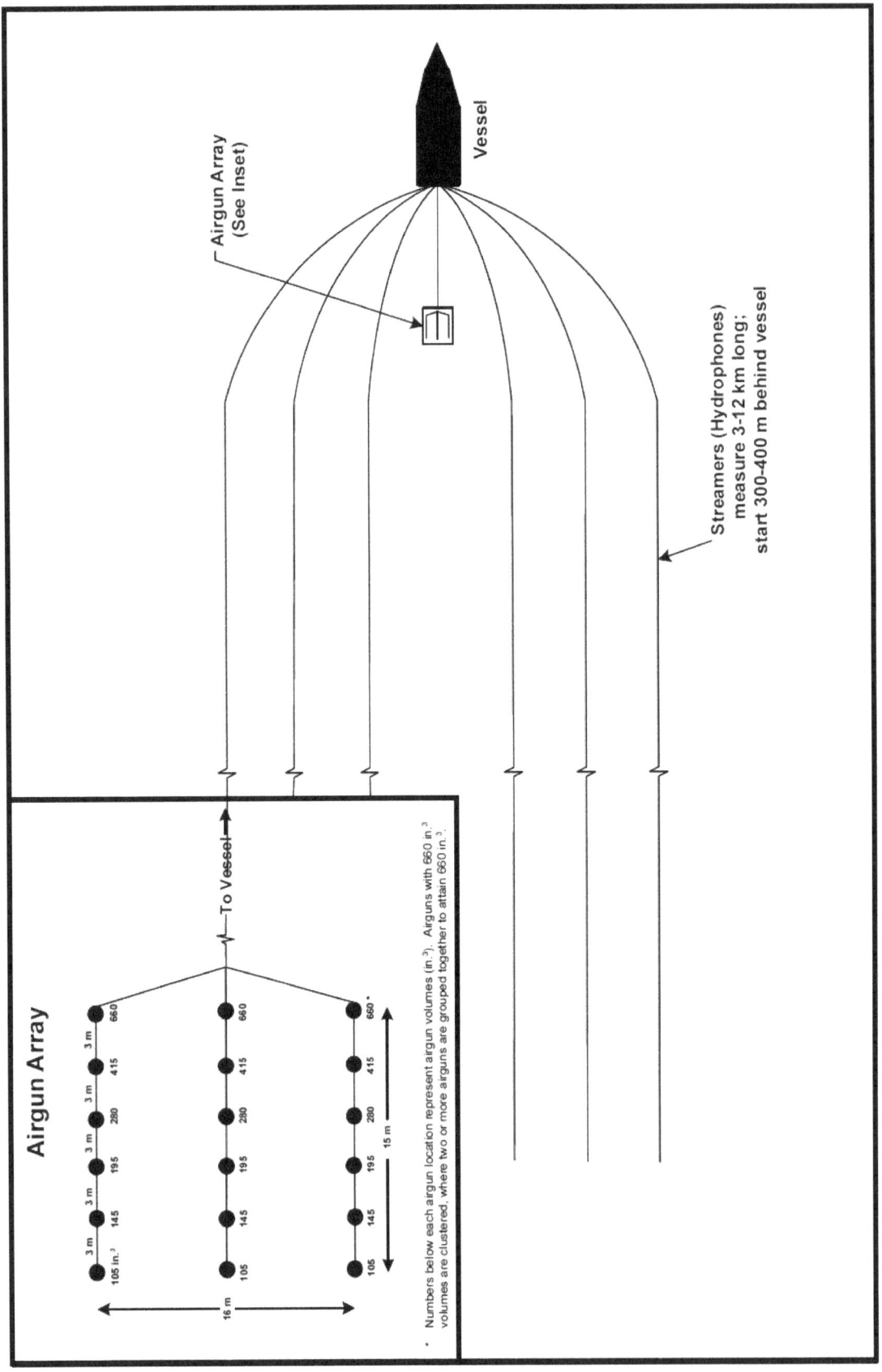

Figure D-2. Typical seismic array configuration.

Table D-1
Airgun Source Levels

Mfr.	Model	Chamber (cu. in.)	Pressure (psi)	Depth (m)	Filter[1] (lo/hi-Hz)	0-Pk Amp (bar-m)	0-Pk Amp (dB re 1 µPa-m)	Mode (if applicable)
Bolt	2800LLX	20	2000	5	0/1000	1.8	225	
	2800LLX	20	2000	2	0/1000	1.8	225	
	2800LLX	30	2000	4	0/512	1.9	226	
	1900LLX	150	2000	6	0/392	3	230	
	1900LLX	160	2000	5	0/1000	4.2	232	
	1900LLX	160	2000	5	0/128	2.05	226	
	1900LLXT	160	2000	5	0/128	2.25	227	
	1500LL	195	2000	4	0/128	2.5	228	
	1500LL	195	2000	5	0/1000	4.1	232	
	1500LL	195	2000	5	0/128	2.6	228	
	1500LL	200	2000	5	0/1000	4.1	232	
	1500LL	300	2000	6	0/392	4.35	233	
	1500LL	460	2000	5	0/1000	5	234	
	1900LL	2x40	2000	5	0/1000	3.8	232	
	1900LL	2x100	2000	5	0/1000	5.6	235	
	1500LL	2x200	2000	5	0/128	4.6	233	
	1500LL	2x200	2000	5	0/1000	7.6	238	
	1900LL	3x40	2000	5	0/1000	5	234	
	1900LL	3x40	2000	5	0/128	2.4	228	
	1900LL	3x100	2000	5	0/1000	7.2	237	
	1900LL	3x100	2000	5	0/128	4.25	233	
	1500LL	3x200	2000	5	0/128	6.1	236	
	1500LL	3x200	2000	5	0/1000	10.5	240	
Sodera	S15		2000	0.22	0/1000	1.5	224	Water Gun
	G	150	2000	6	0/128	2.45	228	
	G	2x250	3000	5	0/128	7.3	237	
	G	2x250	3000	5	0/128	6.5	236	
	GI	150	2000	6	0/128	1.85	225	Harmonic Mode
	GI	150	2000	6	0/128	1.95	226	GI Mode
	GI	90	2000	6	0/128	1.5	224	Harmonic Mode
	GI	90	2000	1.5	0/1000	2.8	229	High Resolution
	GI	90	2000	1.5	0/256	1.7	225	High Resolution
	GI	210	2000	6	0/128	2	226	Harmonic Mode
	GI	2x(45+105)[2]	2000	6	0/256	4.3	233	Deep Penetration
	GI	2x(45+105)	2000	6	0/128	3.5	231	Deep Penetration
Sodera	Mini-G	2x60	3000	1.5	0/1000	5.7	235	
	Mini-G	12	2000	1.5	0/1000	1.4	223	
	Mini-G	12	2000	1.5	0/256	0.8	218	
	Mini-G	12	3000	1.5	0/1000	1.8	225	
	Mini-G	12	3000	1.5	0/256	1.1	221	

Table D-1
Airgun Source Levels
(Continued)

Mfr.	Model	Chamber (cu. in.)	Pressure (psi)	Depth (m)	Filter[1] (lo/hi-Hz)	0-Pk Amp (bar-m)	0-Pk Amp (dB re 1 µPa-m)	Mode (if applicable)
	Mini-G	24	2000	1.5	0/1000	1.7	225	
	Mini-G	24	2000	1.5	0/256	1.1	221	
	Mini-G	24	3000	1.5	0/1000	2.3	227	
	Mini-G	24	3000	1.5	0/256	1.4	223	
	Mini-G	40	2000	1.5	0/1000	2	226	
	Mini-G	40	2000	1.5	0/256	1.3	222	
	Mini-G	40	3000	1.5	0/1000	2.6	228	
	Mini-G	40	3000	1.5	0/256	1.7	225	
	Mini-G	60	2000	1.5	0/1000	2.2	227	
	Mini-G	60	2000	1.5	0/256	1.6	224	
	Mini-G	60	3000	1.5	0/1000	2.9	229	
	Mini-G	60	3000	1.5	0/256	2.1	226	
	Mini-GI	40	2000	1.5	0/256	1.1	221	

[1] Filter is the filter applied to the signal prior to recording the signal on digital tape (i.e., prior to measurement of the pulse amplitudes).

[2] Indicates two guns in a cluster, each with a total volume of 150 cubic inches (i.e., generator volume = 45 cubic inches; injector volume = 105 cubic inches).

source level is 10 dB less or 230 dB re 1 µPa at 1 m. The rise time of the pulse is on the order of 2 ms or less from the first onset to the first peak.

The spectral content of the airgun pressure pulse is dependent upon the number, geometrical layout, and volumes of the individual airguns as well as the depth at which the airgun/airgun array is towed. The primary characteristic of the amplitude spectrum is a series of notches (see **Figure D-3**), which are dependent upon the depth the array is towed. The lowest frequency notch occurs at a frequency that is the inverse of the travel time from the airgun to the surface to the airgun. (For example, for a tow depth of 7.65 m and a water velocity of 1,506.9 m/s, the travel time is 7.65 x 2/1,506.9 = 0.010 sec, or a frequency of 100 Hz). Subsequent notches occur at uniform intervals in the amplitude spectrum. Sound at the frequencies of notches is thus greatly attenuated and may not penetrate the seafloor. The spectrum will usually have significant energy at 10 Hz, reaches its peak frequency about 50 Hz, and falls off roughly at 0.06 dB/Hz to 50 dB down at 900 Hz.

Although the typical pressure in the airgun is 2,000 psi (1.38×10^7 Pa), the pressure can be increased to 3,000 psi (2.07×10^7 Pa) in some models of airguns, which increases the amplitude of the primary pressure pulse by 50%.

Airgun arrays are directive, as can be seen in **Figure D-4** (i.e., broadside [a] and inline [b] directivity). Arrays are designed to focus the energy from the airgun array down toward the seafloor and attenuate the energy directed toward the streamer cables. If one chooses a frequency of 60 Hz (which is about the peak of the amplitude spectrum for a typical array; see **Figure D-3** for the purpose of illustration), then one can simplify, somewhat, the array effects of the array. For example, for energy propagated directly forward and backward from the array along the water surface (±90° in **Figure D-4[a]**), the attenuation of the seismic energy pulse is approximately 60 dB. For energy propagated at an angle of 30° below the surface, the attenuation is about 20 dB. For energy propagated perpendicular to the array along the water surface, the attenuation of the seismic energy pulse is again approximately 60 dB (**Figure D-4[b]**), while for energy directed 30° below the surface, the attenuation is also about 20 dB. For the purposes of determining where the sound reaches a specific level (e.g., 180 dB), one can conservatively use 20 dB to account for the "array effect." At a distance of 200 m from the array center that would be a depth of 100 m, plus a tow depth of the array of 10 m gives a total depth of 110 m. Therefore, at water depths above this surface, the seismic energy is attenuated at levels of 20 dB or greater (Hrvoic, 1999). IAGC has noted that the array effect is more pronounced at higher frequencies. Nevertheless, use of 20 dB for the nominal array effect in the horizontal direction remains conservative. Additional discussion of the array effect is provided in **Appendix C, Section IV - Polar Properties of Airgun Arrays in the Environment**.

B.   HIGH-RESOLUTION AIRGUNS

Airguns used in high-resolution site surveys are manufactured by Seismic Systems Inc. (Sodera) and are its GI series guns. The GI series guns are made up of two guns within the same casing. One gun, the generator, is fired, creating the primary pulse and an air bubble. The second airgun, the injector, is fired into the bubble when the bubble encompasses the ports of the injector gun, which injects air into the bubble produced by the generator. This allows for better control and suppression of the water bubble oscillations. GI airgun volumes for high-resolution work are typically 90- to 150-cubic inches (90 cubic inches is one 45-cubic inch generator gun and one 45-inch injector). Output from a 90-cubic inch airgun is about 229 dB re 1 µPa at 1 m with a spectral content from 40 to 300 Hz (6 dB down points from peak at 130 Hz). Output from the 150-cubic inch cluster (one 45 cubic inches + one 105 cubic inches) GI airgun is 226 dB re 1 µPa at 1 m (high cut filter at 128 Hz). Airgun pressures are typically 2,000 psi, although they can be used at 3,000 psi for more output. For 3D high-resolution work, a tri-cluster of 8- to 10-cubic inch GI airguns is used.

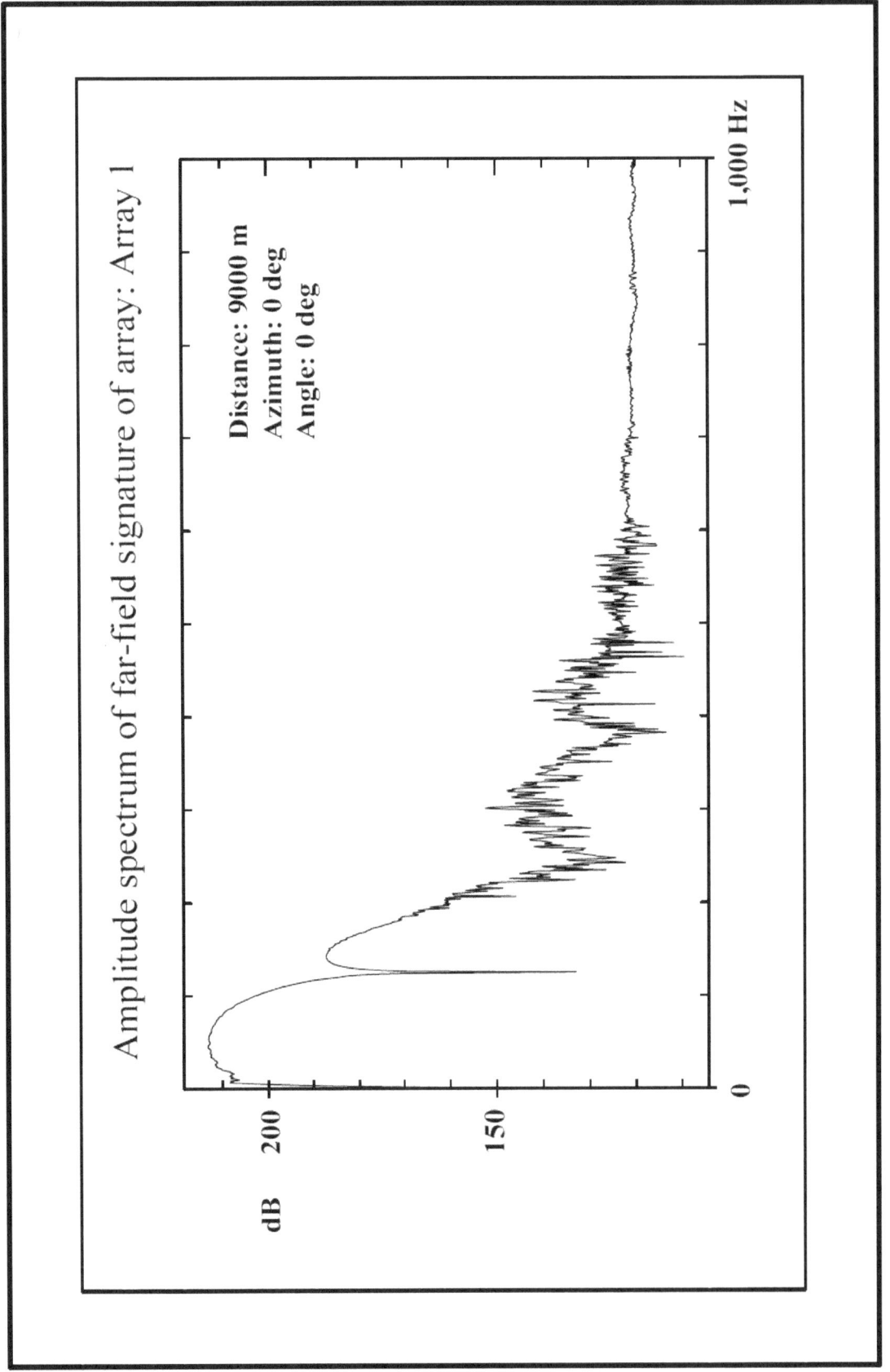

Figure D-3. Amplitude spectrum of far-field signature of a seismic array.

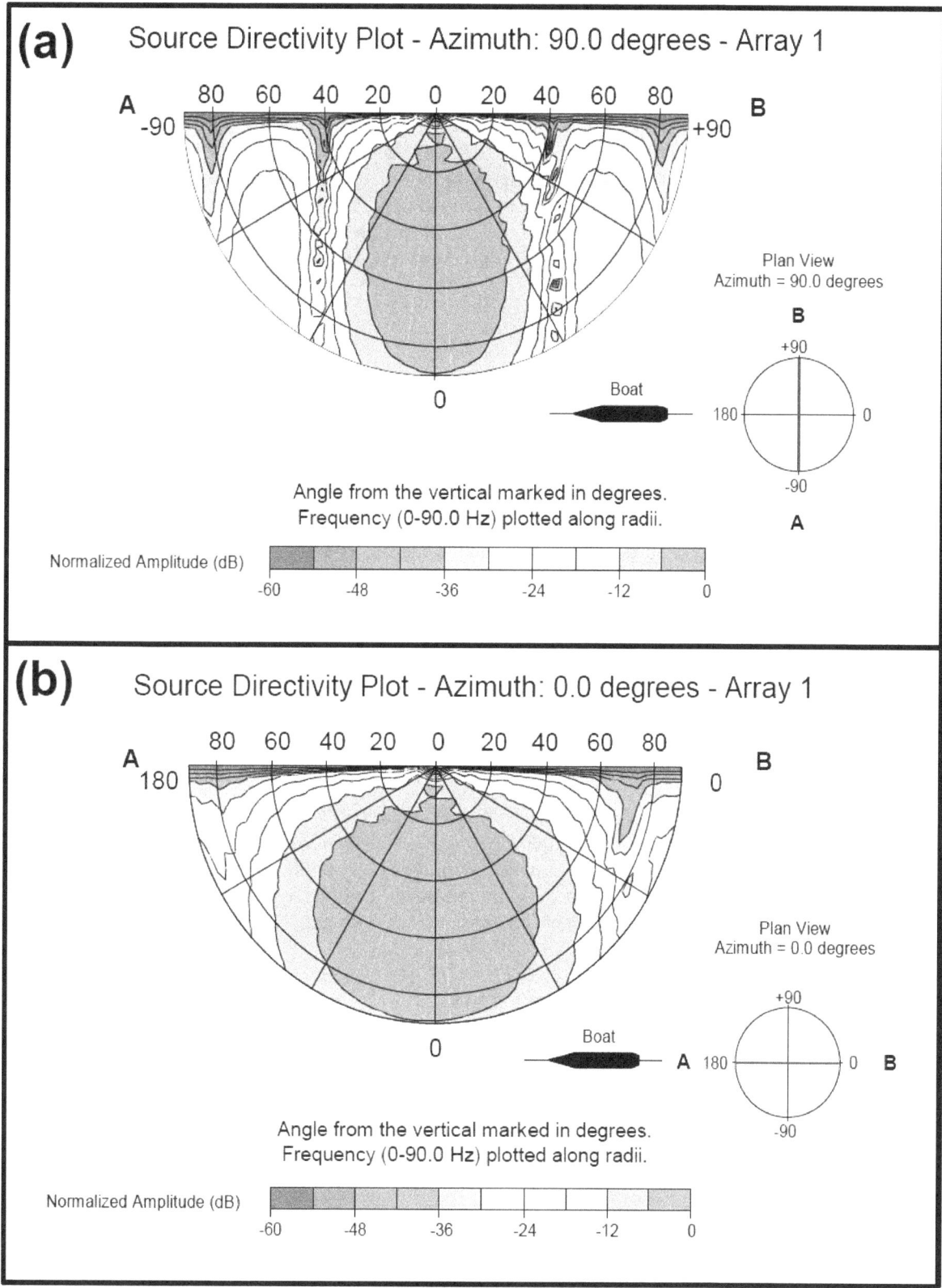

Figure D-4. Source directivity plots showing attenuation (a) broadside to and (b) inline with a seismic array.

C.   WATER GUN SOURCES

Water gun sources have been used in the past when exceptionally high frequencies are desired to be recorded. Water guns are rather small with a length of 546 mm, width of 152 mm, and weight of 13 kg. Water guns use compressed air as do airguns, but instead of releasing the air into the water, the compressed air is used to propel a water jet that creates vacuum cavities, which when imploded by the surrounding hydrostatic pressure, emit a strong, bubble free, high frequency pulse. A typical signature has a source strength of 224 dB re 1 µPa at 1 m (zero-to-peak) and a spectrum with maximum amplitude at around 1,000 Hz, with -6 dB points at 380 and 2,250 Hz. Because of the superior performance of the generator and GI guns (see **Section II.A. - Seismic Sources, Airguns** of this appendix and **Table D-1**), the water gun is being phased out.

D.   VIBROSEIS

Vibroseis sources in a marine environment are rarely used. Marine vibrators rely on changes in volume created by differential displacement over a period of time rather than one explosive moment. One system was developed by Seismograph Service Co. and deployed in the early 1970's, and a second system was developed by Industrial Vehicles Inc. in the 1980's and deployed in the mid-1980's through the mid-1990's. The second system was used in the North Sea and saw limited use in the Gulf of Mexico.

There are few published details on this second marine vibrator system. One test of the marine vibrator source was conducted in Seneca Lake in New York. In that test they found the energy level of a vibrator emitting a single 8- to 120-Hz sweep with fixed amplitude control was comparable to a 750-cubic inch airgun with a modeled output of 24-Bar-m (247 dB re 1 µPa at 1 m) through a high cut filter of 125 Hz (Smith and Jenkerson, 1998). This high output seems hard to believe (e.g., output expressed as peak to peak; zero-to-peak would be 6 dB less), and actual data would have to be checked before using this figure. In a later test in the transition zone offshore Louisiana, a four-vibrator array output was 223 dB re 1 µPa at 1 m, which seems a more likely level. This output, one would assume, is the total energy output of the vibrator and not the distributed output over the sweep length.

Since the vibrator source emits vibrations for a nominal 10 s then is off for 7 s or more, the signal is continuous rather than impulsive. To compare a vibroseis signal on a one-to-one basis, one would have to take the sum of squares of the vibroseis sweep amplitudes and divide by the length of the sweep and compare that to the sum of squares of the airgun impulse divided by the time duration of the airgun pulse. This ratio would compare the powers of each source. When such data become available, this comparison should be made.

III. SEISMIC DETECTOR SYSTEMS

A.   STREAMER CABLES

Towed streamer cables are used to house the hydrophones, which record the seismic signals initiated by the seismic source. In a typical scenario, 12 streamer cables, normally 100 m apart, are towed at a depth of 7 to 10 m behind the ship for 3D surveys; only one cable is towed for 2D surveys. The number of streamer cables varies depending upon ship size and streamer technology. The cables are on the order of 4 to 8 km long, are 50 to 70 mm in diameter, and are made up in sections 75 to 100 m long. Contained within these sections are 6 to 8 groups of hydrophones to detect the seismic energy and a kerosene-like fluid for cable buoyancy. Normally, each section contains roughly 100 to 200 L of fluid. Breaks in the cable are rare and usually occur when currents whip the cables around some installation such as a producing platform. Fish bites from large fishes may also occasionally puncture towed streamer cables. When a break does occur, it is only for one section of cable, and the released fluid evaporates rapidly. Newer cables do not contain buoyancy fluids but instead are filled with a polymer for flotation so cable breaks do not result in a loss of fluid.

The number of streamer cables, their separation, and their length cover an area immediately below the sea surface known as its footprint. For example, a ship towing five streamers, 8 km long with each streamer separated from the other by 100 m leaves a footprint of 3.2 square kilometers. Footprints can be as large as 8 square kilometers and as small as 2 square kilometers.

As the cable travels through the water, it is subject to mechanical vibrations (strumming) and flow noise as well as environmental effects such as sea state and background noise. With a group sensitivity of 14 to 20 volts per bar, noise levels on the cable can be as high as 12 micro-bars but are usually on the order of 1 to 4 micro-bars. Of course, the noise levels are higher on the groups closest to the ship. Noise levels are routinely measured at the beginning and end of each line.

B.   OCEAN BOTTOM CABLES

Ocean bottom cables are designed to lie on the ocean bottom rather than be towed behind a ship. The main difference between the (fluid filled) streamer cables and bottom drag cables is the lack of enclosed fluid for buoyancy, since the bottom drag cables sit on the bottom. These solid cables are 4 to 5 cm in diameter and come in 2,500-m sections. The sensors are plugged into the cable, wrapped with tape, and sealed with silicone gel.

The hydrophone sensors, used with bottom drag cables, have (until recently) limited the depth of operations for these cables to about 180-m water depth. However, new technology has been developed that allows bottom cables to be deployed at depths up to 2,500 m or more. This technology involves newly designed hydrophones, geophones, and capsules, the latter of which enclose the hydrophone and geophone. New hydrophones use a large area, cylindrical, piezoelectric crystal that allows the crystal to expand and contract in response to pressure variations. The wall of the cylindrical capsule serves as a pressure diaphragm for the outer wall of the piezoelectric cylinder to which it is coupled. The inner wall of the piezoelectric crystal is air-backed and free to contract. The pressure variations from outside the capsule are thus transferred to the piezoelectric crystal, producing a piezoelectric signal proportional to the pressure variations outside the capsule. The new geophones use three, mutually perpendicular, long throw coils that are specially constructed. A long throw of the coil surrounding the magnet is required to enable the geophone to respond, regardless of the orientation of the geophone axis. Sensors inside the capsule measure angle of tilt of the capsule, but the exact orientation of the geophone must be determined by recording seismic waves from a known source location. Typically, the sensors and the transmission cables are bundled together. This requires analog converters and telemetry electronics to be included in the capsule. The capsules are linked together with special armor wrapped cable to withstand the pressures at 2,500 m, and constructed to support its own weight and survive the demands of deployment and retrieval. It is noteworthy that, as these systems are deployed at deeper water depths, both the physical demands upon the system and the concern for the integrity of the system increase.

C.   HIGH-RESOLUTION STREAMER CABLES

Teledyne analog streamer cables are used for high-resolution work. The clusters of hydrophones are spaced 12.5 m apart. The streamer cables are filled with fluid for buoyancy and are towed at a depth of 3 m, although in calm water conditions the cables can be towed at 1.5 m.

D. VERTICAL CABLES

Vertical cables used by the exploration industry were originally developed by the Navy as passive listening devices. Each consists of an active section and lead-in section. At the bottom of the active section are a 680-kg steel weight and a transponder for locating the position of the cable. In the active section are 16 specially designed hydrophones at 25-m intervals, which makes the active section 375 m long from first to last hydrophone. At the top are buoyant floats to keep the cable as vertical as possible. Attached also to the top of the active section is the lead-in section leading to the surface recording package.

IV. SHIPS

A. GENERAL

Most ships used on seismic surveys are of modern construction (usually built in Norway) and have the latest safety and communications equipment on board. Since the ships are used for seismic work, they use the very latest propeller designs to ensure low noise generation by the ship. In addition, the diesel electric engines commonly used are very efficient, with low gas emissions from the stack. In most cases the actual gases emitted are measured by the manufacturer at the time of installation and are re-measured during scheduled maintenance. Some ships have continuous monitoring of smoke stack emissions.

B. 3D OPERATIONS

Ships used for 3D seismic streamer work in the Gulf of Mexico are usually on the order of 80 to 90 m long with a beam of 10 to 20 m and a gross tonnage of about 4,000 tons. Some of the larger ships, like the Ramform type, are 86.2 m long, 39.6-m beam, with a gross tonnage of 9,000 tons. Other large ships like the Geco Beta are 92.3 m long, 19.6-m beam, with a gross tonnage of 4,404 tons. A typical ship size would be 85 m long, 20-m beam, with a gross tonnage of 6,354 tons.

C. 2D OPERATIONS

Ships used for 2D seismic streamer work, although rare, are usually smaller ships, although a ship outfitted for 3D work can do 2D work as well. A typical ship size is length 70 m, beam 18 m, with a gross tonnage of 1,573 tons. Both 3D and 2D streamer work are conducted at ship speeds of 4 to 5 kn (typically 4.5 kn). Since streamers are routinely retrieved at the end of a project, cruise speeds between projects are considerably higher and are about 12 to 14 kn.

D. OCEAN BOTTOM CABLE OPERATIONS

Ships used for ocean bottom cable work are even smaller, with lengths of 67 m, 14 m-beam, and a gross tonnage of 1,771 tons. Although streamer work can be done with one vessel, ocean bottom cable work requires the use of up to six ships (i.e., one shooting; one recording; two cable layout/pickup; two smaller, all purpose boats for trouble shooting, survey, etc.)

E. HIGH-RESOLUTION OPERATIONS

Ships used in high-resolution operations are generally smaller vessels, as the equipment used is less complex than that used in the normal seismic 2D and 3D surveys. Ships are generally 37 to 47 m long, with the longer ships used for the 3D high-resolution work. The ships are designed to be ultra-quiet as the higher frequencies used in high-resolution work are easily lost in the noise if special attention is not paid to keeping the ships quiet.

F.   VERTICAL CABLE OPERATIONS

Two identical ships are used in vertical cable operations. These ships are 60 to 80 m long with a 14 to 16-m beam.

V.   LEVEL OF ACTIVITY

A.   CHARACTERIZATION OF ACTIVITY

Although gravity, magnetic, and electromagnetic surveys are conducted in the Gulf of Mexico, by far the greatest activity is seismic data acquisition. The principal seismic technology is 3D seismic, using either streamers or bottom cables. Time-lapse (4D) surveys are becoming more frequent as the technology for analyzing the data is developed. Four-component (4C) seismic surveys conducted with bottom cables and vertical cable surveys are in their infancy.

In developing a level of activity, only the seismic techniques are considered, as they are the main focus of the environmental analysis in this PEA. The number of lease blocks surveyed is one measure of activity. Seismic survey activity is typically measured by the number of active "crews." Each crew is a complete entity with the ability to collect seismic data. Crews can be ships towing streamer cables, or a fleet of ships collecting either bottom cable or vertical cable seismic data. For the purposes of describing the level of activity, the period from 1988 to the present adequately characterizes the recent level of G&G activity in the Gulf of Mexico.

Another measure of activity is permits issued by MMS for 3D seismic surveys. While permits do not cover seismic surveys conducted under lease terms, most 3D seismic surveys are conducted under permit. Permit records for proposed 3D seismic operations from 2002 to date (August 2003) were reviewed by MMS. A summary of 3D seismic survey, by permit, is outlined in **Table D-2** and graphically depicted in **Figures D-5** and **D-6**. There is considerable variability in the size and location of 3D surveys conducted in the Gulf of Mexico, with most seismic activity occurring in the Central Planning Area, whether in shallow (<200 m) or deeper waters (≥200 m).

While it is of interest to project future levels of G&G seismic activity in the Gulf of Mexico, such predictions must be viewed as gross approximations because many factors influence activity levels, as detailed below. For the purposes of this analysis, recent trends in permitting levels were evaluated, current driving factors considered, and estimates of future survey activity projected.

B.   FACTORS DRIVING ACTIVITY

1.   Economics

The price of oil and natural gas is the predominant driver for exploration activity. Prices for oil are determined on the world market and are primarily set by the Organization of Petroleum Exporting Countries (OPEC). Over the period from August 1985 to May 2003, the closing futures price of Light Sweet Crude Oil has varied from a low in 1998 of $11.21 per barrel to a high of $36.09 per barrel in 1990 (Haver Analytics, written comm., 2003). Over a 10-year period from October 1988 to December 1998, the price of oil began at a low of $11.63 and rose gradually to an average price of about $20 (with one spike at $36.09 in October 1990), then declined to another low of $11.21 in December 1998. Since then,

Table D-2

Summary of 3D Seismic Survey Activities Throughout the U.S. Gulf of Mexico Outer Continental Shelf from January 2002 through August 2003, as Compiled from Minerals Management Service (MMS) Permit Records (From: R. Brinkman, MMS, oral comm., 2003)

Permit Number	Tracklines Covered [line miles (km)]		Line Miles (and Percentage) by Survey, Water Depth, and Planning Area (PA)					
			Shallow (<200 m)			Deep (≥200 m)		
			Western PA	Central PA	Eastern PA	Western PA	Central PA	Eastern PA
2002 Permits								
L02-001	9,150	(14,722)	-	458 (5)	-	-	8,692 (95)	-
L02-007	24,900	(40,064)	-	-	-	-	22,410 (90)	2,490 (10)
L02-008	6,050	(9,734)	-	6,050 (100)	-	-	-	-
L02-016	2,800	(4,505)	-	-	-	-	2,800 (100)	-
L02-019	5,850	(9,413)	-	-	-	-	5,850 (100)	-
L02-027	9,350	(15,044)	-	9,350 (100)	-	-	-	-
L02-039	22,500	(36,202)	-	-	-	-	13,500 (60)	9,000 (40)
L02-049	2,100	(3,379)	-	420 (20)	-	-	-	1,680 (80)
L02-054	3,900	(6,275)	-	-	-	-	3,900 (100)	-
L02-057	4,250	(6,838)	-	-	-	-	4,038 (95)	212 (5)
L02-058	14,700	(23,652)	-	4,410 (30)	-	-	10,290 (70)	-
T02-001[1]	8,000	(12,872)	200 (2)	-	-	7,800 (98)	-	-
T02-002	11,450	(18,423)	-	-	-	11,450 (100)	-	-
T02-007	2,800	(4,505)	-	-	-	2,800 (100)	-	-
T02-009	800	(1,287)	-	-	-	-	800 (100)	-
T02-018	1,000	(1,609)	1,000 (100)	-	-	-	-	-
T02-022	15,400	(24,779)	-	-	-	15,400 (100)	-	-
Total Tracklines (2002) [miles (km)]	145,000	(233,305)	1,200 (1,931)	20,688 (33,287)	0 (0)	37,450 (60,257)	72,280 (116,299)	13,382 (21,532)
% of Total, 2002	-		0.8	14.3	0	25.8	49.8	9.2
2003 Permits								
L03-001	1,400	(2,253)	-	1,400 (100)	-	-	-	-
L03-007	5,650	(9,091)	-	5,650 (100)	-	-	-	-
L03-008	16,150	(25,985)	-	4,845 (30)	-	-	11,305 (70)	-
L03-012	800	(1,287)	-	-	-	800 (100)	-	-
L03-013	6,300	(10,137)	-	-	-	-	6,300 (100)	-
L03-028	9,950	(16,010)	-	-	-	-	9,950 (100)	-

Table D-2

Summary of 3D Seismic Survey Activities Throughout the U.S. Gulf of Mexico Outer Continental Shelf from January 2002 through August 2003, as Compiled from Minerals Management Service (MMS) Permit Records (From: R. Brinkman, MMS, oral comm., 2003)

(Continued)

Permit Number	Tracklines Covered [line miles (km)]		Lines Miles (and Percentage) by Survey, Water Depth, and Planning Area (PA)						
			Shallow (<200 m)			Deep (≥200 m)			
			Western PA	Central PA	Eastern PA	Western PA	Central PA	Eastern PA	
L03-029	4,650	(7,842)	-	1,860 (40)	-	-	2,790 (60)	-	
T03-002	7,750	(12,470)	-	-	-	7,750 (100)	-	-	
T03-003	3,850	(6,195)	-	3,850 (100)	-	-	-	-	
T03-007	2,350	(3,781)	2,350 (100)	-	-	-	-	-	
Total tracklines, 2003[2] [miles (km)]	58,850	(94,690)	2,350 (3,781)	17,605 (28,326)	0 (0)	8,550 (13,757)	30,345 (48,825)	0 (0)	
% of Total, 2003	-		4.0	29.9	0	14.5	51.6	0	

[1] Same permit as T03-002; continuation of permit from 2002 into 2003; estimated line miles/line kilometers adjusted accordingly.
[2] 3D permits through August 2003.

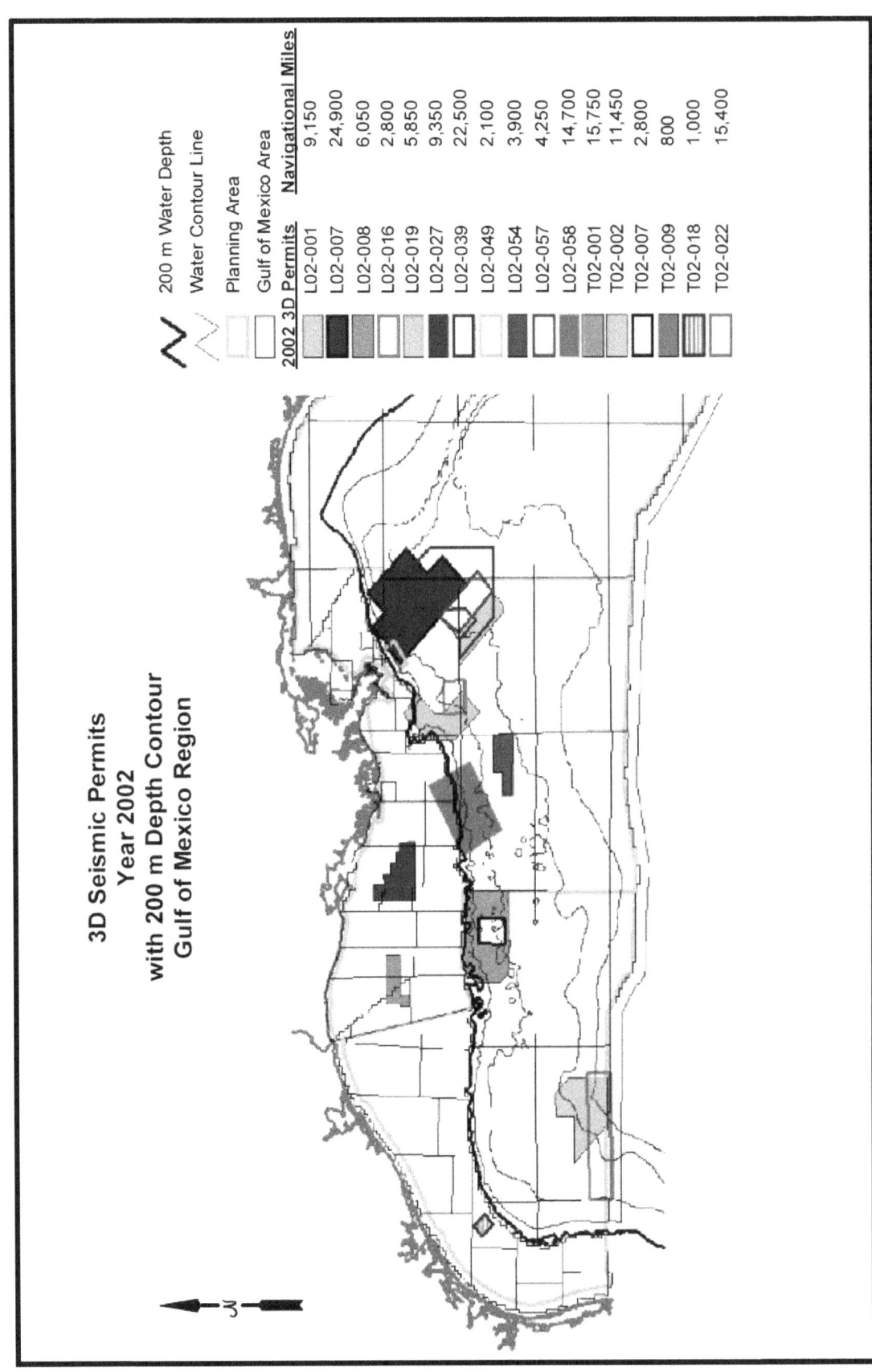

Figure D-5. Location of 3D seismic surveys permitted by the Minerals Management Service in the Gulf of Mexico, 2002.

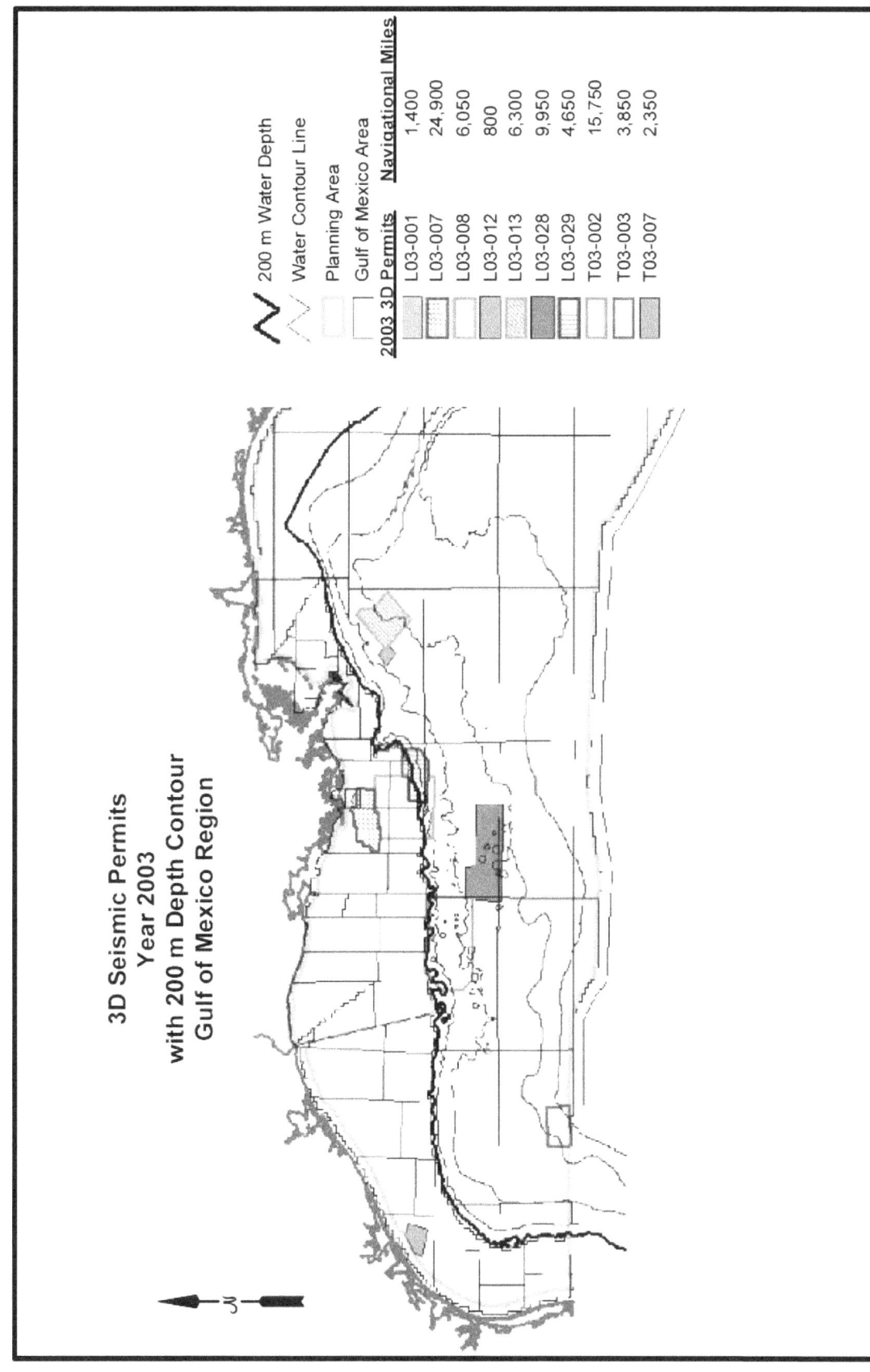

Figure D-6. Location of 3D seismic surveys permitted by the Minerals Management Service in the Gulf of Mexico, through August 2003.

there have been two peaks (i.e., one in November 2000 at $34.16, a second in February 2003 at $35.78) interspersed with a low of $19.53 in December 2001. Most recently (May 2004), the price of oil has reached historic high levels of around $40 per barrel. The general trend seems to be prices increasing over the next few years.

The price of gas is less dependent upon the world market, as gas is usually delivered from field to market by gas pipelines. Gas prices have shown a gradual increase during the decade of the 1990's from a price of about $1.80 per million British thermal units (MMBTU) (closing futures price New York Mercantile Exchange [NYMEX]) to a price of about $2.70 per MMBTU in December 2001 (Haver Analytics, written comm., 2003). One price spike occurred during that time in December 2000 when the price rose to $8.32 per MMBTU. Recently (June 2003), prices seem to be on another rise; however no trend in prices can be discerned. During the projection period, relatively short term price spikes for natural gas may be expected to occur (e.g., as occurred during winter 2000-2001), and the price can be expected to increase over the levels in the 1990's.

This oil and gas price scenario assumes a relatively stable economic environment over the long term with regard to oil and gas producing prices. Offsetting this optimism is the current oversupply of marine seismic vessels for exploration for new and exploitation of discovered fields, and the economic viability of the seismic contracting business. This has caused some vessels to sit idle, and profits for geophysical exploration companies to become elusive. Some reduction in vessel capacity will occur along with potentially drastic reductions in crew size (see **Section III.V.C. - Correlation of Factors to Activity** of this appendix).

2.   Geography

Oil and gas exploration on the continental shelf of the northern Gulf of Mexico is in a mature state, although large discoveries are expected in deeper waters. From a seismic exploration view, about 900 blocks in the Western and Central Planning Areas have not yet been surveyed with 3D seismic techniques (R. Brinkman, MMS Gulf of Mexico OCS Region, oral comm., 2003). For the purpose of this analysis, several factors were considered in the development of activity level projections. Annual survey activity levels during the past several years were considered (i.e., 1993-2002), as well as the number of permits issued through August 2003. On the basis of trends evident in the past several years, it has been assumed that remaining blocks would most likely be surveyed over the next several years, with a gradual tapering off of activity. Resurveying of blocks has also been integrated into the projections. It is assumed that a lower level of new seismic survey activity will occur in the Eastern Planning Area relative to the remaining two Planning Areas (i.e., the vast majority of survey activities are expected in the Central and Western Planning Areas). Industry interest in the Eastern Gulf has historically been limited to the westernmost portions of the planning area (e.g., Destin Dome 56 Unit; Lease Sale 181 area).

3.   Technology

The development of new technology has been a key factor in continuing new seismic activity. Technology improvements to the 3D method are expected to cause previously surveyed blocks to be resurveyed. In the past, improvements in technology have caused resurveying at approximately 8-year intervals and reprocessing of data every 5 years. However, current economic conditions have caused seismic contractors to decrease their investments in new technology. Indeed, these contractors are reducing the number of crews in order to reduce cash flow. Because of these events, new technology will take longer to come into the mainstream, and the resurvey time will be on the order of 10 to 15 years. 4C bottom cable and vertical cable surveys will also be done over older 3D surveys in order to collect additional or new information. Time lapse surveys will be done over existing fields. On occasion, these time lapse surveys will be 4C.

C.   CORRELATION OF FACTORS TO ACTIVITY

The economic factor will be the most important driving factor over the next several years, an assumption based on the relative stability or instability of oil and gas prices and the economic health of the exploration industry. Over the past 10-12 years, periods of relative price stability have been interspersed with rapid fluctuations due to regional and worldwide instabilities and conflicts. The oil and gas pricing scenario assumes a relatively stable economic environment over the long term. The geographic factor is projected to cause the survey of blocks of interest in the Gulf of Mexico that have not been surveyed. On the basis of a review of permit history and MMS consultation, it has been estimated that 900 blocks remain to be surveyed; such new surveys are expected to occur between 2004 and 2008, with highest activity levels expected in the near term. Technology will be the driving factor that will require each block (i.e., each block already surveyed) to be re-shot with a new survey using either streamer cables, water bottom cables with hydrophones and vertical geophones (2C), water bottom cables with 4C, or vertical cables. Normally, new technology would prompt the resurveying of blocks with an approximate 8-year delay, but with the decrease in investment in new technology, that period should expand to approximately 14 years. All blocks without production will likely be resurveyed; those with production will most likely be resurveyed with time lapse surveys. Adding these two numbers together (newly surveyed and resurveyed blocks) gives the number of blocks available for crew activity. This figure varies from 1,041 in 2003 to 6,520 blocks in 2011. The numbers of lease blocks surveyed since 1988 and the numbers of seismic crews working are presented in **Table D-3**.

The methodologies used to compile **Table D-3** included

- Line miles surveyed were derived from the MMS database;
- Actual blocks surveyed also were provided via the MMS database;
- Estimated blocks surveyed were calculated by
    - dividing line miles by actual blocks surveyed during the period 1992 – 1996;
    - assuming an estimate of 230 line miles per block; and
    - dividing line miles by 230 to get estimated number of blocks.
- Estimated blocks remaining to be surveyed were calculated by
    - using original MMS estimate of 1,800 blocks, with 900 blocks being surveyed in 2001 and 2002; and
    - assuming that 900 blocks were surveyed in 2001 and 2002, leaving 900 blocks remaining to be surveyed over a 6-year period.[1]
- First repeat survey has been extended to 14 years because the technology has not improved at the rapid pace displayed in the 1990's and the economic condition of the contractors is poor.
- "Total blocks" represents the sum of estimated blocks surveyed, remaining blocks to be surveyed, and numbers of blocks in the first repeat survey.
- "Actual crews total," 3D and 2D were obtained from World Geophysical News.
- Estimated number of 3D crews was calculated based on
    - using an estimate that 3D surveys were 10 times as efficient as 2D crews in covering a given area, implying that 90% of the blocks surveyed were done by 3D crews;
    - using total blocks, multiplying by 0.9 to get estimate of the number of blocks surveyed by 3D, then multiplying that number by .009 (i.e., crew efficiency); and
    - calculating crew efficiency by dividing crews by 0.9 total blocks for the years 1994-1999, resulting in a median value of .0097 and an average value of .0088. Analyst selected an average value between these two, or 0.009.

---

[1] On the basis of earlier survey activity levels, it has been estimated that approximately 230 line miles were covered per block. With the evolution of G&G survey operations and considering that 6 to 12 streamers are deployed during a 3D survey covering a breadth of 600 m (1,968 ft) or more, a total of eight transects is currently completed per block (24 line miles). MMS estimated future seismic activity, assuming that 16 transects were completed per block, equivalent to ~48 line miles per block, an estimation based on navigation plots from actual surveys. This estimate has been rounded up to 50 miles of survey transects per block.

Table D-3

Lease Blocks Surveyed[1] and Numbers of Seismic Crews under the Proposed Action

Year	Line Miles[2]	Actual Blocks Surveyed	Estimated Blocks Surveyed[3]	MMS Blocks Surveyed[4]	Repeat Blocks[5]	Total Blocks	Crews Total[6]	Crews 3D	Crews 2D	Crews 2D (est.)	Crews 3D (est.)	Crews Total (est.)
1988	193,089	--	841	841	--	841	--	--	--	--	--	--
1989	187,402	--	816	816	--	816	31	--	--	--	--	--
1990	369,540	--	1,610	1,610	--	1,610	25	--	--	--	--	--
1991	363,259	--	1,583	1,583	--	1,583	21	--	--	--	--	--
1992	384,435	1,825	--	1,825	--	1,825	23	--	--	--	--	--
1993	641,514	1,913	--	1,913	--	1,913	31	21	10	--	--	--
1994	566,142	2,488	--	2,488	--	2,488	34	29	5	--	--	--
1995	636,824	3,112	--	3,112	--	3,112	32	28	4	--	--	--
1996	637,311	3,045	--	3,045	--	3,045	37	30	7	--	--	--
1997	--	6,520	--	6,520	--	6,520	50	38	9	--	--	--
1998	--	5,183	--	5,183	--	5,183	41	33	8	--	--	--
1999	--	3,612	--	3,612	--	3,612	15	8	7	--	--	--
2000	--	3,003	--	3,003	--	3,003	17	8	9	--	--	--
2001	--	3,648	--	3,648	--	3,648	15	6	9	--	--	--
2002	--	3,698	--	3,698	841	4,539	--	--	--	--	--	--
2003	--	--	--	225	816	1,041	--	--	--	8	8	16
2004	--	--	--	225	1,610	1,835	--	--	--	8	15	23
2005	--	--	--	113	1,583	1,696	--	--	--	8	14	22
2006	--	--	--	113	1,825	1,938	--	--	--	8	16	24
2007	--	--	--	112	1,913	2,025	--	--	--	8	16	24
2008	--	--	--	112	2,488	2,600	--	--	--	8	21	29
2009	--	--	--	--	3,112	3,112	--	--	--	8	25	33
2010	--	--	--	--	3,045	3,045	--	--	--	8	25	33
2011	--	--	--	--	6,520	6,520	--	--	--	8	53	61
2012	--	--	--	--	5,183	3,270	--	--	--	8	42	50
2013	--	--	--	--	3,612	3,612	--	--	--	8	29	37
2014	--	--	--	--	3,003	3,003	--	--	--	8	24	32
2015	--	--	--	--	450	450	--	--	--	8	4	12

[1] Tabular data represent prelease survey activity levels only; data are unavailable for postlease surveys or other surveys (e.g., pipeline rights of way, scientific research).

[2] Records maintained by MMS (Gulf of Mexico OCS Region) during 1988-1991 were in line miles only regardless of whether the survey was 2D or 3D; during 1992-1996, both line miles and number of blocks surveyed were also recorded, allowing for calculation of blocks/mile. After 1996, only blocks are recorded regardless of survey type (2D, 3D). Line miles of data collected are equal to the length of transects surveyed times the number of streamers. For example, a survey of one mile transect distance by a ship towing six streamers results in 6 line miles of data collected.

[3] Estimated blocks surveyed during the period 1988-1991 based on line miles traveled and blocks/mile data (for the period 1992-2002).

[4] Projections (beyond 2002) are approximations based on the number of remaining unsurveyed blocks (900).

[5] Repeat blocks represent those blocks to be resurveyed.

[6] The term "crews" refers to a seismic data acquisition operation. In most cases, it represents a ship towing streamer cables, but can represent multiple ships (e.g., bottom cable surveys; vertical cable surveys). Crew statistics courtesy of IHS Energy Group, World Geophysical News.

- 2D crew levels were kept constant at eight crews due to 1) evidence that a number of individuals are getting into the contracting business using 2D crews and less expensive boats, and 2) the number of 2D crews appeared to remain relatively constant from 1997 to 2003.

A simplified trend analysis was also used to project the number of seismic surveys expected in the foreseeable future in the Gulf of Mexico, as outlined in **Table D-3**. Recent historical survey activity was considered, as were the number of unsurveyed blocks and projected resurvey activities (i.e., estimated 14-year resurvey cycle). Historical activity levels were derived from MMS permit application records and are graphically presented in **Figures D-7** through **D-12**. **Figures D-7** and **D-8** depict 2D seismic activity levels for pre-1990 and 1990 to present, respectively. In both figures, surveys completed in water depths <200 m in the Western Gulf Planning Area and most of the Central Gulf Planning Area are not shown, as noted. With the initiation of permit tracking within MMS in the early 1990's, coupled with industry movement towards 3D seismic activity, areal coverage by lease blocks became easier. **Figures D-9**, **D-10**, and **D-11** depict prelease 3D seismic survey activity in the Gulf of Mexico during the periods 1993-1995, 1996-1998, and 1999-2001, respectively. A composite and complex figure depicting all prelease 3D seismic activity during the period 1993-2001 is provided in **Figure D-12**.

*Note: Tabular data presented in **Table D-3** represent prelease survey activity levels only; data are unavailable for postlease surveys or other surveys (e.g., pipeline rights of way, scientific research). Further, it is important to recognize that projections (beyond 2003) are only approximations based on the remaining unsurveyed blocks (900) and the projection of a 14-year resurvey cycle. While **Table D-3** projects activity out to 2015, there is no historical basis or implied significance to this endpoint.*

The activity levels realized in the Gulf in recent years (i.e., 1995-2003) are expected to produce a corresponding increase in resurveying activities on a 14-year cycle, assuming other factors remain relatively unchanged. It is apparent from this analysis that a) several factors strongly influence survey activity levels (i.e., projections are only approximations), and b) there is a "general trend" or "cycle" evident, based on remaining unsurveyed blocks and projected resurvey activity tied to an approximate 14-year timeframe. Further, it is unknown how many new, unsurveyed blocks may be identified in the near future (e.g., from future lease sales and leasing activity).

One factor that cannot be easily integrated into these projections is the geology of specific areas of interest. For example, MMS notes that the area seaward of the Sigsbee Escarpment (i.e., Lund and Lund South in the Central Gulf of Mexico Planning Area) contains only a few deeply buried prospects. For the near term in this area, 3D surveys will probably be of limited areal extent and may result in additional blocks being surveyed for the second and third time sooner. Areas such as Mississippi Canyon and northern Atwater, where drilling activity has been high during recent years, contain some blocks that have been surveyed three times during the period 1993 through 2003.

Using data for the period 1994 to 1999, "crew efficiency" was also calculated by dividing the number of 3D crews by the number of blocks surveyed. Crew efficiency varied from 0.004 crews per block to 0.009 crews per block, with a median value of 0.009 (see **Table D-3**). Assuming 0.009 crew efficiency and multiplying by the total number of blocks estimated to be surveyed from 2003 on, the number of active 3D crews is predicted to range from 8 to 53. Dellagiarino et al. (1998, 2000, 2001) indicated that the percentage of 3D surveys (relative to all geophysical surveys permitted by the MMS since 1993) has ranged from a low of 40% (in 1998) to a high of 54% (in 1993). MMS notes that 2D permitting activity has remained relatively static over the past several years. The number of 2D crews is assumed to remain constant at eight crews in the foreseeable future. The resulting predictions for total crew and 3D crew activity are listed in **Table D-3**.

In addition to the 2D or 3D seismic surveys (prevalent in the Gulf of Mexico) are the VSP surveys conducted on a much smaller geographic scale (e.g., within a single lease block). Seismic sources used during a VSP survey are the same as those used in conventional seismic surveys. For

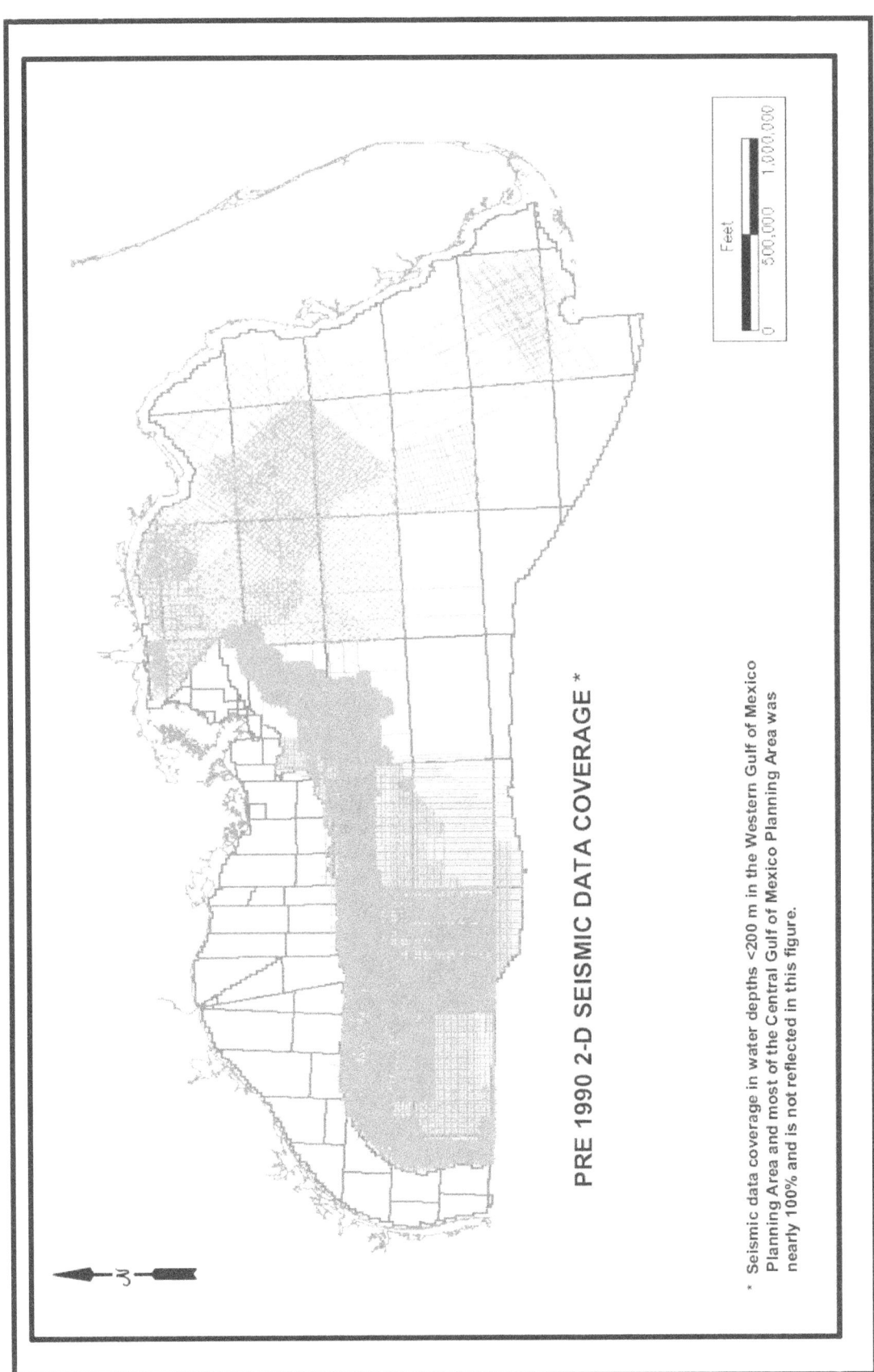

PRE 1990 2-D SEISMIC DATA COVERAGE *

* Seismic data coverage in water depths <200 m in the Western Gulf of Mexico Planning Area and most of the Central Gulf of Mexico Planning Area was nearly 100% and is not reflected in this figure.

Figure D-7. Prelease 2D seismic survey activity in the Gulf of Mexico, pre-1990.

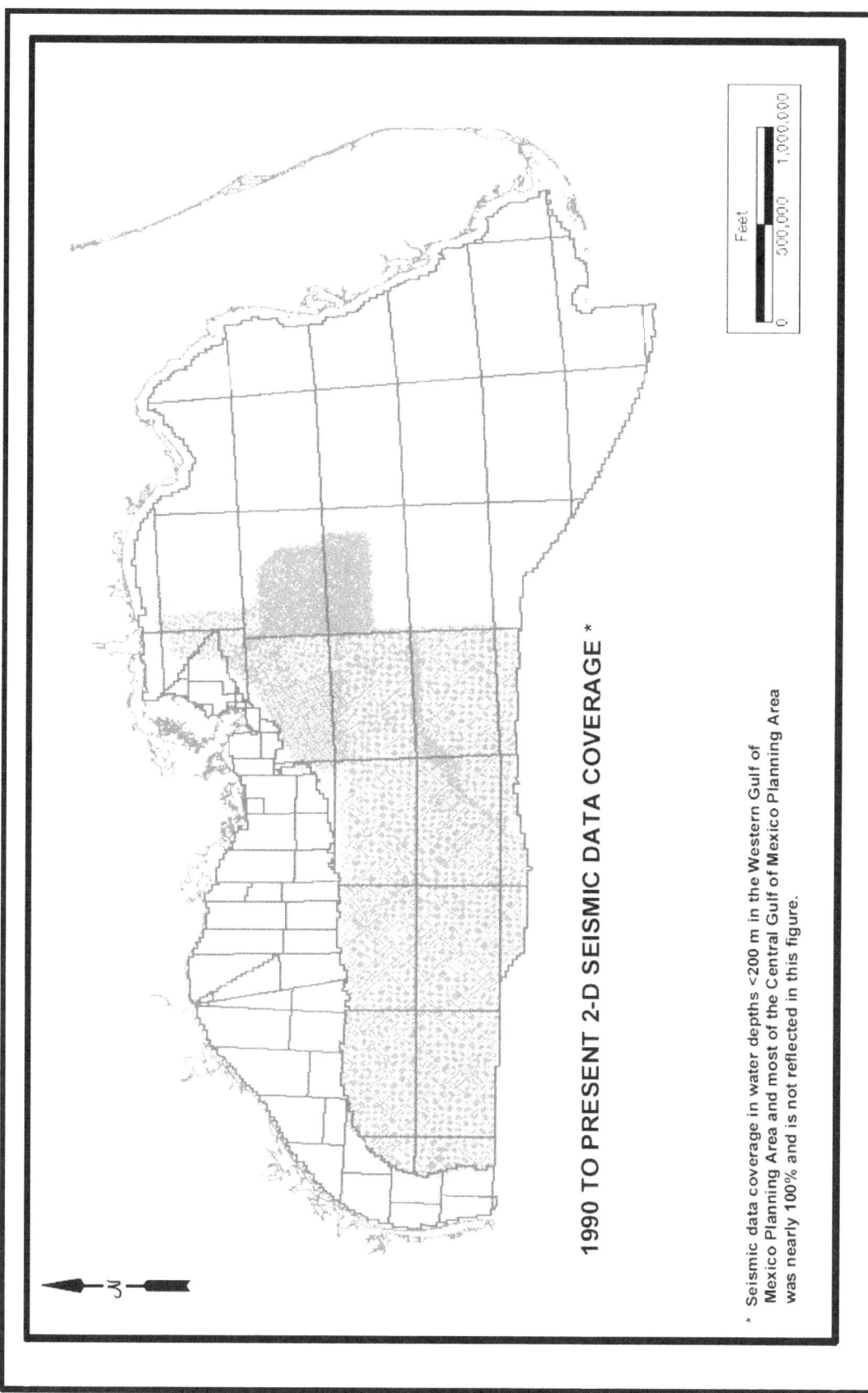

1990 TO PRESENT 2-D SEISMIC DATA COVERAGE *

* Seismic data coverage in water depths <200 m in the Western Gulf of Mexico Planning Area and most of the Central Gulf of Mexico Planning Area was nearly 100% and is not reflected in this figure.

Figure D-8. Prelease 2D seismic survey activity in the Gulf of Mexico, 1990 to Present.

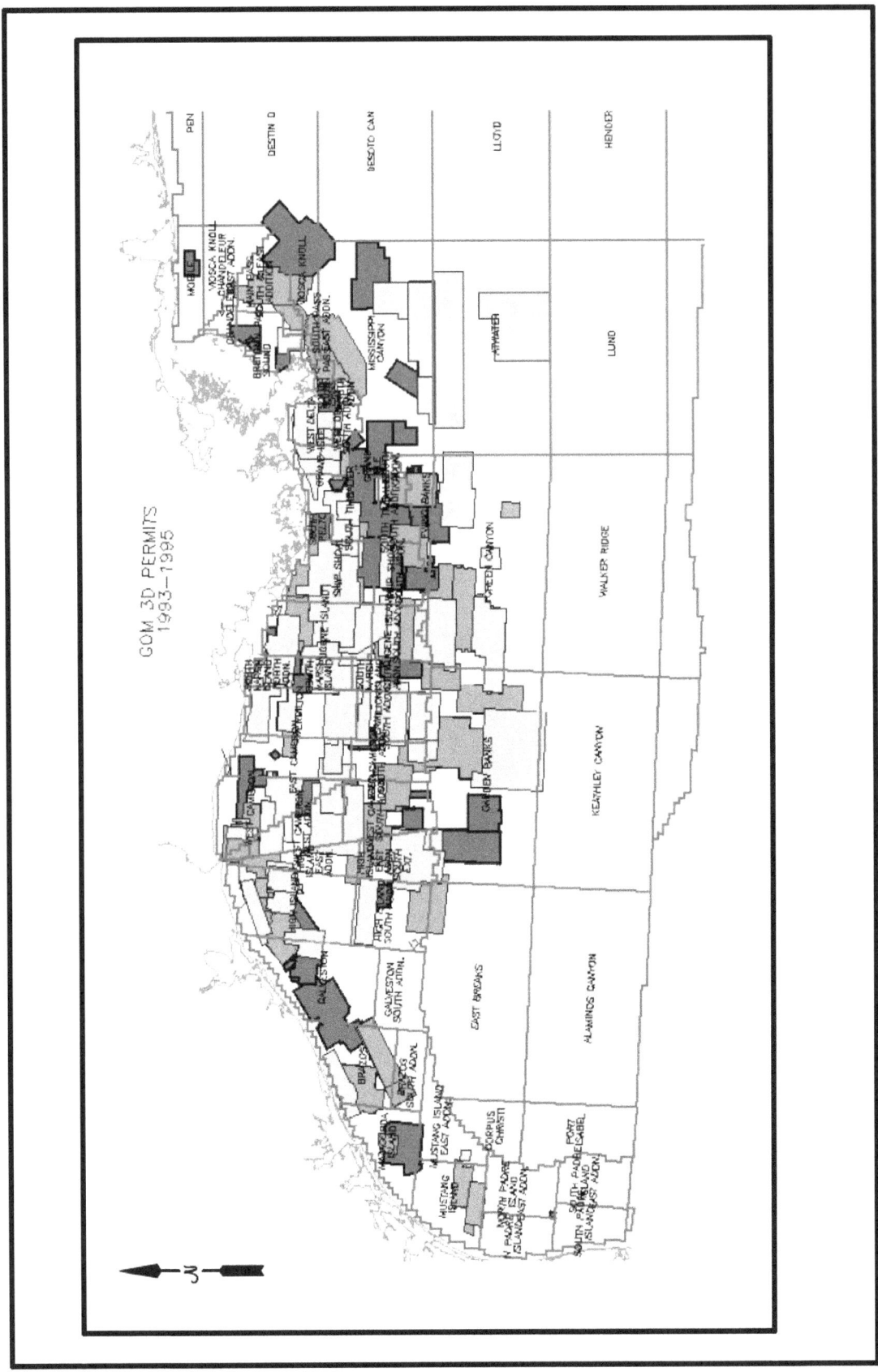

Figure D-9. Prelease 3D seismic survey activity in the Gulf of Mexico, 1993-1995.

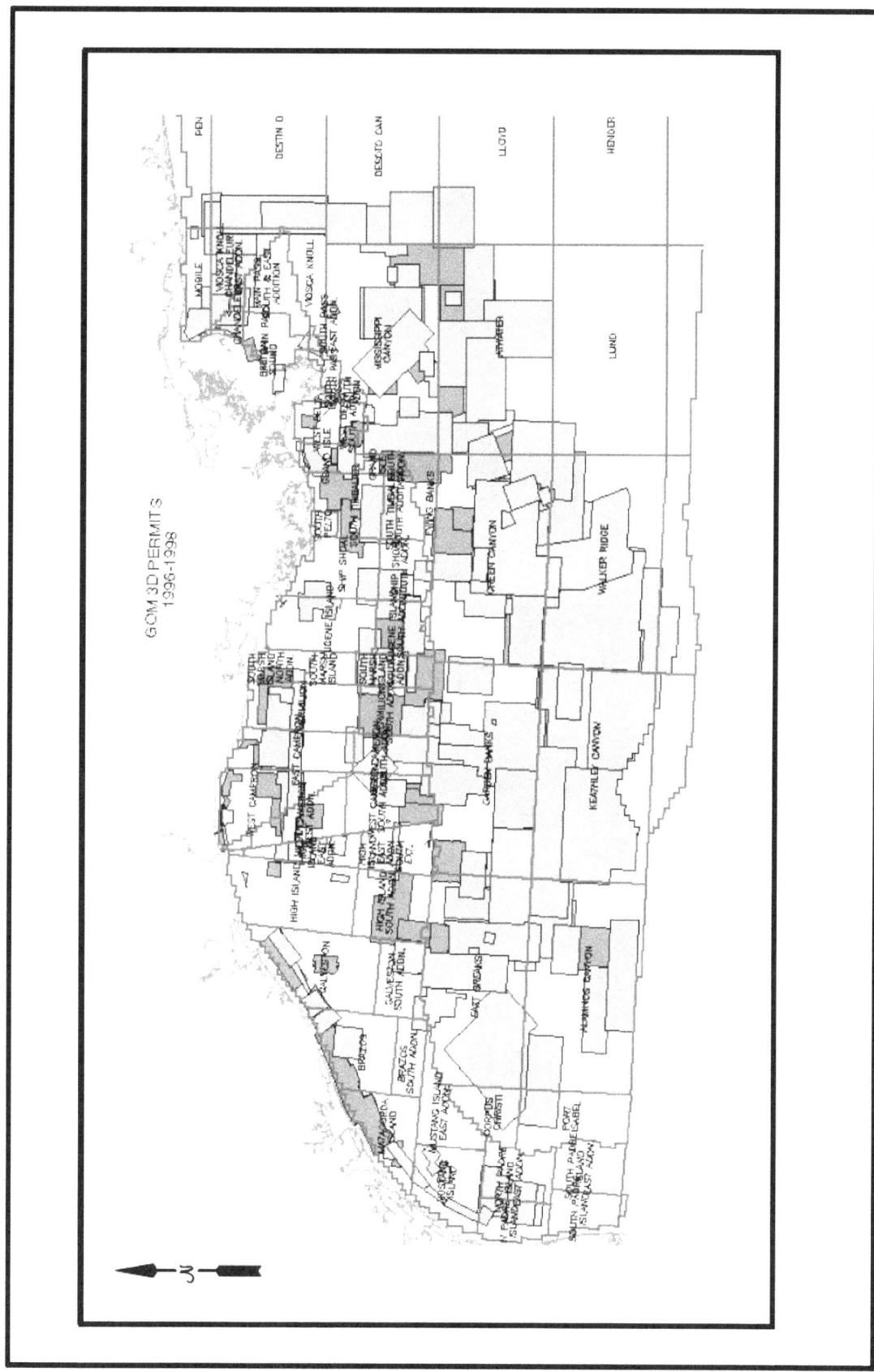

Figure D-10.   Prelease 3D seismic survey activity in the Gulf of Mexico, 1996-1998.

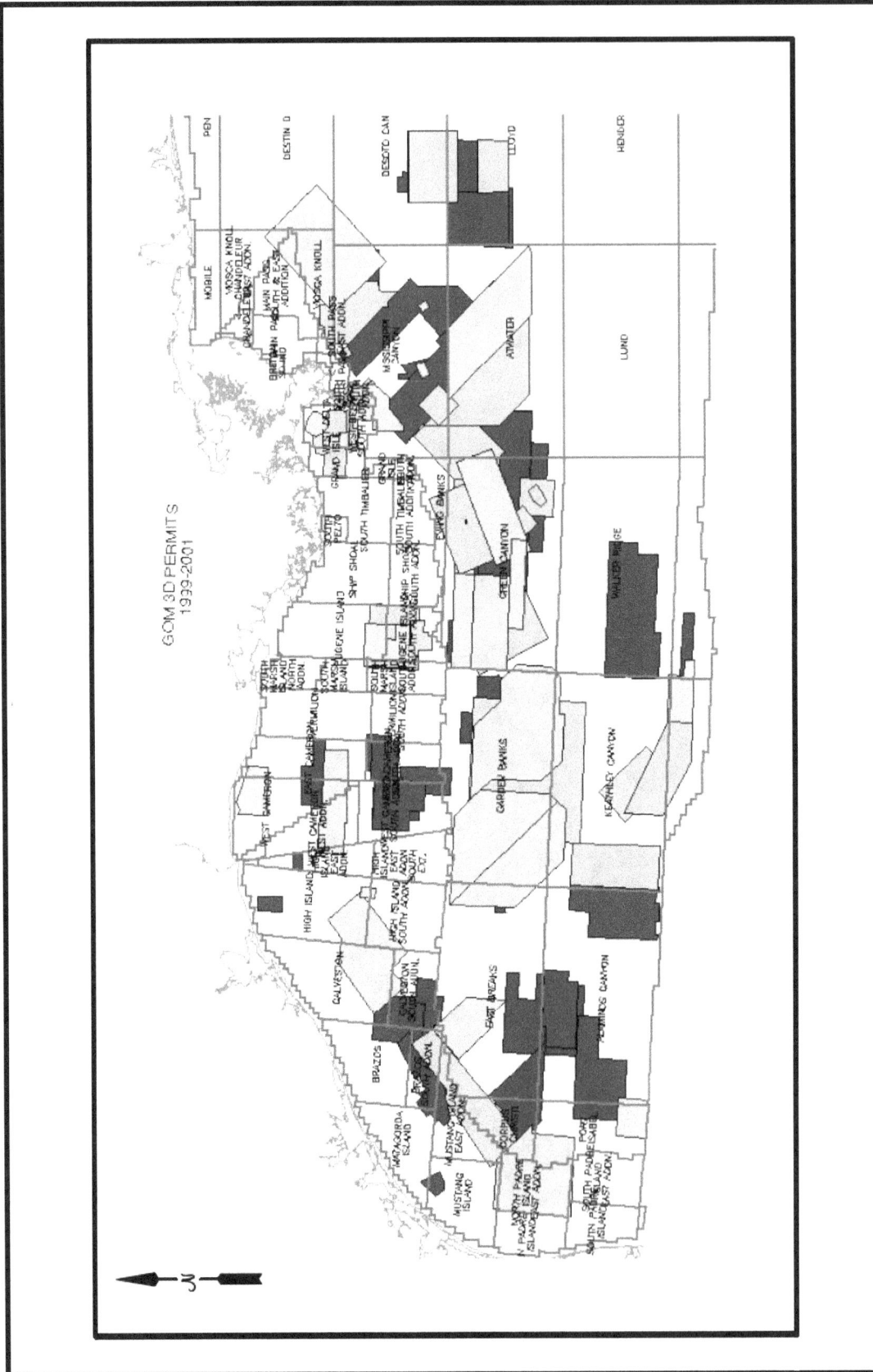

Figure D-11. Prelease 3D seismic survey activity in the Gulf of Mexico, 1999-2001.

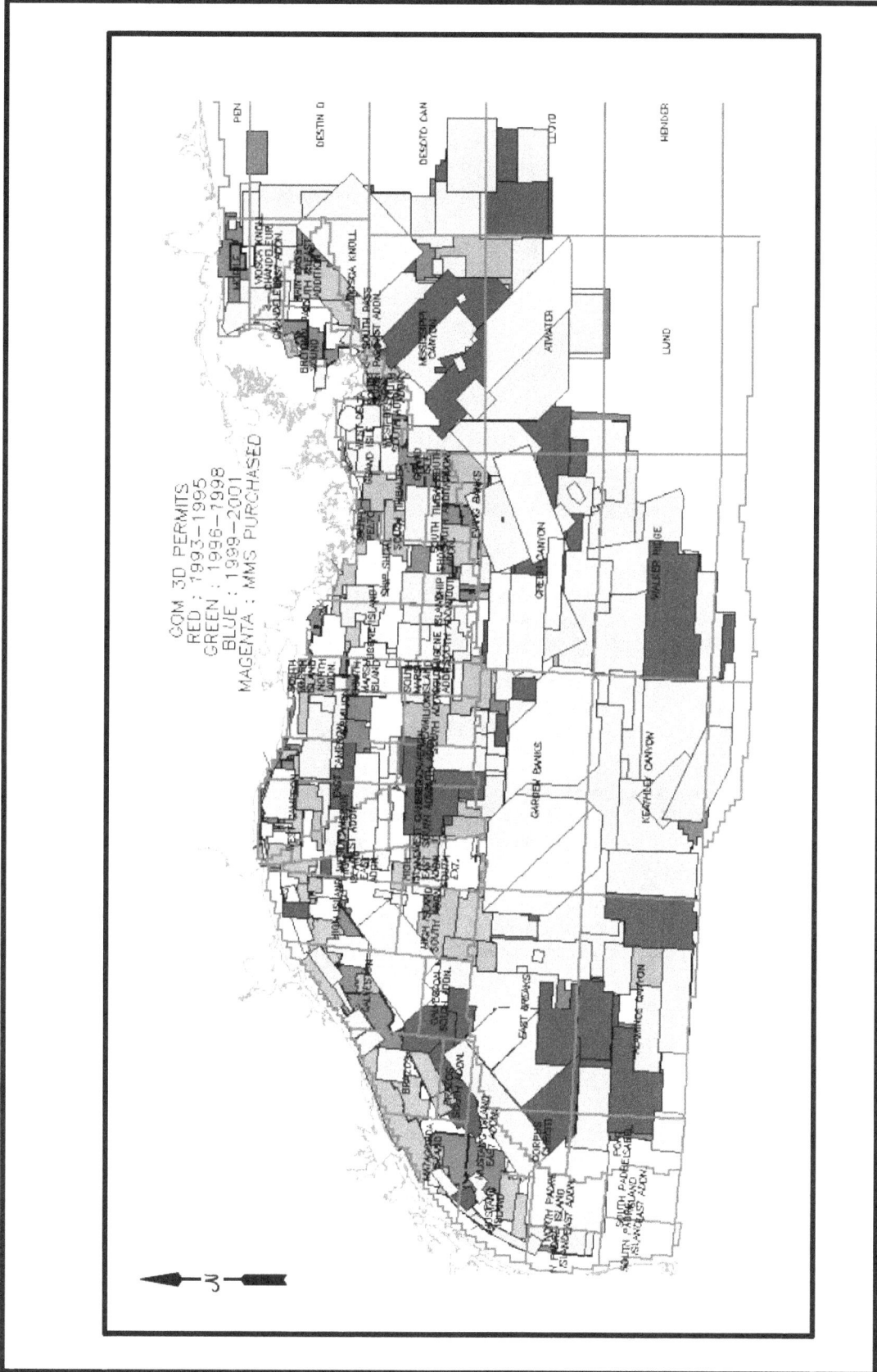

Figure D-12. Prelease 3D seismic survey activity in the Gulf of Mexico, 1993-2001 (composite).

example, zero offset surveys (i.e., also known as check shot surveys) utilize a small volume single airgun suspended by a crane located on the deck of the drilling rig. Walk-away surveys utilize a work boat equipped with four to eight airguns, while 3D VSP surveys use the same airgun arrays as used for conventional 2D and 3D seismic surveys. Check shot surveys are the most common type of borehole seismic survey conducted in the Gulf of Mexico. **Figures D-13** through **D-15** indicate check shot survey activity in the Gulf of Mexico for the periods 1993-1995, 1996-1998, and 1999-2001, respectively. Comparisons of 2D and 3D surveys with check shot survey activity levels can only be made with appropriate caveats, given the inherent differences in survey characteristics. Specifically, the frequency of airgun firing, geographic distribution of survey activity, and duration of survey operations are considerably different in check shot surveys than those characteristic of 2D and 3D seismic surveys, as discussed previously (see **Section I.A.7 - Vertical Seismic Profile Surveys** of this appendix). It is evident, however, that check shot survey activity in recent years has followed the general trend evident in the region's exploration and development sectors - increasing movement into deeper waters of the Gulf.

VI. LITERATURE CITED

American Petroleum Institute. 1999. API Basic Petroleum Data Book, Section VI, Table 1 and Table 2.

Brooks, J.M., M.C. Kennicutt II, and B.D. Carey, Jr. 1986. Offshore surface geochemical exploration. Oil and Gas Journal, 20 October, p. 6.

Brooks, J.M., B.B. Bernard, J.D. Stonebraker, C.F. Schiefelbein, K.A. Allen, and T.J. McDonald. 1997. Design of multidisciplinary surface geochemical exploration surveys to identify active petroleum systems offshore West Africa, p. 3. *In*: Hedberg Research Symposium, "Petroleum Systems of the South Atlantic Margin," Rio de Janeiro, 16-19 November 1997 (extended abstract).

Dellagiarino, G., P. Fulton, K. Meekins, and D. Zinzer. 1998. Geological & geophysical data acquisition, outer continental shelf through 1997. Resource Evaluation Program Report, U.S. Department of the Interior, Minerals Management Service, Resource Evaluation Division, Herndon, VA. OCS Report MMS 98-0027. 32 pp.

Dellagiarino, G., P. Fulton, K. Meekins, and D. Zinzer. 2000. Geological & geophysical data acquisition, outer continental shelf through 1999. Resource Evaluation Program Report, U.S. Department of the Interior, Minerals Management Service, Resource Evaluation Division, Herndon, VA. OCS Report MMS 2000-071. 32 pp.

Dellagiarino, G., P. Fulton, K. Meekins, and D. Zinzer. 2001. Geological & geophysical data acquisition, outer continental shelf through 2000. Resource Evaluation Program Report, U.S. Department of the Interior, Minerals Management Service, Resource Evaluation Division, Herndon, VA. OCS Report MMS 2001-092. 33 pp.

Hrvoic, D. 1999. High resolution magnetics in marine oil exploration. Hydro International 3(2):60-63.

Johnston, R.C., D.H. Reed, and J.F. Desler. 1988. Special Report of the SEG Technical Standards Committee: SEG Standards for specifying marine seismic energy sources. Geophysics 53(4):566-575.

Sassen, R., J.M. Brooks, I.R. MacDonald, M.C. Kennicutt II, N.L. Guinasso, Jr., and A.G. Requejo. 1993. Association of oil seeps and chemosynthetic communities with oil discoveries, upper continental slope, Gulf of Mexico. Trans. Gulf Coast Assoc. Geol. Soc. 43:349-355.

Smith, J.G., and M.R. Jenkerson. 1998. Acquiring and processing marine vibrator data in the transition zone. Abstracts from the 68th Annual Meeting of the Society of Exploration Geophysicists.

U.S. Department of the Interior, Geological Survey. 1976. Final environmental impact statement. Regulations pursuant to geological and geophysical explorations of the outer continental shelf. FES 76-23.

Weber, M. 2001. Advances continue in gravity and magnetic data technology for marine seismic surveys. First Break 19(9):509-514.

Figure D-13. Check shot survey activity in the Gulf of Mexico, 1993-1995.

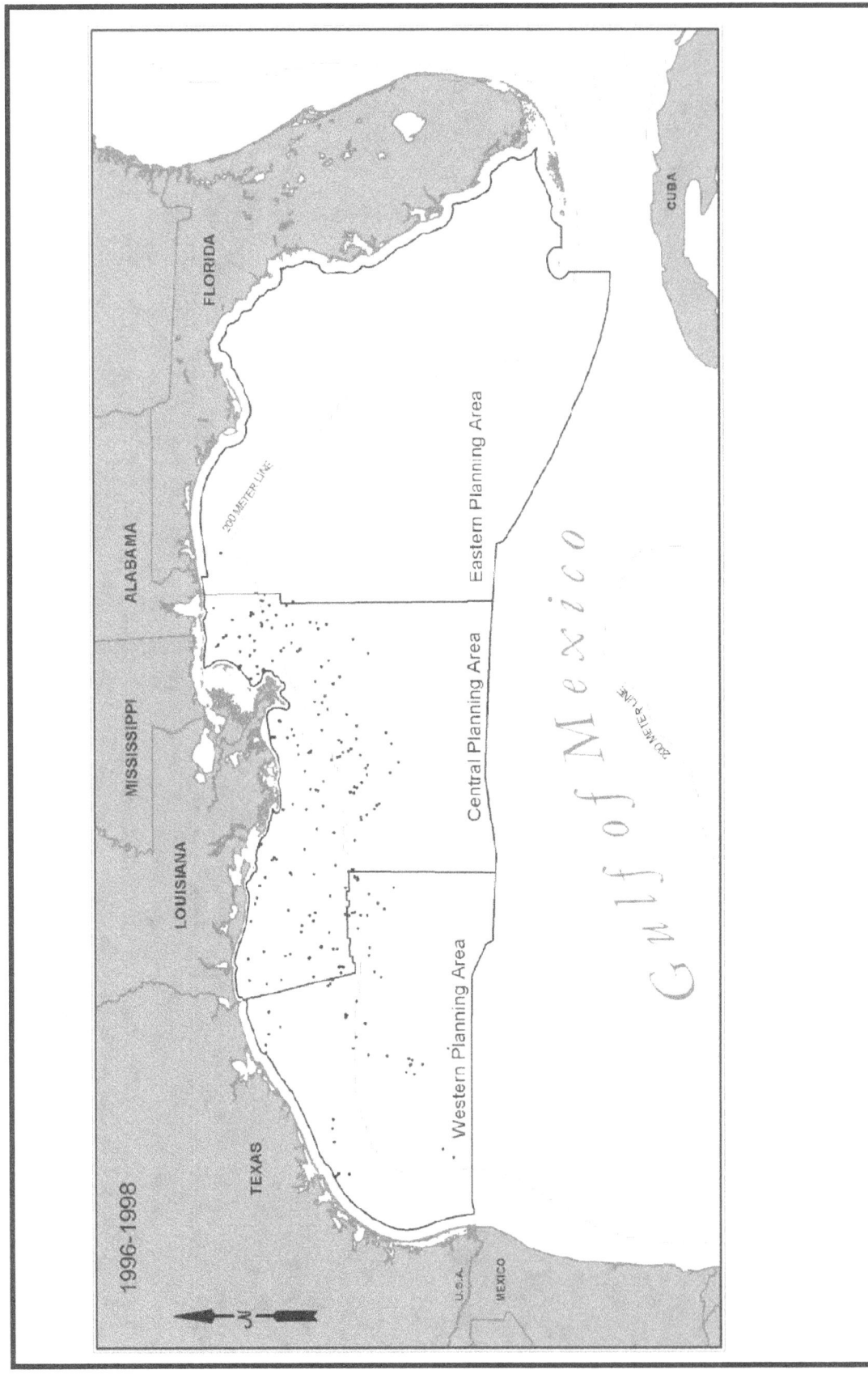

Figure D-14. Check shot survey activity in the Gulf of Mexico, 1996-1998.

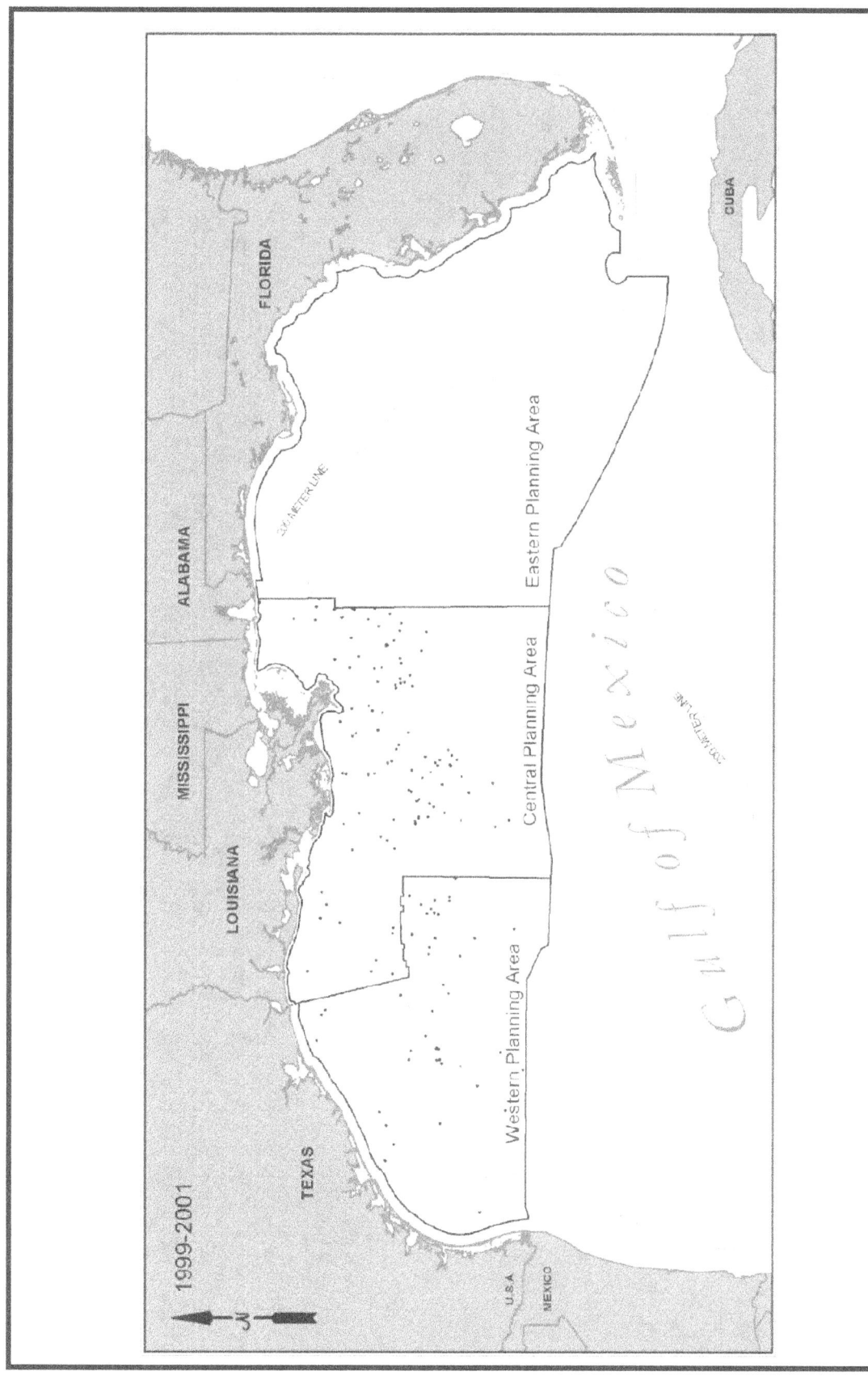

Figure D-15.　Check shot survey activity in the Gulf of Mexico, 1999–2001.

Appendix E
**Alternatives Considered but Not Analyzed, Existing and Proposed
Mitigation Measures, and Potential Operational Restrictions**

Alternatives Considered but Not Analyzed, Existing and Proposed Mitigation Measures, and Potential Operational Restrictions

This appendix 1) describes the alternatives considered during the analysis but not analyzed, and 2) summarizes existing mitigation measures (e.g., permit requirements, lease stipulations, other protective measures) and evaluates additional mitigation measures and potential operational restrictions that may reduce or eliminate impacts associated with geological and geophysical (G&G) operations. Alternatives discussed below were not considered to be viable alternatives, primarily because they failed to meet the purpose and need (see **Section I.B - Purpose and Need** and **Section II - Alternatives** of the Programmatic Environmental Assessment [PEA]). The discussion of mitigation measures was intended to identify current protective measures, to characterize potentially applicable mitigation measures for G&G operations, and to evaluate their feasibility to G&G operations in the Gulf of Mexico (GOM).

I. ALTERNATIVES CONSIDERED BUT NOT ANALYZED

A. REPLACE SEISMIC SURVEYS WITH PASSIVE SEISMIC AND/OR SLIM-HOLE DRILLING

In this alternative, seismic operations would be prohibited. The purpose of this alternative is to eliminate environmental impacts associated with airgun noise from seismic surveys. Impacts of seismic survey noise on marine mammals are the principal environmental issue associated with G&G activities. To meet the underlying need for accurate data on location and extent of hydrocarbon resources, the industry would have to rely on other approaches such as passive seismic and/or slim-hole drilling.

1. Description of Additional G&G Techniques

a. Slim-Hole Drilling

Slim-hole drilling is a technology that has been used on land but not in a marine setting. This technology uses a smaller-than-normal drill stem (i.e., hole sizes <12 inches) and takes many more core samples. Since the drill stem is so small, blowout prevention is a severe problem. Usually areas where abnormal pressures are expected are avoided in order to avoid blowout problems. The idea is to drill a stratigraphic test and not to encounter hydrocarbons. The information from the cores and the wire line logs is correlated from hole to hole, and subsurface maps are drawn from these data. The size of trap is a direct function of the grid size of the stratigraphic tests. Each exploration company would have to decide for themselves the exploration grid size.

In a marine setting, the drilling apparatus would be either ship-mounted, with facilities similar to those of the Ocean Drilling Program, or rig-mounted.

b. Passive Seismic

Passive seismic is a rarely used exploration technique employed in land exploration where other technologies fail. The approach uses earthquakes as its seismic source. Detectors with recorders set to detect the first arrivals of distant earthquakes are placed on the surface to be explored. Iterative use of Marquardt's method on earthquake travel times has been used to estimate seismic velocities (Hawley et al., 1981). Passive seismic data can be jointly interpreted with gravity data, using Birch's Law to provide more accurate information. The resulting information outlines bodies of constant velocity or constant density. Salt bodies in the GOM would be a good example.

In a marine setting, detectors with seismic recorders would have to be placed on the seafloor (probably in a grid similar to vertical cable surveys) and left there for some time. After a sufficient number of earthquakes have been detected, the instrument packages would be recovered and the data interpreted. A technique presently in early development is to record seismic events with bottom cables or in boreholes from microearthquakes that result from active hydrofracturing or injection of fluids in the subbottom during oil and gas development and production. Other passive sources include drilling related (seismic-while-drilling) and other manmade noise (Stephen, 2000; Maxwell and Urbanck, 2001).

2.   Mitigation Measures Included

Existing mitigation measures (protective measures routinely specified in permit requirements and lease stipulations; Notice to Lessees and Operators [NTL] requirements) described previously for the Proposed Action also would be included in this alternative.

3.   Evaluation Relative to Purpose and Need

At this time, neither passive seismic nor slim-hole drilling are viable alternatives to conventional seismic exploration. Passive seismic is a low-resolution technique, and in a mature province like the GOM, would not be able to detect the small features now being drilled. The passive seismic method needs much more development work before it is a viable alternative. Slim-hole drilling suffers from a lack of application because seismic survey techniques are so reliable and highly developed. Much research would have to be done before this technique is proved feasible for GOM exploration. In addition, extensive new drilling programs would raise environmental issues similar to those for current drilling programs (e.g., drilling mud and cuttings discharges) with the addition of increased potential for blowouts due to lack of blowout preventer technology applicable to this method. Slim-hole drilling essentially involves drilling deep stratigraphic test holes, which currently require preparation of an EA due to the potential for environmental impacts.

Under this alternative, the remaining 900 blocks in the Western and Central Planning Areas that have not been surveyed using 3D seismic techniques would not be surveyed, although existing seismic data previously collected on speculation could be analyzed. Blocks with previous 3D surveys would not be resurveyed using more sophisticated technology. An undetermined number of postlease seismic surveys, currently conducted annually to support optimal management of producing fields, would no longer occur. Disallowing seismic surveys would slow outer continental shelf (OCS) exploration and development.

In conclusion, this alternative would not meet the purpose and need specified in **PEA Section I.B - Purpose and Need**. It would not provide the oil and gas industry with sufficiently accurate data on the location, extent, and properties of hydrocarbon resources, as well as information on shallow geologic hazards and seafloor geotechnical properties, in order to explore, develop, produce, and transport hydrocarbons safely and economically. Therefore, this alternative is not considered further in the PEA.

II. MITIGATION MEASURES AND OPERATIONAL RESTRICTIONS

A. EXISTING MITIGATION MEASURES

This section identifies mitigation or protective measures already in place as a result of current G&G permit requirements, including G&G operator compliance with lease stipulations and other protective measures, as well as applicable NTL requirements. Permit requirements and existing mitigation or protective measures are included in the Proposed Action.

1. G&G Permit Requirements

Pursuant to 30 CFR 251.4, a permit must be obtained to conduct prelease geological or geophysical exploration for oil, gas, and sulphur resources. Permits for exploration for other minerals are granted pursuant to requirements outlined in 30 CFR 280.3. Permit applications must be submitted to the Minerals Management Service (MMS) in accordance with the requirements outlined in 30 CFR 251.5 and 30 CFR 251.6 and explained further in applicable Letters to Permittees. The Letter to Permittees dated 20 January 1989 specifies forms and maps, stipulations, and special provisions applicable to most permit activity. The 30 CFR 251 regulations do not apply to G&G activities conducted by, or on behalf of, a lessee on a leased block. Such G&G activities are governed by 30 CFR 250.201 regulations and by applicable Notices to Lessees and Operators. **Table E-1** identifies the appropriate Federal regulations and their applicability to select mineral resources and phase of activity.

Table E-1
Federal Regulations Applicable to Prelease and Postlease Activities,
by Mineral Resource of Interest

Regulatory Citation	Mineral Resource	Activity Phase
30 CFR 250	Oil, gas, and sulphur	Postlease (i.e., on-lease)
30 CFR 251	Oil, gas, and sulphur	Prelease or off-lease exploration or scientific research
30 CFR 280	All minerals exclusive of oil, gas, and sulphur	Prelease (prospecting)

G&G explorations for mineral resources may not be conducted in the OCS without an approved permit unless such activities are being conducted pursuant to a lease issued or maintained under the OCS Lands Act. Separate permits must be obtained for either geological or geophysical explorations for mineral resources.

As noted in the permit requirements, "all geological and geophysical explorations (or scientific research activities) authorized and conducted on the OCS must be performed in accordance with the OCS Lands Act, 30 CFR Part 251, and other applicable Federal statutes and regulations, and amendments thereto." G&G activities associated with mineral exploration authorized under 30 CFR Part 251 cannot

- interfere with or endanger operations under any lease or right-of-way easement, right-of-use, scientific notice, or permit issued or maintained pursuant to the OCS Lands Act;
- cause harm or damage to aquatic life, property, or to the marine, coastal, or human environments;
- cause pollution;
- create hazardous or unsafe conditions;

- unreasonably interfere with or harm other uses of the area;
- disturb archaeological resources; or
- cause hazardous or unsafe conditions.

G&G operators conducting activities under 30 CFR Part 251 must immediately report to the Director, MMS, when

- hydrocarbon occurrences are detected;
- environmental hazards are encountered that constitute an imminent threat to human life or property; or
- activities occur that adversely affect the environment, aquatic life, archaeological resources, or other uses of the area in which the exploration or scientific research activities are conducted.

Any person conducting shallow or deep stratigraphic test drilling activities under a permit for mineral exploration or scientific research under 30 CFR Part 251 must use the best available and safest technologies that the Director, MMS, determines to be economically feasible. Permitted activities approved for a specified period, including requests for extensions, and activities under a notice may not exceed 1 year.

2.   Stipulations and Protective Measures

The MMS currently requires oil and gas operators to comply with a series of stipulations and protective measures during G&G activities. These requirements effectively represent mitigation measures designed to reduce or eliminate impacts to sensitive resources. Such measures are implemented through regulations governing prelease and postlease G&G activities. A total of six stipulations apply to prelease and postlease G&G activities, as well as a series of environmental protective measures incorporated by reference. Key points consist of the following:

- *Explosives Prohibition*: Explosives cannot be used except under written authorization from the Regional Supervisor. Further protective measures (including Endangered Species Act Section 7 consultation with the National Marine Fisheries Service [NMFS]) apply in the event that explosives are used.

- *Protection of the Florida Manatee*: Several measures protect Florida manatees from vessel strikes. These include a separate Endangered Species Act consultation.

- *No-Activity Zone at Topographic, Chemosynthetic, and Pinnacle Features*: Bottom-disturbing activities, including buoys, are prohibited within the No-Activity Zones of certain topographic features, and bottom-disturbing activities are restricted within the "Pinnacle Trend" area of the northeastern GOM and in OCS blocks containing known chemosynthetic communities. Bottom-disturbing activities include, but are not limited to, drilling, anchoring, placing seafloor templates, discharging muds and cuttings, and installing pipelines. Further protections apply for activities within the Flower Garden Banks National Marine Sanctuary.

- *Archaeological Resources*: The permittee must report discovery of any archaeological resource (i.e., shipwreck/prehistoric site) to the MMS and take precautions to protect the resource from operational activities.

- *Seismic Safety*: All pipes, buoys, and other markers used in connection with seismic work must be properly flagged and lighted according to the navigation rules of the U.S. Army Corps of Engineers and the U.S. Coast Guard.

- *Digital Navigation Data*: Digital navigation data must be recorded on tape or other suitable storage media for seismic reflection surveys.

Additional stipulations not covered in this analysis limit activities as a means of avoiding impacts to coastal bird species.

3.   NTL Requirements

NTL No. 2004-G01 (USDOI, MMS, 2004) requires G&G operators conducting seismic operations in all Federal waters >200 m deep in the GOM and all Federal waters of the Eastern Planning Area (regardless of water depth) to 1) employ ramp-up; 2) utilize trained protected species observers; and 3) complete MMS reporting requirements. Ramp-up is to be initiated only during periods of sufficient visibility when observers are able to scan and clear an area (i.e., impact radius, or exclusion zone) several hundred meters around seismic operations. Specifically, the NTL requires that visual observers clear the exclusion zone at and below the sea surface within a radius of 500 m surrounding the center of an airgun array and the area within the immediate vicinity of the survey vessel. Observers must observe no whales within (or approaching) the exclusion zone for a period of 30 minutes, after which ramp-up operations may begin. Once ramp-up has been completed and the seismic array is operating at full power, visual observations are to continue until seismic operations cease or sighting conditions do not allow observation of the sea surface (e.g., fog, rain, darkness). If a whale is sighted either within this exclusion zone or moving towards the exclusion zone, the array must be shut down until the area can be cleared. The seismic array may be powered down to a minimum level of 160 dB re 1 μPa (rms) without reinitiating ramp-up. Procedures for ramp-up, protected species observer training, visual monitoring, and reporting are described in detail in the NTL.

This NTL also details the optional use of passive acoustic monitoring (for vocalizing marine mammals) during periods of poor visibility. Performance of these mitigation measures are a condition of approval of applications for geophysical permits, and are applicable to geophysical activities conducted under lease terms for all seismic survey operations proposed within waters >200 m deep throughout the GOM and all OCS waters of the Eastern Planning Area.

B.   POSSIBLE MITIGATION MEASURES AND OPERATIONAL RESTRICTIONS

There have been several reviews published during the past decade addressing the question of noise and its effects on marine mammals (e.g., National Research Council, 1994, 2003; Richardson et al., 1995). There also have been a series of site-specific EAs dealing with the issue of seismic operations and their effects on marine organisms, with an emphasis on marine mammals, seabirds, and fishes (e.g., Turnpenny and Nedwell, 1994; USDOI, MMS, 1995; Arthur D. Little International, Inc., 1997; Davis et al., 1998; Stone, 1998; LGL, Ltd., environmental research associates, 2003). To varying degrees, these efforts have discussed generic and/or site-specific mitigation designed to reduce the impacts of seismic noise on resident fauna.

Standard mitigation measures pertinent to seismic noise have been summarized by Pierson et al. (1998) as part of the 1998 workshop on seismic operations and marine mammals. Pierson also has summarized worldwide regulatory issues, including mitigation measures, as part of a recent workshop on seismic airgun sources, U.S. regulations, and marine mammals.

Table E-2 summarizes the various mitigation measures used under several separate seismic operations in U.S. waters, including offshore California (i.e., Southern California Bight; Santa Barbara Channel), Beaufort Sea (Alaska), the Puget Sound – Georgia Strait region (Washington/British Columbia), and the northern GOM. Site-specific mitigation, as applied in recent U.S. seismic operations (see **Table E-2**), normally encompasses several different measures (i.e., geographic and seasonal restrictions, ramp-up, impact zones, and real-time monitoring). These measures are typically used in conjunction with one another, although there is one instance where all but one measure has been required. Beyond the mitigation measures noted in **Table E-2**, other mitigation measures are considered in this evaluation. These include reduction of sound source levels and sound baffling.

1.   Geographic and Seasonal Restrictions

Description: Geographic and seasonal restrictions are intended to preclude G&G operations within a prescribed area (e.g., within important species habitat, within known congregation areas, along migratory pathways), thereby avoiding either higher concentrations of marine mammals or individuals that may be more susceptible to acoustic disturbance. Typically, a seasonal restriction is directly associated with a geographic limitation. Such restrictions are intended to reduce or eliminate noise-related impacts to sensitive species associated with seismic operations.

For the purpose of this analysis, geographic and/or seasonal restrictions are covered under Alternative 4 (Restrict G&G Seismic Surveying Operations), as outlined in **PEA Section II.E - Alternative 4: Restrict G&G Seismic Surveying Operations**. Specifically, geographic and/or seasonal restrictions are included under 1) cessation of all types of seismic surveys in the area(s) most frequented by sperm and Bryde's whales (Alternative 4B); and 2) seasonal restrictions on G&G activities in the area(s) most frequented by sperm whales (Alternative 4C).

Examples: Several examples of such mitigating restrictions are found in U.S waters. Seasonal limitations were imposed on Exxon during their 3D seismic survey of the Santa Ynez Unit in the Santa Barbara Channel, offshore southern California (United States Department of the Interior [USDOI], MMS, 1995). Southbound migrating gray whales that are present in the Channel area during early winter (December-January) represented the sensitive resource of concern. While the operator planned to complete the seismic survey prior to 15 December, Exxon was required to place NMFS-approved observers on board the G&G vessel in the event that seismic operations continued past 15 December 1995. Based on an observer's determination of impact, G&G operations could have been either terminated or relocated to a less sensitive location when perceived impacts exceeded a "negligible" level. Further, NMFS required that all G&G operations be completed by the end of December 1995. Because the 3D seismic survey was completed by mid-December, no seasonal limitations were imposed.

Seasonal restrictions also were implemented during geohazards investigations of Puget Sound (U.S. waters) and the Georgia Strait (Canadian waters) in 1998. Geoscientists from various Federal and foreign agencies and academic institutions conducted an extensive seismic survey of the region to assess potential earthquake hazards. As a mitigation measure, the survey was scheduled to occur during a 2-month period when marine mammal abundance levels in the region were lowest.

No seasonal or geographic restrictions were imposed on seismic operations proposed on behalf of British Petroleum Exploration in its Northstar Unit in the Alaskan Beaufort Sea. However, over a 2-year period in 1996 and 1997, seismic operations were conducted using other mitigation measures, including boat-based and aerial monitoring, ramp-up requirements, and adherence to 180- to 190-dB shutdown criteria (LGL Ltd., Inc. and Greeneridge Sciences, Inc., 1997a,b).

Table E-2

Mitigation Measures Used for Seismic Surveys in U.S. Waters (Adapted from: Pierson et al., 1998; Normark, 1999; and National Oceanic and Atmospheric Administration Fisheries, 2003)

Study Area/ Operator	Geographic and Seasonal Restrictions	Ramp-up	Impact Zone	Visual Monitoring	Acoustic Monitoring	Other	Comments and References
California – Santa Barbara Channel/Exxon, Santa Ynez Unit	Yes – with observers aboard from 15 December on; cessation of all operations after 31 December	Yes – maximum increase of 6 dB per minute	Yes – ranging from 152.4 m for pinnipeds and odontocetes to 450 m for sperm whales and mysticetes	Yes – shipboard and aerial	Yes – during nighttime operations and daylight periods of low visibility	None	Impact zones corresponded to the estimated 190-dB and 180-dB isopleths, respectively; U.S. Department of the Interior, Minerals Management Service (1995)
California – southern California, Santa Barbara to San Diego (Federal waters only, >3 mi from shore)	No – survey scheduled for June-July timeframe	Yes – shut down required when animals observed within the impact zone	Yes – ranging from 100 m for dolphins, seals, and sea lions to 250 m for whales	Yes – 24-hr observations	No	Airgun operations restricted to daylight hours only	Observations of marine mammal reactions to seismic noise recorded; Normark (1999)
Alaska – Beaufort Sea/BP Exploration, Northstar	No	Yes – maximum increase of 6 dB per minute	Yes – ranging from 60 to 260 m for pinnipeds, and 640 to 1,020 m for gray, beluga, and bowhead whales	Yes – shipboard and aerial	Yes – via bottom recorders, vessel-based acoustic measurements, and ocean bottom cable acoustic receivers	None	Impact zones corresponded to the estimated 190-dB and 180-dB isopleths for pinnipeds and cetaceans, respectively. Impact zone distances for cetaceans ranged from 650 to 750 m; LGL Ltd., Inc. and Greeneridge Sciences, Inc. (1997a,b)

Table E-2

Mitigation Measures Used for Seismic Surveys in U.S. Waters (Adapted from: Pierson et al., 1998; Normark, 1999; and National Oceanic and Atmospheric Administration Fisheries, 2003)

(Continued)

Study Area/ Operator	Geographic and Seasonal Restrictions	Ramp-up	Impact Zone	Visual Monitoring	Acoustic Monitoring	Other	Comments and References
Washington/ British Columbia – Puget Sound – U.S. Coast Guard, Seismic Hazard Investigations in Puget Sound (SHIPS)	Yes – operations planned during period of low marine mammal abundance	Yes – maximum increase of 6 dB per minute	Yes – 100 m for pinnipeds, 200 m for odontocetes, and 500 m for gray, minke, and humpback whales	Yes – shipboard and aerial	Yes – via hydrophone array	None	National Marine Fisheries Service (1997)
Northern Gulf of Mexico - Lamont-Doherty Earth Observatory (LDEO) *R/V Maurice Ewing* seismic research	No	Yes – maximum increase of 6 dB per 5-minute period over a total duration of ~14 min (6 gun array), 18 to 20 min (10 to 12 gun arrays) or 23 to 25 min (20-gun array)	Yes – ranging from 75 to 1,425 m for cetaceans, depending upon airgun array size	Yes – shipboard	Yes	None	Impact zones corresponded to the calculated 180-dB isopleth times 1.5 (LGL Ltd., environmental research associates, 2003)

Pierson et al. (1998) also noted the use of seasonal restrictions in several international locations (e.g., Australia, Caspian Sea).

Evaluation and Possible Application to the GOM: While there may be justification for limited application of geographic restrictions of G&G operations in the Gulf (i.e., Alternative 4), such restrictions would not have universal benefit (i.e., only two of the three species or species groups of concern would realize potential benefit from such restrictions). Two geographic areas of concern have been identified in **PEA Section III**. These are areas on the continental slope offshore of the Mississippi River mouth, and the Eastern Gulf of Mexico Planning Area, particularly the De Soto Canyon area where Bryde's whales and sperm whales are known to occur. While such geographic restrictions offer the potential for reduced impacts to sperm and Bryde's whales, they offer no mitigation for beaked whales. Beaked whales have shown no apparent site fidelity or affinity for preferred areas within the GOM, nor are seasonal trends in distribution evident.

On the basis of available marine mammal presence and distribution data (e.g., GulfCet I and II), there does not appear to be a basis for seasonal restrictions on G&G operations in the GOM (i.e., Alternative 4C is not a viable alternative).

2. Impact Zones and Real-Time Monitoring

Description: An "impact radius," or impact zone, is established around the seismic vessel based on the nature (i.e., size, output) of the seismic array, ambient oceanographic conditions in the survey area, and the potential for sensitive receptors (i.e., marine mammals) to be present. A discussion of impact zone calculation methods, explicit and inherent assumptions, and data deficiencies is provided in **Appendix C, Sections III** and **IV**. Data indicate that the horizontal transmission from a typical seismic array is not perfectly circular, but rather is more elliptical and irregular in nature. In the absence of modeling results that are directly applicable to this PEA (i.e., specific to a typical seismic array in use in the GOM), a circular geometry is applied in establishing zones of ensonification and potential impact. This approach quantifies the loudest point in a polar field and ascribes a circular zone at that radius. Directivity and the elliptical nature of seismic pulses are discussed further in **Appendix D, Section II.A - Seismic Sources, Airguns**.

Impact zones (also referred to as "safety zones" or "exclusion zones") are generally based on a received sound pressure level, the latter of which varies depending upon the sensitive resource. For example, 180 dB re 1 μPa (rms) has been used as an impact zone/safety zone boundary for baleen and sperm whales, while the 210 dB re 1 μPa (rms) isopleth has been used for odontocetes and pinnipeds. Isopleths are estimated using transmission loss models and are often verified on a site-specific basis (Pierson et al., 1998). Real-time monitoring consists of visual monitoring (from either a vessel or aircraft, or both) and (passive and active) acoustic monitoring. The intent of determining an impact zone and conducting real-time monitoring is to establish and maintain a zone around the seismic vessel that is clear of marine mammals, thereby reducing or eliminating the potential for hearing damage.

Under the Proposed Action (Alternative 1), visual monitoring of a predetermined impact zone (and the optional use of passive acoustic monitoring) is currently required for G&G activities in OCS waters >200 m deep throughout the GOM and in all OCS waters of the Eastern Planning Area.

Examples: It is recognized, though not directly proven, that animals within several hundred meters of a seismic array may be at risk for potential hearing impairment (e.g., temporary or permanent threshold shift, TTS or PTS; see summaries of hearing impairment in NMFS, 2001a,b, and associated references). In the most basic approach, the impact zone is monitored immediately prior to and during ramp-up. Typically, visual monitoring occurs over a 30-minute period prior to ramp-up. In some instances, visual monitoring of the impact zone may continue during the seismic operation. In the event one or more marine mammals are sighted within the impact zone during this period, ramp-up or airgun operations are halted until the animals have left the area. This approach has been employed by most seismic surveys in western U.S. waters – off California, Alaska, and Washington state (Pierson et al., 1998) and during more recent research-based seismic surveys in the northern GOM (LGL Ltd., environmental research associates, 2003).

In a more comprehensive application of impact zone monitoring, continual observations are required during the entire seismic operation. The impact zone is typically monitored visually using one or more shipboard observers; in many cases, aerial observers supplement shipboard observations.

Shipboard surveys generally utilize two teams of three primary observers (Burks et al., 2001; LGL Ltd., 2001). Observers normally have two tasks – to monitor the impact zone for marine mammals, and to observe and document marine mammal behavior. Two observers search with high-powered (25x), pedestal-mounted binoculars. The third observer searches with unaided eyes and occasionally hand-held binoculars, and serves as data recorder. Data are recorded on a lap-top computer that has direct input from the vessel's global positioning system (GPS) navigation system. Observers within a team rotate among the three duty stations at regular intervals, and alternate work and rest periods with the second observer team. Unfortunately, visual observations have definite limitations. Sea state, sun angle, limited visibility, and observer fatigue can all affect observer sighting success (i.e., perception bias). Routine activities of marine mammals (e.g., diving duration patterns, pod size, overt behaviors) show considerable variability between species, thereby affecting whether or not animals are sighted (i.e., availability bias). During nighttime operations or during periods of reduced visibility, several options are available to allow for continual monitoring of the impact zone (e.g., shipboard lighting of waters around the vessel, use of enhanced vision equipment, acoustic monitoring). However, the efficiency of visual monitoring during nighttime hours, using shipboard lighting or enhanced vision equipment, is severely limited when compared with monitoring during daylight hours.

Passive acoustic monitoring (i.e., no acoustic sources are used, only listening devices) can occur either from a vessel-based system, from a directional or non-directional hydrophone or sonobuoy array placed on the seafloor or allowed to drift in the water column, or both (Burks et al., 2001; Swartz et al., 2001). These passive acoustic methods generate data that may provide species-specific signature calls (i.e., presence-absence data) and directional information (i.e., magnetic bearing angles from the survey vessel to the signal[s]). However, passive acoustic monitoring is effective only when an animal or group of animals is actively vocalizing. Furthermore, there are limitations in assessing an accurate enumeration of individuals present in larger groups. Active acoustic monitoring has recently become both of scientific interest and environmental concern (see **Section 4.c - Active Acoustic Monitoring** of this appendix).

In the event a marine mammal is sighted or otherwise detected within the impact zone, seismic operations are suspended until the animal leaves the area. There are definite limitations to real-time monitoring, including 1) limitations on sightability of individuals due to poor visibility (fog, elevated Beaufort sea state, nighttime operations) and/or observer fatigue, and 2) limits to the frequency of vocalization in several marine mammal species, potentially altering the effectiveness of passive acoustic monitoring.

Pierson et al. (1998) summarized available information regarding the variability of impact zones in site-specific situations in U.S. waters. These data are presented in **Table E-2**. By comparison, seismic operations in the United Kingdom (UK) (e.g., within the Atlantic Frontier) use a impact zone of 500 m, applicable to mysticetes, odontocetes, and pinnipeds (Pierson et al., 1998). Australian seismic guidelines require operators to monitor a 3-km area in front and lateral to seismic operations (Environment Australia, 2001). Offshore Sakhalin Island (Russian Federation), monitoring of feeding gray whales and other large cetaceans has been conducted over the past several years in association with offshore oil and gas development (Würsig et al. 1999; Johnson, 2002; Weller et al., 2002), as noted by LGL Ltd. environmental research associates (2003). A comprehensive monitoring program involving vessel- and shore-based observations, aerial surveys, and acoustic measurements was implemented in 2001 to provide information on gray whale reactions to seismic noise, and to facilitate implementation of a mitigation program. Distances to two isopleths (i.e., received sound levels of 180 and 163 dB re 1 µPa rms) were calculated for seismic operations located adjacent to the primary gray whale feeding area. Relative sizes of buffer zones reflected whether non-endangered whales or the gray whale and other endangered whales were expected to be present. There was no indication that gray whales exposed to seismic noise were displaced from their overall feeding grounds near Sakhalin Island during seismic programs in 1997 and 2001, but there were indications of subtle behavioral effects and localized avoidance by some individuals (LGL Ltd., environmental research associates, 2003).

Calculation of an Impact Zone: On the basis of prior case histories of G&G activities conducted in the Santa Barbara Channel and Alaskan Beaufort Sea (**Table E-2**), operators have been required to establish impact zones at 180 and 190 dB re 1 µPa for cetaceans and pinnipeds, respectively. *Note: The lead agency must consider a series of factors in establishing appropriate mitigation measures for a proposed G&G activity, including hearing sensitivities for the species of concern, site-specific marine mammal activity patterns and their significance (e.g., migration pathway, breeding or feeding area), characteristics of the seismic operation and associated equipment, and site-specific physical characteristics, as well as any new scientific data or regulatory requirements that may be applicable.* For the purposes of this analysis and for the cetacean species of concern, calculations of radial distances to the 160- and 180-dB isopleths have been completed[1].

As noted in **Appendix D** (**Section II.A - Seismic Source Airguns**), sound from an airgun array attenuates as it passes through the water column. Sound propagation in water is a complex phenomenon, affected by location (e.g., position and orientation of the sound source in the water column), water depth, bottom type, and water column temperature/density. Received sound levels are a function of both source and receiver depth. Gordon et al. (1998) note that a generalized spreading model (i.e., $n$log[R]) is often used to estimate sound attenuation; such models are further modified for specific operations within a particular area, accounting for site-specific characteristics that may affect transmission loss. For example, spherical spreading (20log[R]) is normally applied to seismic sources in moderate to deep water; such spreading is spherical to a radial distance approximately equivalent to water depth. Beyond this radius, Gordon et al. (1998) suggest that cylindrical spreading (10log[R]) may be more representative. In shallower water, cylindrical spreading provides a closer approximation of sound attenuation (i.e., sound is "trapped" between the sea surface and the seafloor; see LGL Ltd., Inc. and Greeneridge Sciences, Inc., 1997b and Gordon et al., 1998). In a more recent analysis (see **Appendix C**, **Section III - Seismic Source Levels**), Goold considers near-field and far-field noise attenuation, and evaluates the applicability of various sound attenuation models to G&G operations in the GOM. Goold concludes that, for pulsed sound from an airgun array, spherical spreading is appropriate to a horizontal range that is 1.5 times the water depth, after which modified cylindrical spreading may occur (e.g., as a result of seabed reverberation, etc.).

---

[1] Based on MMS High Energy Seismic Survey (HESS) workshop and NMFS Workshop on Acoustics Findings, as well as recent Final Rule determinations issued by NMFS. Received sound pressure levels (SPLs) of impulse noise above 180 dB re 1 µPa (rms) may be of concern, resulting in overt behavioral, physiological, and hearing effects on marine mammals in general, as cited by NMFS (2001a,b). The 160- and 180-dB rms levels are of regulatory and scientific interest (see **Appendix B**, **Table B.1**). Seismic activities produce repetitive, mobile noise that is neither single impulse nor continuous in nature.

The complexity of horizontal or near horizontal transmission of sound from an airgun array is non-trivial in a non-homogeneous environment. Thermal stratification of the water column may have a profound effect on the propagation characteristics of a sound wave. Seiche.com Ltd. (2001) shows a number of ray trace plots from sound sources located at different depths in the water column. It is evident, especially in the near surface examples (analogous to airgun arrays), that "patches" of sound convergence may occur in the near surface waters (i.e., sound levels could be greater at certain points than predicted by simple $n\log[R]$ models). Similar modeling is also shown for prospects in Australian waters (Anonymous, 2001), which demonstrates the complexity of sound ducting. Figure 28 in Anonymous (2001) shows predicted sound attenuation from an airgun after propagation modeling. From the first 10 km of the graph, propagation loss can be crudely estimated by eye to be some $11\log[R]$ (i.e., highly ducted).

However, for the 180- and 160-dB isopleths under consideration here, it is unlikely that ducting will be so extreme, given that the free-field (spherical spreading) ranges necessary to attenuate seismic sources to 180 dB are approximately an order of magnitude lower than the example given above. Depending much upon water depths, bottom type(s), and temperature profiles at specific locations in the GOM, sound propagation could range from close to free-field (i.e., $\sim20\log[R]$) to ducted (approximated by $15\log[R]$). Further details on free-field vs. ducted sound propagation are found in **Appendix C, Section III - Seismic Source Levels**.

Although the validity of the 180- or 160-dB isopleths as a mitigation sound level is not a subject addressed in this document, it should be noted that were the mitigation level to be reduced to 160 dB for example (as some scientific opinion would argue should be the case), then propagation ranges to the isopleth would be far larger and sound ducting would potentially be far more significant. Although $15\log[R]$ is suggested here as a compromise, it has been shown by the work of Anonymous (2001) that cylindrical spreading that approaches the theoretical minimum losses is possible.

For the purposes of this analysis, G&G seismic operations in Federal OCS waters will produce sound that should be assumed (in a conservative approach) to attenuate via spherical spreading (i.e., $20\log[R]$) to a distance that at least matches the water depth. Complications with bottom reverberation may become apparent once a distance of approximately 1.5 times the water depth is reached. Examples are given for both 180- and 160-dB isopleths calculated via this approach.

A rough approximation of radial distance to a given isopleth (e.g., 160 or 180 dB) can be calculated using these assumptions. This approach may serve as the basis for determining an appropriate impact zone for marine mammals.

Calculation of an impact zone radius relative to a seismic array must initially take into consideration how the array affects sound source levels, particularly for surface and near surface waters (i.e., from the horizontal plane to approximately 30° below the horizon). Through the use of multiple airguns, seismic arrays direct the sound downward, with maximum energy directed towards the seafloor. The "array effect" (see **Appendix D, Figure D-4**) reduces sound source levels in surface and near surface waters by 20 to 60 dB or more. Richardson et al. (1995) note that sound pressure levels may propagate differently in the horizontal direction. For seismic arrays with sound source levels of 240 to 250 dB re 1 μPa-m, they state that the effective source level can be 220 to 230 dB perpendicular to the array's long axis, but as much as 20 dB lower along the array axis. Recent modeling conducted by International Association of Geophysical Contractors (IAGC) indicates the difference between vertical and horizontal signatures is higher, in the vicinity of 60 dB (P. Fontana, Veritas Marine Acquisition, oral comm., 2003). Modeling results included in the *R/V Maurice Ewing* seismic EA (LGL, Ltd., environmental research associates, 2003) also indicate the elliptical nature of the sound field. Therefore, the array effect may cause the target isopleth to be ellipsoid in shape (i.e., the isopleth may not be a perfect circle around the array). In the absence of validated sound spreading loss models and associated modeling results (which are applicable to all seismic arrays), a radial impact zone is considered in this analysis, although it is recognized that the sound field and respective isopleths are ellipsoid. In light of these factors, a conservative estimate for the array effect has been established at 20 dB.

Further, airgun array sound source levels are typically expressed as zero-to-peak or peak-to-peak (see **Appendix C, Section III - Seismic Source Levels**). Conversion from zero-to-peak to rms sound levels[2] reduces the sound source level by 10 dB.

Details of impact zone calculations for the 160- and 180-dB isopleths are provided in **Appendix C, Section III - Seismic Source Levels**. The major determinations are

- Horizontal distance of the 180-dB isopleth - nominal range to target isopleth is ~300 m, based on extrapolated sound source level of 260 dB re 1 μPa (zero-to-peak), -10 dB (rms conversion), and -20 dB (array effect); and

- Horizontal distance of the 160-dB isopleth - nominal range to target isopleth is ~3,000 m, based on extrapolated sound source level of 260 dB re 1 μPa (zero-to-peak), -10 dB (rms conversion), and -20 dB (array effect).

Visual monitoring and clearance of a 500-m safety zone has been established as a requirement of operators working in water depths >200 m throughout the GOM and all OCS waters of the Eastern Planning Area under NTL No. 2004-G01.

3. Ramp-Up

Description: Ramp-up (also known as "soft start" or "slow build up") entails the gradual increase in intensity of a sound source (e.g., airgun array) over a period of 15 to 20 minutes or more, until maximum source levels are reached. Ramp-up is considered in the present analysis under Alternative 1. The intent of ramp-up is to either avoid or reduce the potential for instantaneous hearing damage to an animal (from the sudden initiation of an acoustic source at full power) that might be located in close proximity to an airgun array. Increasing sound levels are designed to warn animals of pending seismic operations, and to allow sufficient time for those animals to leave the immediate area. Increasing sound levels (e.g., from an airgun array) are thought to be annoying or aversive to marine mammals. Under optimal conditions, sensitive individuals are expected to move out of the area, beyond the range where hearing damage might occur.

Examples: Ramp-up has become a standard mitigation measure in select U.S. circumstances, and is a broadly accepted measure overseas (e.g., in waters offshore of the UK, within the Atlantic Frontier). In U.S. waters, ramp-up has been used in all five recent cases (**Table E-2**), with increases of 6 dB re 1 μPa per minute. Ramp-up is a current requirement of operators working in water depths >200 m throughout

---

[2] The measurement of rms values is complex, as aperiodic waveforms do not lend themselves to simple calculation (see **Appendix C, Sections II** through **IV**). Although low frequencies dominate, seismic pulses are a mix of sounds at different frequencies and should be characterized through power spectral density analysis. If the correct routines are chosen, this can be used to decompose the pulse into the rms levels of its constituent spectral components. Given the dominance of the low frequency pressure spike, there is a tendency to read peak pressures directly in terms of signal amplitude. Calculations of rms levels and power also should take account of pulse durations, which are typically less than 1 second. Power is defined as energy per second. Strictly speaking, for complex waveforms, the rms levels will depend upon the individual waveform characteristics as devolved through fourier analysis, and cannot be easily generalized (J. Goold, oral comm., 2002).

the GOM and all OCS waters of the Eastern Planning Area under NTL No. 2004-G01. NTL requirements call for ramp-up over a 20 to 40 minute period, starting at 160 dB re 1 µPa (rms). Per IAGC (2003), output sound pressure level for any airgun array is more closely related to the total number of elements (i.e., airguns) in the array than to the total array volume. IAGC has recommended to the MMS that ramp-up procedures specify that signal output should be increased by a factor of two (i.e., increasing output by 6 dB) every 5 minutes.

Ramp-up also has been used in other acoustic-based studies. The Acoustic Thermometry of Ocean Climate (ATOC) study effort was initiated by the Scripps Institute of Oceanography and other academic institutions in the mid-1990's. By introducing continuous sound levels of 195 dB re 1 µPa at depth into several of the world's oceans, ATOC was designed to monitor ocean temperature changes. In response to concerns about further ensonification of the marine environment and the potential for instantaneous hearing damage, ramp-up of the ATOC system was implemented (e.g., see Final Rule, Operation of a Low Frequency Sound Source by the North Pacific Acoustic Laboratory – NMFS, 2001a).

The U.S. Navy also has initiated a low-frequency active sonar program (termed LFA) to detect submarines using low frequency sound. Concern over potential impacts of LFA have prompted preparation of an environmental impact statement (Department of the Navy, 2001). Pierson et al. (1998) have also summarized the application of ramp-up to further mitigate potential impacts of G&G operations.

Evaluation: The effectiveness of ramp-up has yet to be documented, either within U.S. waters or overseas. Common sense, however, suggests that ramp-up may reduce the potential for hearing damage or other injury to marine mammals upon start-up of an airgun array. Slow build up in the output of a seismic array offers a possible warning to marine mammals of the presence of a potentially aversive and injurious sound source, and may allow adequate time for the individual to swim out of the immediate area.

In spite of the absence of data on its effectiveness, a Gulf-wide requirement for ramp-up is currently part of NTL 2004-G01. Further, ramp-up must be combined with visual monitoring (and may be utilized with passive acoustic monitoring) to provide another mechanism to identify when marine mammals are present within the impact zone, and to establish protocols for powering down the seismic array until the impact zone is clear of marine mammals.

4.   Acoustic Monitoring

a.   Introduction

Acoustic monitoring (or sonar) systems detect underwater objects using sound and are categorized by the way in which they operate – either in a passive or active mode. In principle, both of these methods are in widespread and common use for various purposes in the ocean environment every day. The military uses passive acoustics to detect ships and submarines while scientists use it to study natural phenomena (e.g., volcanic and tectonic activity) and to monitor vocalizing marine mammals. Active acoustic systems or sonars are used everyday and all over the world by the military, industry, scientists, and the general public for navigation, finding fish, locating submerged objects, measuring currents, locating and positioning equipment, etc. Acoustic monitoring, in either passive or active mode, represents a possible mitigation for geophysical and geological (G&G) surveys. The technology can be used also to monitor marine mammal impacts, thereby providing regulators data to assess potential impacts on marine mammal stocks.

At this time, both technologies, particularly active acoustics, are under development and are still evolving. Acoustic monitoring for marine mammals is a developing area of science, and answers to a number of fundamental technical questions are still being sought. In addition, operational considerations relative to actual deployment of acoustic monitoring systems in conjunction with seismic survey operations still need to be worked out. Further research, development, validation, and field trials are required before these technologies will be available for routine use, although a few passive systems are commercially available at present.

Either approach has its own advantages and disadvantages. Ultimately, the technical issues for either approach are reduced to a single purpose – the ability to determine three dimensional (3D) range and bearing of a target marine mammal. By providing information on bearing and distance to the target and, if possible, species identification and direction of movement, an acoustic monitoring system should be able to the assess the risk of a marine mammal being in the "impact" zone during a seismic survey and allow decisions on whether the survey may proceed or not.

This section of the appendix discusses passive and active acoustic monitoring, and outlines the merits or problems associated with each technology. The discussion focuses on the current state of these technologies, their application, and their capabilities to identify (and thereby potentially protect) marine mammals at risk within the "impact zone" of seismic operations. Information has been obtained and summarized on the basis of literature and Internet searches, complemented with individual interviews with industry, military, and regulatory experts. Experts were asked about their knowledge of available technologies and if they knew of organizations that were supplying or developing acoustic monitoring systems. Presentations given at the Marine Mammal Workshop (sponsored by the IAGC in May 2002) also were reviewed for relevant information. Sources contacted and respective affiliations are summarized in **Table E-3**.

Table E-3
Summary of Contacts and Respective Affiliations – Acoustic Monitoring Systems

Contact	Affiliation
Jay Barlow	National Oceanic and Atmospheric Administration, Southwest Fisheries Science Center, La Jolla, California
Craig Douglas	Seamap, Inc., Singapore
Phil Fontana	Veritas/International Association of Geophysical Contractors, Houston, Texas
Robert Gisiner	Office of Naval Research, Arlington, Virginia
George Ioup	University of New Orleans, New Orleans, Louisiana
Mark McDonald	Whale Acoustics, Bellvue, Colorado
Robert McCauley	Curtin University, Australia
David Mellinger	Oregon State University, Corvalis, Oregon
Simon Mustoe	Applied Ecology Solutions, Melbourne, Australia
Tim Pinnington	Seamap, Inc., Houston, Texas
Marcus Smith	OceanEar, Ltd., Bude, Cornwall, United Kingdom
Peter Stein	Scientific Solutions, Inc., Nashua, New Hampshire
David Simmons	Department of Trade and Industry, United Kingdom
Aaron Thode	Scripps Institute of Oceanography, La Jolla, California
Peter Tyack	Woods Hole Oceanographic Institution, Woods Hole, Massachusetts
Roy Wyatt	Seiche Measurements, Inc., Anglesey, United Kingdom
Matthew Zimmerman	FarSounder, Inc., Providence, Rhode Island

MMS recently implemented requirements to protect marine mammals such as the sperm whale during G&G activities in the GOM (i.e., MMS NTL No. 2004-G01, Implementation of Seismic Survey Mitigation Measures and Protected Species Observer Program, effective 1 March 2004 and superceding

NTL Nos. 2003-G08 and 2002-G07 and its amendment; see **Appendix A**). Mitigation measures are now required during all seismic survey operations using airgun arrays in waters deeper than 200 m throughout the GOM; further, seismic survey operations conducted in all Federal waters of the Eastern Planning Area (regardless of water depth) also require mitigation. The latest NTL describes the ramp-up procedure, the exclusion zone for visual monitoring, the roles of visual observers, and reporting requirements relevant to marine mammals. The requirements specify that trained marine mammal observers be onboard seismic survey vessels to monitor a 500-m exclusion zone surrounding the airgun array for whales during seismic survey operations except at night and in bad weather. Ramp-up procedures must be followed when a survey is initiated, and visual observers must be able to clear the exclusion zone (i.e., no whales visible for 30 minutes) prior to ramp-up (i.e., daylight and adequate visibility necessary). The latest NTL also provides the opportunity for G&G operators to utilize passive acoustic monitoring in lieu of visual monitoring should ramp-up be implemented at night or during periods of reduced visibility. Ramp-up would otherwise not be permitted at night or during periods of reduced visibility.

Visual observations to detect the presence of whales that may be potentially affected by seismic surveys are possible during ideal conditions but are less effective or ineffective at night and in poor weather. Visual monitoring suffers from target detection limitations as observers will only sight a fraction of the actual numbers of animals along a survey track. This is due to the fact that observers cannot sight every marine mammal present on the surface due to several factors (i.e., perception bias). Secondly, whales are not at the surface all the time for observers to see them (i.e., availability bias). Even under optimal conditions (e.g., with a large whale, an excellent platform, calm seas, and experienced observers), the visual detection rate may be considerably less than 100%; detection rates drop for sperm whales and beaked whales due to the amount of time these species spend underwater (Gisiner, 2002).

Acoustic monitoring has advantages over visual methods because 24-hour operations are possible and are rarely affected by weather. Acoustic monitoring also offers greater detection ranges than visual methods, and the range is less affected by poor meteorological conditions. For many species, acoustic cues are more conspicuous and are produced more often than visual cues. Given the same duration of observation as visual monitoring, acoustic monitoring also is less tiring and onerous since it requires smaller personnel teams (e.g., during daylight hours). Acoustic monitoring has the potential to be partially or wholly automated, unlike visual methods. In addition, passive acoustic monitoring provides a permanent record of effort and detection.

Acoustic monitoring does have some shortcomings. For example, some marine mammal species rarely vocalize and only do so at certain times of the year. Passive acoustic monitoring also requires a quiet vessel and an initial investment in specialized equipment (Gordon et al., 1998; Gordon, 2002). Further, since acoustic monitoring can be done on a continual basis (i.e., 24 hours a day, 7 days a week), it could also require more operators than the two or three observers required for visual monitoring during daylight hours. Although there is already some partial automation of the detection process, wholly automated systems are not very likely at present given that there is insufficient acoustic information on a number of species (S. Mustoe, Applied Ecology Solutions, oral comm., 2003).

In the GOM, cetaceans are found in both shallow and deeper waters. While bottlenose and Atlantic spotted dolphins are prevalent in shallow water, more than 20 species of toothed whales, dolphins, and a few baleen whale species are present in deeper waters of the Gulf. Sperm whales are the species of major concern with 1,349 individuals estimated in the endangered population in the northern GOM (Mullin and Fulling, in review), some of which co-occur in areas of past and future seismic operations. The area of prime concern is located off the Mississippi River delta at 1,000 m depth, an area frequented by mainly females and calves forming the Delta Group. The 2000 and 2001 Sperm Whale Acoustic Monitoring Program (SWAMP) showed that sperm whales in the GOM could be tracked by passive acoustics and demonstrated the advantages of acoustic techniques over purely visual methods. SWAMP established the integrated visual, acoustic, and tagging teams and their methods for tracking sperm whales. Baseline information was collected in Fiscal Years 2000 and 2001, which was the basis for planning a follow-through project, the Sperm Whale Seismic Study (SWSS) (Lang, 2002). The current belief is that the answer to the problem of effective monitoring for marine mammals, particularly for sperm whales in the GOM, lies with acoustic methods.

b.   *Passive Acoustic Monitoring*

Passive acoustic monitoring simply involves "listening" for sounds emitted by a source without putting any energy into the water. Passive sonar (i.e., sound navigation and ranging) detects sounds from the propulsion systems of ships and submarines and was developed for military purposes, allowing listeners to detect targets while not generating any noise themselves. Techniques such as beamforming[3] and Target Motion Analysis (TMA) are used passively to determine over time the target location, speed, and direction of travel. With passive acoustic monitoring, the targets are whales and dolphins, and the sounds detected are the animals' own vocalizations. Passive acoustic monitoring may employ hydrophone arrays on streamers towed behind a seismic acquisition vessel or a chase boat, sonobuoys, ocean bottom cables, and fixed (boat-based, moored) or free floating sound recorders. Seismic survey operations conducted in U.S. waters have used various passive acoustic systems for monitoring, including bottom recorders, ocean bottom cables, sonobuoys, and hydrophones (Pierson et al., 1998).

Monitoring for deep-diving whales such as sperm whales is made possible by passive acoustic monitoring. NMFS uses passive acoustic monitoring in marine mammal surveys to supplement visual observers, at least in the GOM. Passive acoustic monitoring works for sperm whales because they make loud clicks most of the time they are underwater. However, beaked whales are more problematic because they are cryptic in nature and few recordings of their vocalizations are available. Nonetheless, there is good evidence to suggest that these species (e.g., Cuvier's beaked whale) also produce clicks that are consistent with echolocating cetaceans, suggesting that this species does echolocate. The clicks are similar in nature to those produced by sperm whales but at higher frequencies (Frantzis et al., 2002). Passive acoustic monitoring has proven to far exceed the capabilities of visual observers in the GOM as shown during the 2002 SWSS surveys and in surveys in Australia for Santos Ltd. in 2002. However, the efficacy of passive acoustic monitoring will diminish for those species that only infrequently vocalize, or those animals whose acoustic signals cannot be recognized by the operator or are masked by ambient noise. As discussed below, active acoustic monitoring can detect marine mammals with infrequent vocalization, but the technology is still not developed for use as a mitigation tool. It is apparent that no single system can yet provide effective monitoring to mitigate impacts of seismic operations on all species of marine mammals.

Although the hardware and software for passive acoustic monitoring are available and technologically advanced, complete integrated systems specifically designed and validated for use with marine mammals during seismic surveys are not. Systems for detecting and recording sounds from marine mammals and determining their bearing and distance relative to the receiver are readily available. However, systems that can provide real-time information to allow operational decisions to be made during a seismic survey are limited.

---

[3]   The mathematical process of combining signals from each hydrophone in an array is called beamforming. Beamforming enhances the ability to detect sound arriving from a particular direction, while reducing readings from other directions. In this way, a target can be located.

1)   *Fixed Passive Acoustic Systems*

There are available systems that can be deployed to allow passive acoustic monitoring from fixed positions (e.g., sonobuoys, ship-mounted/floating or bottom mounted recorders, or ocean bottom cables) provided the monitoring system adequately covers the survey area. However, these may not be feasible to use for mitigating seismic survey impacts if data and information cannot be processed, analyzed, and evaluated in real time. Although passive systems have more extensive detection ranges than active systems, they still suffer from limited sound propagation ranges, especially in the GOM where there are several sources of ocean noise (e.g., petroleum activity, shipping, fishing, etc.).

Sonobuoys.

Directional Frequency Analysis and Recording (DIFAR) sonobuoys are independent portable hydrophones that transmit ocean sounds by very high frequency (VHF) radio link to the ship from as far away as 20 km. DIFAR buoys are a valuable tool for correlating visual and acoustic behavior of whales. During use, the DIFAR buoy and radio antenna remain on the surface, while the hydrophone can be set to drop to a preset depth (e.g., from surface to 305 m). The direction to each sound from each sonobuoy can be determined through a magnetic compass and a set of acoustic particle motion sensors. The compass bearing to the sounds are displayed in near real time on a computer display on a ship equipped with receivers. The range to which whales can be heard depends on how rough and noisy the sea is near the sonobuoy, but is commonly up to 20 km, a generalization representing typical baleen whales and average to poor propagation. With sperm whales, the typical detection range for females is about 6 km and for males about 12 km (i.e., due to the fact that higher frequencies tend to attenuate more quickly than lower frequencies). In exceptional cases such as the waters of the Antarctic, the blue whale detection range can exceed 100 km. On the basis of experience, DIFAR sonobuoys have no better detection range than typical marine mammal towed arrays for baleen whales, and only a little better range for sperm whales, which are normally detected at a frequency of about 3,500 Hz above the frequency of flow noise on towed arrays. A multimillion dollar military hydrophone array towed at 3 kn and 1,000 m depth will have better detection range than a DIFAR sonobuoy for baleen whales, but a typical marine mammal towed array running 8 or 10 kn at 2 m depth and costing only $10,000 will have a much poorer detection range than a DIFAR sonobuoy. The biggest issues are ship, ambient, and flow noise on towed arrays. (M. McDonald, Whale Acoustics, oral comm., 2003; http://www.whaleacoustics.com).

In the GOM, the primary species of concern include sperm and beaked whales. Sperm whales are acoustically active about 90 percent of the time when in groups of females and thus are very appropriate for towed array detection, while the added effort and cost of sonobuoy detection may not make sense. Beaked whales exhibit limited vocalization and are cryptic. Sonobuoy arrays are more appropriate for baleen whales such as right whales or blue whales, which call at frequencies below 150 Hz, where towed arrays are very poor performers. No single method is perfect (M. McDonald, Whale Acoustics, oral comm., 2003).

DIFAR bearing determination has not been successfully used with sperm whales because of the short duration and relatively high frequency of the sperm whale click. Crossed DIFAR bearings from two sonobuoys spaced appropriately provide very accurate localization, but with many animals, the swim speed and infrequent occurrence of calls make it impractical to monitor an area as small as 500 m from a moving ship (M. McDonald, Whale Acoustics, oral comm., 2003).

There are technical and operational limitations to using sonobuoys for acoustic monitoring of seismic surveys. Although they could conceivably be moored, typically sonobuoys are not stationary and drift with the current. Therefore, reporting a location based on two sonobuoy detections, which can change, could be problematic. Use of sonobuoys equipped with GPS could solve the problem of determining the relative position of the sound source; however, GPS equipped sonobuoys still require development. Hermes Electronics Inc. modified a regular Model AN/SSQ53D(2) DIFAR sonobuoy with a GPS receiver and demonstrated its feasibility. However, to minimize modulation interference with DIFAR pilot tones and power considerations required, the directional channels had to be disabled (Baker and Bonin, 2001).

To cover a seismic survey area adequately, a line of sonobuoys would have to be dropped ahead of the seismic source that may be moving at 2 to 5 kn. The sonobuoys would have to be positioned carefully at some distance from the vessel track. Ambient currents could move the sonobuoys, which float on the surface, to a position that could disrupt the survey (e.g., by collision or entanglement with the seismic array or streamers) or damage the sonobuoys. Sonobuoys operate for only a limited time (the standard military versions scuttle themselves after a preset time up to 8 hours, although this could be disabled), and it is likely that given the limited effective range for passive acoustic monitoring and the large areas involved in a seismic survey in the GOM, a large number of expensive sonobuoys would be required. Given the number of surveys that may be conducted in the GOM, it may be unrealistic to use sonobuoys for passive acoustic monitoring. Further, the varying distances between the detector, the seismic source, and the targets would present larger uncertainties about the accuracy of the location, which can confound the operational decisions that have to be made to mitigate potential impacts.

Sonobuoys could be moored; however, the water depths involved in the GOM (beyond 200 m) may make such an approach impractical. Sonobuoys have to be a few miles from a sound source in order to detect it. At least two sonobuoys are required to be within the effective range of a potential target in order for a location to be established. A second vessel or aircraft would be required to place sonobuoys ahead of a moving seismic vessel, along with the cost of a second vessel or an aircraft for deployment. The cost of sonobuoys alone can be substantial. Further, there is the potential that unretrieved sonobuoys may litter the ocean floor.

Further assessment of the utility and operational effectiveness of sonobuoys for passive acoustic monitoring as a mitigation tool is needed before specific recommendations for their use can be made. It remains uncertain as to whether these systems can provide an effective range and bearing to a possible marine mammal target as a mitigation tool.

Other Fixed Passive Acoustic Systems.

Other fixed passive acoustic monitoring systems are available. Directional Autonomous Seafloor Acoustic Recorders (DASARs) are fixed acoustic monitoring systems developed by Greeneridge Sciences, Inc. initially in 1996 for detecting and locating bowhead whale calling rates at the British Petroleum (BP) Northstar facility in the Beaufort Sea. BP annually uses an array of 11 offshore DASARs, configured as two overlapping hexagons in which the unit-to-unit separation was 5 km (2.7 nmi). Two DASARs are installed at the central point in the array for redundancy. The DASARs incorporate components from DIFAR equipment that permit calculation of the bearings to vocalizing whales, with 3 DIFAR channels recording directly to a disk. When a whale call is detected by two or (preferably) more DASARs, the location of the calling whale can be estimated by triangulation. The DASARs could be modified to provide real-time data (similar to DIFARs), but at this time, are limited to about a 30- to 45-day deployment on the bottom and need to be retrieved, reprogrammed, and then returned to the sea bottom. Underwater recorders such as DASARs are useful for purposes other than mitigation (e.g., under the Marine Mammal Protection Act [MMPA] authorizations); they can provide information on the number of vocalizing whales and an indication of the level of impact through changes in calling rates.

Fixed passive acoustic monitoring sensors such as bottom-mounted sensors, midwater moorings, or ocean bottom cables also may be impractical for use as a mitigation tool for seismic surveys that would require real time information. Data would have to be telemetered to a receiver (e.g., the seismic source vessel) from each sensor, each of which require adequate independent power and bandwidth. Acoustic signal processing is data intensive. It may be impractical to be telemetering sound files in real time due to the bandwidth that would be required. The technology to process sound files from multiple fixed sensors to generate a target range and bearing relative to the position and track of a seismic survey vessel would need to be developed. An even more difficult task would be to track the movement of the target marine mammal relative to the survey vessel.

Fixed passive acoustic monitoring sensors could be deployed at enough points to cover a survey area, and data could be telemetered to the survey vessel. Given that mitigation is required for surveys in waters deeper than 200 m throughout the GOM and in all OCS waters of the Eastern Planning Area, there would be technical and cost limitations to the use of bottom-mounted recorders or ocean bottom cables. This approach can be very costly depending on the unit cost and number of sensors necessary to cover a survey area, the cost of deploying the sensors, the cost of retrieving the sensors, and the cost of the system to receive, process, and evaluate the data in real-time to be used by the survey vessel. A bottom-mounted hydroacoustic recorder is estimated to cost $40,000 per mooring after considering all costs involved (C. Fox, Pacific Marine Environmental Laboratory, oral comm., 2001). The hydrophones for a single mooring alone can cost $12,000 to $25,000 depending on the actual number purchased. This does not include costs for deploying and retrieving the mooring (G. Ioup, University of New Orleans, oral comm., 2003).

Acoustic data have to be received in real time, so there would be a problem with adequate bandwidth. Such a fixed system still may not detect all marine mammals that could be affected by the seismic survey. Detection range and accuracy for a fixed passive acoustic monitoring system also would vary depending on distance from the fixed sensors. Due to limitations in the utility of fixed passive acoustic monitoring systems and the absence of a mature and proven commercially available fixed passive acoustic monitoring system (i.e., one developed for marine mammal detection to mitigate impacts), fixed passive acoustic monitoring systems are not considered as a viable mitigation measure for marine mammals. Unlike towed systems, they are available but have yet to be developed for routine use as a mitigation tool.

2)   *Towed Passive Acoustic Systems*

Stationary passive acoustic monitoring techniques (i.e., sonobuoys and fixed buoy sensors/recorders) may have their utility in some types of studies, but as a mitigation tool for seismic surveys, towed systems are more operationally efficient to use, more likely to provide reliable positions of potential targets, and likely to be less expensive. Passive acoustic monitoring technologies are available at present that can be mounted on the seismic source vessel but are also limited in their ability to provide 3D range and bearing to a target. With the maturation and validation of the technology, passive acoustic monitoring systems mounted on the seismic source vessel will be a more reasonable supplement to shipboard observers, allowing an operator to monitor an impact zone (or exclusion zone) around the seismic source as further mitigation to protect deep diving sperm whales and cryptic beaked whales.

Towed passive acoustic systems are described below to present the current state and use of the technology. Present towed systems are still at an experimental stage because of limitations discussed below and are effective only for certain species (e.g., actively vocalizing cetaceans). Compared with a fixed acoustic monitoring system, a towed passive acoustic monitoring system, especially when mounted on the seismic survey vessel, would be better for monitoring an exclusion or safety zone around an airgun array. Towed systems representative of the technology available at this time are detailed below, followed by a discussion of the main problems and limitations of towed systems.

Representative Passive Acoustic Monitoring Systems.

It should be noted that the towed systems described herein are representative and are not intended as a complete listing of available systems. Passive acoustic monitoring systems have been designed and constructed by a number of academic, research, and commercial organizations using readily available (i.e., "off the shelf") components, but the number of working integrated systems that have been deployed in the field is limited. The systems described are meant to show what is available, how they are used, and problems that may be encountered in their use.

The commercial and "off the shelf" passive acoustic monitoring systems presented below are very similar in specifications and performance. There may be differences in specific hardware components, construction, and software applications used, but the systems are very similar in concept, design, and performance. [*Note: Although the discussion below focuses on the hardware and software aspects, much of the effectiveness of the technology also depends on the training and experience of the acoustic system operators. The ability to recognize marine mammals correctly and reliably from numerous other sound sources in the ocean requires extensive training and experience. Understanding of the limitations of the technology in different environmental or acoustic conditions and possible sources of errors is essential for acoustic system operators. The development of passive acoustic monitoring as a mitigation tool should include guidelines for training and experience of acoustic system operators. Companies such as OceanEar offer clients options on the hardware and software systems (e.g., Seamap or Seiche) to use along with working teams of acoustic operators and marine mammal observers. Other companies such as Applied Ecology Services and Ecologic offer their own passive acoustic monitoring systems and trained acoustic operators.*]

The Ecologic passive acoustic monitoring array was developed by Jonathan Gordon of the Sea Mammal Research Unit, University of St. Andrews, and is essentially the same as the Seiche systems. Ecologic provides both acoustic monitoring and mitigation services for clients, assembles and sells customized acoustic monitoring systems, and has experience using passive acoustic monitoring for mitigation, research, and population assessment. Ecologic was part of a collaboration, funded by Shell and coordinated by Birmingham University, that developed the first passive acoustic monitoring system specifically for monitoring and mitigation during offshore seismic surveys (J. Gordon, Ecologic, oral comm., 2003).

The Ecologic Integrated Acoustic Detection System or Integrated Cetacean Monitor System is composed of a medium (200 Hz to 22 kHz) and high (30 kHz to 150 kHz) frequency hydrophone array, electronics, and software. This system was designed to detect high frequency clicks from dolphins and porpoises, medium frequency clicks from sperm whales, and medium frequency tonals (whistles) from many dolphins and toothed whales. Each detector assesses background noise and sets trigger thresholds to optimum levels, and the system calculates the bearing to target from the arrival time on two hydrophone elements. All three detection routines run under a monitor program on the same personal computer. Modular systems further facilitate future development (Gordon, 2002).

During the 2002 SWSS cruise, an Ecologic acoustic array with two hydrophone elements mounted about 3 m apart and housed in a ~10 m long polyurethane tube was used on the *R/V Gyre*. The Ecologic array had no depressor (heavy ballast) and was allowed to stream out horizontally behind the vessel. For most of the time, the Ecologic array was deployed at the maximum cable length (approximately 400 m). With this amount of cable out, the hydrophone elements were towed at an estimated depth of about 50 m. During surveys, the vessel maintained a speed of 6 kn, representing a good compromise between reducing noise and covering ground. At a tow speed of 3.3 kn, the Ecologic array had a detection range of about 3.5 km when tracking clicks from a stationary sperm whale group. The limited detection range may have been due to ship noise because the *R/V Gyre* was not designed to be acoustically silent when commissioned in 1973 (SWSS, 2002).

The Ecologic system utilizes a suite of acoustic detection and data collection programs that were developed as part of the Shell/Birmingham project by Oliver Chappell and Douglas Gillespie (Chappell et al., 1999; Lewis et al., 2000). These shareware programs are based on routines initiated by the International Fund for Animal Welfare (IFAW), and the IFAW continues to support the core programs. Acoustic signals are fed into *RainbowClick* and *Ishmael*. These programs have many features, but primarily allow the acoustic team to determine the relative bearing to many whales at once. There is a left-right ambiguity in the relative bearing, which is resolved also by turning the ship to different bearings and following the change in relative bearing. Two of the detection programs, *RainbowClick* and *Whistle*, run more or less continuously, and their detection records are stored as computer files. Ishmael also can run for long periods in real-time spectrogram mode. *RainbowClick*, which automatically detects sperm whale clicks, calculates and displays their bearings. *Whistle* automatically detects and displays mid-high frequency tonal sounds. *Ishmael* also provides a variety of acoustic detection and display functions (SWSS, 2002).

The system has proven very effective in detecting small odontocetes and sperm whales, increasing detection reliability by an order of magnitude during trials conducted off the UK. Passive acoustic monitoring probably makes the biggest contribution in detection of sperm whales principally because they can be detected at significant ranges, they are consistently vocal for a large proportion of the time, and the *RainbowClick* software is so well developed as a sperm whale detection and analysis tool. Smaller odontocetes such as dolphins, killer whales, pilot whales, and dolphins can be detected at useful ranges, but they often move so quickly that target motion analysis is typically not effective. The effective range for harbor porpoise (~400 m) is usually within most mitigation ranges so that any reliable detection should lead to action (e.g., seismic system shutdown). Beaked whales produce promising signals, but there is insufficient knowledge of their vocal behavior to determine how passive acoustic monitoring can best be used with these species (J. Gordon, Ecologic, oral comm., 2003).

NMFS Norris Array.

The *Stenella* Abundance Research Project (STAR) Survey is a multi-year study designed to assess the status of dolphin stocks that have been taken as incidental catch by the yellowfin tuna purse-seine fishery in the eastern tropical Pacific. Acoustic recordings supplement visual survey data and provide information for on-going studies of cetacean vocalizations. NMFS Southwest Fisheries Science Center (SWFSC) originally constructed its own hydrophone arrays with the help of Don Norris of Sonatech, Santa Barbara, CA. NMFS currently constructs its own arrays. For the 2003 STAR cruise, the main array was a new 5-element Norris array with sensitivity from 15 Hz to 40 kHz. The cruise also planned a 3-element Norris array with sensitivity from 500 Hz to 150 kHz as the first backup array; the second backup array was the 5-element Innovative Transducers, Inc. (ITI) array with sensitivity from 10 Hz to 15 kHz (J. Barlow, NMFS, oral comm., 2003).

Shareware software programs are used for processing the sound records. *Ishmael*, written by Dave Mellinger of Oregon State University, is the main software for estimating bearing angles and provides a variety of acoustic detection and display functions. *WhalTrak*, written by Oliver Chappel and Doug Gillespie, records ship position, date and time, comments, and other survey information sent to it by *Ishmael* (SWSS, 2002).

Seamap, Inc. Cetacean Monitoring System.

The Seamap Cetacean Monitoring System (CMS), also called the Passive Marine Cetacean Detection System, is described as an integrated solution for detecting marine mammal vocalizations and provides direction finding and distance estimation. It combines in-sea equipment composed of a hydrophone array with a 1 Hz to more than 25 kHz frequency response, a tow cable, and deck cable with a computer running proprietary software (Seamap CMS). The wide bandwidth coupled with exceptional low frequency response and variable hydrophone spacing allows detection of almost all vocalizing species including baleen whales. The system, interfaced to GPS, can provide the latitude and longitude of a detected vocalization, as well as range and bearing information, in real time. The information is presented using an intuitive map-based database viewer. It is self contained for rack mounting and can be deployed in several configurations for use on a chase vessel or acquisition vessel. The system is expandable to record, analyze, and process up to 150-kHz signals. A trained operator is required for accurate detection of marine mammals. Details about their system are available on their website at http://www.seamap.com.

The Seamap towed array has low levels of self-generated noise, essential to its ability to detect baleen whales vocalizing at infrasonic frequencies (typically means frequencies below human hearing, but in this case refers to frequencies below 100 Hz). Low noise is not essential for sperm whales or other toothed whales because their vocalizations are usually at higher frequencies and are loud enough that they can be detected out to 500 to 2,000 m with any detector. However, it is very useful for baleen whale detection as well as for monitoring seismic signals (P. Tyack, Woods Hole Oceanographic Institution [WHOI], oral comm., 2003).

Seamap CMS does not provide a 3D location of a target. To position animals in three dimensions, algorithms and software are required to solve the passive acoustic problems for such positioning. This solution is not just for the special case where the depth of the animal can be derived by measuring the arrival times of the surface reflection and bottom reflection relative to the direct arrival (coupled with knowledge of the water depth and hydrophone depth), but a general case solution that would work for any vocalization and without knowledge of the water depth or bottom shape (C. Douglas, Seamap, oral comm., 2003).

Seiche Measurements Ltd. Cetacean Monitoring System.

The Seiche Cetacean Monitoring System is based on the IFAW cetacean monitoring system developed over the last 6 years and used extensively around the world to monitor cetaceans. The system, similar to those developed by Jonathan Gordon of Ecologic (UK), is composed of a short towed hydrophone array section, a tow cable, a deck cable, and a data processing system, which processes, displays, and stores selected data. A depth gauge is fitted within the array. The in-water component of the system can be stored on a small winch or storage reel. The data processing system has electronics and a suite of software programs written by Douglas Gillespie, designed to detect different types of whistles and clicks made by cetaceans. The system also automatically collects navigation information. Summary information on detections and navigation data is stored in a database for data archiving and analysis. The system combines a low frequency detection system (100 Hz to 44 kHz) with a high frequency system for detection of porpoise and dolphin clicks (50 to 150 kHz). A very low frequency (VLF) hydrophone is also available to monitor the frequency band from 0.5 to 150 Hz. Details about their system are available on their website at http://www.seiche.com.

The system gives an estimated range and a bearing, although it cannot resolve the left-right ambiguity without a direction change of the towing vessel (this would be a problem if mounted on a seismic survey vessel). Seiche is evaluating their 2D system to resolve this ambiguity during the course of 2003 (R. Wyatt, Seiche Measurements, Ltd., oral comm., 2003).

Seiche systems have been operated from the back deck of seismic survey vessels, although the normal operating method is to use a chase boat. However, this approach has a disadvantage in that the measurement system and the noise source are not coincident, so relating cetacean activity to the noise source may be problematic. Although towing from the back deck is not a common occurrence, it is likely that it will become the standard operating method when satisfactory deployment methods are achieved. The operation from the back deck of a seismic vessel has been carried out in two ways: 1) by towing from a cable attached to one of the paravane lines, and 2) by towing from the back deck but short of and deeper than the gun arrays. Seiche expects to be doing further experimental deployments from the back deck of various vessels to establish the best towing method from the seismic vessel (R. Wyatt, Seiche Measurements, Ltd., oral comm., 2003).

WHOI Array.

During the 2002 SWSS cruises, the *R/V Gyre* was outfitted with a second acoustic monitoring array provided by WHOI. The WHOI array, developed for research purposes, had three, 1 to 24 kHz hydrophone elements mounted about 3 m apart and housed in a ~10-m long polyurethane tube. Ship noise and tow speed affect hydrophone array performance. The WHOI array was attached to a batfish depressor to pull the hydrophone elements down and allow the WHOI array to monitor at greater depth than the Ecologic array. Unfortunately, above 5 kn, the WHOI hydrophone array/batfish combination was found to rattle and was generally noisier at lower speeds compared with the Ecologic array. Tests revealed an effective detection range of 4 km. The WHOI array, in particular, suffered from ship noise contamination, as the depressor pulled the array not only deeper, but closer to the ship. In addition, whenever the ship traveled slower than 2 kn, a large amount of noise was created when one of the engines was disengaged. The WHOI array was effectively awash in prop noise at higher speeds (SWSS, 2002). An acoustically quiet platform is needed for passive acoustic detection of baleen whales.

Problems and Limitations.

Present towed systems have the following problems and limitations:

- difficulty in determining range;
- array blindspot and left/right ambiguity;
- deployment on chase boats; and
- deployment on seismic vessel.

When a towed array is used for passive acoustic monitoring, it is often difficult to determine the range to the vocalizing animals. Determination of range to a transient sound source on a towed array is based on how accurately the time of arrival difference on the hydrophones can be determined. This depends on the environment; deep water may result in many multipath arrivals and perhaps a very weak or apparently absent direct path arrival, sometimes causing large errors. As a rule of thumb, range can be estimated out to 10 times the array length when the sound source is broadside of the array. A typical marine mammal array is only perhaps 100 m long, and as the angle goes off broadside, the ability to determine range diminishes rapidly. Although 100-km ranges for humpbacks and 10-km ranges for dolphins are possible (e.g., with military arrays towed at very slow speeds at depth with optimal acoustic conditions), the mitigation (i.e., shutdown) range will be much less and will be more difficult to estimate. Hydrophone arrays towed behind the survey vessel (or from an additional chase boat) are affected by tow speed (which affects flow noise) and the ship's acoustic characteristics, thereby limiting detection range. Noise generated by the ship, the towed seismic array, and the seismic pulses also could mask sounds from marine mammals at the same range of frequencies. The airgun array itself will reverberate in the ocean following each shot, and the shots are typically spaced only about 8 to 15 seconds apart, depending on water depth and the depth of the targeted geologic structures. This means that typically the background ambient noise of an airgun vessel is pushed up by about the order of 6 dB at frequencies below 1,000 Hz, making passive localization more difficult, and detection ranges will be reduced to some undetermined degree (M. McDonald, Whale Acoustics, oral comm., 2003).

At present, towed systems also have problems related to determining and measuring the bearing to a potential target marine mammal: a blindspot and left/right ambiguity. Single analog arrays (i.e., a single line of hydrophones) have a blindspot at each end of the array, along the axis. At 0° from the axis of the array, a position cannot be triangulated effectively because the hydrophones are behind one another. The best and most accurate positioning capability is directly perpendicular to the line of the hydrophone array because the two hydrophone elements are side by side. As the angle becomes more acute (towards the blindspot), the accuracy decreases. A bearing to the sound producer is calculated by the difference in the time it takes a signal to reach each hydrophone element. Logically, this is the same for any single angle, irrespective of whether it is coming from the right or the left hand side of the hydrophone array. This is referred to as left/right ambiguity (S. Mustoe, Applied Ecology Solutions, oral comm., 2003). For example, if the sound arrives at both hydrophone elements on a dual element array, the source is located at 90° (bearing angle) to the axis of the array, but it cannot be determined whether the sound came from the left or right side of the hydrophone array. If the sound arrives first at the lead element of the hydrophone array, then the sound source is located in front of the array on the left or right side. Similarly, if the sound arrives first at the trailing element of the hydrophone array, then the source is behind the array on the left or right side.

Single hydrophone arrays are used primarily on chase boats or guard vessels, but there are some problems. Chase boats or guard vessels are routinely used on seismic surveys in the UK. The use of guard vessels in the UK is an arrangement that has developed over many years but is purely a commercial arrangement between the company carrying out the survey and fishermen's organizations. Typically, a guard vessel (also known as a chase boat) contacts fishing vessels, or guards them away from the streamers. Chase or guard vessels, equipped with hydrophone arrays, are typically used as platforms for passive acoustic monitoring in seismic surveys in the UK and elsewhere. This places the passive acoustic monitoring system at some distance ahead of the airgun array, although it also is possible that the chase boat is positioned alongside or behind the seismic vessel. Passive acoustic monitoring equipment is most often deployed from chase vessels because of the perception that it is likely to cause operational difficulties on a seismic vessel and can be easier to deploy from the guard vessel. One drawback to such a deployment rests with the primary role of guard vessels – chasing other vessels. It is possible that a guard vessel could be many miles from the mother vessel and unable to provide useful information on acoustic detections when needed. Communications between visual observers on the seismic vessel and passive acoustic monitoring operators on the chase boat also are made more difficult (D. Simmons, Department of Trade and Industry, oral comm., 2003).

Petroleum Geo-Services (PGS) deployed a passive acoustic monitoring system from a guard vessel positioned 4,050 m ahead of the seismic source during surveys in Brazil. Passive acoustic monitoring was used in 24-hour operations for 90 days, representing 57% of the survey duration of 150 days. Passive acoustic monitoring detected 36 cetaceans in encounters primarily in the shallow water area and performed much better than visual methods. However, passive acoustic monitoring suffered from poor species identification and the inability to detect distances. The lack of automatic flagging of detections, high noise levels, and the dual source/acoustic vessel geometry caused problems resulting in less than satisfactory performance (Walker, 2002).

In the GOM, at least two chase boats or guard vessels generally are used during 3D surveys, while one or more chase boats may be used for 2D surveys depending on the area where the survey is conducted (P. Fontana, IAGC, oral comm., 2003). However, there may be situations and areas when chase boats or guard vessels may not always be used, or are not always available as a platform for acoustic monitoring. If chase or guard vessels are used, they are required generally to perform other duties, and from time to time may have to return to port for provisions (S. Mustoe, Applied Ecology Solutions, oral comm., 2003). Noisy and unhelpful guard vessels are among the issues to be resolved in developing effective acoustic monitoring guidelines (Gordon, 2002).

Deploying the hydrophone array from the back deck of the main seismic vessel solves most of the problems associated with using chase boats. If passive acoustic monitoring arrays are deployed from the seismic vessel, the receivers can be positioned almost at the airguns. This has two major advantages. First, it is simpler to translate distance into a radius from the guns; and second, in the absence of a support vessel or dedicated research boat to accompany the main seismic vessel, passive acoustic monitoring cannot be used otherwise. This has been tried, and the results showed that vocalizing sperm whales and other cetaceans, including pilot whales and smaller dolphins, could still be detected at distances in excess of 500 m from the airgun array. One of the biggest difficulties with single-streamer systems is significant inherent imprecision when it comes to pinpointing animals. For example, assuming the animals, the acoustic vessel, and the seismic vessel are cruising at about 4 kn, any one of these can cover a distance of about 500 m in about 4 minutes. Under certain conditions (e.g., if the seismic vessel and animals were headed toward each other), the closure rate almost doubles (i.e., the 500-m distance could be covered in 2 minutes). In all cases, there may be very little time to locate animals. Furthermore, over a 2-minute period, the animals may have moved significantly with respect to the 500-m distance used for mitigation (S. Mustoe, Applied Ecology Solutions, oral comm., 2003).

There are two ways to get around the blindspot and left/right ambiguity problem. One is by moving the axis of the array, and this is currently what is done when the hydrophone arrays are deployed off chase boats. By trailing the array from a small boat, the boat and therefore the array can be turned a few degrees to either side of the survey track (this means leaving the survey track). Turning the array allows the acoustic operator to a) explore the blindspot; and b) remove left/right ambiguity. Due to the fact that bearing is determined from just the difference in sound arrival times at two hydrophone elements, an acoustic operator cannot determine from which side of the array the sound emanated (left/right ambiguity). If a sound comes from ahead of a hydrophone array at 45° off the bow and an object a fixed distance away, due to left/right ambiguity, two possible lines emanate from the hydrophone array, one at 45° from the left and one at 45° from the right. If the hydrophone array is turned 20° to the left and is moved along some distance and if the object has not moved significantly, the line emanating from the left is now at a much greater angle and is converging with the original line to the left while the line to the right is not diverging from the original line to the right (S. Mustoe, Applied Ecology Solutions, oral comm., 2003).

However, if the hydrophone array is deployed from the seismic vessel, the technique of turning the vessel a few degrees to either side to explore the blindspot and precisely locate animals ahead of the main vessel (an inherent problem with single streamers) and also remove the left/right ambiguity cannot be used. The seismic vessel cannot turn more than a degree or two per minute or otherwise the gear gets tangled (S. Mustoe, Applied Ecology Solutions, oral comm., 2003).

The alternative to addressing the left/right ambiguity is to use a stereo array composed of two arrays deployed side by side that are independently triangulating. A stereo array can be used on either chase boats or the seismic source vessel. With four hydrophones arranged in a square rather than a line, there is no blindspot and no left/right ambiguity. Although this is only a simple hardware step, there is as of now no software specifically designed for interpreting such data (S. Mustoe, Applied Ecology

Solutions, oral comm., 2003). It is possible to use existing software such as *RainbowClick* to resolve the left/right ambiguity problem when towing two arrays (S. Tsoflias, MMS, oral comm., 2003).

In addition to technical limitations (such as the blind spot off the axis of the array), the left/right ambiguity, and difficulty in estimating range, operational problems still have to be addressed before towed arrays can be demonstrated to be an effective mitigation tool. The technical and operational issues regarding the use of a second vessel as a monitoring platform (e.g., noisy guard vessels, distance between seismic source and detector), working from seismic vessels, and the question of incorporating additional hydrophones into a seismic array still have to be considered and resolved. Optimum protocols and working procedures still need to be developed and adapted into regulatory guidelines. In order to develop acoustic monitoring technology, it is important to measure its efficiency and effectiveness. For passive acoustic monitoring, detection efficiency can be measured using independent platform methods; however, defining what is effective mitigation still needs to be considered and acceptable measures or metrics be developed. These are essential for guiding the industry's development of acoustic monitoring systems and allowing regulators to assess the effectiveness of mitigation in minimizing impacts on species of concern.

Although the previous discussion focused on the hardware and software aspects of representative systems, the effectiveness of the technology also depends on the training and experience of the acoustic system operators. The ability to correctly and reliably recognize marine mammals from numerous other sound sources in the ocean requires extensive training and experience. Understanding of the limitations of the technology in different environmental or acoustic conditions and possible sources of errors is essential for acoustic system operators. The development of passive acoustic monitoring as a mitigation tool should include guidelines for training and experience of acoustic system operators.

c. *Active Acoustic Monitoring*

Active sonar places a short sound pulse (energy) from a high power source (transducer) that travels through the water, reflects off objects, and travels back to a hydrophone receiver. The time it takes for the sound to travel to and from the target is easily computed from the difference in time that the source "ping" was sent and the time the reflected returning sound is measured. This travel time multiplied by the speed of sound in water divided by two is the approximate distance to the target. Bearing and range from the ship (or some other platform) can be converted to an absolute position on a map, given the ship position and some simple geometry. This is used, for example, to map seabed features, or to discriminate among different objects on the seafloor and in the water.

Along with passive acoustics, the U.S. Navy has been the major developer of active acoustic technologies. The Sound Surveillance System (SOSUS), now the Integrated Undersea Surveillance System (IUSS) Group, in essence, began development of passive acoustic monitoring (Gisiner, 2002). These military assets that were effective for detecting submarines also have turned out to be valuable tools for studying marine mammals. However, the use of Navy-developed acoustic systems for commercial purposes is limited because much of the technology may be classified due to national security concerns. The technology for active sonar systems is well understood and available in Navy tactical sonars, but the legal issues will likely block these existing sonars from being used for marine mammal detection (M. McDonald, Whale Acoustics, oral comm., 2003).

Analysis of the responses from experts shows there was common consensus that at present there are no commercially available active acoustic monitoring systems for detecting marine mammals. Industry contacts indicated that two systems are currently under development in the U.S. and may be available within the next few years. Given the recent experience in the blocking of the Marine Mammal Active

Sonar Test (MAST) project in California, development could be further delayed. MAST was enjoined from being conducted in 2003.

Development of an effective active sonar system requires consideration of the differences in behavior of various types of marine mammals in the ocean. As such, it may be difficult to find a single approach that will work well with all species. Of the 82 species currently listed by the International Whaling Commission, there are 20 beaked whale species, about 40 ocean-going dolphin and porpoise species, and 13 baleen whale species. Remaining marine mammal species include river dolphins and narwhals (http://www.iwcoffice.org/cetacea.htm).

At present, the species most likely to be injured by mid-range frequencies of man-made sound in the ocean are the beaked whales. Several stranding incidents that occurred following sonar testing suggest that such operations may have prompted these responses. For example, the stranding of 18 cetaceans in the Bahamas (mostly beaked whales) was coincident with the use of Navy sonar. There are several similar but less well documented examples – sonar operations in the Mediterranean Sea, and seismic operations in the Gulf of California. Beaked whales are very elusive animals that normally dive to depths exceeding 1,000 m for 40 minutes or more. Beaked whales are virtually impossible to observe at the surface when winds (and corresponding waves) are greater than about 10 kn. A sonar will likely have difficulty distinguishing them from other animals (e.g., schooling fish). Beaked whales are known to produce clicks and whistles but are difficult to survey with passive acoustics. In the GOM, most of the beaked whale sightings are in the western portion of the Gulf. Beaked whales are rarely found in continental shelf waters (Würsig et al., 2000), but typically are concentrated on the deep-water side of shelf margins, where they feed on deep-water squid. A speculation offered as one possible explanation for the scarcity of beaked whales within U.S. GOM waters may be their sensitivity to seismic surveys (M. McDonald, Whale Acoustics, oral comm., 2003).

Other cetaceans such as dolphin species generally live within water depths of 200 m or less. At the surface, dolphins tend to be found in large schools making them easier to observe. These species can usually be detected with passive acoustics such as a towed array; use of active sonar provides little added benefit.

In the case of baleen whales, there is a risk the active detection system will startle the whales. Whale movement away from an active acoustic source is desirable; however, the initial response to a sudden sound may be towards the surface. Such movement is not advantageous if the vessel is rapidly approaching, creating a ship strike issue (M. McDonald, Whale Acoustics, oral comm., 2003).

Apart from the use of a specialized active sonar (i.e., U.S. Navy HF/M3) to find whales, seals, and sea turtles in order to mitigate impacts during testing of new military sonars (e.g., the U.S. Navy's Surveillance Towed Array Sensor System, Low Frequency Active, or SURTASS LFA – a long-range, low frequency [between 100 and 500 Hz] sonar system composed of both active and passive components), there has been little or no technology and experience in applying active acoustics to locate and track marine mammals. Currently, there are no known efforts to develop active sonar systems for marine mammal monitoring in Asia, Europe, or Australia where there would be potential market interest and manufacturers. At present, there have not been any active acoustic monitoring or sonar systems that have been brought to market except for a system developed for avoiding underwater obstacles that may have limited applicability to mitigation efforts in the GOM due to its limited detection distance. The challenges to active acoustic technology development involve not only technical issues with regard to the design frequency of the sonar, but also testing and validation to establish performance metrics.

If the sonar is to be kept above the hearing range of any marine mammal, it would have to operate at 150 to 200 kHz or higher, greatly limiting the range of the sonar. Most fish finding sonars (e.g., Furuno) operate at around 30 kHz and are probably good for detecting whales out to about 2 km and dolphins out to about 1 km. However, it is probably also audible to all the small marine mammals and some of the larger whales. If the whale detection sonar is operated at frequencies the animals might hear, the detection sonar also would need to be assessed whether it by itself is a source of some disturbance. For any acoustic system, performance metrics need to be established (e.g., percent detections at various ranges, false alarm rate, sensitivity to environmental conditions like ducting, entrained air bubbles, fish schools, etc.; R. Gisiner, Office of Naval Research, oral comm., 2003).

The use of active sonar to monitor marine mammals is not new. Active sonars were developed in the 1950's for use on catcher boats to find and kill whales (e.g., Kelvin-Hughes "whalefinder" sonar). There have been relatively few recent attempts to develop an active sonar for whale detection and monitoring. An active acoustic monitoring system was being developed by the SOLMAR program at SACLANT Centre, a North Atlantic Treaty Organization naval research center, in La Spezia, Italy. This system was designed to obtain target strength measurements of sperm and fin whales in the Ligurian Sea using a research sound source of relatively lower frequency (i.e., about 10 kHz). Plans to develop a whale detection sonar are currently suspended. Pepe Nuzzolo of SeaQuest, Inc. tested a very high frequency sonar (233 kHz) on a New England Aquarium whalewatching ship in 2000, but the high frequency limited the range of detection to less than 100 m. No further information is available regarding plans for use of a lower frequency sonar (60 to 80 kHz) (R. Gisiner, Office of Naval Research, oral comm., 2003).

Jim Miller of the University of Rhode Island, operating as a corporation (Pyrcon/FarSounder), tested a forward-looking high frequency sonar in the 20 to 40 kHz range. Miller was able to obtain several target strength measurements from the broad side of a humpback whale at about 300 to 500 m. This effort was undertaken using limited NMFS funding (Northeast Fisheries Science Center, Woods Hole, MA). The goal was to develop a research sonar and possible protection sonar from ship collisions for right whales (R. Gisiner, Office of Naval Research, oral comm., 2003; Miller and Potter, undated).

Apart from the systems mentioned above, only the systems described below have been developed to near testing stage. Both of these systems have good potential for application to seismic survey mitigation, although further research and development are required. The timeframe for bringing these systems to the point of testing them for use in seismic mitigation ranges from several months to a few years depending on what specific capabilities are needed.

Science Solutions, Inc. Integrated Marine Mammal Monitoring and Protection System (IMAPS).

One of the active acoustic systems currently available is the IMAPS developed by Scientific Solutions, Inc. (SSI). However, as currently configured, IMAPS is not presently applicable for acoustic monitoring for seismic surveys. The IMAPS system operates at 20 to 40 kHz, with a maximum range of 2.5 km. It has a vertical line array source with a source level of up to 210 dB re 1 µPa-m with 60 electronically steered receivers and self-contained power amplifiers and receiver electronics. It has safety features built in that shut down the sonar pulses if a marine mammal is detected within 100 m of the sonar. There are also integrated passive receivers mounted on the system. Although the current system is essentially a fixed or moored system, being deployed over the side of a ship, SSI intends to develop a towed system that has good potential for use as a mitigation tool for seismic surveys. Towed body development and plans for testing the system are under discussion (P. Stein, Science Solutions, Inc., oral comm., 2003). Details about their system are available on their website at http://www.scisol.com.

The U.S. Navy's Office of Naval Research (ONR) attempted to test the effectiveness of the IMAPS sonar and any response the animals might have to it but were blocked by a lawsuit (Hawaii County Green Party vs. Evans, C-99-3719 WHA, U.S. District Court, Northern District of California). A similar sonar used on SURTASS LFA has been demonstrated to be effective in detecting whales within 1 km of the

LFA source, and thus provides a measure of protection that would not be achievable from visual, passive acoustic, or combined visual-passive monitoring. The HF/M3 sonar used with SURTASS LFA detected close to 100% of the whales in a 1-km vicinity. In comparison, visual surveys would usually detect about 50%, and combined visual plus passive might get up to 75%, depending on how active the whales are vocally (R. Gisiner, Office of Naval Research, oral comm., 2003).

FarSounder FS-3.

FarSounder will soon release a forward looking 60-kHz sonar system with 3D capability (i.e., range, depth, and bearing for in-water targets). The system was developed for obstacle detection, seafloor mapping, search and rescue, and whale avoidance. For navigational applications, the software only processes sonar returns above 50 m. The core technology is very flexible and can be customized to meet needs if >50 m water depths are required. FarSounder's active sonar system also may have limitations in seeing a target near the surface due to surface reflections.

FarSounder technology is capable of creating a 3D image for a large field of view with a single ping in deep or shallow water. The system has a 90% probability of detecting a mid-sized whale at 275 m from a single ping with a 3-second update rate; detection capabilities for the FarSounder system depend upon the size of the target (e.g., shipping container and navigation buoy detectable at 300 and 183 m, respectively). The system can detect and display in-water targets; however, the system does not yet have the ability to classify its targets. This system was designed to image navigation hazards including whales. The system includes a transducer, power module, and graphical user interface. The wet-end (transducer) is forward facing and either external or through-hull mounted. The system also is capable of incorporating National Marine Electronics Association Global Positioning System (NMEA GPS), speed, depth, and heading data. Details about the system are available on their website at http://www.farsounder.com.

The FarSounder FS-3 operates at frequencies well above the hearing range of the great whales. The FS-3 sound levels at under 200 dB are quieter than those of many mammals, including the bottlenose dolphin. Through "adaptive ensonification," the system automatically raises and lowers sound levels depending on sea state and distance to objects.

At present, the FarSounder system locates and displays objects in the water column. Classification of targets is a technology being considered for integration into future developments of the system. FarSounder plans to team with one of the various groups who have developed classification technology (M. Zimmermann, Farsounder, Inc., oral comm., 2003).

d.   *Feasibility of Acoustic Monitoring*

Much of the Navy-towed arrays for detecting and tracking enemy vessels utilize the same technology that can be used for detecting whales; the acoustic technology used for seismic survey operations is very much the same technology that can be used for passive acoustic monitoring of marine mammals. Similar Navy systems work very well to detect and track whales although sperm whales would require listening above 1 kHz. If arrays of sensors are being towed anyway, it seems reasonable to make sure that broadband sensors are included, and that these are monitored for marine mammal sounds. However, there has been little effort to adapt seismic receivers for passive acoustic monitoring of marine mammals. This may be due to operational difficulties rather than technical limitations. It may be that some sensors for intense seismic sources are of lower sensitivity than needed to optimize listening for whales, although any whale near enough to require mitigation should be loud enough to detect on a seismic receiver (P. Tyack, WHOI, oral comm., 2003).

For deep marine seismic surveys, the seismic source is fired generally at 8- to 12-second intervals (firing intervals may be as short as 5 seconds and as long as 15 seconds depending on the targeted seismic formations; P. Fontana, IAGC, oral comm., 2003), with a broadband energy content that is very different compared to the longer tonal vocalizations of the baleen whales. Baleen whales can be heard vocalizing "through" the momentary flash of low frequency seismic noise. Passive systems typically only need to detect one distinct vocalization to obtain a bearing. While the seismic hydrophones are technically capable of hearing the vocalizations from the baleen whales, there are a few problems with this approach. Seismic hydrophones are generally "grouped," being summed over anywhere from 12.5 to 40 m to increase the perpendicular group gain. The group is much less sensitive in the inline direction. In addition, the seismic array is usually towed at shallow depths (i.e., from 5 to 10 m subsurface). This is not ideal for horizontal low frequency reception. In this case, the ray path length of the direct and ghost (sea surface) arrivals are almost the same. The ghost arrival has a 180° phase inversion and similar amplitude to the direct arrival, and hence there is a cancellation of the received low frequency signals. The situation is improved by towing the array deeper (e.g., around 40 m subsurface). Deeper towing depth increases the differences in the ghost and direct ray path lengths for horizontally propagating low frequency noise, and hence there is less "canceling effect," therefore more signal is received, which can be detected. There are several problems in fitting new hydrophones suitable for marine mammal detections into a seismic streamer. Essentially, seismic arrays are typically operated too shallow and normally behind the source; they also are noisy in the acoustical as well as electrical sense, especially when looking at the much higher frequency signals that are of interest (C. Douglas, Seamap, oral comm., 2003).

The knowledge and technology for using sound to detect marine mammals by passive or active means is currently evolving. Until recently, there has been limited regulatory agency interest, and thus limited market interest to drive full development of acoustic technology for mitigating impacts to marine mammals. Unlike technologies for military and commercial purposes, there has been no impetus for developing technology for monitoring marine mammals until recently. In the case of active acoustics, there is the additional concern of impact from the system itself that could block development of the technology as has been the recent experience with the MAST proposed in California.

This situation may change in the future if additional regulations are implemented for mitigation of seismic activities. The Department of Trade and Industry (DTI) in the UK is looking at improving the effectiveness of mitigation measures such as passive and active acoustic monitoring in implementing the Joint Nature Conservation Committee (JNCC) Guidelines (i.e., detecting marine mammals within 500 m of seismic operations prior to ramp-up) (JNCC, 2002). The DTI has commissioned a study to evaluate the effectiveness of active techniques in seismic survey mitigation. The study will look at specialized cetacean systems as well as investigating the applicability of fish finding sonars, an existing and well-developed technology, to mitigation. Fish finding sonars are being considered because experience suggests they may be capable of detections at considerable range (D. Simmons, DTI, oral comm., 2003). Some fish finding sonars operate at very high frequencies, suggesting that the risk of behavioral effects would be reduced. However, it is doubtful that fish finding sonars will be optimized enough to both detect marine mammals at sufficient range and not affect their behavior (P. Tyack, WHOI, oral comm., 2003). There is anecdotal evidence that Norwegian whalers could startle and chase minke whales to the surface using fish finding sonars, suggesting that minke whales can hear them (M. McDonald, Whale Acoustics, oral comm., 2003).

Apart from the difficulties of measuring sound in the marine environment, accurately determining marine mammal position through acoustic means is a problem due to the signal complexity, since the process involves receiving both low frequency sound and high sonic clicks. Signal detection involves signal conditioning to reduce the signal to noise ratio, and signal classification (differentiating the signal versus noise). It also requires setting criteria to automate the process, especially if the monitoring is done to gather and evaluate data in real time as would be required for mitigation for marine surveys. Receiver Operating Criteria (ROC) are needed to specify how much noise should be tolerated for the gain in sensitivity in detecting marine mammals. Localization and tracking of the signal sources (i.e., mapping the data) is the next step. Neural networks are thought to present the best possibility of dealing with or processing complex data from sound recordings (Gisiner, 2002).

Although the technology for detecting and locating underwater sounds and their sources in general is well developed, integrated hardware and software systems using acoustics specifically designed to locate and track marine mammals as mitigation for seismic surveys are relatively new and have only been commercially available in recent years. Recognizing that passive acoustic monitoring technology is still under development, there is no country currently requiring their use in seismic surveys, although various government agencies have advocated its optional use. For example, the MMS advocates the optional use of passive acoustic monitoring under NTL No. 2004-G01. The UK also recommends their use in certain sensitive waters such as the West of Shetlands (JNCC, 2002). Australia and Canada recognize their potential usefulness as a mitigation tool and recommend their use but may not require them (Environment Australia, 2001; J. Matthews, Petroleum Research Atlantic Canada, oral comm., 2003). In Australia, acoustic monitoring may be required due to proximity of a seismic survey to habitats of protected species, e.g., blue whale (A. Bigwood, Environment Australia, oral comm., 2003).

While the UK and Australia have developed guidelines on acoustic monitoring, Canada has no specific regulations or guidelines, although monitoring requirements may be stipulated as a license condition or as part of a mitigation plan arising from an environmental assessment process (i.e., by non-regulatory or procedural means; P. Macnab, Canada Department of Fisheries and Oceans, oral comm., 2003). Recently, the MMS implemented NTL No. 2004-G01, which allows G&G operators to utilize passive acoustic monitoring techniques in lieu of visual observers during nighttime or periods of reduced visibility.

Present passive acoustic monitoring systems still require technical development and proving in the field. Minimal experience has been acquired in the application of passive acoustic monitoring as a mitigation during G&G activities. Towed arrays have been used primarily, for sperm whale work, although they have the disadvantage of not being able to see straight ahead or through the ship unless the array is towed deeper than the bottom of the vessel. In addition to technical limitations, operational issues still have to be addressed before towed arrays are effective as a mitigation tool. Areas where seismic activities have involved passive acoustic monitoring are summarized in **Table E-4**.

Due to proprietary concerns, information on future seismic programs employing acoustic monitoring is not available except for some general information. Discussions regarding the use of acoustic monitoring for a seismic survey in the UK are underway (D. Simmons, DTI, oral comm., 2003). For the first time, two seismic programs off Nova Scotia using acoustic monitoring as mitigation are being conducted in 2003. The seismic programs are near a proposed marine protected area and near habitat for the endangered northern bottlenose whale (E. Theirault, Canada - Nova Scotia Offshore Petroleum Board, oral comm., 2003).

Table E-4
Summary of the Use of Passive Acoustic Monitoring as a Mitigation Tool During Seismic Operations

Country	Location/Field	Company	Dates
United States	Alaska (Beaufort Sea), Northstar Unit	British Petroleum	1997-2003
United States	Southern California (Santa Barbara Channel) Santa Ynez Unit	Exxon	1995
United Kingdom	West Wales	Chevron	1994-1995
United Kingdom	Shetlands, West of Shetland	Shell	1996
United Kingdom	Shetlands, Brendan's Dome	Shell	1997
United Kingdom	Shetlands, Brendan's Dome	Shell	1998
United Kingdom	Shetlands, West Cormorant	Shell	1998
Brazil	Southeastern Brazil	Petroleum GeoServices	2001/2002
Canada	Beaufort Sea, Exploration License 420	Devon Canada	2001/2002
United Kingdom	Shetlands, West of Shetlands Schiehallion/Foinaven Field 4-D	British Petroleum	2002
United Kingdom	Shetlands, UK quadrants 204, 176 (UK) and 6004, 6005 (Faroes)	Amerada Hess	2002
Australia	Victoria, Southeast Australia, Bonney Upwelling, Otway 3D program	Santos Limited	2002
Australia	Victoria, Southeast Australia, Bass Strait, Gippsland Basin Tuskfish 3D	Esso Australia	2002/2003

Although there is no current requirement for passive acoustic monitoring in the U.S., the G&G industry is looking to develop operational specifications and procedures for using passive acoustic monitoring (e.g., by towed arrays, in seismic surveys) (P. Fontana, Veritas/IAGC, oral comm., 2003). It appears that regulatory guidance will increase in the future. MMS presently outlines the optional use of passive acoustic monitoring as an element of NTL No. 2004-G01. In Australia, the required use of passive acoustic monitoring in the future will be considered in conjunction with industry and conservation organizations within the next 12 to 18 months, during which time the current guidelines also will be reviewed (A. Bigwood, Environment Australia, oral comm., 2003). In the UK, the DTI is revising its existing seismic surveys guidance to include an operational specification for passive acoustic monitoring systems and their deployment. This is because recent experience suggests that when passive acoustic monitoring has been used it has not been very useful in implementing JNCC guidelines (JNCC, 2002). Revised guidance will encourage companies to seek out and use the best system available to implement the JNCC guidelines as far as possible; of particular interest is the detection of cetaceans within 500 m of the source in the period immediately prior to ramp-up. To date, systems have been installed primarily on guard or chase vessels, not seismic vessels. Guard or chase vessels are often some distance from the seismic vessel so the detections made are difficult to correlate with sightings. Using systems without effective range finding and bearing capabilities means the detections cannot be translated into cetacean location relative to the seismic source; as such, they are often of little use in implementing mitigation. DTI will look at the practicality of the specifications on a case-by-case basis but hopes to drive the usefulness of passive acoustic monitoring on surveys in the right direction (D. Simmons, DTI, oral comm., 2003).

The only dedicated trial of a passive acoustic monitoring array directly from a seismic vessel during a complete seismic survey occurred off of southeast Australia. The passive acoustic monitoring during the Tuskfish 3D survey in the Gippsland Basin was conducted on behalf of Esso Australia in the summer of 2002/2003, using a standard hydrophone array configuration similar in design to all commercially available systems, with an emphasis on location of sperm whales and an effective frequency range of about 300 Hz to 15 kHz. The deployment of the array from a seismic vessel was aimed at providing a cost effective and demonstrably beneficial means of mitigating impacts on marine mammals and testing the efficacy of the equipment as a tool for legal compliance. The trial demonstrated that the system could be effectively deployed off the seismic vessel. Unfortunately, the limited number of encounters with species such as sperm whale (which the system is most effective at detecting) limited the chance of accurately quantifying detection distances. Preliminary results suggest that although it is effective, more work is required before it is possible to quantify its use in strict legal compliance terms (e.g., pinpointing cetaceans at specific distances; S. Mustoe, Applied Ecology Solutions, oral comm., 2003).

Further evaluation and trials on the use of passive acoustic monitoring in the GOM are likely warranted to understand better its effectiveness because propagation conditions in this region are poor, and simplified acoustic propagation assumptions (e.g., cylindrical spreading) may be inappropriate for the region. Numerical modeling results suggest that propagation distances in the Gulf for a well-designed towed array deployed from a quiet vessel may extend the detection range to about 6 km. Detection range could be improved further not by increasing array depth but by increasing the array gain by incorporating additional hydrophones into an array system. For example, a quiet ship such as the *R/V Alliance* using a hydrophone array with 128 elements is able to achieve detection ranges in excess of 20 nmi in summer-stratified conditions in the Mediterranean Sea (SWSS, 2002).

e.   *Passive Versus Active Acoustic Monitoring*

The use of passive acoustics has advantages over that of active acoustics:

- longer ranges can be achieved;
- omnidirectional;
- species can be potentially identified by their vocalization signature;
- no acoustic footprint that could affect the target animals; and
- more mature and affordable technology.

However, passive acoustic monitoring has the following disadvantages:

- only works for vocalizing whales, many animals are quiet much of the time, especially when disturbed;
- difficult to determine the range to the vocalizing animals when a towed array is used;
- cannot readily determine depth to vocalizing animals;
- for fixed hydrophones (ship or bottom mounted recorders, sonobuoys, ocean bottom cables), area of coverage/detection range may be limited by noise, requiring more sensors to cover a seismic survey area;
- requires hydrophone arrays be towed behind the survey vessel (or from an additional chase boat); and
- hydrophone performance may be affected by tow speed and the ship's acoustic characteristics, limiting detection range.

The advantages of active acoustic monitoring compared with passive acoustic are as follows:

- works with non-vocalizing or cryptic whales and those species that exhibit only limited vocalization;
- can in some cases determine 3D range and bearing, including depth of vocalizing animals;
- avoids having to stream behind survey vessel if sound source and hydrophone/receiver are hull-mounted; and
- may involve less bulky equipment, minimizing personnel required for handling and operation.

Active acoustic monitoring has the following disadvantages:

- active source may be more harmful than the sound source it is being used to mitigate;
- limited detection ranges depending on power and frequency;
- inability to identify species on the basis purely of size;
- limited beam width and associated problems seeing deep diving whales at close range;
- active systems could potentially affect the behavior of the animals themselves;
- towfish would be required, possibly larger than passive acoustic monitoring array; and
- current costs for development and deployment are higher.

f. *Summary*

Acoustic signal processing is the major difficulty in the technology, but there have been decades of research in acoustics for scientific, military, and commercial purposes leading to significant developments and a mature technology applicable to acoustic monitoring. The recent availability of inexpensive powerful computers, with larger memory and storage capacities at lower cost, and more simplified programming approaches that provide much better user interfaces, have facilitated solutions to the problem, which should allow development of more affordable systems for use in acoustic monitoring of marine mammals.

Although there are presently two commercially available passive acoustic monitoring systems, there are other acoustic systems that are also available, purpose-built for research and monitoring using "off the shelf" hardware and public domain software that are very similar in specifications and performance. The use of "off-the-shelf" parts allows the evolution of systems to meet particular requirements in the use of passive acoustic monitoring technology. All the present systems in general have the same technical limitations. Present acoustic monitoring systems still require operational testing and validation as a mitigation tool. There are no acoustic systems at present that provide a complete answer for mitigating potential impacts of seismic surveys on marine mammals. Operational procedures and specifications for use of passive acoustic monitoring of marine mammals are still being developed. Clearly, the concern with sperm whales and beaked whales in the GOM warrants consideration of passive acoustic monitoring as a mitigation measure once their effectiveness is validated.

There are questions and concerns about the feasibility of developing active acoustic monitoring systems for marine mammal monitoring since it means exposing marine mammals to high intensity sound from the sonar, which is intended to find the animals so that they will not be exposed to high intensity sound from the airguns. A detection sonar that produces some disturbance is not necessarily useless, especially if it is being used to shut off a potentially much more disturbing or injurious sound source or other activity. This is a decision-making process involving both regulators and risk managers. In any case, it is a consideration that extends beyond the simple question of whether the whale detection sonar works as an effective means of detecting cetaceans (R. Gisiner, Office of Naval Research, oral comm., 2003).

Active mitigation sonar may be intense enough to cause potential TTS. In order for active sonar to be an effective mitigation tool and not be another potentially injurious noise source, mitigation sonar needs to be ramped up at a rate that allows at least one full sweep to ensure that no marine mammals or turtles are in the area before increasing the source level. This ensures that marine mammals do not receive damaging levels by the mitigation sonar, although the effectiveness of this approach also needs to be determined.

The Navy tactical sonars (the AN/SQS-53C array) represent what at present would be the most sophisticated active sonar for detecting whales. Sonars used by civilians for the purpose of targeting whales will undoubtedly require take permits incidental to sonar use through the NMFS, which enforces the Marine Mammal Protection Act (e.g., see **Appendix B**, **Sections I** and **IV**). Permits would not likely be issued for proposed activities approaching the power and low frequency of Navy sonars and may not be granted even for smaller and higher frequency sonars such as the FarSounder, given the present political climate on this issue (M. McDonald, Whale Acoustics, oral comm., 2003).

Active acoustic systems (whalefinders) are becoming sophisticated, but have not undergone validation testing and still require research and development for application to seismic survey mitigation. The MAST project would have provided critical validation for the basic technology. Until validation is done, no one can answer the question about usefulness of active whalefinder sonars (P. Tyack, WHOI, oral comm., 2003). Given the careful scrutiny of current permit applications covering research on the effects of seismic noise on sperm whales, the permitting process for testing an active acoustic system for detecting whales will be subject to even closer scrutiny and will likely be slower to approve. Opposition to the permit for testing active acoustic systems will likely arise from similar groups objecting to the MAST. Opposition to the MAST (resulting in successfully enjoining the test) was based primarily on procedural deficiencies, so there is a chance that applications for future permits for tests may be allowed to proceed.

In light of the probable opposition to such testing, development of active systems may not be practical or probable at this time. Given that there is no current regulatory pressure in the U.S., it is uncertain as to how many other sonar engineering firms might be willing to take on this task even if funding were made available. Present acoustic systems generally remain in development, especially with regard to 3D range and bearing, although the industry may not drive development too rapidly in a time of opposition.

Despite the opposition to the mitigation sonar, it is NMFS' view that it may be a useful tool for seismic surveys. The recent judgment against the MAST project was related to how NMFS met its NEPA requirements, not on the technical aspects of the sonar itself (Hawaii County Green Party vs. Evans, C-99-3719 WHA, U.S. District Court, Northern District of California). NMFS' web page (http://www.nmfs.noaa.gov/pr/readingrm/Applications/whale_cons_setback.htm) provides further information on their position on the issue. The IMAPS sonar has not been field tested, but NMFS recommends that the industry should try using it in the GOM as soon as it is available (R. Gentry, NMFS, oral comm., 2003).

The current state-of-the-art with passive acoustics is to give bearing accurately to radial distances, which are well beyond typical mitigation ranges. Range is more difficult to determine and can only be estimated with current equipment in specific conditions when multipaths can be resolved. By towing two arrays with good aperture (i.e., a stereo array), it should be possible to develop a passive system that can solve the 3D problem (P. Tyack, WHOI, oral comm., 2003).

While academic and research use of acoustic monitoring has shown encouraging results, the offshore use on a commercial basis in not completely satisfactory based on the experience in seismic surveys in the UK and Brazil. There are still technical limitations to passive acoustic monitoring that need to be understood and addressed before being adopted for routine mitigation. The present technology is not adequately developed (i.e., it is not a "plug and play" technology), since it requires substantial user inputs for it to work. The technology is not mature, and there is still a need to find operating standards (Walker, 2002).

The lack of standardization in technical requirements and procedures needs to be addressed for the technology to be implemented routinely. Unless there are standards for the equipment, its calibration, and use, then the data may be of variable quality depending on how the data were gathered. Standards are required so a seismic survey operator can differentiate between good and bad equipment. Furthermore, acceptable standard operating procedures for the equipment and software used need to be defined. Regulators and the oil and gas industry might be concerned that one geophysical operator might have less sensitive equipment than another, and that the training/experience of the acoustic monitoring operator might be inferior.

Therefore, specifications, procedures, and protocols still need to be developed for employing acoustic monitoring and decision-making in the field. For both types of acoustic monitoring, there remain few established procedures and protocols on what to do when a whale is detected. Minimum specifications and performance levels should be discussed and established through a coordinated effort between industry, academia, and regulators. The inclusion of practical operational specifications covering the use of acoustic monitoring on seismic surveys is important. Apart from these, the practicality of the systems from the standpoint of the operator should be considered to ensure that turn key systems are developed with user friendly software that allows surveyors the latitude to make decisions in the field on the basis of accurate and effective monitoring information. Guidelines for the training of operators to detect marine mammal sounds should also be included during technology development efforts.

5. Operational Restrictions

Description: Operational restrictions would reduce or eliminate seismic survey sound sources in certain areas of concern within the GOM. These restrictions on G&G activities are encompassed under Alternative 4. Operational restrictions considered included 1) a prohibition of simultaneous surveys by more than one seismic source vessel in the area(s) frequented by sperm and Bryde's whales (Alternative 4A); 2) cessation of permitting of all types of seismic surveys in the area(s) most frequented by sperm whales (Alternative 4B); and 3) seasonal restrictions on G&G activities in the area(s) most frequented by sperm whales (Alternative 4C). Effectively, several of these restrictions represent seasonal and/or geographic restrictions. Such operational restrictions are being considered as a means of reducing or eliminating the potential for hearing damage to sensitive marine mammal species. Specifically, MMS and other Federal agencies have expressed concern that marine mammals may be trapped between seismic survey vessels, causing unavoidable exposure to injurious or disruptive sound levels or eliciting panic in exposed individuals.

Cessation of all seismic surveys in sperm or Bryde's whale areas (Alternative 4B) and seasonal restrictions on G&G operations in the areas frequented by sperm whales (Alternative 4C) have been summarized previously under the discussion of seasonal and geographic restrictions. In summary, there does appear to be justification for limited application of operational restrictions in certain Gulf areas, with potential benefits to be realized by sperm and Bryde's whales. Beaked whales, unless co-occurring with either sperm or Bryde's whales, would not realize benefits from these restrictions.

Examples: There are no recent applications of operational restrictions either in U.S. waters or overseas, beyond those noted previously under seasonal and geographic restrictions.

Evaluation and Possible Application to the GOM: Cessation of all seismic surveys in areas frequented by sperm or Bryde's whales (Alternative 4B) and seasonal restrictions on G&G operations in the area(s) most frequented by sperm whales (Alternative 4C) have been summarized previously under the discussion of seasonal and geographic restrictions. Alternative 4B does not fully meet the underlying purpose and need as described in **PEA Sections I.B** and **II.E**. It has been included to provide a complete and balanced discussion of what the impacts would be if industry was not allowed to use these survey techniques (see **PEA Section II.E**; significant impacts to the G&G industry and the Gulf's socioeconomic environment are predicted). Alternative 4C is not considered viable (i.e., no seasonality trends are evident for species of concern) and will not be considered further in this analysis.

Alternative 4A (i.e., restrictions on concurrent seismic operations in the same general area) remains a viable alternative, with qualification. Current industry practice effectively eliminates concurrent seismic operations in the same general area. Because of potential acoustic interference created when simultaneous surveys are conducted too close to one another, G&G operators attempt to maintain sufficient separation distance in order to acquire the best data set possible. IAGC helps its members coordinate their respective surveys so as not to interfere with one another. Minimum separation distance requirements are dependant upon various factors (e.g., output of each array, physical characteristics of the environment, relative orientation of each survey vessel, etc.). It has been noted that under optimal conditions, the minimum distance between two seismic survey vessels should be sufficient to reach background (ambient) noise levels (G. Greve, oral comm., 2001). For the GOM noise environment (see **Appendix F, Section IV - Noise Environment**), background noise levels are assumed to be on the order of 140 dB re 1 µPa. Adopting Goold's near-field/far-field model (see **Appendix C, Section III - Seismic Source Levels**), a 50 dB decrease in output from a typical array would be encountered 3.1 to 4.0 km from the source, depending upon water depth (see **Appendix C, Figures C-8** and **C-9**).

One consideration not easily factored into these calculations is the reflected sounds that bounce off the seafloor and may be propagated greater distances (e.g., due to higher sound levels being directed at the seafloor; due to the possible presence of "sound channels").

Under optimal conditions, seismic survey vessels and their arrays should be separated by >3.1 to 4 km, the approximate dimensions of a single OCS lease block. It is highly unlikely that a marine mammal would realize acoustic impact from the simultaneous firing of airguns during concurrent operations located >3.1 to 4 km apart. For the purposes of Alternative 4A, concurrent seismic operations in water depths >200 m should not occur in adjacent OCS lease blocks.

Multiple vessels (e.g., multi-ship surveys) may be employed, as described in **Appendix D** (**Section I.A.8 - Multi-ship Surveys**), to conduct ocean bottom surveys and vertical cable surveys. Generally, multi-ship surveys are conducted to either obtain converted wave data (shear wave) or to penetrate hard seafloor layers. Depending upon whether 2D or 3D operations are being conducted, two ships are placed one behind the other, towing single or multiple streamer cables. Streamer length is sufficient to record seismic data at distances of 8 to 12 km from the source. In both of these cases, only one airgun array is operational. Therefore, it would not be possible for a marine mammal to be positioned between two survey vessels that are both firing their airguns at the same time.

In summary, marine mammal exposure to seismic survey noise will be similar, regardless of whether the sound source originates from concurrent single or multi-ship surveys, or concurrent multiple surveys.

6.   Reduction of Sound Source Levels

Description: Reduction of sound source levels would lower the peak pressures from an airgun array, and would reduce the size of the impact zone, further reducing the potential for hearing damage.

Examples: Reduction of sound source levels has been evaluated on a limited number of occasions in the GOM. Lowering the peak pressure reduces not only the peak frequency but also the entire spectrum of frequencies emitted from a typical airgun array. A typical amplitude spectrum is shown in **Appendix D, Figure D-3**. In general, the spectrum extends from a low frequency of 10 Hz, reaches a peak frequency at about 50 Hz, then falls off. However, frequencies above 50 Hz are extremely important as they provide data on the finer details of a hydrocarbon deposit. Additional information pertinent to airgun output may be found in **Appendices C (Section III - Seismic Source Levels), D (Section II.A - Seismic Source, Airguns), and G (Section I - Introduction)**.

Evaluation and Possible Application to the GOM: G&G operators attempt to optimize their systems to use the broadest spectrum of energy, including higher frequencies. Reduction of source levels will reduce higher frequency output with the subsequent loss of these detailed data.

Because of the amount of G&G data collected (i.e., current state of knowledge for a large area of the Gulf) and continuing interest in re-evaluating producing reservoirs, there is an increasing need to evaluate structures (e.g., potential or producing reservoirs) in detail. The ability to interpret finer detail will be adversely affected should this mitigation measure be implemented. Therefore, G&G data acquired using lower sound source levels are considered unacceptable. Such data are extremely difficult to interpret. If data cannot be properly interpreted, there is also the possibility that areas would have to be re-surveyed, with the additional potential for adverse impacts to marine mammals that might be present.

7.   Sound Baffling

Description: Sound baffling is accomplished using a screen of air bubbles that surround the seismic array at a prescribed distance. Air bubbles create a dynamic barrier that serves to reflect the sound waves from the array.

Examples: While baffling has been used on at least one occasion from a stationary source with some degree of success, its application to a towed array is considered problematic. Amoco (now British Petroleum [BP]) evaluated the feasibility of bubble screens in the early 1980's (G. Greve, oral comm., 2000). Bolt also experimented with bubble screens in the mid-1990's, with the same results. As noted by both Amoco and Bolt, bubble systems proved to be extremely fragile, difficult to deploy, and difficult to maintain. Operational considerations aside, bubble screens as sound baffles were determined by Amoco to be effective; however, the effectiveness of baffling was not as high as theoretical calculations had indicated.

Evaluation and Possible Application to the GOM: A reliable mobile system has yet to be successfully used. Sound baffling cannot be considered a feasible mitigation measure for G&G operations in the GOM.

8.   Conclusions

The following additional mitigation measures for seismic surveys were evaluated:

- Geographic and seasonal restrictions;
- Impact zones and real-time monitoring;
- Ramp-up;
- Acoustic monitoring;
- Operational restrictions;
- Reduction of sound source levels; and
- Sound baffling.

Reduction in sound source levels and sound baffling have been determined to be infeasible for use in the GOM. Reduction in sound source levels would produce inadequate data for interpretation purposes. Sound baffling presents major logistical problems when applied to a moving sound source.

Ramp-up offers the possibility of reducing the risk of auditory damage or other injury to marine mammals and turtles during start-up of an airgun array. While its effectiveness has yet to be demonstrated, its use in other hydrocarbon-producing areas (e.g., North Sea, Alaskan Beaufort Sea) is interpreted as a common sense measure that may reduce the potential for acoustic impact. Ramp-up is currently required for all seismic operations in all GOM waters >200 m deep and all OCS waters (regardless of water depth) in the Eastern Planning Area (NTL No. 2004-G01).

On the basis of distribution information in **Appendix F (Section I.A - Marine Mammals)** and impact analyses in **PEA Section III**, two geographic areas have been identified where species of concern are known to congregate. One is offshore of the Mississippi River delta between the 500- and 2,000-m isobaths where localized aggregations of sperm whales appear to be persistent. Sperm whales are frequently sighted there and in adjacent portions of the Eastern Planning Area (De Soto Canyon area) (Davis et al., 2000). Sperm whale congregations also have been documented in the southeastern Gulf, west of the Dry Tortugas. The second location is the Eastern Planning Area where Bryde's whales (the only mysticete likely to occur in the Gulf) are most likely to occur. The third species group of concern, the beaked whales, apparently do not exhibit site affinity. None of the species of concern exhibit seasonality preferences of sufficient magnitude to warrant seasonal restrictions. In these areas, mitigation measures to reduce impacts and/or restrictions on seismic operations are considered to be feasible (e.g., determination of impact zones, use of real-time monitoring and/or ramp-up).

The following specific alternatives (including several that contain one or more mitigation measures) are outlined further in **PEA Section II** and evaluated on a resource by resource basis in **PEA Section III**. The purpose of these mitigation measures is to establish and maintain a 500-m exclusion (or impact) zone around the seismic array; this zone is to remain free of marine mammals. When a marine mammal is located within (or approaching) the exclusion zone, the array is to be shut down until the animal has cleared the zone. Resumption of survey activities requires verification that the exclusion zone is clear of marine mammals. Alternatives evaluated in this environmental assessment include

- Alternative 1 – Proposed Action, or a continuation of the status quo, consisting of the various G&G activities currently occurring and estimated to occur in the foreseeable future (i.e., over the next several decades) in the Western, Central, and Eastern Planning Areas of the GOM, operating under current MMS regulations and requirements (e.g., applicable NTLs). Specifically, G&G operators must adhere to the requirements of NTL No. 2004-G01 when operating in waters >200 m deep in the GOM (and all Federal waters of the Eastern Planning Area) by utilizing ramp-up, employing visual monitoring using trained observers, completing reporting requirements, and having the option of initiating seismic operations during nighttime and periods of limited visibility using passive acoustic monitoring techniques.

- Alternative 2 – Addition of Vessel-based Passive Acoustic Monitoring as a Requirement. Specifically, G&G operators working in water depths >200 m throughout the GOM and all OCS waters of the Eastern Planning Area must adhere to the status quo (i.e., compliance with the requirements of NTL No. 2004-G01: ramp-up, visual monitoring using trained observers, reporting requirements) and must utilize passive acoustic monitoring techniques prior to ramp-up in conjunction with visual monitoring.

- Alternative 3 – Addition of Both Passive and Active Acoustic Monitoring as a Requirement. Specifically, G&G operators working in water depths >200 m throughout the GOM and all OCS waters of the Eastern Planning Area must adhere to the status quo (i.e., compliance with the requirements of NTL No. 2004-G01) and must utilize both passive acoustic monitoring and active acoustic monitoring techniques prior to ramp-up in conjunction with visual monitoring.

- Alternative 4 – Restrict G&G Seismic Surveying Operations. Under this alternative, the existing suite of G&G activities would continue but with the implementation of additional restrictions on G&G seismic operations. Specifically, G&G operators would be precluded from conducting simultaneous seismic operations (i.e., within 4 km of one another; within an adjacent OCS lease block) in those portions of the GOM most frequented by sperm whales and Bryde's whales (i.e., water depths >200 m). The purpose of this measure is to remove the potential for simultaneous exposure to seismic noise from concurrent surveys in the same general area.

III. LITERATURE CITED

Anonymous. 2001. Marine Acoustic Effects Study, blue whale feeding aggregations, Otway Basin, Bass Strait Victoria. Report No R2001-7, Centre for Marine Science & Technology (CMST), Curtin University, GPO Box U 1987, Perth 6845, WA.

Arthur D. Little International, Inc. 1997. Environmental impact assessment for a seismic survey in Offshore Block S11(b), Republic of Trinidad and Tobago. A report to ELF Exploration Trinidad. 50 pp.

Baker, G.J., and Y.R.M. Bonin. 2001. GPS equipped sonobuoy. www.waypnt.com/html/sonobuoy.pdf.

Burks, C., K. Mullin, S. Swartz, and A. Martinez. 2001. Cruise Results, NOAA Ship *Gordon Gunter* Cruise GU-01-01(11), 6 February - 3 April 2001, Marine Mammal Survey of Puerto Rico and the Virgin Islands, and a Study of Sperm Whales in the Southeastern Gulf of Mexico. NOAA Technical Memorandum NMFS-SEFSC-462. 58 pp.

Chappell, O., T. Lewis, D. Gillespie, R. Swift, O. Boisseau, and J. Gordon. 1999. The development of acoustic techniques for cetacean monitoring on the Guard Vessel Hunter, 1998. A report prepared by Birmingham Research and Development Ltd. for Shell UK Ltd. Contract C10563 - Monitoring Cetacean Presence. 45 pp.

Davis, R.A., D.H. Thomson, and C.I. Malme. 1998. Environmental assessment of seismic exploration on the Scotian Shelf. Prepared by LGL Limited, environmental research associates and Engineering and Scientific Services for Mobil Oil Canada Properties Ltd., Calgary, Canada. 334 pp.

Davis, R.W., W.E. Evans, and B. Würsig (eds.). 2000. Cetaceans, Sea Turtles and Seabirds in the Northern Gulf of Mexico: Distribution, Abundance and Habitat Associations. Volume II: Technical Report. Prepared by Texas A&M University at Galveston and the National Marine Fisheries Service. U.S. Department of the Interior, U.S. Geological Survey, Biological Resources Division, USGS/BRD/CR-1999-005 and Minerals Management Service, Gulf of Mexico OCS Region, New Orleans, LA. OCS Study MMS 2000-003.

Department of the Navy. 2001. Final Overseas Environmental Impact Statement and Environmental Impact Statement for Surveillance Towed Array Sensor System Low Frequency Active (SURTASS LFA) Sonar. Prepared for the Department of the Navy, Chief of Naval Operations, Washington, D.C. January 2001. 2 vols.

Environment Australia. 2001. Guidelines on the application of the Environment Protection and Biodiversity Conservation Act to interactions between offshore seismic operations and larger cetaceans. Environment Australia, Canberra. October 2001. http://www.ea.gov.au/epbc/assessmentsapprovals/guidelines/seismic/pubs/seismic-guidelines-nomaps.pdf.

Frantzis, A., J.C. Goold, E.K. Skarsoulis, M.I. Taroudakis, and V. Kandia. 2002. Clicks from Cuvier's beaked whales, *Ziphius cavirostris*. J. Acoust. Soc. Am. 112(1):34-37.

Gisiner, R. 2002. Acoustic monitoring. Presentation at the Workshop on Passive Acoustic Monitoring. Sponsored by the International Association of Geophysical Contractors. Boston, MA. May 2002.

Gordon, J. 2002. Practical application of acoustic monitoring during seismic survey mitigation procedures. Presentation at the Workshop on Passive Acoustic Monitoring. Sponsored by the International Association of Geophysical Contractors. Boston, MA. May 2002.

Gordon, J.C.D., D. Gillespie, J. Potter, A. Frantzis, M. Simmonds, and R. Swift. 1998. The effects of seismic surveys on marine mammals. Proceedings of the Seismic and Marine Mammals Workshop, London, 23-25 June 1998.

Hawley, B.W., G. Zandt, and R.B. Smith. 1981. Simultaneous inversion for hypocenters and lateral velocity variations: an interactive solution with a layered model. J. Geophys. Res. 86:7073-7086.

International Association of Geophysical Contractors. 2003. Soft start or slow build up process. http://www.iagc.org/public/gom/gom13.doc

Joint Nature Conservation Committee. 2002. Guidelines for minimising acoustic disturbance to marine mammals. Aberdeen, Scotland, UK.

Johnson, S.R. 2002. Marine mammal mitigation and monitoring program for the 2001 Odoptu 3-D seismic survey, Sakhalin Island Russia: Executive summary. Rep. from LGL Ltd, Sidney, B.C., for Exxon Neftegas Ltd., Yuzhno-Sakhalinsk, Russia. 49 pp. Also available as Working Paper SC/02/WGW/19, Int. Whal. Comm., Western Gray Whale Working Group Meeting, Ulsan, South Korea, 22-25 October 2002. 48 pp.

Lang, W. 2002. Sperm Whale Acoustic Monitoring Program. Presentation at the Workshop on Passive Acoustic Monitoring. Sponsored by the International Association of Geophysical Contractors. Boston, MA. May 2002.

Lewis, T., D. Gillespie, J. Gordon, and O. Chappell. 2000. Acoustic cetacean monitoring, 1996 to 1999 - summary report: towards the development of an automated system. A report prepared by Birmingham Research and Development Ltd. for Shell UK Ltd. Contract C10563. 55 pp.

LGL Ltd. 2001. Marine mammal observers' [sic] handbook – marine mammal monitoring during the 2001 Odoptu open-water 3D seismic exploration program. Report prepared for Exxon Neftegas Ltd. by LGL Ltd., Toronto, Canada. 41 pp.

LGL, Ltd., environmental research associates. 2003. Environmental assessment of marine seismic testing conducted by the *R/V Maurice Ewing* in the northern Gulf of Mexico, May – June 2003. Prepared for Lamont-Doherty Earth Observatory and the National Science Foundation by LGL Ltd., environmental research associates, King City, Ontario. 7 March 2003. LGL Report TA2822-3. 99 pp.

LGL Ltd., Inc. and Greeneridge Sciences, Inc. 1997a. Proposed technical plan for marine mammal monitoring during BPXA's 1997 open-water seismic program. Prepared for BP Exploration (Alaska), Inc. LGL Project TA2150. 39 pp.

LGL Ltd., Inc. and Greeneridge Sciences, Inc. 1997b. Northstar marine mammal monitoring program, 1996: Marine mammal and acoustical monitoring of a seismic program in the Alaskan Beaufort Sea. Prepared for BP Exploration (Alaska), Inc., Anchorage, AK and National Marine Fisheries Service, Anchorage, AK and Silver Spring, MD. LGL Report TA2121-2. August 1997. 261 pp.

Maxwell, S.C., and T.I. Urbanck. 2001. The role of passive microseismic monitoring in the instrumented oil field. The Leading Edge 20(6):636-639.

Miller, J.H., and D.C. Potter. Undated. Active high frequency phased-array sonar for whale shipstrike avoidance: target strength measurements. White Paper. http://www.farsounder.com.

Mullin, K.D., and G.L. Fulling. In review. Abundance of cetaceans in the oceanic northern Gulf of Mexico, 1996-2001. Submitted to Marine Mammal Science. NMFS, Southeast Fisheries Science Center, Pascagoula, MS.

National Marine Fisheries Service. 1997. Environmental assessment on the issuance of an Incidental Harassment Authorization for a seismic survey in Puget Sound, Washington. U.S. Department of Commerce, National Oceanic and Atmospheric Administration, Office of Protected Resources, Silver Spring, MD. 46 pp.

National Marine Fisheries Service. 2001a. Taking Marine Mammals Incidental to Operation of a Low Frequency Sound Source by the North Pacific Acoustic Laboratory. Federal Register 66(160):43442-43459.

National Marine Fisheries Service. 2001b. Taking Marine Mammals Incidental to Naval Activities. Federal Register 66(87):22450-22467.

National Oceanic and Atmospheric Administration Fisheries. 2003. Taking marine mammals incidental to conducting oil and gas exploration activities in the Gulf of Mexico. Notice of receipt of application for a small take authorization and request for comments and information. Federal Register 68(41):9991-9996. March 3, 2003.

National Research Council. 1994. Low-frequency sound and marine mammals: current knowledge and research needs. National Academy Press, Washington, D.C. 75 pp.

National Research Council. 2003. Ocean noise and marine mammals. The National Academies Press, Washington, D.C. ISBN 0-309-08563-5. 192 pp. + plates.

Normark, W. 1999. Southern California earthquake hazards, pp. 3-7. *In:* Sound Waves – Coastal and Marine Geology News, August 1999. USGS, St. Petersburg, FL. 11 pp.

Pierson, M.O., J.P. Wagner, V. Langford, P. Birnie, and M.L. Tasker. 1998. Protection from, and mitigation of, the potential effects of seismic exploration on marine mammals. *In:* M.L. Tasker and C. Weir (eds.), Proceedings of the Seismic and Marine Mammals Workshop, London, 23-25 June 1998. http://www.smru.st-and.ac.uk/seismic/index html. (Also found in: Workshop proceedings – Seismic airgun sources, U.S. regulations, and marine mammals. Workshop sponsored by the Energy Research Clearinghouse, Schlumberger, and the International Association of Geophysical Contractors, The Woodlands, TX, 14 March 2000).

Richardson, W.J., C.R. Greene, Jr., C.I. Malme, and D.H. Thomson. 1995. Marine mammals and noise. Academic Press, San Diego, CA. 576 pp.

Seiche.com Ltd. 2001. The Sonar Course. www.seiche.com. ISBN 1-904055-00-1.

Sperm Whale Seismic Study. 2002. Final D-tag Cruise Report (02G11). M. Howard and A. Jochens (compilers). Prepared in consultation with science teams aboard the R/V *Gyre* and M/V *Rylan T*. Funded through MMS, ONR, and the International Association of Geophysical Contractors.

Stephen, R.A. 2000. Ambient seismic noise below the deep seafloor. The Leading Edge 19(3):276-281.

Stone, C.J. 1998. Cetacean observations during seismic surveys in 1997. JNCC Reports, No. 278.

Swartz, S., A. Martinez, T., Cole, P. Clapham, M. McDonald, J. Hildebrand, E. Oleson, C. Burks, and J. Barlow. 2001. Visual and acoustic survey of humpback whales (*Megaptera novaeangliae*) in the eastern and southern Caribbean Sea. NOAA Technical Memorandum NMFS-SEFSC-456. 37 pp.

Turnpenny, A.W.H., and J.R. Nedwell. 1994. The effects on marine fish, diving mammals, and birds of underwater sound generated by seismic surveys. Report by Fawley Aquatic Research Laboratories, Ltd. for the United Kingdom Offshore Operators Association. 40 pp. + apps.

U.S. Department of the Interior, Minerals Management Service. 1995. OCS Environmental Assessment of a Proposed 3-Dimensional Seismic Survey, Santa Ynez Unit, Exxon Company, U.S.A., August 1995. Prepared by the Regional Supervisor, Office of Environmental Evaluation, Minerals Management Service, Pacific OCS Region, Camarillo, CA. 78 pp. + apps.

U.S. Department of the Interior, Minerals Management Service. 2004. MMS Notice to Lessees No. 2004-G01, Implementation of Seismic Survey Mitigation Measures and Protected Species Observer Program. Effective 1 March 2004.

Walker, C. 2002. Passive acoustic monitoring seismic industry experience. Presentation at the Workshop on Passive Acoustic Monitoring. Sponsored by the International Association of Geophysical Contractors. Boston, MA. May 2002.

Weller, D.W., Y.V. Ivashchenko, G.A. Tsidulko, A.M. Burdin, and R.L. Brownell, Jr. 2002. Influence of seismic surveys on western gray whales off Sakhalin Island, Russia in 2001. Working Paper SC/54/BRG14, Int. Whal. Comm., Western Gray Whale Working Group Meeting, Ulsan, South Korea, 22-25 October 2002. 12 pp.

Würsig, B.G., D.W. Weller, A.M. Burdin, S.H. Reeve, A.L Bradford, S.A. Blokhin, and R.L Brownell, Jr. 1999. Gray whales summering off Sakhalin Island, Far East Russia: July-October 1997: A joint U.S.-Russian scientific investigation. Final Report by Texas A&M University, College Station, TX, and Kamchatka Inst. Ecol. and Nature Manage., Russian Acad. Sci., Kamchatka, Russia, for Sakhalin Energy Investment Co. Ltd. and Exxon Neftegaz Ltd., Yuzhno-Sakhalinsk, Russian Federation. 101 pp.

Würsig, B., T.A. Jefferson, and D.J. Schmidly. 2000. The Marine Mammals of the Gulf of Mexico. Texas A&M University Press, Galveston, TX. 232 pp.

**Appendix F
Affected Environment**

Affected Environment

This appendix describes the characteristics of the biological and physical environments of Gulf of Mexico outer continental shelf (OCS) waters, followed by a summary of the Gulf's noise environment. Individual resources (e.g., marine mammals, sea turtles, fish resources, etc.) may be affected by geological and geophysical (G&G) activity. Summary information provided in this appendix is intended to highlight those resource characteristics that may be potentially adversely affected by G&G activity (see **Section III - Environmental Impacts** of the Programmatic Environmental Assessment [PEA]).

I.   BIOLOGICAL ENVIRONMENT

A.   MARINE MAMMALS

1.   Species Occurrence and Status

Twenty-nine species of marine mammals are known to occur in the Gulf of Mexico (**Table F-1**) (Davis et al., 2000). These include 28 species of the order Cetacea, which consists of seven mysticetes (baleen whales) and 21 odontocetes (toothed whales and dolphins); and a single species (with two subspecies) of the order Sirenia, which includes the manatees and dugong (Jefferson et al., 1992). Life history information for marine mammals is summarized in Jefferson and Schiro (1997) and Würsig et al. (2000).

The current population status of Gulf of Mexico marine mammal species is also indicated in **Table F-1**. The following categories are used (adapted from Würsig et al., 2000):

- *Common*: a species that is abundant and widespread throughout the region in which it occurs.
- *Uncommon*: a species that does not occur in large numbers, and may or may not be widely distributed throughout the region in which it occurs.
- *Rare*: a species present in such small numbers in the region that it is seldom seen.
- *Extralimital*: a species known on the basis of few records that are probably the result of unusual movements of few individuals into the region.

Baleen whales (other than Bryde's whale) are considered extralimital or rare in the Gulf (Würsig et al., 2000). The Bryde's whale is the most frequently sighted baleen species in the Gulf, though considered uncommon. Strandings and sightings data suggest that this species may be present throughout the year, generally in the northeastern Gulf near the 100-m isobath between the Mississippi River delta and southern Florida (Davis et al., 2000; Würsig et al., 2000).

Most of the odontocetes are considered common. Beaked whales are the main exception, with most being rare[1] or extralimital. Dwarf and pygmy sperm whales are also considered

---

[1] Determining the population status of Blainville's and Cuvier's beaked whales, which occur in the Gulf of Mexico, is problematic. Würsig et al. (2000) classify their presence in the Gulf as rare. In contrast, National Marine Fisheries Service (NMFS) notes that beaked whales are difficult to identify to species, they are hard to see, and they occur in small groups (see **Table F-1** footnote). NMFS suggests that Blainville's and Cuvier's beaked whales are at least uncommon, and depending on how abundance is viewed (group sightings or number of individuals), may in fact, along with Gervais' beaked whale, be common. Because of the difficulties distinguishing Gervais' and Blainville's beaked whales, it may be that if one species is truly rare, the other is without doubt common or uncommon.

Table F-1
Marine Mammals of the Gulf of Mexico

Scientific Name	Common Name	Management Status[1]	Population Status[2]
ORDER CETACEA	**WHALES AND DOLPHINS**		
SUBORDER MYSTICETI	**BALEEN WHALES**		
Family Balaenidae	Right whales		
Eubalaena glacialis	Northern right whale	E, S	1
Family Balaenopteridae	Rorquals		
Balaenoptera musculus	Blue whale	E, S	1
Balaenoptera edeni	Bryde's whale	none	3
Balaenoptera physalus	Fin whale	E, S	2
Megaptera novaeangliae	Humpback whale	E, S	2
Balaenoptera acutorostrata	Minke whale	none	2
Balaenoptera borealis	Sei whale	E, S	2
SUBORDER ODONTOCETI	**TOOTHED WHALES / DOLPHINS**		
Family Physeteridae	Sperm whales		
Physeter macrocephalus	Sperm whale	E, S	4
Family Kogiidae	Pygmy and dwarf sperm whales		
Kogia breviceps	Pygmy sperm whale	none	4[3]
Kogia simus	Dwarf sperm whale	none	4[3]
Family Ziphiidae	Beaked whales		
Mesoplodon densirostris	Blainville's beaked whale	S	2-4[3]
Ziphius cavirostris	Cuvier's beaked whale	S	2-4[3]
Mesoplodon europaeus	Gervais' beaked whale	S	3
Mesoplodon bidens	Sowerby's beaked whale	S	1

Scientific Name	Common Name	Management Status[1]	Population Status[2]
Family Delphinidae	Dolphins (Delphinids)		
Stenella frontalis	Atlantic spotted dolphin	none	4
Tursiops truncatus	Bottlenose dolphin	none	4
Stenella clymene	Clymene dolphin	none	4
Pseudorca crassidens	False killer whale	none	3
Lagenodelphis hosei	Fraser's dolphin	none	4
Orcinus orca	Killer whale	none	3
Peponocephala electra	Melon-headed whale	none	4
Stenella attenuata	Pantropical spotted dolphin	none	4
Feresa attenuata	Pygmy killer whale	none	3
Globicephala macrorhynchus	Short-finned pilot whale	S	4
Grampus griseus	Risso's dolphin	none	4
Steno bredanensis	Rough-toothed dolphin	none	4
Stenella longirostris	Spinner dolphin	none	4
Stenella coeruleoalba	Striped dolphin		
ORDER SIRENIA	**DUGONGS AND MANATEES**		
Family Trichechidae	Manatees		
Trichechus manatus latirostris	Florida manatee	E	2[4]
Trichechus manatus manatus	Antillean manatee	E	2

[1] Management status: E = endangered under the Endangered Species Act of 1973; S = strategic stock under the Marine Mammal Protection Act of 1972, as indicated by Waring et al. (1999).

[2] Population status: 1 = extralimital; 2 = rare; 3 = uncommon; 4 = common (adapted from Würsig et al., 2000).

[3] Determining the population status of Blainville's and Cuvier's beaked whales (*Kogia*), which occur in the Gulf of Mexico, is problematic. Würsig et al. (2000) classify the presence of Blainville's and Cuvier's beaked whales and dwarf and pygmy sperm whales (*Kogia*) in the Gulf as rare. The National Marine Fisheries Service (NMFS) notes that beaked whales are difficult to identify to species, they are hard to see, and they occur in small groups. In general, only Cuvier's beaked whales and adult male Blainville's beaked whales can be identified in the field. Nevertheless, NMFS suggests that sightings of beaked whales and *Kogia* in the Gulf are not rare or that uncommon. During all NMFS aerial and ship surveys combined, there have been sightings of about 75 beaked whale groups (15 as Cuvier's beaked whale, 36 as *Mesoplodon* spp., 2 as Blainville's beaked whale, and 22 as unidentified ziphiids). While these sightings are widely distributed in the deep waters of the northern Gulf, because they occur in small groups (usually <4 to 6), the abundance of each beaked whale category is low compared with species with a similar number of sightings that occur in much larger groups. Another factor to consider is the sightability of beaked whales and *Kogia*; they rarely leap out of the water or splash at the surface and are difficult to see unless seas are very calm (Beaufort sea state 0, 1). While a quantitative analysis has not been performed, in general, as the sea state decreases, the number of beaked whale sightings increases. The majority of NMFS surveys have been conducted in sea states that are not optimal for sighting beaked whales. Therefore, NMFS suggests that Blainville's and Cuvier's beaked whales are at least uncommon, and depending on how abundance is viewed (group sightings or number of individuals), may in fact, along with Gervais' beaked whale, be common. Because of the difficulties distinguishing Gervais' and Blainville's beaked whale, it may be that if one species is truly rare, the other is without doubt common or uncommon. On the basis of the frequency of their sightings, the Marine Mammal Commission considers *Kogia* as common in the northern Gulf.

[4] Excluding the Florida coast, the Florida manatee is considered rare in the northern Gulf of Mexico.

common. The frequency of occurrence of both beaked whales and dwarf and pygmy sperm whales is most likely underestimated because these "cryptic" species are submerged much of the time and avoid aircraft and ships.

2.   Listed Species and Critical Habitat

Five baleen whales (northern right, blue, fin, humpback, and sei), one toothed whale (sperm whale), and two subspecies of the West Indian manatee that occur or have been reported in the Gulf of Mexico are listed as endangered species under the Endangered Species Act of 1973 (ESA). No listed mysticete whale normally occurs in the Gulf (Jefferson, 1995). Sperm whales are common in certain deep-water areas of the Gulf. The West Indian manatee (*Trichechus manatus*) inhabits only coastal marine, brackish, and freshwater habitats.

The sperm whale (*Physeter macrocephalus*) is the largest toothed whale and is distributed from the tropics to polar zones in both hemispheres. They are deep diving mammals and inhabit oceanic waters, although they may come close to shore in certain areas where deep water approaches the coast. Sperm whales are known to feed on cephalopods, demersal fishes, and benthic invertebrates (Rice, 1989; Jefferson et al., 1993). The sperm whale is the only great whale that is considered to be common in the Gulf of Mexico (Jefferson, 1995). Sightings data (i.e., observations of individuals on the ocean surface) suggest a Gulf-wide distribution over the continental slope. Groups of sperm whales are common in the vicinity of the Mississippi River Delta between the 500 to 2,000 m isobaths (Davis et al., 2000). These groups of sperm whales in the Gulf are thought to consist of adult females, calves, and immature individuals (Brandon and Fargion, 1993; Mullin et al., 1994a; Sparks et al., 1996; Jefferson and Schiro, 1997). Minimum population estimates of sperm whales in the Gulf of Mexico in the 1990's totaled 411 individuals (Waring et al., 1997, 2002); more recent draft estimates indicate a minimum population estimate of 1,035 sperm whales in the northern Gulf of Mexico (NMFS, 2004).

The West Indian manatee, which in the Gulf of Mexico includes two subspecies, the Florida and Antillean manatees, normally inhabits only coastal marine, brackish, and freshwater habitats. Both subspecies are currently federally designated as endangered species. The range of the Florida manatee in the Gulf is confined largely to peninsular Florida, where they are considered to be generally uncommon but sighted occasionally in localized aggregations. Outside of Florida, their presence in the northern Gulf is considered rare, though widespread within coastal habitats as far north and west as Texas, and sightings have been increasing over the past 15 years (Würsig et al., 2000). The Antillean manatee may range as far north and east within the northern Gulf as Texas and perhaps Louisiana. Their presence within the northern Gulf is also considered rare.

No areas of critical habitat have been designated for the sperm whale or any baleen whales in the Gulf of Mexico. The Florida manatee has two critical habitats in the Gulf of Mexico: Crystal River and southwest Florida from around Tampa Bay south to the southwestern tip of the state (U.S. Department of the Interior [USDOI], U.S. Fish and Wildlife Service [USFWS], 1996). In addition, nearshore areas from Crystal River to Apalachicola, and from Sarasota south to the southwestern tip of Florida, are identified as migratory routes for Florida manatees. These areas include nearshore waters from the shoreline to the outer margins of seagrass beds.

3.   Strategic Stocks

NMFS is responsible for managing stocks of cetaceans and pinnipeds under the Marine Mammal Protection Act of 1972 (MMPA). In the MMPA, the term "stock" means a group of marine mammals of the same species or smaller taxa in a common spatial arrangement that interbreed when mature. Some Gulf of Mexico cetacean stocks are classified as *strategic*. Strategic stocks are those that are 1) subject to a level of direct human-caused mortality that exceeds their Potential Biological Removal level (defined later in this section); 2) declining and thus likely to be listed as threatened species under the ESA within the foreseeable future; and/or 3) currently listed as endangered or threatened species under the ESA. The latter would include the sperm whale and manatee, which have already been described as endangered species. Other strategic stocks within the Gulf are four species of beaked whales; the short-finned pilot whale; and bay, sound, and estuarine communities of the bottlenose dolphin (Waring et al., 1997, 1999). The latter are unlikely to come in contact with offshore G&G activities and are not considered further.

Beaked whales classified as strategic stocks in the Gulf of Mexico include three species in the genus *Mesoplodon* (Sowerby's beaked whale [*M. bidens*], Blainville's beaked whale [*M. densirostris*], and Gervais' beaked whale [*M. europaeus*]) and Cuvier's beaked whale (*Ziphius cavirostris*). Generally, beaked whales appear to prefer deep water, though little is known of their respective life histories. Stomach content analyses suggest that these whales feed primarily on deep water cephalopods, although they will also take fish and some benthic invertebrates (Leatherwood and Reeves, 1983; Jefferson et al., 1993). In the Gulf, beaked whales have been sighted at the ocean surface in water depths between approximately 700 and 2,000 m. Cuvier's beaked whale is probably the most commonly sighted beaked whale in the Gulf (Davis and Fargion, 1996; Jefferson and Schiro, 1997; Davis et al., 2000). However, Würsig et al. (2000) state that Gervais' beaked whale is the most frequently stranded species (historically) in the Gulf of Mexico and is possibly the most widely distributed beaked whale in this region.

The short-finned pilot whale (*Globicephala macrorhynchus*) is found in warm temperate to tropical waters of the world. Short-finned pilot whales feed primarily on squids and fishes. In the Gulf, it is most commonly sighted on the ocean surface along the mid- to upper-slope, in water depths of 250 to 2,000 m and often in areas of steep bottom topography (Davis and Fargion, 1996; Jefferson and Schiro, 1997; Davis et al., 2000).

4.   Factors Influencing Cetacean Distribution and Abundance

The distribution and abundance of most cetaceans within the northern Gulf of Mexico is strongly influenced by mesoscale oceanographic circulation patterns, as summarized in Davis et al. (2000). These patterns are primarily driven by river discharge (e.g., the Mississippi-Atchafalaya River), wind stress, and the Loop Current (LC) and its derived circulation phenomena. Circulation on the continental shelf is largely wind-driven, with localized effects from freshwater (i.e., river) discharge. Beyond the shelf, mesoscale circulation is largely driven by the LC in the eastern Gulf. Approximately once or twice annually, the LC sheds anticyclonic eddies (also called warm-core rings). Anticyclones are long-lived, dynamic features that generally migrate westward and transport large quantities of high-salinity, nutrient-poor water across the near-surface waters of the northern Gulf. These anticyclones, in turn, spawn cyclonic eddies (also called cold-core rings) during interaction with one another and upon contact with topographic features of the continental slope and shelf edge. The cyclones contain high concentrations of nutrients and stimulate localized production.

In the north-central Gulf, the relatively narrow continental shelf south of the Mississippi River delta may be an additional factor affecting cetacean distribution (Davis et al., 2000). Outflow from the Mississippi River mouth transports large volumes of low salinity, nutrient-rich water southward across the continental shelf and over the slope. River outflow may also be entrained within the confluence of a cyclone-anticyclone eddy pair and transported beyond the continental slope. In either case, this input of nutrient-rich water leads to a localized deep-water environment with enhanced productivity and may explain the presence of groups of sperm whales within 50 km of the Mississippi River delta in the vicinity of the Mississippi Canyon (**Figure F-1**).[2]

---

[2] Two surveys were conducted under GulfCet I and II (August 1994; October 1996) to obtain behavioral observations and photo-identification of sperm whales. Areas surveyed centered around the Mississippi Canyon, a region previously recognized for high sperm whale densities. NMFS surveys in the GOM have been extensive (see **Figure F-1**), with sperm whale sightings throughout the region, and aggregations evident off the Mississippi River delta and near the Dry Tortugas.

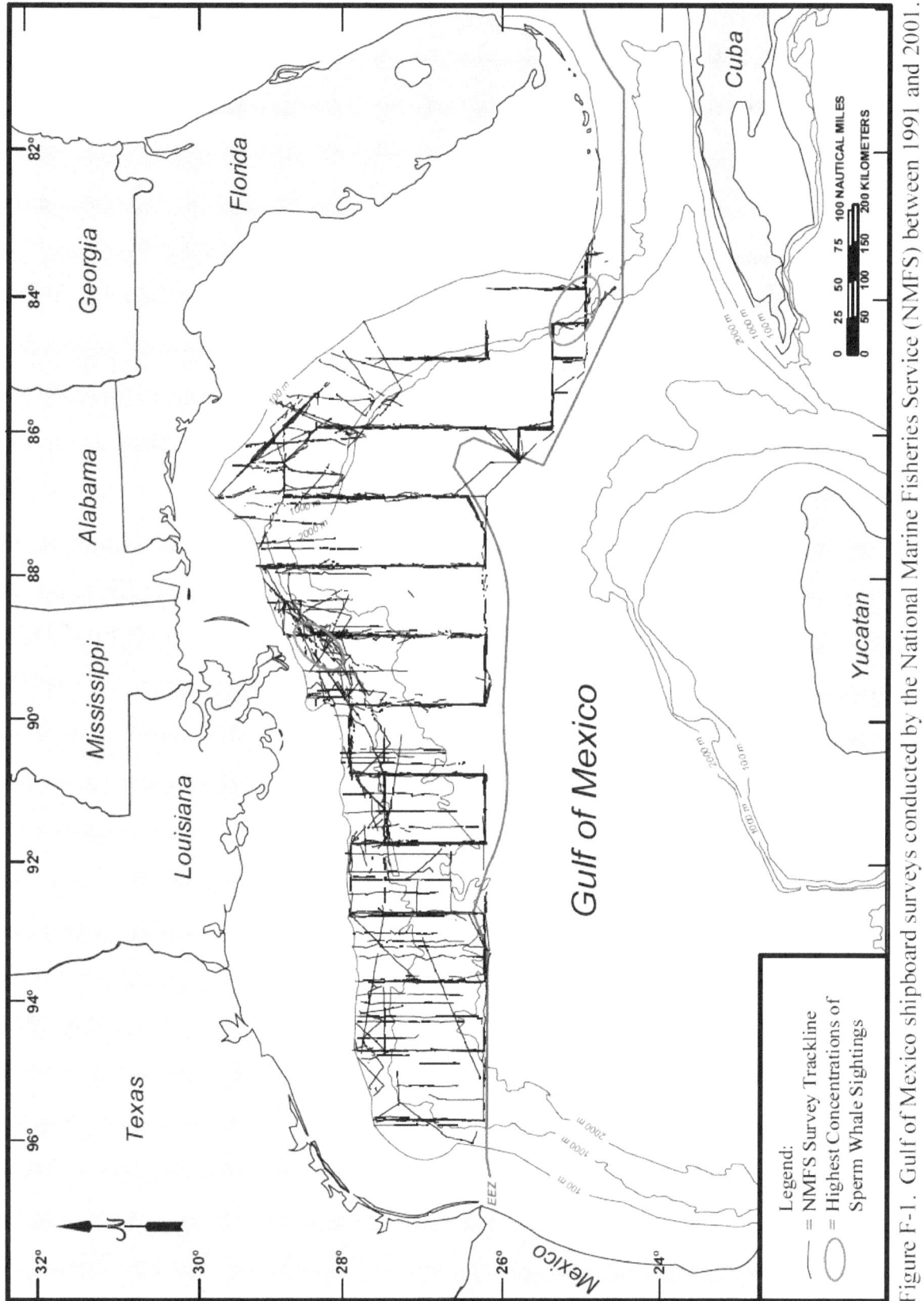

Figure F-1. Gulf of Mexico shipboard surveys conducted by the National Marine Fisheries Service (NMFS) between 1991 and 2001. Areas of highest sperm whale sightings (i.e., aggregations) denoted (Adapted from: K. Mullin, NMFS, written comm., 2003).

The GulfCet II program (Davis et al., 2000) analyzed correlations of environmental factors, such as physical and biological oceanographic variables, with seasonal cetacean sightings data. The sighted odontocete species were then subdivided into five groups, based on relative body size, similar behaviors and dietary preferences, and phylogenetic similarities (congeners). The five groups and their distribution are summarized below:

1) Cetaceans in general (all odontocete species included) were concentrated along the continental slope, especially within areas of cyclonic circulation where chlorophyll concentrations were elevated. They were less likely to occur in depths greater than 2,000 m and within anticyclones. Although cetaceans were sighted throughout the entire oceanic northern Gulf, sighting rates of cetacean groups were approximately 80 percent larger east of 90.0°W as compared with farther west.

2) Sperm whales (based on large body size and deep-diving ability) were found throughout the northern Gulf. They were typically sighted in surface waters over the lower slope and, in some areas, in eddies with highly productive cyclonic circulation. There was usually an aggregation of sperm whale sightings at the ocean surface along the 1,000-m isobath in proximity to the Mississippi River delta, which may provide preferred habitat for this stock.

3) Squid-eaters (i.e., a general category based on intermediate body size and apparent dietary preference for squid, for most marine mammal species included herein) occurred most frequently in waters over the upper slope in areas outside of anticyclonic circulation. This category included dwarf and pygmy sperm whales, false killer whales, melon-headed whales, pilot whales, pygmy killer whales, Risso's dolphins, rough-toothed dolphins, and beaked whales. Short-finned pilot whales and melon-headed whales were sighted almost exclusively in the north-central and northeastern Gulf. Killer whales, while not squid eaters, were included within this group, typically sighted throughout the northern Gulf, with highest densities in a broad but distinct region southwest of the Mississippi River delta. Rough-toothed dolphins were primarily sighted west of the Mississippi River delta during studies prior to GulfCet II and east of the delta during GulfCet II. Dwarf and pygmy sperm whales also were sighted throughout the northern Gulf. Risso's dolphins were concentrated in surface waters over the upper continental slope, but sightings were made throughout the northern Gulf. False and pygmy killer whales were uncommon but widely distributed. Beaked whales also were widely distributed in waters greater than 1,000 m deep.

4) Oceanic stenellids (all oceanic dolphins from the genus *Stenella*) occurred more often in surface waters over the lower slope and abyssal regions (at depths greater than 2,000 m) within areas of cyclonic circulation or within the confluence of adjacent eddies. This category included the Clymene dolphin, pantropical spotted dolphin, spinner dolphin, and striped dolphin. Spinner dolphins were found primarily in the north-central and northeastern Gulf. Pantropical and striped dolphins were sighted throughout the northern Gulf but rarely in the extreme northwestern section. As in the case of rough-toothed dolphins, Clymene dolphins also were primarily sighted west of the Mississippi River delta during studies prior to GulfCet II and east of the delta during GulfCet II. Fraser's dolphins were uncommon but widely distributed.

5) Bottlenose dolphins and Atlantic spotted dolphins were most frequently sighted in waters over the continental shelf and along the upper slope in association with the shelf edge, but outside of deep-water hydrographic features such as cyclones and anticyclones.

5.   Abundance Estimates

Table F-2 lists current minimum population estimates (N_{min}) and Potential Biological Removal (PBR) values for all but nine of the marine mammal species found in the Gulf of Mexico. The values were adapted from current U.S. Atlantic and Gulf of Mexico Marine Mammal Stock Assessment Reports (Waring et al., 1997, 1999, 2002), NMFS survey efforts (USDOI, USFWS, 1997a,b), and recent draft assessments (NMFS, 2004).

Table F-2
Minimum Population (N_{min}) and Potential Biological Removal (PBR)
Estimates for Gulf of Mexico Marine Mammal Stocks

Species	Stock	N_{min}	PBR
Northern right whale	N/A	N/A	N/A
Blue whale	N/A	N/A	N/A
Bryde's whale	Northern Gulf of Mexico	17	0.2
Fin whale	N/A	N/A	N/A
Humpback whale	N/A	N/A	N/A
Minke whale	N/A	N/A	N/A
Sei whale	N/A	N/A	N/A
Dwarf/pygmy sperm whales	Northern Gulf of Mexico	733*	N/A
Sperm whale	Northern Gulf of Mexico	411-1,035**	0.8-2.1**
Blainville's beaked whale	Northern Gulf of Mexico	N/A	N/A
Cuvier's beaked whale	Northern Gulf of Mexico	20-58**	0.2-0.6**
Gervais' beaked whale	Northern Gulf of Mexico	N/A	N/A
Sowerby's beaked whale	Northern Gulf of Mexico	N/A	N/A
Atlantic spotted dolphin	Northern Gulf of Mexico	2,255	23
Bottlenose dolphin	Gulf of Mexico Outer Continental Shelf	43,233	432
Bottlenose dolphin	Gulf of Mexico Continental Shelf Edge and Slope	4,530	45
Bottlenose dolphin	West Gulf of Mexico	2,938	29
Bottlenose dolphin	Northern Gulf of Mexico	3,518	35
Bottlenose dolphin	Eastern Gulf of Mexico	8,963	90
Clymene dolphin	Northern Gulf of Mexico	4,120	41
False killer whale	Northern Gulf of Mexico	236	2.4
Fraser's dolphin	Northern Gulf of Mexico	66	0.7
Killer whale	Northern Gulf of Mexico	197	2
Melon-headed whale	Northern Gulf of Mexico	2,888	29
Pantropical spotted dolphin	Northern Gulf of Mexico	26,510	265
Pygmy killer whale	Northern Gulf of Mexico	285	2.8
Short-finned pilot whale	Northern Gulf of Mexico	186	1.9
Risso's dolphin	Northern Gulf of Mexico	2,199	22
Rough-toothed dolphin	Northern Gulf of Mexico	660	6.6
Spinner dolphin	Northern Gulf of Mexico	4,465	45
Striped dolphin	Northern Gulf of Mexico	3,409	34
West Indian manatee	Florida	2,229***	4.5***
West Indian manatee	Antillean	101***	0.2***

Sources: Waring et al., 2002 (except as indicated below)
    * = Waring et al., 1999
    ** = Higher values from NMFS, draft 2003 marine mammal stock assessment report (NMFS, 2004)
    *** = U.S. Department of the Interior, U.S. Fish and Wildlife Service, 1997a,b
N/A = Not applicable.

The term PBR (16 USC 1362) is defined as the maximum number of individuals, not including natural mortalities, that may be removed from a marine mammal stock while allowing that stock to reach or maintain its Optimum Sustainable Population (OSP, defined as the number of animals that will result in the maximum productivity of the population or the species, considering the carrying capacity of the habitat and the health of the ecosystem where they are found). The PBR is the product of the following factors:

1) The minimum population estimate of the stock;
2) One-half of the maximum or theoretical or estimated net productivity of the stock at a small population size; and
3) An assigned "recovery" factor, relative to the OSP, of between 0.1 and 1.0.

According to Waring et al. (1997, 1999, 2002), the most abundant cetacean within the Gulf of Mexico was the bottlenose dolphin. From GulfCet II surveys, the most commonly sighted cetaceans on the continental shelf were bottlenose dolphins and Atlantic spotted dolphins. On the continental slope, the most commonly sighted cetaceans included the bottlenose dolphin (pelagic form), pantropical spotted dolphins, Risso's dolphins, and dwarf/pygmy sperm whales. The most abundant species on the slope were pantropical spotted and spinner dolphins (Davis et al., 2000).

B.   SEA TURTLES

1.   Species Occurrence and Status

Five species of sea turtles are known to inhabit the waters of the Gulf of Mexico (Pritchard, 1997). These are the loggerhead, leatherback, Kemp's ridley, green, and hawksbill turtles (**Table F-3**). All five are listed as either endangered or threatened species under the ESA (Pritchard, 1997). Brief life history descriptions for these species may be found in recent EISs (e.g., USDOI, Minerals Management Service [MMS], 1997a, 2002a,b).

Table F-3
Sea Turtles of the Gulf of Mexico

Common and Scientific Name	Status
Family Cheloniidae	
Loggerhead turtle (*Caretta caretta*)	T
Green turtle (*Chelonia mydas*)	T/E*
Hawksbill turtle (*Eretmochelys imbricata*)	E
Kemp's ridley turtle (*Lepidochelys kempi*)	E
Family Dermochelyidae	
Leatherback turtle (*Dermochelys coriacea*)	E

E = Currently listed as endangered under the Endangered Species Act of 1973
T = Currently listed as threatened under the Endangered Species Act of 1973
E* = Listed as endangered in the State of Florida

The loggerhead turtle (*Caretta caretta*), a threatened species, is the most abundant sea turtle in the Gulf of Mexico (Dodd, 1988). Loggerhead nesting along the Gulf Coast occurs primarily along the Florida panhandle, although some nesting has been reported from Texas through Alabama as well (NMFS and USFWS, 1991a). Loggerhead turtles have been primarily sighted in continental shelf waters, although many surface sightings of this species have also been made over the deeper waters of the slope (i.e., in waters greater than 1,000 m deep). Sightings of loggerheads over the continental slope suggest that they may be in transit through these waters to distant foraging sites or while seeking warmer waters during winter. Although loggerheads are widely distributed across the shelf during both summer and winter, their abundance in slope waters is significantly greater during winter than summer (Mullin and Hoggard, 2000).

The leatherback turtle (*Dermochelys coriacea*), an endangered species, is the most abundant turtle in waters over the northern Gulf of Mexico continental slope (Mullin and Hoggard, 2000). It is the most pelagic and wide-ranging sea turtle, undertaking extensive migrations from the tropics to boreal waters. Leatherback nesting within the continental U.S. is limited to eastern Florida (NMFS and USFWS, 1992a; Ernst et al., 1994; Meylan et al., 1995). Leatherbacks appear to use spatially both continental shelf and slope waters in the Gulf of Mexico (Fritts et al., 1983a,b; Collard, 1990; Davis and Fargion, 1996). GulfCet I and II surveys suggest that the region from Mississippi Canyon to De Soto Canyon, especially near the shelf edge, appears to be an important habitat for leatherbacks (Mullin and Hoggard, 2000). Temporal variability in leatherback distribution and abundance suggest that specific areas may be important to this species, either seasonally or for short periods of time. During the GulfCet I and II programs, leatherbacks were sighted in significant numbers during both summer and winter surveys. High variability recorded in the relative numbers of individual leatherbacks sighted within specific areas of the Gulf suggests that their distribution and density patterns may have been associated with favorable changes in the environment (Mullin and Hoggard, 2000).

The Kemp's ridley turtle (*Lepidochelys kempi*) is an endangered species and the smallest sea turtle. Survey data from the Gulf of Mexico suggest that Kemp's ridley turtles occur mainly in waters over the continental shelf. Juvenile and adult Kemp's ridleys are typically found in shallow areas and especially in areas of seagrass habitat (Marquez, 1990; NMFS and USFWS, 1992b; Ernst et al., 1994). The major nesting area for this species is near Rancho Nuevo, along the northeastern coast of Mexico (Tamaulipas), although scattered nesting has also been reported in other areas of Mexico and Texas, Colombia, Florida, and South Carolina (Ernst et al., 1994). Adult Kemp's ridleys exhibit extensive internesting movements, although they appear to travel near the coast, especially within shallow waters along the Louisiana coast.

The green sea turtle (*Chelonia mydas*) is listed as threatened, except for the Florida breeding population, which is listed as endangered. This species is found throughout the Gulf of Mexico. They occur in small numbers over grass beds along the south Texas coast and the Florida Gulf coast. Reports of green turtle nesting along the Gulf of Mexico coast are infrequent, and the closest significant nesting aggregations are along the Florida East coast and the Yucatan Peninsula (NMFS and USFWS, 1991b).

The hawksbill turtle (*Eretmochelys imbricata*), an endangered species, has been recorded in all the Gulf states. However, sightings north of Florida are rare. This is the least common sea turtle in the Gulf of Mexico (Marquez, 1990; Hildebrand, 1995). Hawksbill nesting within the continental U.S. is limited to southeastern Florida and the Florida Keys.

2.   Sea Turtle Distribution and Critical Habitat

Historical surveys of sea turtles within the northern Gulf of Mexico suggest that (with the exception of leatherbacks) they are largely distributed in waters less than 100 m in depth. Surveys conducted by Fritts et al. (1983a,b) identified loggerhead, green, Kemp's ridley, and leatherback turtles across the northern Gulf. Surveys conducted during the GulfCet I and II programs sighted only loggerheads, leatherbacks, and Kemp's ridleys (Davis and Fargion, 1996; Mullin and Hoggard, 2000). These surveys found the greatest abundance of sea turtles in the Gulf of Mexico along the continental shelf, east of Mobile Bay (Lohoefener et al., 1990; Mullin and Hoggard, 2000).

Most sea turtles (except perhaps the leatherback) exhibit differential distributions among their various life stages (hatchling, juvenile, adult) (Marquez, 1990; Hirth, 1997; Musick and Limpus, 1997). Hatchling turtles may be found within zones of water mass convergence and/or sargassum rafts, which are rich in prey and provide shelter (NMFS and USFWS, 1991b; Hirth, 1997). These hatchlings may have originated from nesting sites along Gulf of Mexico shores or adjacent areas such as the Caribbean Sea. Juvenile and subadult turtles may actively move across the Gulf to neritic developmental habitats and adult foraging habitats, respectively. Adult foraging habitats may be, in some populations, geographically distinct from juvenile developmental habitats (Musick and Limpus, 1997).

There are no designated critical habitats for sea turtles in the northern Gulf of Mexico. NMFS does recognize many coastal areas of the Gulf as preferred habitat (important, sensitive habitats that are essential for the species within a specific geographic area) – e.g., seagrass beds in Texas lagoons and other nearshore or inshore areas (including jetties) for green turtles; bays and lakes, especially in Louisiana and Texas for ridleys. Sargassum mats are also recognized as preferred habitat for hatchlings. There are no designated migratory routes for turtles in the Gulf.

C.   FISH RESOURCES

The Gulf of Mexico's marine habitats, ranging from coastal marshes to the deep-sea abyssal plain, support a varied and abundant fish fauna. Distinctive fish assemblages can be recognized within broad habitat classes for the continental shelf and oceanic waters as follows: soft bottom fishes, hard bottom fishes, and coastal pelagic fishes on the continental shelf; and epipelagic, midwater fishes, and demersal fishes in oceanic waters (>200 m water depths). One endangered fish species, the smalltooth sawfish (*Pristis pectinata*), occurs in the open Gulf of Mexico.

1.   Continental Shelf Fishes

a.   *Soft Bottom Fishes*

The soft bottom or demersal shelf fish fauna can be generally characterized by substrate composition and water depth. Chittenden and McEachran (1976); Darnell et al. (1983); and Darnell and Kleypas (1987) have described this fauna in detail. From Rio Grande to the Florida Keys, a total of 372 species of demersal fishes was recorded by Darnell and Kleypas (1987). Of these, 164 occurred in the northwestern Gulf and 347 in the northeastern Gulf. While some species are widespread, the number of species is much higher in the northeastern Gulf of Mexico. The number of unique species is also higher in the northeastern Gulf. These patterns were influenced by the high diversity of fishes found on the southwest Florida shelf. Sediment composition, rainfall, river discharge, and isolation all contribute to these observed patterns. As with the common shrimp species of the Gulf, soft bottom fishes generally prefer certain types of sediments over others. This fact led to the naming of three primary fish assemblages by the dominant shrimp species found in the same sediment/depth regime. These assemblages are as follows:

- Pink shrimp assemblage (carbonate sediments, east of De Soto Canyon, 10 to 41 m);
- Brown shrimp assemblage (coarse sediments, west of De Soto Canyon, 22 to 91 m); and
- White shrimp assemblage (fine sediments, west of De Soto Canyon, 3.5 to 22 m).

Common members of the pink shrimp assemblage include Atlantic bumper, sand perch, silver jenny, dusky flounder, and pigfish. This assemblage occurs on the west Florida shelf. Longspine porgy, leopard searobin, horned searobin, and dwarf goatfish characterize the brown shrimp assemblage. Most of these species spend their entire life cycle in marine waters. The white shrimp assemblage consists of species such as Atlantic croaker, star drum, Atlantic cutlassfish, sand seatrout, silver seatrout, Atlantic threadfin, and hardhead catfish. Most of these species spawn in shelf waters and spend their early life stages in estuarine waters.

In some areas offshore of west Florida, particularly the Big Bend area and Florida Bay, soft bottom areas are vegetated with seagrasses and macroalgae. These vegetated bottoms support numerous fishes, including red drum, pinfish, spotted seatrout, filefishes, and spot. Both adults and juveniles of these species utilize the vegetated habitats (Gulf of Mexico Fishery Management Council [GMFMC], 2003).

One endangered fish species, the smalltooth sawfish (Pristis pectinata), occurs in the open waters of the Gulf of Mexico. This species normally inhabits shallow waters (<10 m), often near river mouths or in estuarine lagoons over sandy or muddy substrates, but may also occur in deeper waters (20 m) of the continental shelf. It is considered rare in the northern Gulf of Mexico, with its known range at present reduced to the coastal waters of Everglades National Park in extreme southern Florida. Critical habitat has not been defined, and data are being collected on the life history and biology of this species.

Only one threatened fish species occurs in the Gulf of Mexico: the Gulf sturgeon. This species occurs primarily off Florida and Alabama, where it spends winter months in estuaries and inner shelf waters (over soft bottoms). The biology and status of this species is discussed in detail by USDOI, MMS (1999, 2002a).

b. *Hard Bottom Fishes*

Another important habitat for fishes on the continental shelf is hard bottom. The term hard bottom generally refers to exposed rock, but can refer to other substrata such as coral and clay, or even artificial structures. The estimated areal extent of natural hard bottom in the Gulf of Mexico is 4,772,600 ha, and 94 percent of this exists on the west Florida shelf from Dry Tortugas to Pensacola (GMFMC, 2003). Outside of Florida, hard bottom occurs on the Mississippi/Alabama shelf, the Texas/Louisiana shelf, and the south Texas shelf. Colonized by stony corals, sea whips, sponges, tunicates, and algae, these structures provide shelter, food, and spawning sites for fishes. Fishes found over hard bottom habitats in middle (10 to 25 m) and outer (25 to 50 m) shelf waters include reef and coastal pelagic forms. Reef fishes such as snappers, groupers, grunts, porgies, squirrelfishes, angelfishes, damselfishes, butterflyfishes, surgeonfishes, parrotfishes, and wrasses inhabit hard bottom habitats in the Gulf of Mexico (Dennis and Bright, 1988). In water depths exceeding 50 m, a distinctive deep reef assemblage mixes with depth tolerant members of the shallow reef assemblages. Deep reef species in the Gulf of Mexico include roughtongue bass, yellowtail reeffish, short bigeye, and wrasse bass. Deep reef fishes occur on hard bottom features in water depths of 50 to 105 m off southwest Florida, Mississippi/Alabama Pinnacle trend (Brooks, 1991), Texas/Louisiana shelf edge, and the south Texas carbonate banks (GMFMC, 2003).

Some species use hard-bottom habitat as adults and juveniles, whereas others undergo ontogenetic migrations from adjacent habitats such as seagrass meadows. Some species such as gag grouper aggregate to spawn on hard bottom sites that may be used by the population for many generations (GMFMC, 2003). Other species deposit demersal eggs on the substrate, whereas others shed eggs and sperm into the water column where they are fertilized and then transported to other areas often many kilometers from the spawning site.

Artificial hard bottom, including sunken vessels, oil and gas platforms, and other debris, represents 1.3 percent of all hard bottom in the Gulf of Mexico (GMFMC, 2003). Nevertheless, these structures support abundant fish populations in the shelf waters of all Gulf coast states (GMFMC, 2003).

2.   Coastal Pelagic Fishes

The basic pelagic fish assemblage found in shelf waters of the Gulf of Mexico is usually termed coastal pelagic. The major coastal pelagic families occurring in the Gulf are requiem sharks, eagle rays (*Aetobatus narinari*), cownose rays (*Rhinoptera bonasus*), ladyfish, anchovies, herrings, mackerels, jacks, mullets, bluefish, and cobia. Coastal pelagic species traverse shelf waters of the region throughout the year. Some species form large schools (e.g., Spanish mackerel), while others travel singly or in smaller groups (e.g., cobia). The distribution of most species depends upon water column structure, which varies spatially and seasonally.

King mackerel exist in two populations in the Gulf of Mexico, an eastern group and a western group. The eastern population migrates from near the Mississippi River delta eastward, then southward around the Florida peninsula, wintering off southeastern Florida (Sutter et al., 1991). The western population travels to waters off the Yucatan Peninsula during winter. In summer, both populations migrate to the northern Gulf of Mexico, where they intermix to an unknown extent (Johnson et al., 1994). Cownose rays form dense schools in the summer in the eastern Gulf of Mexico, whereas eagle rays are common during the winter in shelf waters of the western Gulf. Spanish mackerel, cobia, bluefish, crevalle jack, and coastal sharks (*Carcharhinus* spp.) are migratory, but their routes have not been studied. Spanish mackerel, bluefish, and crevalle jack generally migrate westward along the shelf in warm months and back eastward towards Florida during cold months (Barry A. Vittor & Associates, Inc., 1985). All of these species are predatory, feeding upon a range of fishes and invertebrates.

Coastal pelagic fishes can be divided into two ecological groups. The first group includes larger predatory species such as king and Spanish mackerel, bluefish, cobia, jacks, and little tunny. These species typically form schools, undergo migrations, grow rapidly, mature early, and exhibit high fecundity. Each of these species is important to some extent to regional fisheries. The second group exhibits similar life history characteristics, but the species are smaller in body size and are planktivorous. This group is composed of Gulf menhaden, thread herring, Spanish sardine, round scad, and anchovies (USDOI, MMS, 1999, 2002a).

Some coastal pelagic species are found in the nearshore environment along sandy beaches from the shoreline to the swash zone (USDOI, MMS, 1999, 2002a). This habitat occurs along the seaward shore of barrier islands off all Gulf coast states. Commonly occurring species include scaled sardine, Florida pompano, and various anchovies. Nearshore fish assemblages show considerable seasonal structuring in the northern Gulf of Mexico. The lowest abundance of all species occurs in winter, with peak numbers found during summer and fall. Larger predatory species (particularly bluefish, Spanish mackerel, and blue runner) may be attracted to large concentrations of anchovies, herrings, and silversides, which congregate in nearshore areas.

3.   Oceanic Fishes (Beyond the Continental Shelf Edge)

a.   *Epipelagic Fishes*

Epipelagic fishes inhabit the upper 200 m of the water column and include several shark species (mako, silky, oceanic whitetip, whale shark), billfishes (marlins, sailfish, and swordfish), herrings, flyingfishes, halfbeaks, opahs, oarfishes, bluefish, scads, jacks, pilotfishes, dolphin, remoras, pomfrets, tunas, butterfishes, and tetraodontiform fishes (molas and triggerfishes). A number of these species such as dolphin, sailfish, white marlin, blue marlin, and tunas are important to commercial and recreational fisheries. Many of these species, such as bluefin tuna and swordfish, spawn in the eastern Gulf of Mexico in relation to the LC boundary (USDOI, MMS, 1999, 2002a). All of the epipelagic species are migratory, but specific patterns are not well understood. Many of the oceanic species associate with flotsam, which provides forage areas and/or nursery refuge.

Floating seaweed (*Sargassum*), jellyfishes, siphonophores, and driftwood attract juvenile and adult epipelagic fishes. Many species will associate with drifting objects. Larger predators forage around flotsam. As many as 54 fish species are closely associated with floating *Sargassum* at some point in their life cycle, but only two spend their entire lives there: the sargassumfish and the sargassum pipefish (USDOI, MMS, 1999, 2002a). Most fish associated with *Sargassum* are temporary residents, such as juveniles of species that reside in shelf or coastal waters as adults (USDOI, MMS, 1999, 2002a). However, several larger species of recreational or commercial importance, including dolphinfish, yellowfin tuna, blackfin tuna, skipjack tuna, Atlantic bonito, little tunny, and wahoo, feed on the small fishes and invertebrates attracted to *Sargassum* (USDOI, MMS, 1999, 2002a).

b. *Midwater Fishes*

Below the epipelagic zone, the water column may be layered into mesopelagic (200 to 1,000 m) and bathypelagic (>1,000 m) zones. Taken together, these two zones and their inhabitants may be referred to as midwater. In the mesopelagic zone of the Gulf of Mexico, fish assemblages are numerically dominated by lanternfishes, bristlemouths, and hatchetfishes (USDOI, MMS, 1999, 2002a). Lanternfishes are small silvery fishes that can be extremely abundant, often responsible for the deep scattering layer in sonar images of the deep sea. Lanternfishes, and other mesopelagic fishes, spend the daytime in depths of 200 to 1,000 m, but migrate vertically at night into food rich, near-surface waters. Mesopelagic fishes, while less commonly known, are important ecologically because they transfer significant amounts of energy between mesopelagic and epipelagic zones over each daily cycle. The lanternfishes are important prey for meso- and epipelagic predators (e.g., tunas), and particularly the mesopelagic dragonfishes (Hopkins et al., 1997).

Deeper dwelling bathypelagic fishes inhabit the water column at depths greater than 1,000 m. This group is composed of strange, little known species such as snipe eels, slickheads, deep-sea anglers, bigscales, and whalefishes (McEachran and Fechhelm, 1998). Most species are capable of producing and emitting light (bioluminescence) to aid in communicating in an environment devoid of sunlight. Little scientific information is available on bathypelagic fishes of the Gulf of Mexico.

c. *Demersal Fishes*

Demersal fishes are those that are either in direct contact with the substrate or hover above it from the shelf-slope transition down to the abyssal plain. The deep-sea demersal fish fauna in the Gulf of Mexico includes about 300 species. The most diverse group is the cod-like fishes such as hakes and grenadiers, followed by eels, cusk-eels, sharks, and flatfishes. Members of these groups were collected during MMS-sponsored demersal sampling programs summarized by Pequegnat (1983) and Gallaway and Kennicutt (1988). In general, fish species diversity decreases with increasing water depth. The highest diversity and density of demersal fishes was found along the continental slope in the eastern Gulf. Deep-sea demersal fishes consume a wide range of organisms, including fishes and epifaunal, infaunal, meiofaunal, and planktonic invertebrates.

4.   Ichthyoplankton

Larval fishes occur in the upper water column throughout the Gulf of Mexico; the best known assemblages of ichthyoplankton occur in nearshore, shelf, and oceanic waters as well as the Mississippi River plume and the LC boundary (USDOI, MMS, 1999, 2002a).

Ichthyoplankton assemblages in nearshore shelf waters of the region are composed of species that are also common as adults (Ditty, 1986; Ditty et al., 1988). The temporal occurrence of these taxa in ichthyoplankton samples reflects the spawning times of adults. In the northern Gulf of Mexico, spawning activity can be broadly classified into cold water and warm water periods that parallel the seasons (Barry A. Vittor & Associates, Inc., 1985). Because expected seasonal patterns of fish egg and larval occurrence can be inferred from a knowledge of the known adult spawning times, this information is presented to augment information on the temporal patterns of ichthyoplankton occurrence. **Table F-4** gives the spawning times for economically important species from the region.

Ditty et al. (1988) summarized information from over 80 ichthyoplankton studies from the northern Gulf of Mexico (north of 26°N) and reported 200 coastal and oceanic fishes from 61 families. Many taxa were only collected over waters within certain depth ranges. Species found exclusively in water depths shallower than 25 m were mostly inshore demersal species such as Atlantic bumper (*Chloroscombrus chrysurus*), spotted seatrout (*Cynoscion nebulosus*), pigfish (*Orthopristis chrysoptera*), and black drum (*Pogonias cromis*). At depths <100 m, several clupeids (*Brevoortia patronus*, *Opisthonema oglinum*, and *Sardinella aurita*), several serranids (*Centropristis striata*, *Diplectrum formosum*, and *Serraniculus pumilio*), Atlantic croaker (*Micropogonias undulatus*), and spot (*Leiostomus xanthurus*) were most common in collections. More recent data collected from around oil and gas platforms in the Gulf of Mexico support Ditty's general finding of depth-related larval assemblages across the shelf (Hernandez et al., 2002; Shaw et al., 2002).

Species such as Atlantic croaker, spot, and Gulf menhaden (*Brevoortia patronus*) migrate to the outer shelf during winter months to spawn. Consequently, larvae of these species are often numerically dominant during winter months. Larvae of speciose families such as engraulids (*Anchoa* spp.), searobins (*Prionotus* spp.), tonguefishes (*Symphurus* spp.), and pufferfishes (*Sphoeroides* spp.) were collected during all months.

Ichthyoplankton distribution is greatly influenced by mesoscale hydrographic features such as eddies, fronts, and rings. In the eastern Gulf of Mexico, the LC boundary is recognized as an important area of spawning and larval fish aggregation for many species (Richards et al., 1989, 1993; Lamkin, 1997). Recently, Lamkin (1997) used 6 years of Southeast Area Management and Assessment Program (SEAMAP) data, 1983-1988, in an investigation of the frontal zones associated with the northern excursions of the LC. Lamkin found a positive correlation between the abundance of larval nomeid fish and the location of the northern edge of the LC. In particular, bigeye cigarfish (*Cubiceps pauciradiatus*) has adult spawning grounds and larval habitats closely related to sharp temperature gradients. Other fish larvae appear to vary in abundance in relation to mesoscale hydrographic features, as well. Larvae of apex predators like bluefin and yellowfin tuna seem to be most abundant along LC frontal zones and within eddy peripheries (Richards et al., 1989), and the adults, as well, can be caught in such frontal zones.

Another area where ichthyoplankton aggregate is the frontal zone of the Mississippi River discharge plume (Govoni et al., 1989; Grimes and Finucane, 1991; Govoni and Grimes, 1992; USDOI, MMS, 1999, 2002a). Hydrodynamic convergence and the continually reforming turbidity fronts associated with the discharge plume probably accounted for the concentration of larval fishes at the front. These investigators hypothesized that frontal waters provide feeding and growth opportunities for larvae.

Table F-4
Spawning Times of Economically Important Fishes and Invertebrates
in the Northern Gulf of Mexico

Species	Month											
	J	F	M	A	M	J	J	A	S	O	N	D
Cold-Water Spawners												
Sheepshead		•	•	•	•	•				•	•	•
Gulf menhaden	•	•	•							•	•	•
Spot	•	•	•									•
Atlantic croaker	•	•	•	•						•	•	•
Striped mullet	•	•	•	•	•							
Gulf flounder	•	•									•	•
Southern flounder	•	•	•	•					•	•	•	•
Gulf butterfish		•	•	•	•				•	•	•	
Black drum		•	•	•								
Bluefish	•	•	•	•				•	•	•	•	•
Brown shrimp	•	•	•	•						•	•	•
Warm-Water Spawners												
Hardhead catfish					•	•	•	•				
Crevalle jack			•	•	•	•	•	•	•			
Sand seatrout			•	•	•	•	•	•	•			
Silver seatrout									•	•	•	
Red snapper						•	•	•	•	•		
Lane snapper			•	•	•	•	•	•	•			
Harvestfish			•	•	•							
Cobia				•	•	•	•	•				
Red drum									•	•	•	
Spanish mackerel			•	•	•	•	•	•	•			
Tarpon				•	•	•	•	•				
Pink shrimp				•	•	•	•	•	•	•	•	
White shrimp			•	•	•	•	•	•	•	•		
Year-Round Spawners												
Bay anchovy	•	•	•	•	•	•	•	•	•	•	•	•
Blue runner	•	•	•	•	•	•	•	•	•	•	•	•

Source: adapted from Barry A. Vittor & Associates, Inc., 1985.

Bothids (lefteye flounders), carangids, engraulids, exocoetids (flying fishes and halfbeaks), gobiids (gobies), sciaenids, scombrids (mackerels and tunas), synodontids (lizardfishes), and tetraodontids (pufferfishes) were the 10 most frequently caught taxa in the plume/shelf samples off the Mississippi River delta (Grimes and Finucane, 1991).

Like many groupers, the gag aggregates to spawn during a late winter-early spring time period (February to April). Two aggregation sites on the west Florida shelf were identified by researchers (Coleman et al., 1996). One site is south of Panama City, Florida, and the other is west of Tarpon Springs, Florida. Recently, the GMFMC proposed closing these two sites to all fishing.

5.   Endangered Fish Species

The endangered status of the smalltooth sawfish (*Pristis pectinata*) was finalized on 1 May 2003 (50 CFR Part 224). Critical habitat has not been defined, and data are being collected on the life history and biology of this species. The information on biology and distribution provided below was obtained from NMFS (2000).

The smalltooth sawfish is distributed in tropical and subtropical waters worldwide. Within U.S. waters it was historically distributed throughout the GOM and along the Atlantic coast to North Carolina. This species has become rare in the northern GOM during the past 30 years, and its known range is now reduced to the coastal waters of Everglades National Park in extreme southern Florida. Fishing and habitat degradation have extirpated the smalltooth sawfish from much of its former range.

The smalltooth sawfish normally inhabits shallow waters (10 m or less), often near river mouths or in estuarine lagoons over sandy or muddy substrates, but may also occur in deeper waters (20 m) of the continental shelf. Shallow water less than 1 m appears to be important nursery area for young smalltooth sawfish. Smalltooth sawfish grow slowly and mature at about 10 years of age. Females bear live young, and the litters reportedly range from 15 to 20 embryos requiring a year of gestation.

The diet of the smalltooth sawfish consists of macroinvertebrates and fishes such as herrings and mullets. The saw is reportedly used to rake surficial sediments in search of crustaceans and benthic fishes, or to slash through schools of herrings and mullets.

D.   COASTAL AND MARINE BIRDS

The waters and adjacent coastal landforms of the northern Gulf of Mexico are inhabited by a diverse assemblage of resident and migratory birds. These include primarily seabirds, shorebirds, wetlands birds, and waterfowl (**Table F-5**). Brief descriptions of these species groups may be found in USDOI, MMS (1997a, 2002a). Generally, noise produced from activities associated with the Proposed Action might impact only those offshore species of birds that spend quantities of time underwater, either swimming or plunge diving while foraging for food. Offshore Gulf of Mexico birds that may be classified as underwater swimmers include certain waterfowl (some diving ducks) and seabirds (loons and cormorants). Generally, these species are limited to waters of the inner continental shelf. Waterfowl and loons are both seasonal migrants (winter), whereas cormorants are resident species. Gulf of Mexico plunge diving birds include only certain seabirds (primarily brown pelicans, gannets, and boobies). The brown pelican is a resident species that also is limited to waters of the inner continental shelf. Gannets and boobies, however, are seasonal migrants (gannets in winter and boobies primarily in summer) that may range throughout the Gulf of Mexico.

Birds of the Gulf of Mexico and adjacent landforms that are currently listed as endangered or threatened species under the ESA comprise primarily coastal or inshore species. These include the piping

Table F-5
Coastal and Marine Birds of the Gulf of Mexico, by Order and Family

Seabirds
Order Charadriiformes
  Family Laridae – Gulls and Terns
  Family Scolopacidae – Phalaropes
Order Gaviiformes
  Family Gaviidae – Loons
Order Pelicaniformes
  Family Fregatidae – Frigatebirds
  Family Pelicanidae – Pelicans
  Family Phaethontidae – Tropicbirds
  Family Phalacrocoracidae – Cormorants
  Family Sulidae – Gannets and Boobies
Order Procellariiformes
  Family Diomedeidae – Albatrosses
  Family Hydrobatidae – Storm-Petrels
  Family Procellariidae – Petrels and Shearwaters

Shorebirds
Order Charadriiformes
  Family Charadriidae – Plovers
  Family Haematopodidae – Oystercatchers
  Family Recurvirostridae – Stilts and Avocets
  Family Scolopacidae – Sandpipers, Snipes, and Allies

Wetland Birds
Order Charadriiformes
  Family Jacanidae – Jacanas
Order Ciconiiformes
  Family Aramidae – Limkins
  Family Ardeidae – Bitterns, Egrets, and Herons
  Family Ciconiidae – Storks
  Family Threskiornithidae – Ibises and Spoonbills
Order Gruiformes
  Family Gruidae – Cranes
  Family Rallidae – Rails and Coots, Moorhens, and Gallinules
Order Pelicaniformes
  Family Anhingidae – Darters and Anhingas
Order Podicipediformes
  Family Podicipedidae – Grebes

Waterfowl
Order Anseriformes
  Family Anatidae – Ducks, Geese, and Swans

plover, whooping crane, eskimo curlew, brown pelican, wood stork, and bald eagle (USDOI, USFWS, 1998). Of these, the brown pelican is the only endangered or threatened bird species that may be affected by the Proposed Action.

In the northern Gulf of Mexico, brown pelicans occur regularly, but in small numbers. Mississippi, Louisiana, and Caribbean populations of the brown pelican are endangered, but this designation does not apply to populations in Alabama or Florida. In addition, Louisiana has requested the removal of its brown pelican population from the endangered species list (USDOI, MMS, 1997a, 2002a). Brown pelicans generally do not venture far offshore. Fritts and Reynolds (1981) observed them only within about 40 km (22 nmi) of the shoreline, and only one was seen over the northern Gulf of Mexico continental slope during the GulfCet surveys (Peake, 1996).

E.   PLANKTON AND PRIMARY PRODUCTIVITY

In general, primary productivity of the Texas/Louisiana continental shelf proceeds at moderate rates ($0.2 - 1$ g C m^{-2} d^{-1}) when not enhanced by high "new" nitrogen nutrient loads from upwelling or from riverine/estuarine outflows (Biggs and Sanchez, 1997). Maximum values of biomass and primary productivity are typically observed at intermediate salinities in coastal waters impacted by riverine plumes, where these coincide with non-conservative decreases in nutrients along the salinity gradient (Lohrenz et al., 1999). Highest values of productivity (>10 g C m^{-2} d^{-1}) and biomass (>30 mg chlorophyll m^{-3}) are apparently constrained by low irradiance and mixing in the more turbid, low salinity regions of the plume, and by nutrient limitation outside the plume. In other words, both mixing and advection are important in determining the location and magnitude of primary production maxima and nutrient depletion. The major river system influencing the continental margin of the Gulf of Mexico is the Mississippi-Atchafalaya River (MAR) complex, as detailed by Biggs (2000). Local aggregations of ichthyoplankton (fish larvae) were documented in the MAR discharge plume by Grimes and Finucane (1991), and Dagg (1995) documented elevated stocks of meso- and microzooplankton as well. Dagg concluded that the daily grazing impact of zooplankton and fish larvae in such coastal plumes was less than the total daily primary production. As a result, a portion of the primary production in the plume should sink out of surface waters or be otherwise exported. The mass flux of, and the biogeochemical characterization of, the particulate organic matter sinking out of near surface water in the MAR plume were reported by Redalje et al. (1994). The potential this creates for increased hypoxia and anoxia in near-bottom waters is under continuing study (i.e., Rabalais et al., 1996).

In contrast, the standing stocks and biological productivity of plant and animal communities living in the water column of the Gulf of Mexico (GOM) away from upwelling and riverine/estuarine outflows are in general those that might be expected in a nutrient-limited ecosystem. In 1970, as part of a review of plankton primary productivity of the world ocean, Soviet scientists characterized the deep-water GOM by mean primary productivity of just 100 to 150 mg C m^{-2} d^{-1} (Koblenz-Mishke et al., 1970). A few years later, extensive surveys of phytoplankton chlorophyll and primary production that span the period 1964-1971 were summarized by El-Sayed et al. (1972) in atlas format, as averages within $2°$ squares of latitude and longitude. These atlas maps show that surface chlorophyll-*a* generally ranges from 0.06 to 0.32 mg m^{-3} in the deep-water portions of the central and western GOM, equivalent to just 3 to 21 mg m^{-2} when integrated from the surface to the base of the photic zone. Low values of primary production (<0.25 mg C m^{-3} h^{-1}) characterize the majority of the oceanic stations in this atlas, equivalent to <10 mg C m^{-2} h^{-1} when integrated from the surface to the base of the photic zone. If there are on average 12 hours of sunlight per day, this is equivalent to ≤120 mg C m^{-2} d^{-1} and so is in good agreement with the summary by Koblenz-Mishke et al. (1970). Allowing for primary production to proceed 365 days a year in the GOM because of its subtropical climate, this rate of primary productivity nevertheless totals <50 g C m^{-2} y^{-1}. As a consequence, the deep-water GOM average falls at the low end of the estimated range of 50-160 g C m^{-2} y^{-1} that is generally accepted for the annual primary production in open-ocean ecosystems (Smith and Hollibaugh, 1993).

Even in a subtropical ocean, though, there are seasonal changes. Pigment concentration at the surface in the deep-water GOM undergoes a well-defined seasonal cycle that is generally synchronous throughout the region. Müller-Karger et al. (1991) reviewed monthly climatologies of near-surface phytoplankton pigment concentration from multiyear series of coastal zone color scanner (CZCS) images for the period 1978-1985. They reported that highest surface concentrations of chlorophyll occur between December and February, and lowest values occur between May and July. Over deep water, there is about a three-fold variation between the lowest (~0.06 mg m^{-3}) and highest (0.2 mg m^{-3}) surface pigment concentrations. Model simulations show that the single most important factor controlling the seasonal cycle in surface pigment concentration over deep water is the depth of the mixed layer (Walsh et al., 1989). Müller-Karger et al. (1991) concluded that because of this dependence, annual cycles of phytoplankton biomass are usually out of phase relative to the seasonal sea surface temperature (SST) cycle.

Biggs and Müller-Karger (1994) combined CZCS data with ship data to document that another way in which high-chlorophyll plumes may form in the central and western GOM occurs when a seaward-moving surface flow confluence is created by deep-water cyclone-anticyclone circulation pairs. Analogous to a pair of anticlockwise-rotating and clockwise rotating gears, these circulations entrain coastal water from the central and western continental shelf and draw this offshore when the cyclone (anticlockwise circulation) lies immediately to the north or east of the anticyclone (clockwise circulation). Recent fieldwork has shown these mesoscale oceanographic features have additional impacts upon deep-water plankton and micronekton communities, for locally high nutrients are also introduced to the surface of deep-water ocean regions at eddy edges where there is enhanced vertical mixing. In fact, the periphery region of high velocity surface currents that surrounds both the cyclonic and the anticyclonic eddies is a zone of locally high vertical shear. In the CZCS ocean color climatology from 1978-1985, and in imagery from the current generation ocean color sensor (the Sea Wide-Field Scanner, or SeaWiFS, in orbit since November 1997), the periphery of the LC and of the anticyclonic LC eddies (LCEs) of diameter 200 to 300 km that are shed from the LC is often seen to be outlined by surface pigment concentrations that are two-fold to three-fold higher than the extremely low concentrations (0.04 to 0.06 mg m^{-3}) in the interior of these circulations.

NMFS has studied the "fertile fisheries crescent" that is maintained by the transport of nutrients onto the continental shelf of the northern GOM. NMFS has 50$^+$ years of data on menhaden catch (the largest volume fishery in the U.S.) as well as on landings of other commercial target species; Dagg et al. (1991) gave a useful review. In addition, since 1982 the NMFS SEAMAP has made over 2,000 deep-water collections of zooplankton and micronekton in the GOM to survey for ichthyoplankton (eggs and larvae of commercially important fish species). Grimes and Finucane (1991) illustrated how such data can be applied to investigate regional questions (see also http://www.gsmfc.org/seamap.html). In addition, the presence of multiple cyclonic and anticyclonic features in the Gulf of Mexico can set up strong frontal gradients between these features. Lee et al. (1991) have shown that meanders and eddies in the Gulf Stream are often marked by local aggregations of phytoplankton, and elevated fish stocks appear to concentrate in such areas (Atkinson and Targett, 1983).

Because the interiors of the anticyclones are areas of convergence, the upper 100 m or so of the water column in both LC and LCEs are areas in which surface waters are infrequently renewed and so they are impoverished in nitrogen and phosphorus nutrients (Biggs, 1992). The interiors of these regions of convergence are generally regarded as biological "ocean deserts." However, the cyclonic cold-core eddies (local areas of divergence) that are frequently associated with these anticyclones represent areas of higher biological productivity. Both types of these mesoscale features can be detected by the topography of the 15°C-isotherm; this is domed upward in the cyclones and pushed locally deep within the anticyclones. Both types of features can now be located with satellite altimetry since GOM cold-core eddies (15°C-isotherm domed) show up as 10- to 20-cm local depressions in sea surface height, whereas warm-core eddies (15°C-isotherm pushed locally deep) show up as 20- to 70-cm local elevations in sea surface height (Leben et al., 1993).

Subsurface sampling of these GOM eddies from ships showed there was a highly predictable negative first order relationship between temperature <22°C and nitrate concentration. Temperature could thus be used as a proxy for nitrate concentration, and in particular, the depth of the 19°C-isotherm was a good estimation of the depth of the 10-μM nitrate concentration (Biggs et al., 1988). Within one cyclone sampled in 1996, the nitracline was domed 40 to 60 m shallower than within the LCE that was sampled concurrently (see Figure 6 in Zimmerman and Biggs, 1999). Because this doming facilitated a higher flux of new nitrogen into surface waters in the cyclone than in the anticyclone, the deep chlorophyll maximum (DCM) was locally shallower, and chlorophyll reached higher maximum concentration in the cyclone than in the LCE. Because this resulted in higher standing stocks of chlorophyll in the upper 100 m in the cyclone, the cyclones are regarded as biological "oases," while the interior of the LCEs are biological "deserts." During the recently completed GulfCet II research program, which was co-sponsored by the U.S. Geological Survey and MMS, trawling and bioacoustic survey work showed the cyclone, but not the LCE, had locally higher standing stocks of zooplankton and nekton (Wormuth et al., 2000).

In summary, the interaction of freshwater on the shelf, with circulation regimes over the outer shelf and slope, is a key mechanism that can produce "hot spots" of biological production in the deep-water GOM. Another key mechanism that can create locally high plankton resources is the doming of nutrients within the cyclonic slope eddies, and within the high current velocity shear zones that surround these cyclones and the anticyclonic eddies. So, when and where anticyclonic and cyclonic hydrographic features occur over deep water in the western, central, or eastern GOM, these will play an important role in determining biogeographic patterns and controlling population ecology in the GOM. The fisheries potential of the cyclones and of the frontal zones of both types of eddies is becoming better understood, now that these have been identified as deep water concentrating mechanisms for higher trophic levels and apex predators.

F.   BENTHIC COMMUNITIES

The benthic environment has both a floral and faunal component. The floral component consists of the algae and seagrasses generally seen in shallower waters. The faunal component may be divided into those animals that live in the sediments (infauna), such as burrowing worms, and mollusks, and those animals that live on the sediments (epifauna), such as sponges, corals, starfishes, shrimps, crabs, sea urchins, etc. Substrate type is the single most important factor controlling the distribution of benthic animals. For a more complete description of the benthic flora and fauna in the Gulf of Mexico OCS, see the series of recent lease sale environmental impact statements prepared by MMS (e.g., USDOI, MMS, 1996, 1997a,b, 2002a). Species distribution patterns are summarized in Gallaway (1988) and Pequegnat et al. (1990).

Most of the Gulf of Mexico has a soft, muddy bottom in which burrowing worms are the most abundant animals present. Benthic habitats of environmental concern include low-relief hard bottom and seagrass/algal areas on the inner and middle continental shelves, and areas of high relief hard bottom (topographic features, pinnacles, and ridges) on the outer continental shelf and upper slope. Hard bottom and seagrass areas are described as "live bottom" habitats because they are characterized by high productivity and a diverse animal community. Plants and attached or sessile animals forming these live bottom communities include seagrass, algae, sponges, hydroids, corals, seafans, and anemones. These attached animals and plants provide a habitat that attracts fishes and mobile invertebrates such as crabs and lobster.

West of the mouth of the Mississippi, the Louisiana and Texas continental shelves and shelf edges show many, relatively small, topographic features known to be hard or live bottom habitats. While these features are significant in terms of their vertical relief above the surrounding seafloor, and the great variety of fish and invertebrate species they support, they are small in terms of their overall area. Essentially these features are isolated islands of high epibenthic biomass on the sediment covered continental shelf.

Immediately east of the mouth of the Mississippi, there is an extended area of concentrated high and low relief hard bottom known as the "Pinnacle Trend" along the outer shelf and upper continental slope. This area shows a variety of features from low relief hard bottom up to major pinnacles, as well as scarps, ridges, and relic patch reefs.

The continental shelf to the east of mouth of the Mississippi does not show the isolated topographic features seen along the Texas/Louisiana shelf. In the northeastern Gulf off the Florida panhandle, there are some significant areas of high and low relief hard bottom near the edge of the De Soto Canyon (Shipp and Hopkins, 1978; Barry A. Vittor & Associates, Inc., 1995; Benson et al. 1997; Thompson et al., 1999), but these features have a different geologic origin from the topographic features seen west of the Mississippi. Eastward from the head of the De Soto Canyon and southeastward from Cape San Blas, some of the most extensive seagrass beds in the world are seen in the shallow waters of Florida's Big Bend area. Farther offshore, the west Florida continental shelf is covered by extensive areas of low relief live bottom. The extent of the west Florida continental shelf actually covered by hard or live bottom has been estimated to range from 20 percent (Continental Shelf Associates, Inc., 1992) up to 38 percent (Parker et al., 1983). Data analyzed by Continental Shelf Associates, Inc. (1992) indicated approximately 85 percent of the live bottom seen on the west Florida shelf had a vertical relief of less than 1 m.

There are distinctive differences between the deep-water benthic fauna seen in the eastern and western Gulf of Mexico. Animals seen east of the Lower Mississippi Fan in the north and the eastern Campeche Shelf in the south are representative of the eastern Atlantic and Caribbean faunal regions. Animal groups found west of these features form a unique assemblage and have been called the true Gulf of Mexico benthic community (Pequegnat, 1983; LGL Ecological Research Associates, Inc. and Texas A&M University, 1986). Most benthic animals found on the deep continental slope and the abyssal plain of the Gulf of Mexico are limited to specific depth ranges. These animal groups can be divided into six faunal assemblages, as follows (Pequegnat et al., 1990):

- The Shelf/Slope Transition Zone – 118 to 475 m depth;
- The Archibenthal Zone Horizon A – 500 to 775 m depth;
- The Archibenthal Zone Horizon B – 800 to 975 m depth;
- The Upper Abyssal Zone – 1,000 to 2,275 m depth;
- The Mesoabyssal Zone – 2,300 to 3,225 m depth; and
- The Lower Abyssal Zone – 3,250 to 3,850 m depth.

Chemosynthetic communities are defined as persistent groups of largely attached animals that depend on chemosynthetic bacteria as their primary food source. These communities are unique because they are not dependent upon photosynthesis as the first step in the food chain. Originally discovered around "hot vents" along the mid-oceanic ridges, these communities are generally composed of clams, mussels, and tube worms. Unlike hot vent communities, the chemosynthetic communities of the Gulf of

Mexico are associated with hydrocarbon seeps rather than geothermal vents. Hydrocarbons provide the energy source for these communities rather than superheated mineral rich water. First discovered in 1986, these "cool vent" communities have now been found along much of the continental slope off Texas, Louisiana, and Alabama in depths ranging from 350 to 2,200 m (MacDonald et al., 1990, 1995). Another type of chemosynthetic community has been discovered in the eastern Gulf off the west Florida shelf in approximately 3,270 m of water (Paull et al., 1984). These communities are supported by a cold, deep brine seep coming from the porous limestone carbonate rocks underlying the Florida peninsula. A complete description of these scarp communities may be found in Hecker (1985) and Paull et al. (1988).

II.   SOCIAL AND CULTURAL ENVIRONMENT

A.   COMMERCIAL FISHERIES

Commercial fisheries are very important to the economies of the Gulf coastal states (Browder et al., 1991). The Gulf of Mexico leads all other U.S. regions in fishery production. In 2002, commercial fishery landings in the Gulf of Mexico, which includes western Florida, Alabama, Mississippi, Louisiana, and Texas, were over 1.7 billion pounds worth over $700 million (NMFS, 2003a). Of the individual states, Louisiana led in total landings and value in 2002 with 1.3 billion pounds landed worth $305 million (**Table F-6**). Mississippi was second with landings exceeding 217 million pounds worth $48 million, followed by Texas (93 million pounds, $174 million), Florida's west coast (81 million pounds, $142 million), and Alabama (23 million pounds, $34 million). Pounds landed and dollar value of landings are characteristics that are both used to rank and compare landings by state. Differences in rank (i.e., pounds landed vs. dollar value) reflect differential market prices for the species targeted. For example, while Mississippi landings in 2003 ranked second among the Gulf states by total pounds, the value of that catch ranked fourth.

Table F-6
Pounds and Dollar Values of Landings for Gulf of Mexico Coastal States in 2002
(Source: National Marine Fisheries Service, 2003a)

State	Pounds	Dollars
Florida	81,284,932	142,456,198
Alabama	23,156,340	34,976,372
Mississippi	217,959,716	47,547,418
Louisiana	1,308,530,987	305,534,493
Texas	93,483,598	174,249,714
	1,724,415,573	704,764,195

1.   Target Species

Many species are caught and landed in the Gulf of Mexico commercial fisheries. Browder et al. (1991) stated that the fishery includes at least 97 species from 33 families. They considered the most important species groups to be oceanic pelagic (epipelagic) fishes, reef (hard bottom) fishes, coastal pelagic species, and estuarine dependent species. Primary estuarine dependent species targeted are menhaden, penaeid shrimps (brown, white, and pink), and blue crab; oysters are important, but are not considered here because they are harvested exclusively in inshore waters. Targeted species from the other groups include yellowfin tuna and swordfish (epipelagic); king and Spanish mackerel (coastal pelagic); and spiny lobster, red snapper, red grouper, and gag (reef/hard bottom).

2.   Types of Activity

Each species or species group is caught using various methods and gear types. Shrimps are taken by bottom trawling, menhaden are caught in purse nets, yellowfin tuna are caught on surface longlines, and snapper and grouper are caught by hook and line. **Table F-7** summarizes main fishing practices and seasons in the Gulf of Mexico. **Figure F-2** shows locations of longline sets in the Gulf of Mexico.

3.   Commercial Landings

The top 20 species in terms of pounds landed in 2002 are given by state in **Table F-8**. Menhaden, a small coastal pelagic species, contributed the highest (74.5 percent) proportion of the landings. Shrimps and blue crab were also important, collectively representing over 15 percent of the total 2002 landings. Other species comprising the list included reef fishes (red snapper and red grouper), epipelagic fishes (yellowfin tuna), demersal soft bottom (black drum), and coastal pelagic (mullets and sharks).

Dollar values of Gulf coast 2002 landings are given for the top 20 species in **Table F-9**. The three penaeid shrimp species (brown, white, and pink) were the most valuable species landed in 2002. Other invertebrates such as blue crab, spiny lobster, stone crab, and rock shrimp contributed significantly to the landings value. Valuable finfish landed during 2002 were menhaden, yellowfin tuna, red snapper, red grouper, gag, and striped mullet.

4.   Essential Fish Habitat in the Gulf of Mexico

Most fishery species in the Gulf of Mexico are managed by the GMFMC. This council has prepared fishery management plans (FMPs) for corals and coral reefs, shrimps, stone crab, spiny lobster, reef fishes, coastal pelagic fishes, and red drum. All of these FMPs were recently amended to address Essential Fish Habitat (EFH) for the managed species (GMFMC, 2003). This addendum provides maps and tabular information on the geographical distribution of various life stages of the managed species. Another group of exploited species, the highly migratory pelagic fishes, are managed by NMFS Highly Migratory Species (HMS) Management Division, Office of Sustainable Fisheries. This office recently prepared an FMP (NMFS, 1999a) for Atlantic tunas, swordfish, and sharks that inhabit a broad geographic region that encompasses the Gulf of Mexico. This document also addressed EFH for the managed highly migratory species.

These two documents were consulted to gather information on EFH for the Federal waters of the Gulf of Mexico. Tables were prepared listing those species and life stages whose EFH occurred within the Federal waters of the Gulf. For each species, the tables also indicate whether the habitat for the appropriate life stage is pelagic (oceanic or coastal) or benthic (soft bottom or hard bottom). In some cases (such as corals and some sharks and reef fishes), there was insufficient information available to accurately describe EFH.

Table F-10 presents invertebrate and reef fish species managed by the GMFMC for which EFH has been identified. Corals were not included in the table as there are many soft and hard coral species in the Gulf, but formal EFH descriptions have yet to be made by GMFMC. **Table F-11** presents EFH information for managed species (coastal pelagic fishes and red drum). **Table F-12** gives EFH for HMS such as swordfish, tunas, and sharks managed by NMFS.

Table F-7

Commercial Fishing Methods, Seasons, and Locations in the Gulf of Mexico

Fishing Method	Species Sought	Primary Fishing Season	Primary Fishing Area
Commercial Fishing			
Bottom trawling	Brown shrimp, pink shrimp, white shrimp, seabob, royal red shrimp, and groundfish	Year-round, depending on species and seasonal closures	Soft bottom, shelf waters offshore all Gulf states
Purse netting	Menhaden, butterfish, scads, blue runner, and Spanish sardines	Spring and summer months	Menhaden off Louisiana and Mississippi; scads and sardines off Florida panhandle
Gillnetting	Coastal sharks, mullet, black drum	Spring and summer, depending on species and seasonal closures	
Hook-and-lining (bottom fishing and trolling)	Snappers, groupers, amberjacks, triggerfishes, sharks, king mackerel, Spanish mackerel, and cobia	Year-round; effort varies with species-specific closures	Oil platforms, artificial reefs, and natural hard bottom areas throughout the Gulf
Surface longlining	Sharks, swordfish, tunas, and dolphinfish	Year-round with summer peaks	Open Gulf seaward of 200 m
Bottom longlining	Groupers, snappers, tilefishes, and sharks	Year-round; effort varies with species-specific closures	Outer shelf waters from Florida to Texas on suitable bottom type
Trapping	Spiny lobster, stone crab, and reef fishes	Stone crab (Oct to Mar); spiny lobster (July to March); fish (year round)	Florida shelf waters

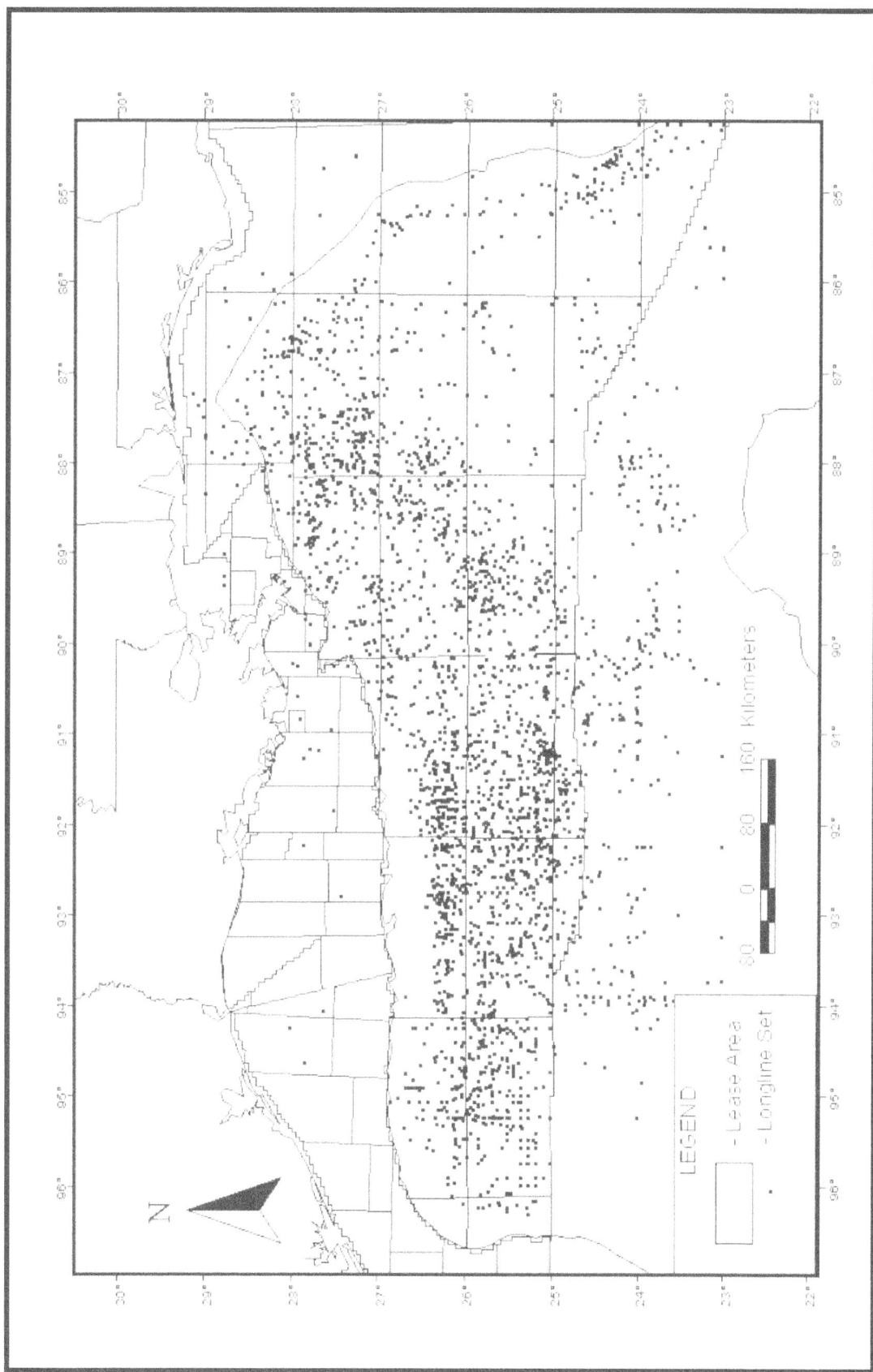

Figure F-2. Spatial distribution of surface longline sets made in the Gulf of Mexico during 1998 (Source: NMFS, 1999b).

Table F-8
Top 20 Species Landed in the Gulf of Mexico during 2002, Ranked by Total Pounds

Species	State					Total	Percent
	Florida	Alabama	Mississippi	Louisiana	Texas		
Menhaden	58,512	979,645	195,370,513	1,087,761,750		1,284,170,420	74.5
Shrimp, brown	639,344	9,378,656	11,135,492	52,685,041	45,835,088	119,673,621	6.9
Shrimp, white	436,538	3,508,688	5,494,536	46,627,069	25,885,312	81,952,143	4.8
Crab, blue	5,485,544	2,572,155	715,814	53,976,318	7,037,012	69,786,843	4.0
Oyster	1,890,946	761,223	2,737,839	13,955,644	5,111,045	24,456,697	1.4
Shrimp, pink	14,953,896	735,330	166,000	45,629	510,261	16,411,116	1.0
Mullet, striped	7,871,642	1,776,050	64,169	2,555,221		12,267,082	0.7
Shrimp, seabob	191,049		1,960	6,801,625	753,721	7,748,355	0.4
Grouper, red	6,937,776	389				6,938,165	0.4
Crab, Florida stone claws	6,383,959			8,130	39,381	6,431,470	0.4
Drum, black	5,963	69,903	8,713	3,117,166	2,329,063	5,530,808	0.3
Snapper, red	946,726	150,664	45,799	2,164,099	1,477,374	4,784,662	0.3
Tuna, yellowfin	410,458			3,409,830	425,645	4,245,933	0.2
Lobster, Caribbean spiny	4,066,671					4,066,671	0.2
Gag	3,055,826	1,129		29,017		3,085,972	0.2
Shrimp, marine, other	1,605,922	123			1,407,105	3,013,150	0.2
Herring, Atlantic thread	2,945,017					2,945,017	0.2
Finfishes, unclassified general	2,295,609				20,906	2,316,515	0.1
Sheepshead	153,801	400,211	46,683	1,583,335	93,705	2,277,735	0.1
Snapper, vermilion	973,874	28,049		754,845	216,569	1,973,337	0.1
TOTAL LANDINGS (Top 20)	61,309,073	20,362,215	215,787,518	1,275,474,719	91,142,187	1,664,075,712	96.5
TOTAL LANDINGS (All species)	81,284,932	23,156,340	217,959,716	1,308,530,987	93,483,598	1,724,415,573	

Source: National Marine Fisheries Service, 2003a.

Table F-9
Top 20 Species Landed in the Gulf of Mexico during 2002 Ranked by Total Dollar Value

Species	State					Total	Percent
	Florida	Alabama	Mississippi	Louisiana	Texas		
Shrimp, brown	1,192,818	17,634,596	17,600,154	60,478,249	86,287,323	183,193,140	26.0
Shrimp, white	815,042	8,386,820	12,020,997	76,202,595	55,661,872	153,087,326	21.7
Menhaden, Atlantic	11,134	101,662	11,625,203	66,419,446		78,157,445	11.1
Oyster, eastern	2,963,802	1,606,308	4,455,647	30,275,812	12,162,706	51,464,275	7.3
Crab, blue	5,139,260	1,489,356	568,419	30,679,950	4,522,532	42,399,517	6.0
Shrimp, pink	28,174,545	1,204,230	243,584	62,144	989,078	30,673,581	4.4
Crab, Florida stone claws	22,868,661			16,915	200,922	23,086,498	3.3
Lobster, Caribbean spiny	18,876,383					18,876,383	2.7
Grouper, red	12,765,652	642				12,766,294	1.8
Tuna, yellowfin	1,077,652			10,340,125	1,179,740	12,597,517	1.8
Snapper, red	2,185,000	363,876	99,567	4,637,149	3,360,825	10,646,417	1.5
Shrimp, marine, other	4,686,711	43			3,898,592	8,585,346	1.2
Mullet, striped (liza)	5,894,306	888,046	21,696	1,680,807		8,484,855	1.2
Gag	7,197,105	2,557		65,704		7,265,366	1.0
Clam, quahog	3,524,322					3,524,322	0.5
Snapper, vermilion	1,772,582	53,961		1,306,307	385,993	3,518,843	0.5
Drum, black	3,647	33,095	5,786	1,615,296	1,815,980	3,473,804	0.5
Shrimp, seabob	56,206		823	2,487,861	297,821	2,842,711	0.4
Snapper, yellowtail	2,586,007			279		2,586,286	0.4
Grouper, yellowedge	1,378,719	61,190		375,950	445,661	2,261,520	0.3
TOTAL LANDINGS (Top 20)	123,169,554	31,826,382	46,641,876	286,644,589	171,209,045	659,491,446	93.6
TOTAL LANDINGS (All Species)	142,456,198	34,976,372	47,547,418	305,534,493	174,249,714	704,764,195	

Source: National Marine Fisheries Service, 2003a.

Table F-10
Managed Species for which Essential Fish Habitat has been Identified in the Gulf of Mexico
(From: Gulf of Mexico Fishery Management Council, 2003)

Species	Life Stages (Reproductive Activity)	Habitat
INVERTEBRATES		
Brown shrimp (*Penaeus aztecus*)	Adults; larvae	Soft bottom; pelagic
White shrimp (*Penaeus setiferus*)	Adults; larvae	Soft bottom; pelagic
Pink shrimp (*Penaeus duorarum*)	Adults; larvae	Soft bottom; pelagic
Stone crab (*Menippe* spp.)	Adults; larvae	Soft bottom; pelagic
Spiny lobster (*Panulirus argus*)	Adults; larvae	Hard bottom; pelagic
REEF FISHES		
Red grouper (*Epinephelus morio*)	Adults and juveniles; eggs and larvae	Hard bottom; pelagic
Gag (*Mycteroperca microlepis*)	Adults and juveniles; eggs and larvae	Hard bottom; pelagic
Scamp (*Mycteroperca phenax*)	Adults and juveniles; eggs and larvae	Hard bottom; pelagic
Red snapper (*Lutjanus campechanus*)	Adults; juveniles; eggs and larvae	Hard bottom; soft bottom; pelagic
Lane snapper (*Lutjanus synagris*)	Adults and juveniles; eggs and larvae	Hard bottom; pelagic
Yellowtail snapper (*Ocyurus chrysurus*)	Adults and juveniles; eggs and larvae	Hard bottom; pelagic
Tilefish (*Lopholatilus chamaeleonticeps*)	Adults and juveniles; eggs and larvae	Soft bottom; pelagic
Greater amberjack (*Seriola dumerili*)	Adults and juveniles; eggs and larvae	Hard bottom; pelagic
Lesser amberjack (*Seriola fasciata*)	Adults and juveniles; eggs and larvae	Hard bottom; pelagic
Gray triggerfish (*Balistes capriscus*)	Adults; eggs; larvae; and juveniles	Hard bottom; pelagic

Table F-11
Managed Species (Coastal Pelagic Fishes and Red Drum) for which Essential Fish Habitat has been
Identified in the Gulf of Mexico (From: Gulf of Mexico Fishery Management Council, 2003)

Species	Life Stages (Reproductive Activity)	Habitat
COASTAL PELAGIC FISHES		
Cobia (*Rachycentron canadum*)	Adults; juveniles/subadults; larvae and eggs	Pelagic
King mackerel (*Scomberomorus cavalla*)	Adults; juveniles/subadults; larvae and eggs (spawning area)	Pelagic
Spanish mackerel (*Scomberomorus maculatus*)	Adults; juveniles/subadults; larvae and eggs (spawning area)	Pelagic
Dolphin (*Coryphaena hippurus*)	Adults; juveniles/subadults; larvae and eggs (spawning area)	Pelagic
RED DRUM		
Red drum (*Sciaenops ocellatus*)	Adults; larvae and eggs (spawning area)	Soft bottom; pelagic

Table F-12
Managed Highly Migratory Species for which Essential Fish Habitat has been Identified in the Gulf of Mexico (National Marine Fisheries Service, 1999a)

Species	Life Stages (Reproductive Activity)	Habitat
SWORDFISH		
Swordfish (*Xiphias gladius*)	Adults; larvae and eggs (spawning area)	Pelagic
TUNAS		
Skipjack tuna (*Katsuwonus pelamis*)	Adults; larvae and eggs (spawning area)	Pelagic
Yellowfin tuna (*Thunnus albacares*)	Adults; juveniles/subadults; larvae and eggs (spawning area)	Pelagic
Bluefin tuna (*Thunnus thynnus*)	Adults; larvae and eggs (spawning area)	Pelagic
SHARKS		
Nurse shark (*Ginglymostoma cirratum*)	Adults; late juveniles/subadults; neonates/early juveniles	Pelagic
Longfin mako shark (*Isurus paucus*)	Adults; late juveniles/subadults; neonates/early juveniles	Pelagic
Blacknose shark (*Carcharhinus acronotus*)	Adults; late juveniles/subadults; neonates/early juveniles	Pelagic
Spinner shark (*Carcharhinus brevipinna*)	Late juveniles/subadults	Pelagic
Silky shark (*Carcharhinus falciformis*)	Adults; late juveniles/subadults; neonates/early juveniles	Pelagic
Bull shark (*Carcharhinus leucas*)	Adults; late juveniles/subadults; neonates/early juveniles	Pelagic
Blacktip shark (*Carcharhinus limbatus*)	Late juveniles/subadults	Pelagic
Dusky shark (*Carcharhinus obscurus*)	Neonates/early juveniles	Pelagic
Caribbean reef shark (*Carcharhinus perezi*)	Adults; late juveniles/subadults	Pelagic
Sandbar shark (*Carcharhinus plumbeus*)	Adults; late juveniles/subadults; neonates/early juveniles	Pelagic
Tiger shark (*Galeocerdo cuvier*)	Adults; late juveniles/subadults; neonates/early juveniles	Pelagic
Lemon shark (*Negaprion brevirostris*)	Adults; late juveniles/subadults; neonates/early juveniles	Pelagic
Scalloped hammerhead (*Sphyrna lewini*)	Adults; late juveniles/subadults	Pelagic
Great hammerhead (*Sphyrna mokarran*)	Adults; late juveniles/subadults	Pelagic
Bonnethead (*Sphyrna tiburo*)	Adults; late juveniles/subadults; neonates/early juveniles	Pelagic
Atlantic sharpnose shark (*Rhizoprionodon terraenovae*)	Adults; late juveniles/subadults; neonates/early juveniles	Pelagic

Although billfish (sailfish [*Istiophorus platypterus*], blue marlin [*Makaira nigricans*], white marlin [*Tetrapterus albidus*], and longbill spearfish [*T. pfluegeri*]) are now considered among the species administered by the NMFS HMS Management Division, there were no EFH designations for these species in NMFS (1999a).

Spatially limited EFH, called habitat areas of particular concern (HAPCs), have also been identified in the Gulf of Mexico by the GMFMC. These include Dry Tortugas (Fort Jefferson National Monument), Florida Keys National Marine Sanctuary, Florida Middle Grounds, and Flower Garden Banks National Marine Sanctuary. All of these HAPCs are important with respect to corals and coral reefs and provide habitats for reef species such as snappers, groupers, and spiny lobster. Recently, two hard bottom areas (i.e., Steamboat Lumps and Madison Swanson) offshore of west Florida near the Florida Middle Grounds have been closed for a minimum of 5 years to fishing for reef fishes by the GMFMC and HMS of NMFS.

EFH includes most of the substrate and water column of the Gulf of Mexico where the managed species commonly occur. There will be some impact by the ensonification of the water column during the various types of geophysical surveys expected for the Gulf of Mexico. The potential effects of sound on fishes are described in detail in the PEA (see **PEA Section III.D - Fishes**) and **Appendix I - Fish Hearing and Sensitivity to Acoustic Impact**. These effects include physiological or anatomical effects on auditory systems, potential behavioral alterations, and auditory masking. Highest energy levels produced by seismic airguns fall within the frequency range from 10 to 200 Hz (up to 1,000 Hz), which is within the audible range for Gulf of Mexico fishes previously discussed (i.e., reef fishes, red drum, coastal pelagic fishes, and highly migratory fishes). Although rigorous supporting data are not available, several studies indicate that seismic shooting can temporarily alter the behaviors and movements of several fish species when received sound pressures are sufficiently high.

B. RECREATIONAL FISHERIES

The primary source for marine recreational fisheries in U.S. waters is the NMFS Marine Recreational Fisheries Statistics Survey (MRFSS). This survey combines random telephone interviews with on-site intercept surveys of anglers to estimate recreational catch and effort for inland, state, and Federal waters. In the Gulf of Mexico, surveys are conducted in western Florida, Alabama, Mississippi, and Louisiana. Texas conducts its own surveys, and these data were not available at this time. MRFSS data for 2002 were obtained from NMFS (2003b). Other recreational fishing information is available in USDOI, MMS (1999, 2002a,b).

An estimated 4 million fishers from Florida, Alabama, Mississippi, and Louisiana engaged in some form of recreational fishing during 2002. These anglers fished from shore, piers, jetties, private/rental boats, party boats, and charter boats. Recreational fishing takes place from inland waters to the open Gulf, with most effort concentrated in coastal and inshore waters.

1. Activity

Of the four states, western Florida had the highest number of anglers and saltwater fishing trips in 2002. Following Florida (in descending order of number of trips) were Louisiana, Alabama, and Mississippi. The following provides the estimated numbers of anglers and trips taken during 2002 by state:

State	Number of Anglers	Number of Trips
Florida	3,349,948	14,418,275
Alabama	355,615	1,190,004
Mississippi	264,717	1,038,353
Louisiana	642,807	3,018,946

The mode of fishing that was most common in all states was private/rental boats, comprising over 50 percent of the effort in each state. This was followed closely by fishing from shore and distantly by fishing from charter/party vessels. The percentage of trips by three fishing modes during 2002 were as follows:

Fishing Mode	Florida	Alabama	Mississippi	Louisiana
Shore	32	44	—	38
Private/Rental Boat	61	48	87	56
Charter/Party Boat	7	8	3	6

Party boats operate mostly from ports in Florida and Alabama, whereas charter boats are found in all coastal states (Continental Shelf Associates, Inc., 1997; USDOI, MMS, 1999, 2002a).

In 2002, the percentage of effort expended in inland, state, and Federal waters varied by state. For Mississippi and Louisiana anglers, most trips were made in inland waters as opposed to state and Federal waters. In Florida and Alabama, the percentage of trips made in state waters was much higher than the other states. Percentage of trips made in inland, state, and Federal waters in 2002 was as follows:

Fishing Area	Florida	Alabama	Mississippi	Louisiana
Inland Waters	53	32	93	88
State Waters	38	44	2	9
Federal Waters	9	24	5	3

Fishing in state and offshore shelf waters often occurs around artificial structures. Off Alabama, Mississippi, Louisiana, and Texas, these structures are oil and gas platforms (Stanley and Wilson, 1990; Continental Shelf Associates, Inc., 1997; USDOI, MMS, 1999, 2002a).

2.   Target Species and Estimated Catches

The top five species commonly caught by recreational fishers in the Gulf coast states are listed in **Table F-13** (NMFS, 2003b). Spotted seatrout, an inshore species, was the most common fish caught by recreational anglers in the Gulf of Mexico during 2002. The estimated catch of spotted seatrout for 2002 was over 22 million fish. The target species varied among states, with Florida being somewhat different than the other three states. This difference reflected the prevalence of hard bottom species such as white grunt in the Florida catches. Recreational fishers in the other three states caught soft bottom species such as red drum and sand seatrout (NMFS, 2003b). The species in **Table F-13** are primarily inshore forms. In offshore oceanic waters of the Gulf of Mexico, commonly sought species include yellowfin tuna, sailfish, blue marlin, dolphin, wahoo, and sharks (Continental Shelf Associates, Inc., 1997). Catch and effort for these epipelagic fishes are much less than for the inshore and shelf species (Continental Shelf Associates, Inc., 1997).

Table F-13
Numbers (in thousands) of the Five Most Frequently Encountered Recreational Fishes
Caught in Gulf Coastal States during 2002

Species	State				Total	Percent
	Florida	Alabama	Mississippi	Louisiana		
Spotted seatrout	12,241,851	359,891	931,348	9,131,830	22,664,920	30.4
Pinfishes	11,014,949	405,401	176,151	246,596	11,843,097	15.9
Saltwater catfishes	4,331,832	534,646	759,604	3,033,847	8,659,929	11.6
Red drum	1,667,636	187,744	177,272	5,319,247	7,351,899	9.9
White grunt	5,529,179				5,529,179	7.4
Sand seatrout	1,488,730	557,934	976,729	1,105,672	4,129,065	5.5
Spanish mackerel	3,675,189	122,158	32,604	52,242	3,882,193	5.2
Atlantic croaker	455,610	654,503	1,143,264	1,336,148	3,589,525	4.8
Red snapper	1,444,231	1,456,339	208,744	87,540	3,196,854	4.3
Kingfishes	762,242	573,733	395,977	278,266	2,010,218	2.7
Black drum	135,775	31,760	22,460	1,396,289	1,586,284	2.1
TOTAL	42,747,224	4,884,109	4,824,153	21,987,677	74,443,163	

Source: National Marine Fisheries Service, 2003b.

C. CULTURAL RESOURCES

1. Prehistoric

Evidence indicates that at the end of the last ice age, approximately 18,000 years before present (YBP), sea level in the Gulf of Mexico was approximately 45 m lower than it is today. Eighteen thousand years ago, large areas of today's continental shelves were dry land. Paleo-indian settlements discovered along the Gulf of Mexico's present coastline have been dated to 12,000 YBP, and it is felt there is a high probability of prehistoric archaeological sites existing on today's continental shelves inshore of the 45-m isobath, and perhaps as deep as the 60-m isobath (USDOI, MMS, 1996, 1997a,b, 2002a).

2. Historic

With the exception of unique specific situations such as the Ship Shoal Lighthouse, historic resources found on the outer continental shelves consist of shipwrecks. These shipwrecks may be modern (20th century), or they may be historic, dating back to the 1500's. Coastal Environments, Inc. (1977) reported 1,589 documented shipwrecks from the Gulf of Mexico, based on literature surveys. Of these reported shipwrecks, only 110, or less than 10 percent, were precisely located. This figure dropped to less than 2 percent for pre-20th century wrecks (Coastal Environments, Inc., 1977). Using additional literature searches, unpublished historical logs, navigational and commercial fishing "hang" data, as well as other sources, Texas A&M University (Garrison et al., 1989) expanded the Coastal Environments, Inc. (1977) list to over 4,000 possible shipwreck locations in the Gulf. Fifteen hundred of the potential wreck sites were located in OCS waters.

Analysis shows that shipwrecks on the Gulf continental shelf cluster in patterns related to harbor entrances and navigational hazards. Once a ship has gone down, the spatial distribution of material or "site integrity," and therefore the potential for preservation, is governed by sea state, water depth, type of bottom, nature of the adjacent coast, strength and direction of currents, and the size and type of construction of the vessel. Garrison et al. (1989) investigated how these factors affect the potential preservation of the wreck sites in the Gulf of Mexico. They concluded that near and to the west of the Mississippi River delta, the potential for wreck preservation was moderate to high. To the east of the delta and along the west Florida shelf, the preservation potential was lower. The preservation potential varied in this manner because of the thick blanket of organically rich sediments deposited by the Mississippi River in the western Gulf that protects and covers wrecks once they settle. Low sedimentation rates are the rule on the west Florida shelf, and wrecks settling there would remain on the seabed, exposed to decay and storm damage from the tropical storms that regularly traverse the area. Considering the problems of inaccurate wreck location reporting, drift and break up of wrecks once on the bottom, and the number of ships that have been lost without being reported sunk, it is difficult to assess the true value of Gulf shipwrecks as an archaeological resource (USDOI, MMS, 1996).

D.   MILITARY USES

The surface and airspace of large areas of the Gulf of Mexico are used extensively by the military for testing, evaluation, training, and qualification of aircraft, vessels, weapon systems, and personnel. Military activities in the Gulf of Mexico have been extensively summarized previously in USDOI, MMS (1997a, 1998, 1999), with the most recent updates outlined in USDOI, MMS (2002a). These activities normally consist of various air-to-air, air-to-surface, and surface-to-surface fleet-training and air force exercises. In addition to carrier operations, air-to-air, air-to-surface, and subsurface operations, the Navy uses the Gulf for shakedown cruises on newly built ships, and for ships completing overhaul or extensive repairs in Gulf shipyards, such as those located in Pascagoula, Mississippi. No aircraft carriers or submarines are currently home-ported or stationed in the Gulf, but may from time-to-time conduct operations there. Military Warning Areas (MWAs) designated by the Federal Aviation Administration (FAA) in the Gulf of Mexico are used by various Air Force, Navy, and Air National Guard units. The airspace over MWAs are FAA Special Use Airspace reserved for military use, and FAA air traffic controllers may prohibit unauthorized aircraft from entering MWAs. The specific boundaries, authorized activities, and military command contacts for each MWA are available in appropriate U.S. Air Force Flight Information Publications and Navy Fleet Training Area Range Manuals (e.g., Department of the Navy, 1998).

Under Department of Defense (DOD) and Department of the Interior (DOI) agreements, lease activities in MWAs may be under Military Areas Stipulations, which include hold and save harmless clauses, electromagnetic emissions restrictions, and operational sections to reduce potential multi-use conflicts. The stipulations promote safety and reduce interference between operations but do not reduce or eliminate the actual physical presence of oil and gas operations in areas where military operations are conducted. Except for a few areas and activities, detailed information on the level of present and future levels of military activity in the MWAs is not available. The lease stipulations and MOAs between DOD and MMS have been effective in avoiding multi-use conflicts over a period of many years.

E.   MARINE TRANSPORTATION

The northern Gulf of Mexico supports an extensive maritime industry transporting goods between the major Gulf ports and ports located outside the Gulf. Gulf waters and adjacent waters, including the Gulf Intracoastal Waterway and large rivers and canals, are heavily used for domestic and foreign waterborne commerce within the Gulf. Cargo ships, bulk freighters, and tankers of various sizes and flags ply its waters, transporting a large range of commodities including crude oil, fuel, petroleum products, raw materials, manufactured goods, food and farm products, and chemicals. In 1998, Gulf ports handled approximately 32 percent of the total domestic U.S. waterborne commerce (based on tonnage) and 49 percent of the total U.S. foreign waterborne commerce (U.S. Department of the Army, Corps of Engineers, 2000).

III.   PHYSICAL ENVIRONMENT

Figure F-3 shows the geographical boundary and major bathymetric features in the Gulf of Mexico. A summary of the Physical Environment is presented in the 2002 Final Environmental Impact Statement (FEIS) for the Outer Continental Shelf Oil and Gas Leasing Program: 2003-2007 (USDOI, MMS, 2002a). Environmental data are also summarized in the FEIS for Gulf of Mexico OCS Oil and Gas Lease Sales 169, 172, 175, 178, and 182 Central Planning Area (USDOI, MMS, 1997a), FEIS for Gulf of Mexico OCS Oil and Gas Lease Sales 171, 174, 177, and 180 Western Planning Area (USDOI, MMS, 1997b), and Draft EIS for Gulf of Mexico OCS Oil and Gas Lease Sales 189 and 197 (USDOI, MMS, 2002b).

Geologically there are seven provinces in the Gulf of Mexico:

1) the Gulf Basin, containing thick sediment sequences and underlain by oceanic crust;
2) the shelf and slope areas of the northeastern Gulf, formed from a subsiding carbonate bank;
3) the South Florida carbonate platform;
4) the Yucatan platform and Campeche Bank;
5) the Isthmain Embayment;
6) the eastern Mexican shelf; and
7) the northwestern Gulf coast geosyncline.

Geologic hazards affecting oil and gas development are primarily associated with seafloor instability in the following areas:

1) at the edge of the continental shelf;
2) in areas of high deposition resulting in gravity faulting;
3) in areas where high deposition rates have led to slumping;
4) in areas of deformation, gravity faulting, and slumping;
5) in areas of high biogenic gas; and
6) in the karst features seen on the west Florida shelf.

Mineral resources currently being commercially extracted from the northern Gulf of Mexico include quartz sand, sulfur, and salt, in addition to oil and natural gas and methane hydrates. Other known mineral resource deposits within coastal waters include phosphate, oyster-shell, limestone, sand and gravel, and magnesium.

F-38

Appendix F

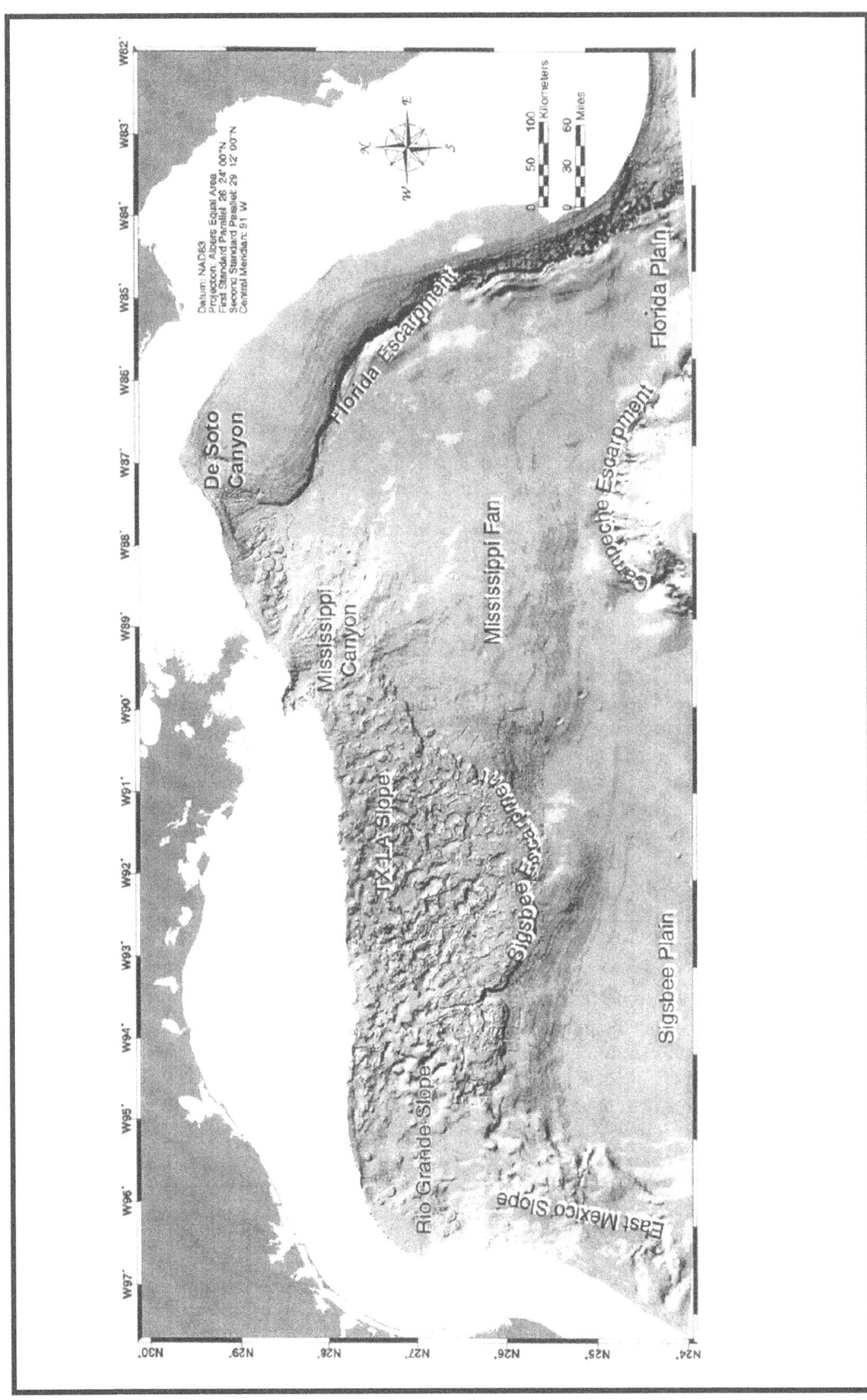

Figure F-3. Geographical boundary and major bathymetric features in the area of the proposed action. Water depths range from the edges of the continental shelves to about 3,700 m in the Sigsbee Plain. This bathymetry is a shaded relief based on multibeam and digitized seismic data (Modified from: Bryant et al., 1990).

Topography, the LC, and the influence of the Mississippi River control currents within the Gulf of Mexico. Coastal and marine water quality is influenced primarily by the extensive river discharge seen here, and to a lesser degree by discharges from offshore activities, primarily OCS oil and gas development and marine transportation.

Climate in the Gulf of Mexico region is controlled by clockwise wind circulation patterns formed around the semipermanent high alternating between the Azores and Bermuda, coupled with the persistent high pressure seen across North America in the winter. Humidity, temperature, precipitation, cloudiness, and visibility are all typical of maritime climates and show little variation from season to season.

Air quality in the coastal areas of the northern Gulf is measured against the National Ambient Air Quality Standards (NAAQS) resulting from the Clean Air Act, as amended, or against the more restrictive standards adopted by some coastal states. For a detailed summary of air quality considerations in the northern Gulf of Mexico, see either the results of specific air quality studies (e.g., USDOI, MMS, 1995) or the Air Quality section of the Final EIS addressing the Outer Continental Shelf Oil and Gas Leasing Program: 2003-2007 (USDOI, MMS, 2002a). Operations to the west of 87.5°W longitude fall under the jurisdiction of the MMS for enforcement of the Clean Air Act, while operations to the east are subject to U.S. Environmental Protection Agency (USEPA) air quality regulations (USDOI, MMS, 2002a).

IV.   NOISE ENVIRONMENT

Ambient noise is defined as typical or persistent environmental background noise, lacking a single source or point. Ambient noise has both horizontal and vertical directionality. In the ocean, there are numerous sources of ambient noise, both natural and manmade, which are variable with respect to season, location, time of day, and noise characteristics (e.g., frequency). Generally, the ambient noise spectral level is about 140 dB re 1 μPa^2 per Hz at 1 Hz and decreases at the rate of 5 to 10 dB per octave to a level of approximately 20 dB re 1 μPa^2 per Hz at 100 kHz (Office of Naval Research [ONR], 1999). Higher frequencies are attenuated with distance from the source more rapidly than lower frequencies. **Figure F-4** illustrates the ambient noise spectra (from Wenz, 1962) in the marine environment. Due to its importance to the sensitivity of instrumentation for research and military applications, ambient noise has been of considerable interest to oceanographers and naval forces. Recent concerns over potential impacts of strong sources of sound from scientific and military activities have driven considerable public and political interest in the issue of noise in the marine environment (National Research Council, 1994, 2000, 2003; Richardson et al., 1995; ONR, 1999).

Natural sources of ambient noise include wind and waves and surf noise, produced by waves breaking on shore. Volcanic and tectonic noise generated by earthquakes on land or in water propagates as low frequency, locally generated "T-phase" waves, with energy levels generally below 100 Hz (Richardson et al., 1995). Biological noises from fishes, certain shrimps (Myrberg, 1978; Dahlheim, 1987; Cato, 1992), and marine mammals can produce sounds at frequencies ranging from approximately 12 to over 100,000 Hz (Richardson et al., 1995).

Sources of ambient noise in the Gulf of Mexico include wind and wave activity, including surf noise near the land-sea interface; precipitation noise from rain and hail; lightning; biological noise from marine mammals, fishes, and crustaceans; and distant shipping traffic (Richardson et al., 1995). Several of these sources may contribute significantly to the total ambient noise at any one place and time, though ambient noise levels above 500 Hz are usually dominated by wind and wave noise. Consequently, ambient noise levels at a given frequency and location may vary widely on a daily basis. A wider range of ambient noise levels occurs in water depths less than 200 m (shallow water) than in deep water. Ambient noise levels in shallow waters are directly related to wind speed and indirectly to sea state (Wille and Geyer, 1984). Bottom conditions also have a strong effect on shallow water ambient noise, with

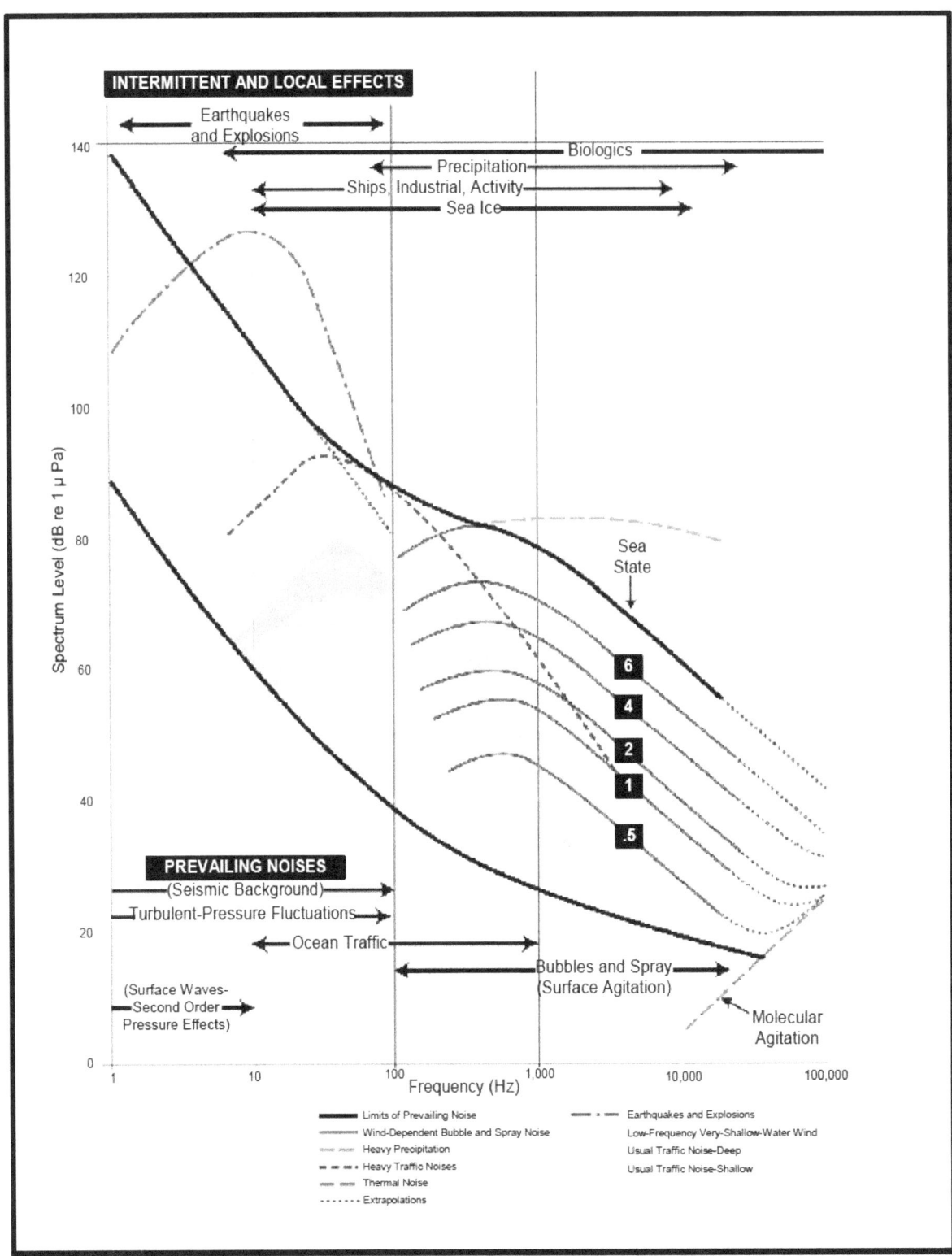

Figure F-4. Wenz curves describing pressure spectral density levels of marine ambient noise
        from weather, wind, geologic activity, and commercial shipping (Adapted from:
        Wenz, 1962 and National Research Council, 2003).

generally higher levels of ambient noise where the bottom is very reflective and low where it is absorptive (Urick, 1983). Ship traffic is a major source of low-frequency ambient noise in the deep ocean, generally dominating frequencies below 500 Hz (frequencies from 10 to 200 Hz).

Table F-14 summarizes the various types of manmade noises in the ocean. Sources include transportation, dredging, construction, hydrocarbon and mineral exploration, geophysical surveys, sonars, explosions, and ocean science studies. Noise levels from most human activities are greatest at relatively low frequencies (less than 500 Hz). Several manmade noise sources may contribute to the total noise at any one place and time (Richardson et al., 1995).

Within the Gulf of Mexico, transportation-derived noise sources include aircraft (both helicopters and fixed-wing aircraft), and surface and subsurface vessels. Underwater sounds from aircraft are transient. The primary sources of aircraft noise are their engine(s) (either reciprocating or turbine) and rotating rotors or propellers. Sound levels from both helicopters and fixed-wing aircraft are at relatively low frequencies (usually below 500 Hz) and are dominated by harmonics associated with the rotating propellers and rotors (Smith, 1989; Hubbard, 1995).

The propagation and levels of underwater noise from passing aircraft are influenced by the altitude and incident angle of the aircraft, water depth, sound receiver depth, bottom conditions, source duration, and aircraft size and type. Peak received noise level in the water, as an aircraft passes overhead, decreases with increasing altitude and increasing receiver depth. At incident angles greater than 13° from the vertical, much of the incident noise from passing aircraft is reflected and does not penetrate the water (Urick, 1972). As mentioned previously, bottom type may strongly affect the reflectivity or absorption of sound. The duration of sound from a passing aircraft is variable, depending on the aircraft type, direction of travel, receiver depth, and altitude of the source (Greene, 1985). Large, multi-engine aircraft tend to be noisier than small aircraft. A four-engine P-3 Orion with multi-bladed propellers has estimated source levels of 160 to 162 dB re 1 μPa-m in the 56-80 Hz range and 148 to 158 dB re 1 μPa-m in the 890 to 1,120 Hz band. A twin-engine Twin Otter generates source levels of 147 to 150 dB re 1 μPa-m at 82 Hz. Helicopters are typically noisier and produce a larger number of acoustic tones and higher broadband noise levels than do fixed-wing aircraft of similar size. Estimated source levels for a Bell 212 helicopter range from 149 to 151 dB re 1 μPa-m (Richardson et al., 1995).

Vessels are the greatest contributors to overall noise in the sea. Sound levels and frequency characteristics of vessel noises underwater are generally related to vessel size and speed. Larger vessels generally emit more sound than smaller vessels do, and those underway with a full load, or those pushing or towing a load, are noisier than unladen vessels. The primary sources of sounds from all machine-powered vessels are related to their machinery and rotating propellers. The frequency of propeller sounds is inversely related to their size. Propeller cavitation is usually the dominant underwater noise source of many vessels (Ross, 1976). Propeller "singing," typically a result of resonant vibration of the propeller blade(s), is an additional source of propeller noise. Noise from propulsion machinery is generated by engines, transmissions, rotating propeller shafts, and mechanical friction. These sources reach the water through the vessel hull. Other sources of vessel noise include a diverse array of auxiliary machinery, flow noise from water dragging along a vessel's hull, and bubbles breaking in the vessel's wake.

In shallow water, shipping traffic more than 10 km away from a receiver generally contributes only to background noise. However, in deep water, traffic noise up to 4,000 km away may contribute to background noise levels (Richardson et al., 1995). Shipping traffic is most significant at frequencies from 20 to 300 Hz. Source levels from a freighter can be 172 dB re 1 μPa-m in the dominant tone of 41 Hz. Large vessels such as tankers, bulk carriers, and containerships can generate 169 to 181 dB re 1 μPa-m, while a very large containership generates as much as 181 to 198 dB re 1 μPa-m. Supertankers generate

Table F-14
General Types of Manmade Sounds in the Ocean and Estimated Level of Maritime Activity
(Adapted from: Richardson et al., 1995 and various sources, as noted)

Activity	Sources	Source Level (dB re 1 μPa-m)	Frequency Range (Hz)	Gulf of Mexico Level of Activity
Transportation	Aircraft (fixed-wing and helicopters)	156-165	45-7,070	Moderate flight activity, estimated to be in the range of several hundred flights annually (most low level flights for oil and gas support, aerial surveys)
	Small vessels (ships, boats)	145-170	37-6,300	High activity level; hundreds to thousands of fishing vessels, pleasure craft, small ships daily; millions of angler trips per year (**Section II.B - Recreational Fisheries** of this appendix); oil and gas support vessel activity estimated to be 304,807 to 319,921 trips per year, with most concentrated in the Central Planning Area.
	Large vessels	169-198	6.8-428	In the U.S. Gulf of Mexico in 1999, tankers and other freight vessels completed a total of approximately 279,000 vessel trips in Gulf and Gulf Intracoastal Waterway (GIWW) waters
	Ice breakers	171-191	10-1,000	None
	Hovercraft and vehicles on ice	130	224-7,070	None; related watercraft would include "jet skis," whose numbers are estimated to range into the thousands
Dredging and Construction	Dredging	150-180	10-1,000	Precise levels unknown, although harbor maintenance activity is very common for major Gulf ports; very limited in shipping channels
	Tunnel boring	Low	10-500	Unknown; expected to be rare in the Gulf of Mexico
	Other construction operations	Low	<1,000	Unknown; expected to be limited in the Gulf of Mexico
Oil and Gas Drilling and Production	Drilling from islands and caissons	70-100	10-160	None in the Gulf of Mexico
	Drilling from bottom-founded platforms	119-127 (received)	5-1,200	Variable; may range from tens to hundreds of wells drilled from Gulf platforms annually; January 2001 drilling activity levels: 61 wells. MMS notes 40,361 approved applications to drill in Gulf of Mexico Federal waters
	Drilling from vessels	154-191	10-10,000	Low level of activity, on the order of tens of drill ships operating in Gulf waters annually
	Offshore oil and gas production	Low	50-500	4,019 production platforms on 7,564 active leases in Federal waters of the Gulf of Mexico, as of 2 September 2003; as of 31 July 2001, there were 3,476 active offshore production platforms in GOM Federal waters.
	Support activity	See small vessels	See small vessels	304,807 to 319,921 trips per year, with most (~90%) concentrated in the Central Planning Area; ~10% of support vessel activity occurs in the Western Planning Area, while 0.2% to 0.3% is projected for the Eastern Planning Area

Table F-14
General Types of Manmade Sounds in the Ocean and Estimated Level of Maritime Activity
(Adapted from: Richardson et al., 1995 and various sources, as noted)
(Continued)

Activity	Sources	Source Level (dB re 1 µPa-m)	Frequency Range (Hz)	Gulf of Mexico Level of Activity
Geophysical Surveys	Airguns	216-259	<120	Tens to 30+ surveys per year; may have as many as five surveys running concurrently (see **Appendix D, Section V - Level of Activity**)
	Sleeve exploders and gas guns	217	Low	Unknown; expected to be very rare
	Vibroseis	187->210; instantaneous level dependent upon sweep length (i.e., ~18-22 dB less than an airgun pulse)	10-70	Estimated to be rare (see **Appendix D, Section II.D - Vibroseis**)
	Other techniques (sparkers, boomers)	212-221	N/A	Estimated to be rare
Navigation and Target Detection (sonars, pingers)	Fathometers	180+	12,000+	Potentially high, given the presence of thousands of ships and boats in the Gulf
	Military active sonars	230+	4-1,000	Unknown; expected to be periodic, infrequent (e.g., tens to a hundred or more annually)
	Transponders	180-200	7,000-60,000	Unknown; expected to be periodic, infrequent (e.g., several hundred per year)
Explosions	Military ordnance Ship and weapons testing	>279 >294 (10,000 lb charge)	Peak Broadband	Low; live fire testing very limited in the Gulf of Mexico Periodic, infrequent
	Offshore demolition (structure removals)	267-279 (based on charge weights)	Peak	53-130 removals per year
Ocean Science Studies	Seismology	N/A	N/A	Unknown, expected to be very limited
	Acoustic propagation	220	Broadband	None
	Acoustic tomography	N/A	N/A	Unknown, expected to be very limited
	Acoustic thermometry	195	Broadband	None

peak sources levels of 185 to 190 dB re 1 µPa-m at about 7 Hz. At frequencies of 20 to 60 Hz, supertankers generate a source level of 160 dB re 1 µPa-m (Richardson et al., 1995).

Coastal commercial shipping traffic is also a source of noise, producing noise of 150 to 170 dB re 1 µPa-m at frequencies below 1,000 Hz. A tug pulling a barge generates 164 dB re 1 µPa-m when empty and 170 dB re 1 µPa-m loaded. A tug and barge underway at 18 km/h can generate broadband source levels of 171 dB re 1 µPa-m. A small crewboat produces 156 dB re 1 µPa-m at 90 Hz. A small boat with an outboard engine generates 156 dB re 1 µPa-m at 630 Hz, while an inflatable boat with a 25-hp outboard engine produces 152 dB re 1 µPa-m at 6,300 Hz (Richardson et al., 1995).

Fishing in coastal regions also contributes sound to the overall ambient noise. Sound produced by these smaller boats is typically at a higher frequency, around 300 Hz. A 12-m long fishing boat, underway at 7 knots, generates 151 dB re 1 µPa-m in the 250 to 1,000 Hz range. Trawlers generate source levels of 158 dB re 1 µPa-m at 100 Hz (Richardson et al., 1995).

Marine dredging and construction activities are common within the coastal waters of the Gulf of Mexico. Underwater noises from dredge vessels are typically continuous in duration (for periods of days or weeks at a time) and strongest at low frequencies. Marine dredging sound levels vary greatly, depending upon the type of dredge (Greene, 1985, 1987). Source levels from marine dredging operations range from 150 to 180 dB re 1 µPa-m between 10 to 1,000 Hz. A clamshell dredge generates broadband source levels of ~167 dB re 1 µPa-m while pulling a loaded clamshell back to the surface. Sounds from various onshore construction activities vary greatly in levels and characteristics. These sounds are most likely within shallow water. Onshore construction activities may also propagate into coastal waters, depending upon the source and ground material (Richardson et al., 1995).

Offshore drilling and production involves a variety of activities that produce underwater noises. Noises emanating from drilling activities from fixed, metal-legged platforms are considered not very intense and generally are at very low frequencies, near 5 Hz. Gales (1982) reported received levels of 119 to 127 dB re 1 µPa-m at near-field measurements. Noises from semisubmersible platforms also show rather low sound source levels. Drillships show somewhat higher noise levels than semisubmersibles as a result of mechanical noises generated through the drillship hull. The drillship *Canmar Explorer II* generates broadband source levels of 174 dB re 1 µPa-m. Noises associated with offshore oil and gas production are generally weak and typically at very low frequencies (~4.5 to 38 Hz) (Gales, 1982). The specialized ice-strengthened floating platform *Kulluck* produced broadband (10 to 10,000 Hz) source levels of 191 dB re 1 µPa-m while drilling and 179 dB re 1 µPa-m while tripping. Support activity associated with oil and gas operations such as supply/anchor handling and crewboats and helicopters also contribute to the noise from offshore activity.

Marine geophysical (seismic) surveys are commonly conducted to delineate oil and gas reservoirs below the surface of the land and seafloor. These operations direct high-intensity, low frequency sound waves through layers of subsurface rock, which are reflected at boundaries between geological layers with different physical and chemical properties. The reflected sound waves are recorded and processed to provide information about the structure and composition of subsurface geological formations (McCauley, 1994). In an offshore seismic survey, a high energy sound source is towed at slow speed behind a survey vessel. The sound source typically used is an airgun, a pneumatic device that produces acoustic output through the rapid release of a volume of compressed air. The airgun is designed to direct the high energy bursts of low-frequency sound (termed a "shot") downward towards the seafloor. Airguns are usually used in sets, or arrays, rather than singly (McCauley, 1994). Reflected sounds from below the seafloor are received by an array of sensitive hydrophones on cables (collectively termed "streamers") that are either towed behind a survey vessel or attached to cables placed on or anchored to the seafloor. A summary of G&G activities being conducted in the Gulf of Mexico is provided in **Appendix D, Section I - Description of Activities**.

Sounds produced by seismic pulses can be detected by mysticetes and odontocetes from 10 to 100 km from the source (Greene and Richardson, 1988; Bowles et al., 1994; Richardson et al., 1995). Airgun arrays are the most common source of seismic survey noise. A typical full-scale array produces a source level of 248 to 255 dB re 1 µPa-m, zero-to-peak (Barger and Hamblen, 1980; Johnston and Cain, 1981). Typical seismic arrays being used in the Gulf produce source levels (sound pressure levels) of approximately 240 dB re 1 µPa (see **Appendix D, Section II.A - Airguns**). While the seismic airgun pulses are directed towards the ocean bottom, sound propagates horizontally for several kilometers (Greene and Richardson, 1988; Hall et al., 1994). In waters 25 to 50 m deep, sound produced by airguns can be detected 50 to 75 km away, and these detection ranges can exceed 100 km in deeper water (Richardson et al., 1995).

Active sonars are used for the detection of objects underwater. These range from depth-finding sonars (fathometers), found on most ships and boats, to powerful and sophisticated units used by the military. Sonars emit transient, and often intense, sounds that vary widely in intensity and frequency. Unlike most other manmade noises, sonar sounds are mainly at moderate to high frequencies that attenuate much more rapidly than lower frequencies (Richardson et al., 1995). Acoustic pingers used for locating and positioning of oceanographic and geophysical equipment also generate noise at high frequencies.

Underwater explosions in open waters are the strongest point sources of anthropogenic sound in the Gulf of Mexico. Sources of explosions include both military testing and non-military activities such as offshore structure removals. Explosives produce rapid onset pulses (shock waves) that change to conventional acoustic pulses as they propagate. Even a small 0.5-kg charge of TNT generates broadband source levels of 267 dB re 1 µPa-m, while a 20-kg charge of TNT produces 279 dB re 1 µPa-m. Detonation of very large charges during ship shock tests produces source levels of more than 294 dB re 1 µPa-m (Richardson et al., 1995).

To assess the availability of data for developing a baseline on ambient noise levels and their sources in the Gulf of Mexico, a literature and information search was conducted. An electronic search for publications on ocean or marine noise was conducted through the Harbor Branch Oceanographic Institution library. Searches were also conducted through Internet search engines. Available literature on ocean noise, including relevant environmental impact statements, environmental assessments, environmental reports, conference proceedings, and National Research Council reports, were reviewed. Researchers and experts on ocean noise and acoustics were consulted for information (see **PEA Section IV**), including Federal agencies (e.g., Marine Mammal Commission [MMC], NMFS, ONR), academic institutions (e.g., Texas A&M University, Massachusetts Institute of Technology [MIT], Woods Hole Oceanographic Institution [WHOI], Cornell Laboratory of Ornithology), and members of the private sector (e.g., LGL, Ltd., Environmental Research Associates).

Published efforts on long-term monitoring of ambient noise in the ocean is limited. The VENTS program of the Pacific Marine Environmental Laboratory (PMEL) has monitored underwater seismic activity and large whale migrations in the East Pacific Rise, the central North Atlantic, and the Gulf of Alaska through hydrophones and acoustic recorders. Some of the East Pacific Rise moorings have been in place since 1996 (C. Fox, PMEL, oral comm., 2001; http://www.pmel.noaa.gov/vents/acoustics/ seismicity/deploy-info.html). The multi-year deployments of hydroacoustic recorders have resulted in some surprising results, such as detection of geophysical activity near Sable Island, Canada, Brazil, and West Africa in hydroacoustic recorders located in the central North Atlantic. The Bioacoustics Research Program of the Cornell Laboratory of Ornithology has deployed bottom-mounted "pop-up" hydroacoustic recorders in several sites in the Pacific and Atlantic Oceans on short-term (up to a few months) deployments (K. Fristrup, Cornell Laboratory of Ornithology, oral comm., 2001; http://www.birds.cornell.edu/brp/popup.html). Deployments were in the following areas:

- Southern California Bight;
- Point Buchon, California;
- North side of Kauai, Hawaii;
- West of the Shetland Islands;
- Ligurian Sea, Italy;
- Gulf of California;
- Onslow Bay, North Carolina;
- Point Barrow, Alaska;
- Offshore Cape Cod, Massachusetts; and
- Monterey Bay, California.

Short-term measurements of ambient noise have also been made in the Stellwagen National Marine Sanctuary and the St. Lawrence Seaway by Peter Scheifele of the National Undersea Research Center at the University of Connecticut. These short-term ambient noise measurements were aimed at studying impacts on humpback whales and beluga whales (http://www.sustainableseas.noaa.gov/missions /stellwagen1/background/sea_sounds.html).

Based on the literature search and interviews, available data and published reports on which to develop baseline information on ambient noise, specifically for the Gulf of Mexico, are extremely limited. There have only been a few efforts to monitor ambient noise in the Gulf of Mexico that were usually associated with acoustic surveys for marine mammals. These include the multi-year study of Gulf of Mexico cetaceans (GulfCet) program sponsored by the MMS that involved acoustic surveys for marine mammals in 1996 and 1997 (Davis and Fargion, 1996; Davis et al., 2000). During GulfCet II cruises, noise from geophysical activity was monitored during acoustic surveys for marine mammals. However, ambient noise from distant, continuous sources was filtered out and is not part of the recordings (S. Rankin, NMFS, Southwest Fisheries Science Center, oral comm., 2001). More recently, MMS has sponsored surveys for marine mammals and sea turtles on board NOAA research vessels conducting annual plankton cruises in 2000 and 2001. The month-long summer 2000 cruise on the *R/V Gordon Gunter* was a pilot study to work out novel approaches and methods to conduct marine mammal surveys using a variety of visual and acoustic techniques and multidisciplinary teams. The 2000 and 2001 cruises included acoustic tracking of endangered sperm whales near the mouth of the Mississippi River (A. Thode, MIT, oral comm., 2001).

Establishing background levels or a baseline on ambient noise for the Gulf of Mexico will be a complex effort. The ambient noise levels in the Gulf vary strongly with position, due to changes in water depth, possibly bottom composition, and due to manmade background activities. Values from one area of the Gulf of Mexico cannot be used somewhere else without accounting for these factors (A. Thode, MIT, oral comm., 2001).

Continuous recordings of background noise in the Gulf of Mexico, using towed arrays at different tow speeds, have been made during the sperm whale surveys, so in theory, the noise levels at "zero" tow speed can be extrapolated. At present, the acoustic data require further processing after the array is calibrated. Once the array is calibrated and the data processed, the data can be used to address many ambient noise questions. Samples of background noise when it is raining and various types of ship traffic in the Gulf of the Mexico have been collected (A. Thode, MIT, oral comm., 2001).

As scientific research efforts have been limited, the U.S. Navy and petroleum industry are likely the main sources of acoustic data for the Gulf of Mexico (R. Gentry, NMFS, oral comm., 2001; C. Fox, PMEL, oral comm., 2001; P. Tyack, WHOI, oral comm., 2001; R. Gisiner, ONR, oral comm., 2001). The Naval Oceanographic Office (NAVOCEANO) in Stennis Space Center, Mississippi is the data repository for a large amount of acoustic data from Navy sonobuoys, submarines, and surface vessels spanning 30 years. Long-term acoustic records from the U.S. Navy's Sound Surveillance System (SOSUS) network were not kept and are not part of the archive. Even if these records had been kept, due to national security concerns, these data would not be available to the public and researchers (C. Fox, PMEL, oral comm., 2001). Access to the acoustic data archive may be possible through negotiations with the U.S. Navy (R. Gisiner, ONR, oral comm., 2001). Similarly, due to the proprietary nature of the data, acoustic data would not be available to the public and researchers from the large number of geophysical seismic surveys that have been done in the Gulf of Mexico.

The National Research Council has created a panel on ambient noise chaired by George Frisk of WHOI to review the issue of ambient noise, define the state of the knowledge, and develop policy and research recommendations. As part of their assessment, the panel will examine the feasibility of using NAVOCEANO's acoustic or ambient noise archives to develop baseline levels or sound budgets in the oceans. The panel has had its first meeting of three planned meetings, and a report from the panel is expected within a year (R. Gentry, NMFS, oral comm., 2001). One of the recommendations arising from the 1998 Workshop on Effects of Anthropogenic Noise in Marine Environment was developing a total sound budget of the ocean and its principal constituent parts (ONR, 1999).

The available measurements of ambient noise in the Gulf of Mexico are limited and inadequate for developing a baseline on ambient noise. Further, currently there is no ambient noise monitoring effort in the Gulf of Mexico that could produce the data required for a reasonable baseline with some degree of temporal and spatial coverage. There have not been any bottom-mounted acoustic recorders reported in the Gulf of Mexico, unlike in the Pacific and Atlantic Oceans where researchers have established bottom-mounted acoustic recorders to monitor oceanic phenomena, such as underwater seismic and volcanic activity and marine mammals. There have been efforts in the Gulf of Mexico to record ambient noise associated with marine mammal surveys such as the GulfCet Program (S. Rankin, NMFS, Southwest Fisheries Science Center, oral comm., 2001) and recent 2000 and 2001 sperm whale surveys aboard the *R/V Gordon Gunter*. Acoustic recordings were made in 2000 and 2001 on a Norris array similar to the towed array used by Jay Barlow (A. Thode, MIT, oral comm., 2001). MMS has been considering the funding of bottom-mounted recorders in the Gulf of Mexico, but none have been deployed to date (P. Tyack, WHOI, oral comm., 2001). The technical issues surrounding the long-term monitoring of ambient noise are generally resolved, and costs are likely the issue. A bottom-mounted hydroacoustic recorder is estimated to cost $40,000 per mooring (C. Fox, PMEL, oral comm., 2001). Due to the large numbers of offshore platforms in the Gulf of Mexico, the cost of installing and maintaining hydroacoustic recorders can be substantially reduced (R. Gisiner, ONR, oral comm., 2001).

At present, ambient noise monitoring in the Pacific and Atlantic Oceans is being conducted by PMEL and Cornell University using bottom-mounted recorders (C. Fox, PMEL, oral comm., 2001; K. Fristrup, Cornell Laboratory of Ornithology, oral comm., 2001). The Coastal Marine Mammal Program in the NMFS Southwest Fisheries Science Center is conducting marine mammals surveys in the Eastern Pacific called ORCAWALE 2001, using a towed hydrophone array system developed by Jay Barlow. The array has been used to estimate sperm whale abundance in the eastern temperate North Pacific, identify tropical Pacific delphinid species from whistles, and examine the fraction of marine mammals not detected by visual observers (see http://swfsc.nmfs.noaa.gov/sars).

V.   LITERATURE CITED

Atkinson, L.P., and T.E. Targett. 1983. Upwelling along the 60-m isobath from Cape Canaveral to Cape Hatteras and its relationship to fish distribution. Deep-Sea Res. 30:221-226.

Barger, J.E., and W.R. Hamblen. 1980. The air gun impulsive underwater transducer. J. Acoust. Soc. Am. 68(4):1038-1045.

Barry A. Vittor & Associates, Inc. 1985. Tuscaloosa Trend regional data search and synthesis study, synthesis report, Volume I. OCS Study MMS 85-0056. U.S. Department of the Interior, Minerals Management Service, Gulf of Mexico OCS Regional Office, New Orleans, LA.

Barry A. Vittor & Associates, Inc. 1995. Photodocumentation report for Destin Dome Area Block 57, wellsite 57 No. 1. Report prepared for Chevron U.S.A. Production Company, New Orleans, LA. 40 pp. + app.

Benson, J.B., W.W. Schroeder, and A.W. Shultz. 1997. Sandstone hardbottoms along the western rim of the De Soto Canyon, northeastern Gulf of Mexico. Gulf Coast Assoc. Geo. Soc. Trans. XLVII:43-48.

Biggs, D.C. 1992. Nutrients, plankton, and productivity in a warm-core ring in the western Gulf of Mexico. J. Geophys. Res. 97:2143-2154.

Biggs, D.C. 2000. Environmental patterns and oceanographic processes: hydrography and mesoscale circulation of the continental margin of the northeastern Gulf of Mexico, 1996 and 1997 (Chapter 2), pp. 15-68. *In:* Davis, R.W., W.E. Evans, and B. Würsig (eds.), Cetaceans, Sea Turtles and Seabirds in the Northern Gulf of Mexico: Distribution, Abundance and Habitat Associations. Volume II: Technical Report. Prepared by Texas A&M University at Galveston and the National Marine Fisheries Service. U.S. Department of the Interior, U.S. Geological Survey, Biological Resources Division, USGS/BRD/CR-1999-005 and Minerals Management Service, Gulf of Mexico OCS Region, New Orleans, LA OCS Study MMS 2000-003.

Biggs, D.C., and F.E. Müller-Karger. 1994. Ship and satellite observations of chlorophyll stocks in interacting cyclone-anticyclone eddy pairs in the western Gulf of Mexico. J. Geophys. Res. 99:7371-7384.

Biggs, D.C., and L.L. Sanchez. 1997. Nutrient enhanced primary productivity of the Texas-Louisiana continental shelf. J. Mar. Sys. 11:237-247.

Biggs, D.C., A.C. Vastano, R.A. Ossinger, A. Gil-Zurita, and A. Perez-Franco. 1988. Multidisciplinary study of warm- and cold-core rings in the Gulf of Mexico. Mem. Soc. Ciencias Nat. Venezuela 48(3):11-31.

Bowles, A.E., M. Smultea, B. Würsig, D.P. DeMaster, and D. Palka. 1994. Relative abundance and behavior of marine mammals exposed to transmissions from the Heard Island Feasibility Test. J. Acoust. Soc. Am. 96(4):2,469-2,484.

Brandon, E.A., and G.S. Fargion. 1993. Mesoscale temperature features and marine mammals in the Gulf of Mexico. Tenth Biennial Conference on the Biology of Marine Mammals, Abstracts, Galveston, TX, p. 31.

Brooks, J.M. (ed). 1991. Mississippi-Alabama continental shelf ecosystem study: data summary and synthesis. Volume II: Technical narrative. OCS Study MMS 91-0063. U.S. Department of the Interior, Minerals Management Service, Gulf of Mexico OCS Regional Office, New Orleans, LA. 862 pp.

Browder, J., B. Brown, W. Nelson, and A. Bane. 1991. Multispecies fisheries in the Gulf of Mexico. ICES mar. Sci. Symp.:194-197.

Bryant, W.R., J.R. Bryant, M.H. Feeley, and G.S. Simmons. 1990. Physiography and bathymetric characteristics of the continental slope, Gulf of Mexico. Geo-Marine Letters 10:182-199.

Cato, D.H. 1992. The biological contribution to the ambient noise in waters near Australia. Acoust. Austral. 20(3):76-80.

Chittenden, M.E., and J.D. McEachran. 1976. Composition, ecology, and dynamics of demersal fish communities on the northwestern Gulf of Mexico continental shelf, with a similar synopsis for the entire Gulf. Texas A&M University Sea Grant Publication TAMU-SG-76-298.

Coastal Environments, Inc. 1977. Cultural resources evaluation of the northern Gulf of Mexico continental shelf. Prepared for the Interagency Archeological Service, Office of Archeology and Historic Preservation, National Park Service, U.S. Department of the Interior. Baton Rouge, LA. 4 vol.

Coleman, F.C., Koenig, C.C., and L.A. Collins. 1996. Reproductive styles of shallow water groupers (Pisces: Serranidae) in the eastern Gulf of Mexico and the consequences of fishing spawning aggregations. Environ. Biol. Fishes 47:129-141.

Collard, S. 1990. Leatherback turtles feeding near a water mass boundary in the eastern Gulf of Mexico. Marine Turtle Newsletter 50:12-14.

Continental Shelf Associates, Inc. 1992. Compilation of existing data on the location and areal extent of reef fish habitat on the Mississippi/Alabama/Florida continental shelf - eastern Gulf of Mexico. Report prepared for the U.S. Dept. of Commerce, National Oceanic and Atmospheric Administration, National Marine Fisheries Service, St. Petersburg, FL. 25 pp.

Continental Shelf Associates, Inc. 1997. Characterization and trends of recreational and commercial fishing from the Florida panhandle. U.S. Department of Interior, Minerals Management Service, Gulf of Mexico OCS Region, New Orleans, LA. USGS/BRD/CR-1997-0001 and OCS Study MMS-97-0020. 333 pp.

Dagg, M.J. 1995. Copepod grazing and the fate of phytoplankton in the northern Gulf of Mexico. Contl. Shelf Res. 15:1303-1310.

Dagg, M.J., C. Grimes, S. Lohrenz, B. McKee, R. Twilley, and W. Wiseman. 1991. Continental shelf food chains of the northern Gulf of Mexico, pp. 329-345. *In:* Food Chains, Yields, Models, and Management of Large Marine Ecosystems (K. Sherman, L.M. Alexander, and B.D. Gold, eds.). Westview Press, Boulder, CO.

Dahlheim, M.E. 1987. Bio-acoustics of the gray whale (*Eschrichtius robustus*). Ph.D. Thesis, University of British Columbia, Vancouver, B.C. 315 pp.

Darnell, R.M., and J.A. Kleypas. 1987. Eastern Gulf shelf bio-atlas, a study of the distribution of demersal fishes and penaeid shrimp of soft bottoms of the continental shelf from the Mississippi River Delta to the Florida Keys. OCS Study MMS 86-0041. U.S. Department of the Interior, Minerals Management Service, Gulf of Mexico OCS Regional Office, New Orleans, LA.

Darnell, R.M., R.E. Defenbaugh, and D. Moore. 1983. Northwestern Gulf shelf bio-atlas, a study of the distribution of demersal fishes and penaeid shrimp of soft bottoms of the continental shelf from the Rio Grande to the Mississippi River Delta. Open File Report 82-04. U.S. Department of the Interior, Minerals Management Service, Gulf of Mexico OCS Regional Office, New Orleans, LA.

Davis, R.W., and G.S. Fargion (eds.). 1996. Distribution and abundance of cetaceans in the north-central and western Gulf of Mexico: Final Report. Volume II: Technical Report. OCS Study MMS 96-0027. Prepared by the Texas Institute of Oceanography and the National Marine Fisheries Service. U.S. Dept. of the Interior, Minerals Management Service, Gulf of Mexico OCS Region, New Orleans, LA. 357 pp.

Davis, R.W., W.E. Evans, and B. Würsig (eds.). 2000. Cetaceans, Sea Turtles and Seabirds in the Northern Gulf of Mexico: Distribution, Abundance and Habitat Associations. Volume II: Technical Report. Prepared by Texas A&M University at Galveston and the National Marine Fisheries Service. U.S. Department of the Interior, U.S. Geological Survey, Biological Resources Division, USGS/BRD/CR-1999-005 and Minerals Management Service, Gulf of Mexico OCS Region, New Orleans, LA OCS Study MMS 2000-003.

Dennis, G.D., and T.J. Bright. 1988. Reef fish assemblages on hard banks in the northwestern Gulf of Mexico. Bull. Mar. Sci. 43(2):280-307.

Department of the Navy. 1998. Fleet Training Area/Range Directory. Naval Warfare Assessment Station, Coronado, CA. 4105 Ser SE 40/122 (CD-ROM).

Ditty, J.G. 1986. Ichthyoplankton in neritic waters of the northern Gulf of Mexico off Louisiana: Composition, relative abundance, and seasonality. Fish. Bull. 84(4):935-946.

Ditty, J.G., G.G. Zieske, and R.F. Shaw. 1988. Seasonality and depth distribution of larval fishes in the northern Gulf of Mexico above 26°00'N. Fish. Bull. 86:811-823.

Dodd, C.K., Jr. 1988. Synopsis of the Biological Data on the Loggerhead Turtle, *Caretta caretta* (Linnaeus, 1758). U.S. Fish and Wildlife Service Biol. Rep. 88. 100 pp.

El-Sayed, S.Z., W.M. Sackett, L.M. Jeffrey, A.D. Fredericks, R.P. Saunders, P.S. Conger, G.A. Fryxell, K.A. Steidinger, and S.A. Pearle. 1972. Primary productivity and standing crop of phytoplankton, pp. 8-13. *In:* V.C. Bushnell (Editor), Chemistry, Primary Productivity, and Benthic Algae of the Gulf of Mexico. American Geographical Society, New York.

Ernst, C.H., R.W. Barbour, and J.E. Lovich. 1994. Turtles of the United States and Canada. Smithsonian Institute Press, Washington, DC. 578 pp.

Fritts, T.H., and R.P. Reynolds. 1981. Pilot study of the marine mammals, birds, and turtles in the OCS areas of the Gulf of Mexico. Rept. FWS/OBS-81/36. U.S. Fish and Wildlife Service, Washington, D.C. 139 pp.

Fritts, T.H., W. Hoffman, and M.A. McGehee. 1983a. The distribution and abundance of marine turtles in the Gulf of Mexico and nearby Atlantic waters. J. Herpetol. 17:327-344.

Fritts, T.H., A.B. Irvine, R.D. Jennings, L.A. Collum, W. Hoffman, and M.A. McGehee. 1983b. Turtles, birds, and mammals in the northern Gulf of Mexico and nearby Atlantic waters. Final report for the U.S. Department of the Interior, U.S. Fish and Wildlife Service, Division of Biological Services, Washington, D.C. FWS/OBS-82/65. 455 pp.

Gales, R.S. 1982. Effects of noise of offshore oil and gas operations on marine mammals-An introductory assessment. U.S. Naval Ocean Systems Center, San Diego, CA. NOSC TR 844, 2 vols. NTIS AD-A123699 and AD-A123700. 79 pp. and 300 pp.

Gallaway, B.J. 1988. Northern Gulf of Mexico Continental Slope Study, Final Report, Year 4. Volume II: Synthesis Report. Prepared for the U.S. Department of the Interior, Minerals Management Service, Gulf of Mexico OCS Regional Office, New Orleans, Louisiana. OCS Study MMS 88-0053.

Gallaway, B.J., and M.C. Kennicutt II. 1988. Characterization of benthic habitats of the northern Gulf of Mexico, Chapter 2. In: B. J. Gallaway (ed.), Northern Gulf of Mexico continental slope study, Final report, Year 4. Volume II: Synthesis Report. U.S. Department of the Interior, Minerals Management Service, Gulf of Mexico OCS Region, New Orleans, LA. OCS Study MMS 88-0053.

Garrison, E.G., C.P. Giammona, F.J. Kelly, A.R. Tripp, and G.A. Wolff. 1989. Historic shipwrecks and magnetic anomalies of the northern Gulf of Mexico: reevaluation of archaeological resource management zone. U.S. Department of the Interior, Minerals Management Service, Gulf of Mexico OCS Region, New Orleans, LA. 3 vol. OCS Study MMS 89-0023, 89-0024, and 89-0025.

Govoni, J.J., D.E. Hoss, and D.R. Colby. 1989. The spatial distribution of larval fishes about the Mississippi River plume. Limnol. Oceanogr. 34:178-187.

Govoni, J.J., and C.B. Grimes. 1992. The surface accumulation of larval fishes by hydrodynamic convergence within the Mississippi River plume front. Cont. Shelf Res. 12(11):1265-1276.

Greene, C.R., Jr. 1985. Characteristics of waterborne industrial noise, pp. 197-253. *In:* W.J. Richardson (ed.), Behavior, disturbance responses, and feeding of bowhead whales *Balaena mysticetus* in the eastern Beaufort Sea, 1980-84. OCS Study MMS 85-0034. Report by LGL Ecological Research Associates Inc., Bryan, TX for the U.S. Department of the Interior, Minerals Management Service, Reston, VA.

Greene, C.R., Jr. 1987. Characteristics of oil industry dredge and drilling sounds in the Beaufort Sea. J. Acoust. Soc. Am. 82(4):1315-1324.

Greene, C.R., Jr., and W.J. Richardson. 1988. Characteristics of marine seismic survey sounds in the Beaufort Sea. J. Acoust. Soc. Am. 83(6):2246-2254.

Grimes, C.B., and J.H. Finucane. 1991. Spatial distribution and abundance of larval and juvenile fish, chlorophyll and macrozooplankton around the Mississippi River discharge plume, and the role of the plume in fish recruitment. Mar. Ecol. Prog. Ser. 75:109-119.

Gulf of Mexico Fishery Management Council. 2003. Draft environmental impact statement for the generic essential fish habitat amendment to the following fishery management plans of the Gulf of Mexico: shrimp fishery of the Gulf of Mexico; red drum fishery of the Gulf of Mexico; reef fish fishery of the Gulf of Mexico; stone crab fishery of the Gulf of Mexico; coral and coral reef fishery of the Gulf of Mexico; spiny lobster fishery of the Gulf of Mexico and South Atlantic; and coastal migratory pelagic resources of the Gulf of Mexico and South Atlantic, Gulf of Mexico Fishery Management Council, Tampa FL.

Hall, J.D., M.L. Gallagher, K.D. Brewer, P.R. Regos, and P.E. Isert. 1994. ARCO Alaska, Inc. 1993 Kuvlum exploration area site specific monitoring program. Final report. Report from Coastal & Offshore Pacific Corp., Walnut Creek, CA for ARCO Alaska, Inc., Anchorage, AK. 219 pp. + data appendix, vols. 1 and 2.

Hecker, B. 1985. Fauna from a cold sulfur-seep in the Gulf of Mexico. Comparison with hydrothermal vent communities and evolutionary implications. Biol. Soc. Wash. Bull. 6:465-473.

Hernandez, F.J., R.F. Shaw, J.S. Cope, J.G. Ditty, T. Farooqi, and M.C. Benfield. 2002. The across-shelf larval, postlarval, and juvenile fish community associated with offshore oil and gas platforms and a coastal rock jetty west of the Mississippi River delta, pp. 185-269. *In:* McKay, M., J. Nides, and D. Vigil (eds.), Proceedings: Gulf of Mexico fish and fisheries: Bringing together new and recent research, October 2000. U.S. Department of the Interior, Minerals Management Service, Gulf of Mexico OCS Region, New Orleans, LA. OCS Study MMS 2002-004. 674 pp.

Hildebrand, H.H. 1995. A historical review of the status of sea turtle populations in the western Gulf of Mexico, pp. 447-453. *In:* Bjorndal, Karen A., ed., Biology and Conservation of Sea Turtles, Second Edition, Smithsonian Institution Press, Washington, D.C.

Hirth, H.H. 1997. Synopsis of biological data on the green turtle, *Chelonia mydas* (Linnaeus 1758). Report to the U.S. Dept. of the Interior, U.S. Fish and Wildlife Service, Washington, D.C. Biological Report 97(1). 120 pp.

Hopkins, T.L., T.T. Sutton, and T.M. Lancraft. 1997. The trophic structure and predation impact of a low latitude midwater fish assemblage. Prog. Oceanogr. 38: 205-239.

Hubbard, H.H. (ed.). 1995. Aeroacoustics of flight vehicles, theory and practice. Vol. 1, Noise Sources. Acoust. Soc. Am. New York. 608 pp.

Jefferson, T.A. 1995. Distribution, abundance, and some aspects of the biology of cetaceans in the offshore Gulf of Mexico. Ph.D. dissertation, Texas A&M University, College Station, TX. 107 pp.

Jefferson, T.A., and A.J. Schiro. 1997. Distributions of cetaceans in the offshore Gulf of Mexico. Mammal Rev. 27(1):27-50.

Jefferson, T.A., S. Leatherwood, L.K.M. Shoda, and R.L. Pitman. 1992. A field guide for aerial and shipboard observers. Texas A&M Univ. Print Ctr., College Station, TX. 92 pp.

Jefferson, T.A., S. Leatherwood, and M.A. Webber. 1993. FAO species identification guide. Marine mammals of the world. Food and Agriculture Organization of the United Nations, Rome. 320 pp.

Johnson, A.G., W.E. Fable, Jr., C.B. Grimes, L. Trent, and J.V. Perez. 1994. Evidence for distinct stocks of king mackerel, *Scomberomorus cavalla*, in the Gulf of Mexico. Fish. Bull. 92:91-101.

Johnston, R.C., and B. Cain. 1981. Marine seismic energy sources: Acoustic performance comparison. Manuscript presented at 102nd Meet. Acoust. Soc. Am., Miami Beach, FL, Dec. 1981. 35 pp.

Koblenz-Mishke, O.J., V.K. Volkovinsky, and J.C. Kabanova. 1970. Plankton primary production of the world ocean, pp. 183-193. *In:* W.W. Wooster (Editor), Scientific Exploration of the South Pacific, National Academy of Sciences, Washington, D.C.

Lamkin, J. 1997. The Loop Current and the abundance of larval *Cubiceps pauciradiatus* (Pisces: Nomeidae) in the Gulf of Mexico: Evidence for physical and biological interaction. Fish. Bull. 95(2):251-267.

Leatherwood, S., and R.R. Reeves. 1983. The Sierra Club handbook of whales and dolphins. Sierra Club Books, San Francisco, CA. 302 pp.

Leben, R.R., G.H. Born, D.C. Biggs, D.R. Johnson, and N.D. Walker. 1993. Verification of TOPEX altimetry in the Gulf of Mexico. TOPEX/POSEIDON Research News 1:3-6.

Lee, T.N., J.A. Yoder, and L.P. Atkinson. 1991. Gulf Stream frontal eddy influence on productivity of the southeast U.S. continental shelf. J. Geophys. Res., 96:22191-22205.

LGL Ecological Research Associates, Inc. and Texas A&M University. 1986. Gulf of Mexico Continental Slope Study, Annual Report: Year 2. Prepared for the Minerals Management Service, Gulf of Mexico OCS Region, Contract No. 14-12-0001-30212. 3 vols. OCS Study MMS 86-0089, 86-0090, and 86-0091.

Lohoefener, R.R., W. Hoggard, K. Mullin, C. Roden, and C. Rogers. 1990. Association of sea turtles with petroleum platforms in the north-central Gulf of Mexico. OCS Study MMS 90-0025. U.S. Dept. of the Interior, Minerals Management Service, Gulf of Mexico OCS Region, New Orleans, LA. 90 pp.

Lohrenz, S.E., G.L. Fahnenstiel, D.G. Redalje, G.A. Lang, M.J. Dag, T.E. Whitledge, and Q. Dortch. 1999. Nutrients, irradiance, and mixing as factors regulating primary production in coastal waters impacted by the Mississippi River plume. Cont. Shelf Res. 19:1113-1141.

MacDonald, I.R., N.L. Guinasso, J.F. Reilly, J.M. Brooks, W.R. Callender, and S.G. Gabrielle. 1990. Gulf of Mexico hydrocarbon seep communities: Patterns in community structure and habitat. Geo-Marine Letters 10(4):244-252.

MacDonald, I.R., W.W. Schroeder, and J.M. Brooks. 1995. Chemosynthetic Ecosystems Studies Final Report. Volume II, Technical Report. U.S. Department of the Interior, Minerals Management Service, Gulf of Mexico OCS Region, New Orleans, LA. OCS Study MMS 95-0022.

Marquez, M.R. 1990. FAO species catalogue. Vol. 11: Sea Turtles of the World. An Annotated and Illustrated Catalogue of Sea Turtle Species Known to Date. FAO Fisheries Synopsis. No. 125, Vol. 11. Rome, FAO. 81 pp.

McCauley, R.D. 1994. Environmental implications of offshore oil and gas development in Australia - seismic surveys, pp. 19-121. *In:* Report by the Australian Inst. of Marine Sci. (Townsville, QLD) for the Australian Petroleum Exploration Association (APEA) and Energy Research and Development Corporation (ERDC).

McEachran, J.D., and J.D. Fechhelm. 1998. Fishes of the Gulf of Mexico, Volume 1. University of Texas Press, Austin, Texas. 1,112 pp.

Meylan, A., B. Schroeder, and A. Mosier. 1995. Sea turtle nesting activity in the State of Florida 1979-1992. Florida Marine Research Publications, State of Florida, Dept. of Environmental Protection, No. 52.

Müller-Karger, F.E., J.J. Walsh, R.H. Evans, and M.B. Meyers. 1991. On the seasonal phytoplankton concentration and sea surface temperature cycles of the Gulf of Mexico as determined by satellites. J. Geophys. Res. 96:12,645-12,665.

Mullin, K.D., and W. Hoggard. 2000. Visual surveys of cetaceans and sea turtles from aircraft and ships (Chapter 4), pp. 111-171. *In:* Davis, R.W., W.E. Evans, and B. Würsig (eds.), Cetaceans, Sea Turtles and Seabirds in the Northern Gulf of Mexico: Distribution, Abundance and Habitat Associations. Volume II: Technical Report. Prepared by Texas A&M University at Galveston and the National Marine Fisheries Service. U.S. Department of the Interior, U.S. Geological Survey, Biological Resources Division, USGS/BRD/CR-1999-005 and Minerals Management Service, Gulf of Mexico OCS Region, New Orleans, LA OCS Study MMS 2000-003.

Mullin, K.D., W. Hoggard, C.L. Roden, R.R. Lohoefener, C.M. Rogers, and B. Taggart. 1994a. Cetaceans on the upper continental slope in the north-central Gulf of Mexico. Fishery Bulletin 92:773-786.

Musick, J.A., and C.J. Limpus. 1997. Habitat Utilization and Migration in Juvenile Sea Turtles, pp. 137-164. *In:* Lutz, P.L. and J.A. Musick, eds., The Biology of Sea Turtles. CRC Press, Boca Raton, FL. 432 pp.

Myrberg, A.A., Jr. 1978. Ocean noise and the behavior of marine animals: Relationships and implications, pp. 169-208. *In:* J.L. Fletcher and R.G. Busnel (eds.), Effects of Noise on Wildlife. Academic Press, New York. 305 pp.

National Marine Fisheries Service. 1999a. Fishery Management Plan for Atlantic tunas, swordfish, and sharks, Volume II. National Marine Fisheries Service, Division of Highly Migratory Species, Office of Sustainable Fisheries, Silver Spring, MD. 302 pp.

National Marine Fisheries Service. 1999b. Pelagic longline logbook data. Southeast Fisheries Science Center, Miami, FL.

National Marine Fisheries Service. 2000. Status review of smalltooth sawfish (*Pristis pectinata*). Retrieved October 2003 from www.nmfs.noaa.gov/endangered htm. 63 pp. + apps.

National Marine Fisheries Service. 2003a. Landings Statistics web page, URL: www.nmfs noaa.gov.

National Marine Fisheries Service. 2003b. Marine Recreational Fisheries Statistics Survey, URL: www nmfs noaa.gov.

National Marine Fisheries Service. 2004. Draft 2003 marine mammal stock assessment report. www nmfs.noaa.gov/prot_res/readingrm/draft_2003_sars/sefsc2003sar_revision_whales_atl.pdf. 35 pp.

National Marine Fisheries Service and U.S. Fish and Wildlife Service. 1991a. Recovery plan for U.S. populations of loggerhead turtle. National Marine Fisheries Service, Washington, DC. 64 pp.

National Marine Fisheries Service and U.S. Fish and Wildlife Service. 1991b. Recovery plan for U.S. population of Atlantic green turtle. National Marine Fisheries Service, Washington, DC. 52 pp.

National Marine Fisheries Service and U.S. Fish and Wildlife Service. 1992a. Recovery plan for leatherback turtles in the U.S. Caribbean, Atlantic, and Gulf of Mexico. National Marine Fisheries Service, Washington, DC. 65 pp.

National Marine Fisheries Service and U.S. Fish and Wildlife Service. 1992b. Recovery plan for the Kemp's ridley sea turtle (*Lepidochelys kempii*). National Marine Fisheries Service, St. Petersburg, FL. 40 pp.

National Research Council. 1994. Low-frequency sound and marine mammals: current knowledge and research needs. National Academy Press, Washington, D.C. 75 pp.

National Research Council. 2000. Marine mammals and low-frequency sound; progress since 1994. Prepublication copy. National Academy Press, Washington, D.C. 134 pp.

National Research Council. 2003. Ocean noise and marine mammals. The National Academies Press, Washington, D.C. ISBN 0-309-08563-5. 192 pp. + plates.

Office of Naval Research. 1999. Proceedings: workshop on the effects of anthropogenic noise in the marine environment, 10-12 February 1998. Arlington, Virginia.

Parker, R.O., D.R. Colby, and T.D. Willis. 1983. Estimated amount of reef habitat on a portion of the U.S. South Atlantic and Gulf of Mexico continental shelf. Bull. Mar. Sci. 33:935-940.

Paull, C.K., B. Hecker, R. Commeau, R.P. Freeman-Lynde, C. Neummann, W.P. Corse, S. Golubic, J.E. Hook, E. Sikes, and J. Curray. 1984. Biological communities at the Florida escarpment resemble hydrothermal vent taxa. Science 226:965-967.

Paull, C.K., F.N. Spiess, J.R. Curray, and D. Twitchell. 1988. Morphology of Florida escarpment chemosynthetic brine seep community sites; deep-tow, Seabeam, and GLORIA surveys. AAPG Bull., 1988-02, 72 (2):233.

Peake, D.E. 1996. Bird surveys, pp. 271-304. *In:* R.W. Davis and G.S. Fargion (eds.), Distribution and abundance of cetaceans in the north-central and western Gulf of Mexico, Final Report. Volume II: Technical Report. U.S. Department of the Interior, Minerals Management Service, Gulf of Mexico OCS Region, New Orleans, LA. OCS Study MMS 96-0027. 357 pp.

Pequegnat, W.E. 1983. The ecological communities of the continental slope and adjacent regimes of the northern Gulf of Mexico. A final report by TerEco Corporation for the Minerals Management Service, Contract No. AA851-CT1-12.

Pequegnat, W.E., B.J. Gallaway, and L.H. Pequegnat. 1990. Aspects of the ecology of the deep-water fauna of the Gulf of Mexico. Am. Zool. 30:45-64.

Pritchard, P.C.H. 1997. Evolution, Phylogeny, and Current Status, pp. 1-28. *In:* Lutz, P.L. and J.A. Musick, eds., The Biology of Sea Turtles. CRC Press, Boca Raton, FL. 432 pp.

Rabalais, N.N., R.E. Turner, D. Justic, Q. Dortch, W.W. Wiseman, Jr., and B.K. Sen Gupta. 1996. Nutrient changes in the Mississippi River and system responses on the adjacent continental shelf. Estuaries 19:386-407.

Redalje, D.G., S.E. Lohrenz, and G.L. Fahnenstiel. 1994. The relationship between primary production and the vertical export of particulate organic matter in a river impacted coastal ecosystem. Estuaries 17:829-838.

Rice, D.W. 1989. Sperm whale- *Physeter macrocephalus* (Linnaeus, 1758). *In:* Ridgway, S.H. and R. Harrison, eds. Handbook of Marine Mammals. Vol. 4: River Dolphins and Larger Toothed Whales. Academic Press, London. pp. 17-234.

Richards, W.J., T. Leming, M.F. McGowan, J.T. Lamkin, and S. Kelley-Farga. 1989. Distribution of fish larvae in relation to hydrographic features of the Loop Current boundary in the Gulf of Mexico. Rapp. P.-v. Reun. Cons. Int. Explor. Mer. 191:169-176.

Richards, W.J., M.F. McGowan, T. Leming, J.T. Lamkin, and S. Kelley. 1993. Larval fish assemblages at the Loop Current boundary in the Gulf of Mexico. Bull. Mar. Sci. 53:475-537.

Richardson, W.J., C.R. Greene, Jr., C.I. Malme, and D.H. Thomson. 1995. Marine Mammals and Noise. Academic Press, San Diego. 576 pp.

Ross, D. 1976. Mechanics of Underwater Noise. Pergammon Press, New York, NY. 375 pp.

Shaw, R.F., D.C. Lindquist, M.C. Benfield, T. Farooqi, and J.T. Plunket. 2002. Offshore petroleum platforms: functional significance for larval fish across longitudinal and latitudinal gradients. Prepared by the Coastal Fisheries Institute, Louisiana State University. U.S. Department of the Interior, Minerals Management Service, Gulf of Mexico OCS Region, New Orleans, LA. OCS Study MMS 2002-077. 107 pp.

Shipp, R.L., and T.L. Hopkins. 1978. Physical and biological observations of the northern rim of the De Soto Canyon made from a research submersible. N.E. Gulf Sci. 2(2):113-121.

Smith, M.J.T. 1989. Aircraft Noise. Cambridge Univ. Press. Cambridge, U.K. 359 pp.

Smith, S., and J.T. Hollibaugh. 1993. Coastal metabolism and the oceanic organic carbon balance. Rev. Geophys. 31:75-89.

Sparks, T.D., J.C. Norris, R. Benson, and W.E. Evans. 1996. Distributions of sperm whales in the northwestern Gulf of Mexico as determined from an acoustic survey. 11th Biennial Conference on the Biology of Marine Mammals (14-18 December 1995, Orlando, FL), p. 108.

Stanley, D.R., and C.A. Wilson. 1990. A fishery-dependent based study of fish species composition and associated catch rates around oil and gas structures off Louisiana. Fish. Bull. 88:719-730.

Sutter, F.C., III, R.O. Williams, and M.F. Godcharles. 1991. Movement patterns of king mackerel in the southeastern United States. Fish. Bull. 89:315-324.

Thompson, M.J., W.W. Schroeder, and N.W. Phillips. 1999. Ecology of Live Bottom Habitats of the Northeastern Gulf of Mexico: A Community Profile. U.S. Department of the Interior, U.S. Geological Survey, Biological Resources Division, USGS/BRD/CR—1999-0001 and Minerals Management Service, Gulf of Mexico OCS Region, New Orleans, LA, OCS Study MMS 99-0004. 74 pp.

Urick, R.J. 1972. Noise signature of an aircraft in level flight over a hydrophone in the sea. J. Acoust. Soc. Am. 52(3, Pt. 2):993-999.

Urick, R.J. 1983. Principles of Underwater Sound. McGraw-Hill, New York. 423 pp.

U.S. Department of the Army, Corps of Engineers. 2000. Waterborne commerce statistics. U.S. Army Corps of Engineers, Waterborne Commerce Statistics Center. Website statistics accessed via www.armycoe.gov.

U.S. Department of the Interior, Minerals Management Service. 1995. Gulf of Mexico Air Quality Study, Final Report. U.S. Department of the Interior, Minerals Management Service, Gulf of Mexico OCS Region, New Orleans, LA.

U.S. Department of the Interior, Minerals Management Service. 1996. Final Environmental Impact Statement, Outer Continental Shelf Oil & Gas Leasing Program 1997-2002. U.S. Department of the Interior, Minerals Management Service, Herndon, VA.

U.S. Department of the Interior, Minerals Management Service. 1997a. Gulf of Mexico OCS Oil and Gas Lease Sales 169, 172, 175, 178, and 182, Central Planning Area. Final Environmental Impact Statement. U.S. Department of the Interior, Minerals Management Service, Gulf of Mexico OCS Region, New Orleans, LA.

U.S. Department of the Interior, Minerals Management Service. 1997b. Gulf of Mexico OCS Oil and Gas Lease Sales 171, 174, 177, and 180, Western Planning Area. Final Environmental Impact Statement. U.S. Department of the Interior, Minerals Management Service, Gulf of Mexico OCS Region, New Orleans, LA.

U.S. Department of the Interior, Minerals Management Service. 1998. Environmental Assessment: Proposed OCS Lease Sale 172, Central Gulf of Mexico. U.S. Department of the Interior, Minerals Management Service, Gulf of Mexico OCS Region, New Orleans, LA.

U.S. Department of the Interior, Minerals Management Service. 1999. Destin Dome 56 unit development and production plan and right-of-way-pipeline application. Draft Environmental Impact Statement. Volume I: Sections I-X. OCS EIS/EA MMS 99-0040. U.S. Department of the Interior, Minerals Management Service, Gulf of Mexico OCS Region, New Orleans, LA.

U.S. Department of the Interior, Minerals Management Service. 2002a. Gulf of Mexico OCS Oil and Gas Lease Sales: 2003-2007. Central Planning Area Sales 185, 190, 194, 198, and 201. Western Planning Area Sales 187, 192, 196, and 200. Final Environmental Impact Statement. U.S. Department of the Interior, Minerals Management Service, Gulf of Mexico OCS Region, New Orleans, LA. OCS EIS/EA MMS 2002-052. 2 vols.

U.S. Department of the Interior, Minerals Management Service. 2002b. Gulf of Mexico OCS Oil and Gas Lease Sales 189 and 197. Eastern Planning Area. Draft Environmental Impact Statement. U.S. Department of the Interior, Minerals Management Service, Gulf of Mexico OCS Region, New Orleans, LA. OCS EIS/EA MMS 2002-056. 2 vols.

U.S. Department of the Interior, U.S. Fish and Wildlife Service. 1996. Florida manatee recovery plan (*Trichechus manatus latirostris*), second revision. Prepared by the Florida Manatee Recovery Team for the Southeast Region, U.S. Fish and Wildlife Service, Atlanta, GA. Approved January 29, 1996.

U.S. Department of the Interior, U.S. Fish and Wildlife Service. 1997a. Draft stock assessment for West Indian manatee (*Trichechus manatus latirostris*): Florida stock. November 5, 1997. 4 pp.

U.S. Department of the Interior, U.S. Fish and Wildlife Service. 1997b. Draft stock assessment for West Indian manatee (*Trichechus manatus manatus*): Antillean stock. November 5, 1997. 3 pp.

U.S. Department of the Interior, U.S. Fish and Wildlife Service. 1998. Endangered and Threatened Wildlife and Plants. 50 CFR 17.11 and 17.12. December 31, 1998. 56 pp.

Walsh, J.J., D.A. Dieterle, M.B. Meyers, and F.E. Müller-Karger. 1989. Nitrogen exchange at the continental margin: A numerical study of the Gulf of Mexico. Prog. Oceanogr. 23:248-301.

Waring, G.T., D.L. Palka, K.D. Mullin, J.H.W. Hain, L.J. Hansen, and K.D. Bisack. 1997. U.S. Atlantic and Gulf of Mexico Marine Mammal Stock Assessments - 1996. NOAA Tech. Memo. NMFS-NE-114. 250 pp.

Waring, G.T., D.L. Palka, P.J. Clapham, S. Swartz, M. Rossman, T. Cole, L.J. Hansen, K.D. Bisack, K. Mullin, R.S. Wells, D.K. Odell, and N.B. Barros. 1999. U.S. Atlantic and Gulf of Mexico marine mammal stock assessments - 1999. NOAA Tech. Memo. NMFS-NE-153. 196 pp.

Waring, G.T., J.M. Quintal, and C.P. Fairfield (eds.) 2002. U.S. Atlantic and Gulf of Mexico Marine Mammal Stock Assessments - 2002. NOAA Tech. Memo. NFMS-NE-169. 318 pp.

Wenz, G.M. 1962. Acoustic ambient noise in the ocean: spectra and sources. J. Acoust. Soc. Am. 34(12):1936-1956.

Wille, P.C. and D. Geyer. 1984. Measurements on the origin of the wind-dependent ambient noise variability in shallow water. J. Acoust. Soc. Am. 75(1):73-185.

Wormuth, J.H., P.H. Ressler, R.B. Cady, and E.J. Harris. 2000. Biological oceanography (Chapter 3), pp. 69-110. *In:* Davis, R.W., W.E. Evans, and B. Würsig (eds.), Cetaceans, Sea Turtles and Seabirds in the Northern Gulf of Mexico: Distribution, Abundance and Habitat Associations. Volume II: Technical Report. Prepared by Texas A&M University at Galveston and the National Marine Fisheries Service. U.S. Department of the Interior, U.S. Geological Survey, Biological Resources Division, USGS/BRD/CR-1999-005 and Minerals Management Service, Gulf of Mexico OCS Region, New Orleans, LA OCS Study MMS 2000-003.

Würsig, B., T.A. Jefferson, and D.J. Schmidley. 2000. The Marine Mammals of the Gulf of Mexico. Texas A&M University Press, College Station, TX. 232 pp.

Zimmerman, R.A., and D.C. Biggs. 1999. Patterns of distribution of sound-scattering zooplankton in warm- and cold-core eddies in the Gulf of Mexico, from a narrowband acoustic Doppler current profiler survey. J. Geophys. Res. 104:5251-5262.

**Appendix G
Marine Mammal Hearing and
Sensitivity to Acoustic Impact**

Marine Mammal Hearing and
Sensitivity to Acoustic Impact

This appendix describes the current state of knowledge regarding marine mammal hearing (and vocalization) capabilities and sensitivity to acoustic impact. This summary serves as a basis for assessing the environmental impact of geological and geophysical (G&G) activities on marine mammals that frequent outer continental shelf (OCS) waters of the Gulf of Mexico (see **Section III - Environmental Impacts** of the Programmatic Environmental Assessment [PEA]).

I. INTRODUCTION

A. MARINE MAMMAL HEARING

Having evolved from ancient land dwelling ancestors, marine mammals have retained the same basic ear structure as used by land mammals, but with some specializations for an aquatic existence. Marine mammal ears have been studied, and detailed accounts given, by Ketten (1992, 1994, 1998). Essentially all mammalian ears have three basic divisions: an outer ear; an air-filled middle ear with bony levers and membranes; and a fluid filled inner ear with mechanical resonators and sensory cells. The outer ear acts as a sound collector, the middle ear transforms acoustic energy into mechanical energy (i.e., acting as a coupling device between the outer and inner ears), and the inner ear detects the energy passed to it via the mechanical apparatus.

In marine mammals, especially cetaceans, the outer ear tends to consist of a waxy plug coupled to the middle ear. The middle ear acts to impedance match acoustic energy, via mechanical advantages of the bony levers (ossicles), from the outer ear to the inner ear. The inner ear is divided into the vestibular system and the spiral cochlea, of which only the cochlea concerns this discussion (see **Figure G-1**). The cochlea has a resonator (the basilar membrane) and a neuroreceptor, the Organ of Corti. Movement of the basilar membrane deflects cilia on hair cells in the Organ of Corti, which in turn stimulate the release of neurotransmitters. The auditory nerve then carries acoustic information from the hair cells to the brain. The information encoded includes frequency, amplitude, and temporal patterning, based on the location, degree of deflection, and sequencing of hair cells that are excited by the basilar membrane motion (Ketten, 1998).

In general, mammalian ears scale with body size. Large marine mammals tend to have large ears and small marine mammals have smaller ears. Large ears (large marine mammals) also tend to be associated with lower frequency hearing, and smaller ears (smaller marine mammals) tend to be associated with higher frequency hearing. The basilar membrane thickness-to-width ratio in mysticetes makes for a relatively loose, low frequency resonator, and suggests that mysticetes are sensitive to infrasonic sounds in the region of 10 Hz upwards. Ketten (1998) supposes an upper functional hearing range in mysticetes extending to some 30 kHz. The basilar membrane thickness-to-width ratio in odontocetes makes for a relatively stiff, high frequency resonator, and suggests that odontocetes are sonic to ultrasonic specialists (100 Hz to 100 kHz), a supposition supported by behavioral audiogram data (see reviews in Au, 1993; Richardson et al., 1995; also see Nachtigal et al., 1996; Popov and Supin, 1997; Popov et al., 1997; Ridgway and Carder, 1997; Erbe and Farmer, 1998; Sauerland and Dehnhardt, 1998; Szymanski et al., 1999).

Tables G-1 and **G-2** identify the major characteristics of baleen and toothed whale vocalization and hearing, including associated threshold levels that have been measured or are inferred.

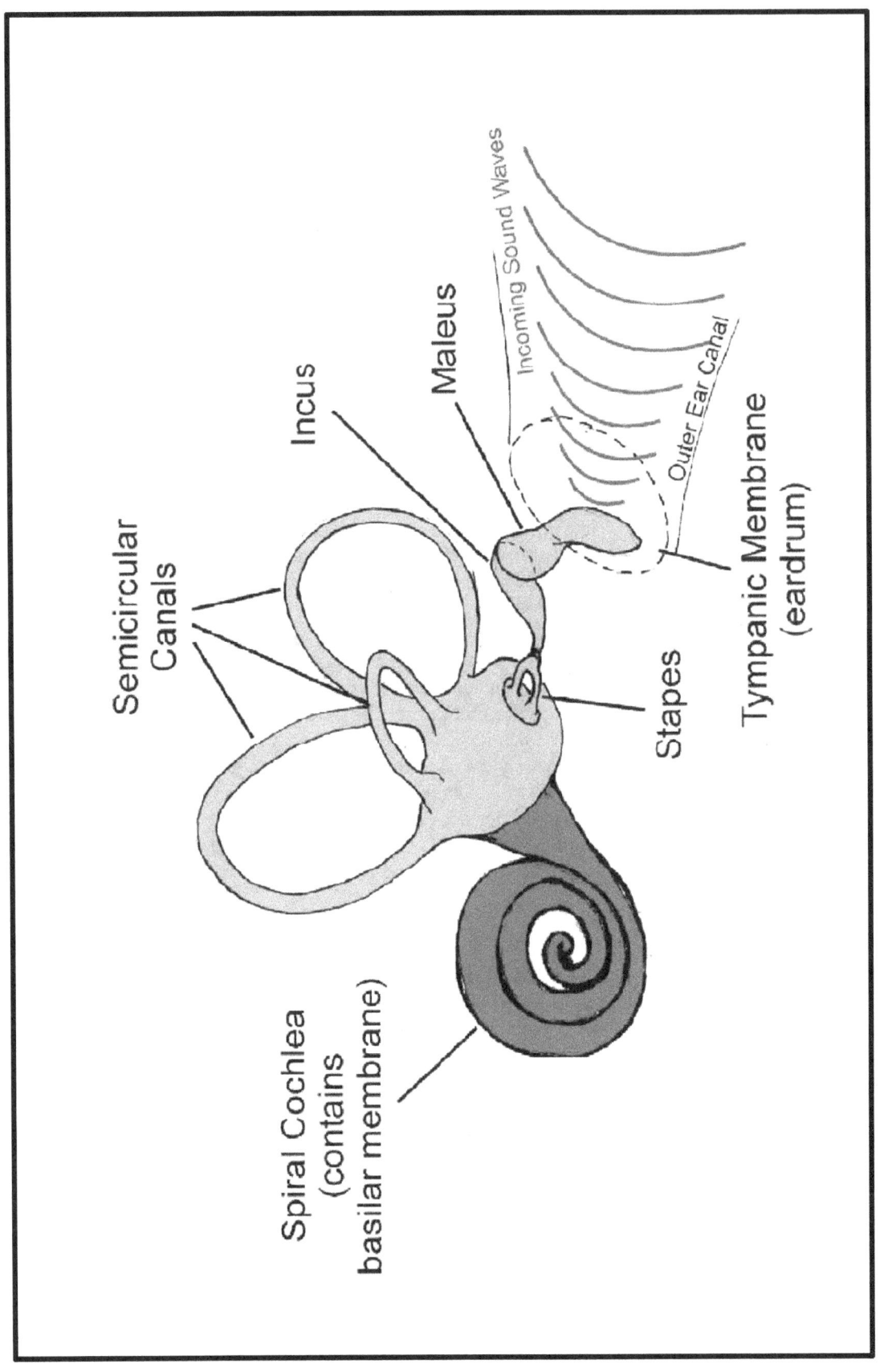

Figure G-1. Ear anatomy of a marine mammal (figure courtesy of John C. Goold).

Table G-1

Vocalization Characteristics of Select Baleen (Mysticete) and Toothed (Odontocete) Whales[1], Including Frequency Ranges and Source Levels (Source: Adapted from Richardson et al., 1995 and Wartzok and Ketten, 1999)

Species	Signal Type	Frequency Range (Hz)	Dominant Frequencies (Hz)	Source Level[2] (dB)	References
Baleen Whales:					
Right whale[3]	tonal	30-1,250	160-500	-	Cummings et al., 1972; Clark, 1982, 1983
	pulsive	30-2,200	50-500	172-187	Cummings et al., 1972; Clark, 1982, 1983
				181-186	Clark in Würsig et al., 1982
	call	<400	<200	-	Watkins & Schevill, 1972; Clark, 1990
	moans	<400	-	-	Watkins & Schevill, 1972; Thompson et al., 1979; Spero, 1981
Fin whale	moans, downsweeps	14-118	20	160-186	Watkins, 1981; Watkins et al., 1987, Edds, 1988, Cummings & Thompson, 1994 Edds, 1988
	constant call	20-40	-	-	Watkins, 1981; Cummings et al., 1986; Edds, 1988
	moans, tones, upsweeps	30-750	-	155-165	Watkins, 1981; Edds, 1988
	rumble	10-30	-	-	Thompson et al., 1979
	whistles?, chirps?	1,500-5,000	1,500-2,500	-	Thompson et al., 1979
	clicks?	16,000-28,000	-	-	Thompson et al., 1979, 1992; Edds, 1988
	moans	16-750	20	160-190	Clark, 1990
	pulse	40-70	-	-	Watkins, 1981a
	pulse	18-25	20	-	Watkins, 1981a
	ragged pulse	<30	-	-	Watkins, 1981a
	rumble	-	<30	-	
Blue whale	moans	12-400	12-25	188	Cummings & Thompson, 1971, 1994, Edds, 1982; Stafford et al., 1994
	clicks?	6,000-8,000, 21,000-31,000	6,000-8,000, 25,000	130, 159	Beamish & Mitchell, 1971; Beamish, 1979
		16-18	16 5		Rivers, 1997
Sei whale	FM sweeps	1,500-3,500	-	-	Thompson et al., 1979; Knowlton et al., 1991
Humpback whale	song components	30-8,000	120-4,000	144-174	Thompson et al., 1979; Payne & Payne, 1985
	shrieks	-	750-1,800	179-181	Thompson et al., 1986
	horn blasts	-	410-420	181-185	Thompson et al., 1986
	moans	20-1,800	35-360	175	Thompson et al., 1986
	grunts	25-1,900+	-	190	Thompson et al., 1986
	pulse trains	25-1,250	25-80	179-181	Thompson et al., 1986
	underwater blows	100-2,000	-	158	Beamish, 1979
	fluke and flipper slap	30-1,200	-	183-192	Thompson et al., 1986
	clicks	2,000-8,200	-	-	Winn et al., 1970; Beamish, 1979
	songs	30-8,000	100-4,000	144-186	Thompson et al., 1979; Watkins, 1981a; Edds, 1982, 1988; Payne et al., 1983; Silber, 1986; Clark, 1990
	social	50-10,000	<3,000	-	Thompson et al., 1979
Minke whale	downsweeps	60-130	-	165	Schevill & Watkins, 1972
	moans, grunts	60-140	60-140	151-175	Schevill & Watkins, 1972; Winn & Perkins, 1976
	ratchet	850-6,000	850	-	Winn & Perkins, 1976
	clicks	3,300-20,000	<12,000	151	Beamish & Mitchell, 1973; Winn & Perkins, 1976
	thump trains	100-2,000	100-200	-	Winn & Perkins, 1976
	sweeps, moans	60-140	-	151-175	Schevill & Watkins, 1972; Winn & Perkins, 1976

Table G-1

Vocalization Characteristics of Select Baleen (Mysticete) and Toothed (Odontocete) Whales[1], Including Frequency Ranges and Source Levels (Source: Adapted from Richardson et al., 1995 and Wartzok and Ketten, 1999) (Continued)

Species	Signal Type	Frequency Range (Hz)	Dominant Frequencies (Hz)	Source Level[2] (dB)	References
Bryde's whale	moans	70-245	124-132	152-174	Cummings et al., 1986
	pulsed moans	100-930	165-900	-	Edds et al., 1993
	discrete pulses	700-950	700-900	-	Edds et al., 1993
Toothed Whales:					
Sperm whale	clicks	100-30,000	2,000-4,000, 10,000-16,000	160-180	Backus & Schevill, 1966; Levenson, 1974; Watkins, 1980a,b; Goold & Jones, 1995; Goold, 1999
	clicks in coda	16,000-30,000	-	-	Watkins, 1980a,b
Pygmy sperm whale	clicks	60,000-200,000	120	-	Santoro et al., 1989
Blainville's beaked whale	chirps/short whistles	<1,000-6,000		-	Caldwell & Caldwell, 1971a
	whistles	2,600-10,700		-	Buerki et al., 1989; Lynn & Reiss, 1992
Killer whale	whistles	1,500-18,000	6,000-12,000	-	Steiner et al., 1979; Ford & Fisher, 1983; Morton et al., 1986
	pulsed calls	500-25,000	1,000-6,000	160	Schevill & Watkins, 1966; Awbrey et al., 1982; Ford & Fisher, 1983; Moore et al., 1988
	echolocation clicks	100-35,000	12,000-25,000	180	Diercks et al., 1971; Diercks, 1972
	clicks	250-500	-	-	Schevill & Watkins, 1966
	scream	2,000	-	-	Schevill & Watkins, 1966
False killer whale	echolocation clicks		25,000-30,000, 95,000-130,000	220-228	Kamminga & van Velden, 1987; Thomas & Turl, 1990
	whistles		4,000-9,500	-	Busnel & Dziedzic, 1968; Kamminga & van Velden, 1987
Pygmy killer whale	growls, blats	-	-	-	Pryor et al., 1965
Long-finned pilot whale	whistles	1,000-8,000	1,600-6,700	-	Busnel & Dziedzic, 1966a
	clicks	1,000-18,000	-	-	Taruski, 1969; Steiner, 1981
	echolocation clicks	-	6,000-11,000	-	McLeod, 1986
Short-finned pilot whale	whistles	500-20,000+	2,000-14,000	180	Caldwell & Caldwell, 1969; Fish & Turl, 1976
	echolocation clicks	-	30,000-60,000	180	Evans, 1973
Bottlenose dolphin	echolocation clicks	-	110,000-130,000	218-228	Au et al., 1974; Au, 1993
	whistles	800-24,000	3,500-14,500	125-173	Lilly & Miller, 1961; Tyack, 1985; Caldwell et al., 1990; Schultz & Corkeron, 1994; Wang Ding et al., 1995
	low frequency narrowband	<2,000	300-900	-	Schultz et al., 1995
	rasp/grate/mew/bark/yelp	-	-	-	Wood, 1953
	click	200-150,000	30,000-60,000	-	Diercks et al., 1971; Evans, 1973
	bark	200-16,000	-	-	Evans & Prescott, 1962
	whistle	4,000-20,000	-	-	Caldwell & Caldwell, 1967; Evans & Prescott, 1962

Table G-1

Vocalization Characteristics of Select Baleen (Mysticete) and Toothed (Odontocete) Whales[1], Including Frequency Ranges and Source Levels (Source: Adapted from Richardson et al., 1995 and Wartzok and Ketten, 1999)

(Continued)

Species	Signal Type	Frequency Range (Hz)	Dominant Frequencies (Hz)	Source Level[2] (dB)	References
Risso's dolphin	whistles	-	3,500-4,500	-	Caldwell et al., 1969
	rasp/pulse burst	100-8,000+?	2,000-5,000	-	Watkins, 1967
	echolocation clicks	-	65,000	~120	Au, 1993
Fraser's dolphin	whistles	7,600-13,400	-	-	Leatherwood et al., 1993
Spotted dolphin[4]	whistles	5,000-19,800	6,700-17,900	-	Caldwell et al., 1973; Steiner, 1981; Wang Ding et al., 1995
	clicks	1,000-8,000	-	-	Caldwell & Caldwell, 1971b
	squawk, barks, growls	100-3,000	-	-	Caldwell & Caldwell, 1971b; Caldwell et al., 1973
	chirps	4,000-8,000	-	-	Caldwell et al., 1973
	echolocation clicks	-	-	-	Caldwell & Caldwell, 1971b
Pantropical spotted dolphin	whistles	3,100-21,400	6,700-17,800	-	Wang Ding et al., 1995
	whistles	-	-	-	Evans, 1967
	pulse	to 150,000	-	-	Diercks, 1972
Striped dolphin	whistles	6,000-24,000+	8,000-12,500	-	Busnel et al., 1968
Spinner dolphin	echolocation clicks	low-65,000+	6,800-16,900	109-125?	Watkins & Schevill, 1974; Norris et al., 1994
	whistles (= squeals?)	1,000-22,500	5,000-60,000	108-115	Watkins & Schevill, 1974; Steiner, 1981; Wang Ding et al., 1995
	pulse bursts	very wide	-	-	Watkins & Schevill, 1974; Norris et al., 1994
	screams	-	-	-	Norris et al., 1994
Clymene dolphin	whistles	6,300-19,200	-	-	Mullin et al., 1994
Rough-toothed dolphin	echolocation clicks	-	5,000-32,000	-	Norris & Evans, 1967
	whistles	-	4,000-7,000	-	Busnel & Dziedzic, 1966b

Notes and Footnotes:

? Indicates data infrequently recorded and/or questionable correlation of sound with species, per Richardson et al. (1995)

[1] Data provided for select species only; several mysticetes (e.g., bowhead, gray, and pygmy right whale) and odontocetes (e.g., beluga, narwhal) not included, based on their expected absence from the study area

[2] Decibel level (dB) using an underwater/in-water reference pressure unit of one microPascal at 1 m distance (re 1 μPa at 1 m), rms

[3] Both northern and southern right whale results presented; and

[4] Spotted dolphin (*Stenella plagiodon*) is a congener to Atlantic spotted (*S. frontalis*), the latter of which is common in the Gulf of Mexico; no sound or vocalization characterization data are available for Atlantic spotted dolphin

Table G-2
Vocalizations and Hearing Sensitivities in Select Species of Baleen and Toothed Whales
(Source: Adapted from Richardson et al., 1995)

Characteristic	Baleen Whales	Toothed Whales
Vocalizations	Frequency-modulated "simple calls": principal energy <1,000 Hz; Amplitude-modulated "complex calls": bandwidths of 500 to 5,000 Hz; Knocks and grunts: 100 to 1,000 Hz, signal duration of 50 to 100 ms; Clicks and pulses: bandwidths of 3 to 3,100 Hz. Summary: <10 to 25,000 Hz, with principal energy <1,000 Hz	Whistles: various narrow-band sounds; energy typically <20,000 Hz; Clicks: echolocation possible; sperm whale clicks from <100 to 30,000 Hz, most energy at 2,000 to 4,000 Hz and 10,000 to 16,000 Hz; click rates of 1 to 90/s noted; Pulses: extremely complex; energy at 600 to 25,000 Hz; pulse repetition rates of up to 5,000/s. Summary: Very high frequency sounds used for echolocation in Delphinidae; moderately high frequency sounds used for communication
Hearing	No direct measurements available; suggested to be maximal below 1 kHz	75 Hz to 150 kHz; best above 10 kHz; sensitivity deteriorates with decreasing frequency below 10 kHz; below 1 kHz, sensitivity appears to be poor
Duration of Vocalizations	0.5 to 1 s; >1 s	Whistles of variable length, but can be extensive. Pulses from 0.05 to 10 s, with most 0.5 to 1.5 s
Frequency Patterns	Moans, grunts, downsweeps, upsweeps, constant calls, growls, song components, shrieks, horn blasts, pulse trains, rumble, ratchet, thump trains, possibly whistles, chirps, and echolocation clicks	Whistles (unmodulated, trilled, ascending, descending, ascending-descending, descending-ascending, or slowly wavering), clicks, and pulses
Direct Measurement of Thresholds	Direct measurements lacking; inferred thresholds: 10 to 31,000 Hz frequency range, with dominant frequencies of 16 to 25,000 Hz; vocalization source levels of 130 to 192 dB for those baleen whale species evaluated	Highly species-specific; 60 to 130,000 Hz frequency range, with dominant frequencies of 120 to 60,000 Hz in Delphinidae; vocalization source levels of 85 to 222 dB for those toothed whale species evaluated

Baleen whale hearing has not been extensively studied. There are no specific data regarding sensitivity, frequency or intensity discrimination, or localization abilities in baleen whales. Baleen whales apparently are more dependent on low frequency sounds than other marine mammals. The lack of specific data on baleen whale hearing abilities remains a major limitation in evaluating the effects of manmade noise on this group.

Toothed whales are probably sensitive to sounds above approximately 10 kHz. Below the 10 kHz level, sensitivity deteriorates with decreasing frequency, with the possible exception of the sperm whale (Carder and Ridgway, 1990). The sensitivity of many toothed whale species to high frequency sounds is attributed to their use of high frequency sound pulses in echolocation and moderately high frequency calls for communication. Low frequency hearing has not been studied extensively in toothed whales, however, some species may be able to detect sound frequencies as low as 60 to 105 Hz. Below 1 kHz, where most industrial noise energy is concentrated, toothed whale hearing sensitivity appears to be relatively poor.

Toothed whales also possess good frequency and intensity discrimination abilities, as well as good directional localization capabilities.

Several environmental factors must be considered when conducting hearing studies or assessing the impacts of manmade noise on free-ranging cetaceans (Dahlheim and Ljungblad, 1990), including determinations of 1) ambient noise levels and the potential for masking; 2) sound propagation characteristics of the medium (e.g., water depth, substrate, temperature, salinity, seasonal fluctuation in characteristics such as stratification); 3) absolute sound levels and frequencies reaching the cetacean; and 4) orientation of the cetacean relative to the sound source.

Biological factors should also be considered when evaluating the results of hearing studies conducted on marine mammals. Hearing may vary among individuals according to age or sex (Awbrey et al., 1988). Indeed, age related hearing loss has been shown in the structure of cetacean ears, which with a restricted sample size of animals for testing and behavioral observation, could be misinterpreted as a hearing injury. The behavioral state of test animals may also influence the responses evoked (Ljungblad et al., 1988). Habituation may also occur under those conditions where a cetacean is repeatedly exposed to a manmade sound (Dahlheim and Ljungblad, 1990). However, with regard to habituation, it is difficult to determine if habituation is behavioral (i.e., the animal is voluntarily tolerating a noise level) or whether the animal has become de-sensitized to repeated noise exposure through either temporary or permanent threshold shift.

B. ACOUSTIC UNITS AND DEFINITIONS

Sound is essentially the transfer of energy, propagated by particle motion in the conducting medium. In fluids, such as air and water, sound is propagated as a compressional wave, and hence it is pressure that is usually sensed by manmade measuring devices. Sound (i.e., waves of compression and rarefaction) is typically sensed in fluids with some sort of piezoelectric device, which converts the fluctuations in pressure (or sometimes particle velocity) to fluctuations in electrical voltage. The microphone and hydrophone are the devices commonly used for sounds transduction in air and water, respectively.

Due to the logarithmic nature of mammalian hearing, a wide range of pressure fluctuations is audible (i.e., mammals have a large dynamic hearing range). Measurement of such a large range of pressure fluctuations generates large and cumbersome numbers when dealing with linear units. Pressure is measured in Pascals (Pa; equivalent to 1 newton per square meter), which is an extremely large unit relative to the typical fluctuations in pressure generated by acoustic waves. The expression of underwater acoustic measurements is therefore typically condensed to a logarithmic scale, *the decibel scale*, using a reference pressure of 1 micropascal (μPa). For instance, a measured acoustic pressure fluctuation of 1,000 micropascals becomes 60 dB re 1 μPa when expressed logarithmically (the conversion formula is $20 \log[p_1/p_0]$, where p_1 is the measured pressure and p_0 is the reference pressure of 1 μPa). Each order of magnitude increase in pressure is an additional 20 dB on the scale. A decibel is a dimensionless unit, and as such it is critical to define the reference unit used.

It is sometimes appropriate to express sound measurements in terms of intensity, especially when making comparisons between air-borne and water-borne measurements. Intensity is the acoustic power incident on a 1-m^2 area at right angles to the direction of propagation, and has linear units of Watts per meter squared. Acoustic power is proportional to the mean square pressure of the acoustic wave $p^2/\rho c$, where "ρ" is the density of the medium and "c" is sound velocity within it. In decibel terms, the intensity is again referenced to 1 micropascal, but it is commonly forgotten that in this definition, 1 μPa is actually the intensity of a plane waveform of rms pressure equal to 1 Pascal.

The term rms, or "root mean square," describes a form of averaging applied to signal amplitudes. In its simplest form, the rms level of a continuous narrowband sinusoidal signal would be the product of the signal amplitude and the reciprocal of the square root of two, or simply dividing signal amplitude by root 2. Unfortunately, this simple relationship is generally not applicable to seismic airgun pulses, due to their pulsed and time-varying nature.

C. ACOUSTIC TERMINOLOGIES IN RELATION TO MARINE MAMMALS AND SEISMIC SURVEYS

In various studies of marine mammal behavioral responses to seismic surveys, it is common to see measurements of received sound pressure levels, quoted in dB re 1 μPa. These measures can be either peak-to-peak, zero-to-peak, or rms, although often the units are not specified. Strictly speaking, it is only possible to obtain direct peak sound pressure levels from narrowband sounds. Peaks can be measured from pure, or nearly pure, sinusoids, and their amplitude expressed in dB relative to the reference unit of 1 μPa. As most studies of seismic surveys have tended to concentrate on the low frequency sound levels of airgun pulses, from which strong narrowband signals can be visualized, such measures have often been used. (Source levels presented in **Appendix D** are zero-to-peak; rms levels are approximately 10 dB lower than zero-to-peak measurements.) This has the advantage of simplicity, and direct comparability with the low frequency end of known cetacean audiograms, which are also derived in response to narrowband tones.

Narrowband measurements of airgun output give an indication of the kinds of over-hearing-threshold sound pressures that may occur. For instance, a received sound pressure level of 180 dB re 1 μPa at 200 Hz would be about 70 dB above hearing threshold (the level at which a sound just becomes audible) of a dolphin at the same frequency. Another terminology, used by Richardson et al. (1995) and Davis et al. (1998), is Sound Exposure Level (SEL). This is essentially a sound dosage, and represents the total acoustic energy in a transient sound. SEL is proportional to $10 \log(p^2 t)$, where "p" is the rms pressure measured over the duration, "t," of the transient. SEL is defined as the constant equivalent sound level that supplies the same acoustic energy dose in one second as provided by the actual time-varying transient sound over duration "t." If t<1 s, SEL is less than the rms level; if t>1 s, SEL is greater than the rms level.

It has become increasingly apparent from recent measurements (Goold and Fish, 1998; Sodal, 1999) that airguns output a broad sound spectrum. When considering impacts of seismic surveys on smaller marine mammal species, such as the delphinids, it becomes increasingly difficult to express seismic sound pressure levels and relate them in any meaningful way to delphinid auditory threshold curves. Taking the holistic view, airgun pulses are broadband transients, although low frequency components below 200 Hz dominate, and those components can be acceptably represented and measured via narrow band filtering. Broadband sounds have continuous frequency spectra (i.e., not a distinct peak at a particular frequency) and must be described with a power density spectrum, where power per unit frequency is plotted against frequency. Modern spectral analysis software can easily produce power spectral density plots, with units of dB re 1 μPa²/Hz. The frequency spread of airgun pulse power extends across the entire human audio range of 20 Hz to 20 kHz (Goold and Fish, 1998), and into the ultrasonic range >20 kHz (Sodal, 1999). Due to the increasing sensitivity of the dolphin auditory system as frequency increases, these high frequency components cannot simply be ignored. Some limited data are available on broadband airgun output, convolved with dolphin auditory thresholds, and this will be discussed further.

D. OTHER CONCEPTS: TTS AND PTS

Exposure to very high level sounds can induce physical and physiological effects that can be either temporary or permanent. Temporary threshold shift (TTS) is the temporary raising of hearing threshold resulting from exposure to high level sounds. The condition results from metabolic exhaustion of the inner ear sensory systems, which reduces sensitivity to sounds until recovery is achieved after a period of time. Permanent threshold shift (PTS) is, again, a raising of the hearing threshold from overexposure to high level sound; but, in this case, permanent damage occurs to the ear inner ear sensory mechanisms and hence the shift is non-reversible. In humans, such threshold shifts are usually the result of prolonged exposure to continuous sounds, which create threshold shift within the critical hearing bandwidth overexposed. Repeated exposure to TTS levels without sufficient recovery time can lead to PTS. Overexposure to explosive or concussive sounds, however, can also induce threshold shift and tends to do so over a broad bandwidth.

Pressure and overpressure effects are phenomena concerned mainly with the near-field around a sound source. Rapid rise-time events such as explosions or seismic pulses create large and rapid pressure differentials at source. Tissue structures in the near-field are first exposed to a rapid pressure rise (over ambient) due to the compressional wave, followed rapidly by a pressure drop caused by the rarefaction. These pressure effects can rupture tissues and cause other physical damage. In the case of chemical explosives, these pressure fluctuations can occur within the space of a few microseconds. The rise and fall times of airgun pulses are not so rapid, but nevertheless have the potential to cause physical damage to sensory structures.

E. RECENT THEORY

1. Resonance Phenomena

a. *Definition and Examples of Resonance Phenomena*

Resonance is usually defined as the condition of a body or system when it is subject to a periodic disturbance of the same frequency as the natural frequency of the body or system, whereupon it displays an enhanced oscillation or vibration. Familiar types of resonators are objects such as organ pipes, which are manufactured to resonate at particular frequencies to give the desired musical note. Somewhat less familiar are helmholtz resonators, which take the form of an air cavity with a narrow neck — like a round bottomed flask. If such a system is compared to a mass oscillating on a spring under the force of gravity, then the air in the cavity of the helmholtz resonator is analogous to the spring, and the air in the narrow neck is analogous to the mass. A practical example of a helmholtz type resonator is the effect obtained when blowing air across the open top of a glass bottle. A distinctive note is heard as the air inside the bottle resonates from the stimulation provided by the passing air flow.

Resonance phenomena occur in many walks of life and can be desirable (e.g., organ pipes), or undesirable (e.g., built structures such as bridges sent into forced oscillations by high winds). Resonance is commonly associated with air cavities, and in the ocean the presence of air cavities in either biological structures (e.g., fish swimbladders, mammalian lungs), or inert forms (such as air bubbles) means that resonance effects can be induced and observed. Bubble resonance has been studied at some length, and many mathematical formulas have been derived to describe them (e.g., see Glotov et al., 1962; Clay and Medwin, 1977; Pace et al., 1997; Kargi et al., 1998; Orris and Nicholas, 2000). At the simplest level, taking an idealized situation without losses from surface tension and thermal conductivity, the resonant frequency of a free floating bubble in water can be accounted for by the compressibility of the enclosed gas, and the liquid mass moved by the bubble as it pulsates (Clay and Medwin, 1977). In reality, losses occur from factors such as surface tension and thermal conductivity, which effectively create damping. Without damping, there would be no limit to the amplitude of the bubble oscillation at resonance, or to the duration of resonance. The actual calculation of resonant frequencies can be quite involved, and may have to be calculated in a variety of different ways. Mathematical derivation of resonance is beyond the scope of this discussion. As but one example, Glotov (1962, from Clay and Medwin, 1977) calculated the resonant frequency of 60 micron bubbles in breaking waves to be 52 kHz. Advances in technology now make possible experiments such as those by Hoff et al. (2000), in which polymer encapsulants are used to vastly alter the resonant characteristics of microbubbles. Such small bubbles tend to have high resonant frequencies, but much larger air cavities, such as lungs, tend to have much lower resonant frequencies. Again, the analogy can be drawn with a set of organ pipes.

When an air bubble resonates, it absorbs energy at the frequency that drives the oscillations. A resonating bubble effectively acts as a notch filter, absorbing energy at a certain frequency. In addition, bubbles will also scatter and diffuse sound energy, due to their large impedance mismatch with the surrounding water. Bubble curtains have been used around manmade noise sources, such as percussive pile driving (Würsig et al., 2000), in order to reduce sound levels escaping to the surrounding environment. One other effect that occurs when water is highly aerated with bubbles is a dramatic reduction in the "stiffness" of the medium (i.e., related to the bulk modulus), leading to a reduction in sound velocity.

Resonance may be viewed as a phenomenon that occurs in response to a lengthy stimulus (e.g., the sustained note from an opera singer that resonates, and ultimately shatters, a wine glass). While it may take several cycles of a stimulus signal to drive a real system into forced oscillation, a mathematically resonant condition may be stimulated by very short input signals. An abstract analogy can be seen in the realms of digital signal processing, by illustrating the impulse response of a narrow bandpass filter. Filters are typically tested with a unit impulse, which delivers all frequencies simultaneously. A filter will resonate in response to the presentation of a unit impulse (i.e., the filter's impulse response will be derived), as depicted in **Figure G-2**.

A somewhat less abstract analogy is that of a church bell. The stimulus is rapid, in the form of a single hammer blow to the metal. The response is prolonged, as the bell resonates at a particular frequency, or a particular set of frequencies, dependent upon its size and construction. It should be noted that while the bell will resonate at a fundamental frequency, there are often multiples of this frequency, or harmonics, that are also stimulated into oscillation. Typically, the harmonics will decay away first, leaving the tail end drone of the fundamental frequency. The waveform and spectrogram of a bell chime are depicted in **Figure G-3**. It is not necessary to rely totally on abstract or familiar examples of resonance induced by short duration stimuli. In a more relevant vein, short pulse excitation of microbubbles in water has also been demonstrated (Pace et al., 1997).

b. *Biological Context*

In a biological context, as evidenced by consideration in recent environmental assessments of noise impacts, much attention has focussed upon the auditory abilities of marine mammals. This is typically in consideration of source and received sound pressure levels in relation to known, or projected, hearing thresholds. Knowledge of marine mammal hearing thresholds is still incomplete, and exists only for the small- to medium-sized odontocetes, among the cetaceans. Projected hearing thresholds for large mysticetes are based upon ear anatomy, the frequencies at which mysticetes vocalize (i.e., typically corresponding to good auditory sensitivity in odontocetes), and background noise levels in the ocean at the projected frequencies of sensitivity (Ketten, 1998). The logistical problems of obtaining hearing threshold data for large mysticetes are considerable, and in the foreseeable future these data are likely only to be obtained by specialized equipment (e.g., portable techniques to measure auditory brainstem response). Further, nothing is known in regards to equal loudness curves for cetaceans as sound pressures build above hearing threshold (these curves typically flatten out in humans), and there is only limited conjecture about broadband noise interactions with weighted hearing thresholds (Goold and Fish, 1998).

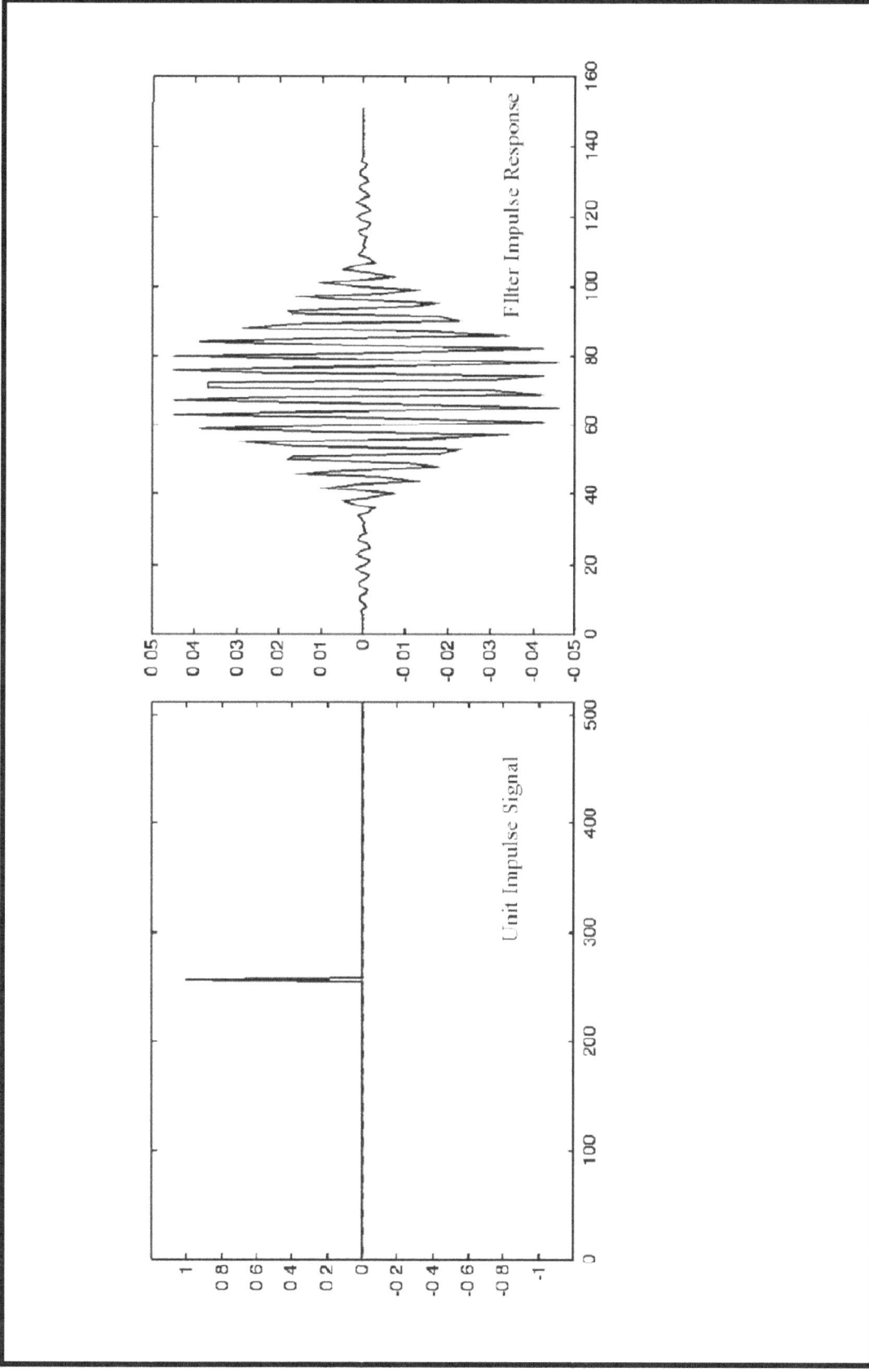

Figure G-2. Unit impulse signal, left, and impulse response of a narrowband filter, right.

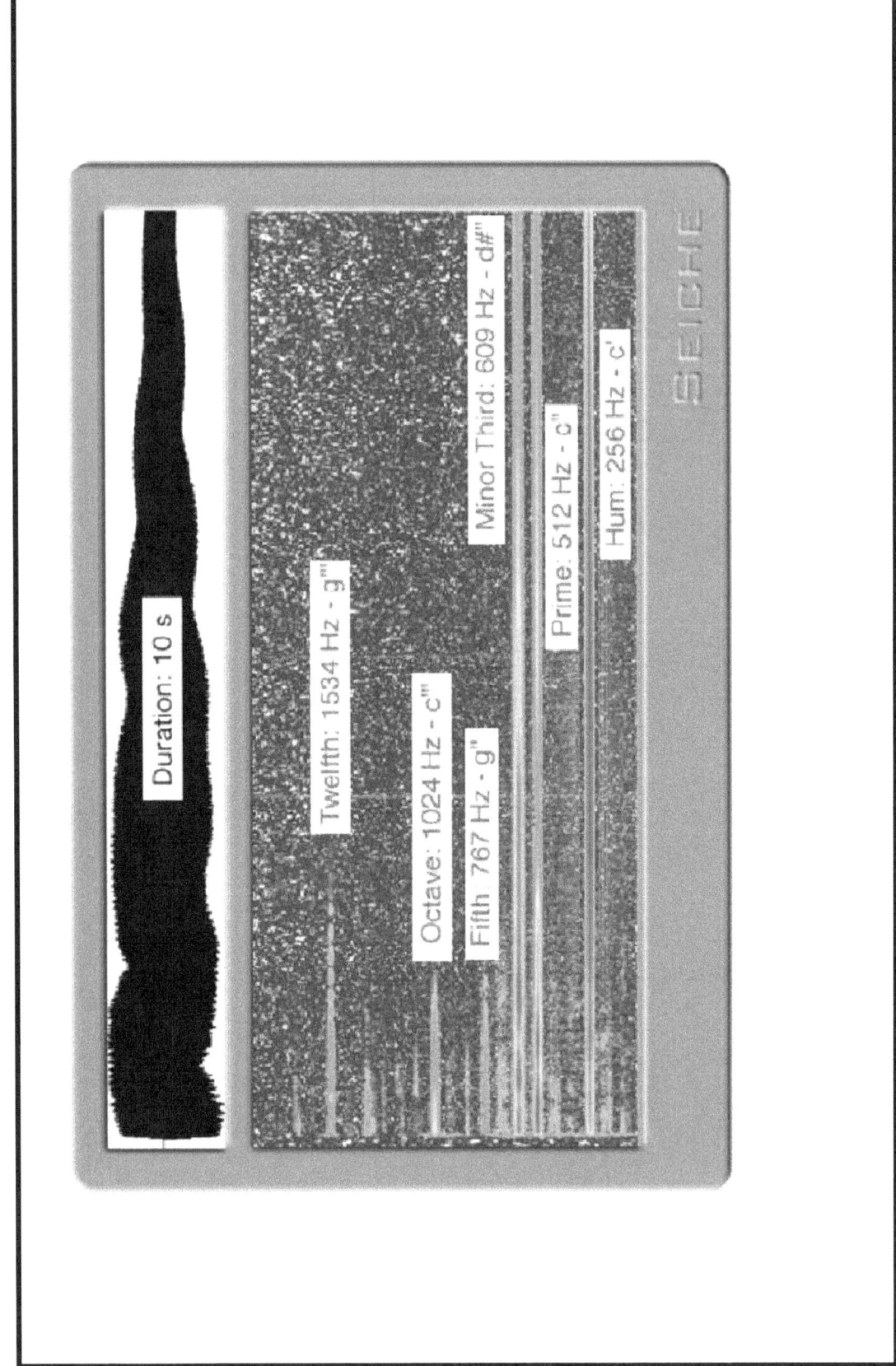

Figure G-3. Wave form and spectrogram of a bell chime, showing prolonged resonant components following an initial impulse stimulus (From: Coates, 2001a, b).

Studies of hearing thresholds, reactions to noise, and noise impacts on marine mammals have, for the most part, not extended to the consideration of resonance phenomena. In fact, there is a paucity of scientific studies of the effects of resonance phenomena on marine mammals. This makes discussion of the subject largely hypothetical, but in the wake of recent mortalities (Frantzis, 1998; Balcomb and Claridge, 2000, 2001; NMFS, 2001) there is mounting evidence of sub-lethal and lethal acoustic impacts on marine mammals as a result of acoustic stimuli. Initial discussion and scientific debate regarding the Bahamas mass stranding incident (Balcomb, 2001; Tepley, 2001) centered around the possibility of resonance phenomena occurring in marine mammal, particularly beaked whale, tissues. Further informed debate since that time (NMFS, 2002) suggests that resonance in tissue structures and air spaces was unlikely to have occurred in the animals involved in the Bahamas mass stranding, given a mismatch of military sonar frequencies and cranial air space volumes and projected resonant frequencies. Other ideas for causative mechanisms of tissue trauma have since emerged, and these, along with resonance in general, will be discussed further in this section.

In a biological context, it seems that many animals may use resonance to their own advantage. Fish may resonate their swimbladders in order to communicate or attract mates. For instance, haddock produce popping sounds by stridulating the swimbladder, and seem to use these in courtship displays (Goold and Coates, 2001). It has been conjectured that the blue whale may employ a helmholtz resonator type effect, using its lungs and tracheal air spaces, in order to produce certain types of low frequency moan (Aroyan et al., 2000). It is further conjectured that the driver for this system in terms of air flow is the change in hydrostatic pressure as the whale ascends and descends. The model proposed is necessarily hypothetical, based upon anatomy and physiology, due to the extreme practical difficulties of actually observing such a mechanism in action, and is not offered to explain the vocalizations of other species. Barham (1973) suggested that the lung and tracheal air spaces of the fin whale may actually be used as a resonant receiver, in order to receive the 20-Hz signals produced by this species.

However, while some animals may use resonance to their advantage, the induction of resonance into a system that is not designed to cope with the associated mechanical stresses and strains might be expected to result in damage. When high level noise sources are introduced into the marine environment, there is potential for resonance to be induced, particularly in the air spaces of biological structures. Much of the public comment and discussion surrounding the phenomenon of resonance in marine mammals has taken place within the context of the U.S. Navy's proposed use of a low frequency active sonar system (LFA; see Department of the Navy, 2001). In the event, it appears that the mass stranding event in the Bahamas resulted from the deployment of mid-frequency, tactical sonars (NMFS, 2001) with a center frequency of approximately 3.5 kHz. Tepley (2001) has calculated the relative severities of LFA versus tactical sonars in terms of particle motion at resonance and concludes that sound at LFA frequencies would likely be more damaging to tissue structures than would sound at mid-frequencies. Ironically, it is the mid-frequencies that have so far been strongly linked to the Bahamas strandings, leaving the suggestion open that low frequencies could be potentially more damaging if resonance were indeed a plausible agent.

c. *Potential Sites of Trauma in Marine Mammals*

As much of the mammalian body is composed of liquid (essentially water) and bone, it is largely incompressible. This applies to marine mammals as well as humans. Solid structures (e.g., bone) may resonate, but given that their densities are of a similar order of magnitude to water, the particle motion differential is not vast. However, both humans and marine mammals also possess air spaces within their bodies. These spaces include the lungs, tracheal cavities, sinuses, and the middle ear. Despite the specialized evolution of marine mammals to an aquatic environment over the past 50 million years or so, they have retained an air-filled middle ear.

When sound waves in the ocean encounter a flexibly delimited air space, they encounter what is described as a pressure release boundary. The biggest such boundary is the sea surface itself, but the same applies to air cavities such as free air bubbles, fish swimbladders, mammalian lungs, and soft bounded air spaces. At these pressure release boundaries, the acoustic pressure wave in the water is translocated as enhanced particle motion into the air cavity. It is this fact that can make acoustic shock waves such as explosives so damaging, as a rapid pulse of compression and rarefaction across the boundary of two media differing greatly in acoustic impedance can literally tear soft tissue structures apart.

It appears that many air cavities within cetaceans may have flexible boundaries, at least along parts of their margins (Fleischer, 1978; Tepley, 2001). Tepley (2001) argues from Fleischer's work that about one-third of the cetacean air-filled middle ear cavity may be up against the pterygoid bone, which is very rigid, but that the remainder could be surrounded by either a flexible tympanic bone or soft tissue, which couples the cavity to the water outside. Similarly, the sinuses in cetaceans may have largely flexible boundaries, which can expand and contract in much the same way as does a bubble oscillating in open water (Tepley, 2001). Tissues surrounding cetacean lungs are usually flexible, hence expansion and contraction can occur. The flexibility of the lung and chest wall, in the context of acoustic resonance, has been demonstrated in humans (Martin et al., 2000).

It therefore seems that, if the correct stimulus were applied, relatively large amplitude oscillations could occur in soft-bounded air spaces of the cetacean middle ear, sinus cavities, and lungs. Balcomb (2001) was of the opinion that resonance in the cranial air spaces of beaked whales, from the Bahamas strandings, was responsible for traumatizing delicate tissues around the brain and ears. Different air spaces will be resonant at different frequencies given their respective volumes. Further, frequency dependence will alter with depth, as air is compressible, and a flexible cavity volume will shrink greatly as depth increases. The delicate tissues to which Balcomb (2001) refers are associated with the beaked whale pterygoid sacs and the laryngeal air space. Below 100 m, virtually all air from the lungs is likely to be compressed into the laryngeal and cranial air spaces. The pterygoid sacs (sinuses) are adjacent to the earbones and the base of the brain. It was conjectured by Balcomb (2001) that as the sacs resonate in response to external stimuli (i.e., the military sonar signals), trauma is caused to the surrounding tissue structures. However, a subsequent scientific workshop convened to discuss resonance phenomena (NMFS, 2002) reached the conclusion that air space volumes in beaked whales and dolphins were too mismatched with incident sound frequencies to support resonance in response to tactical military sonar signals, and that particle displacements were too small to have been damaging.

Although the discussions of marine mammal air-cavity resonance are to an extent hypothetical, resonance effects have nonetheless been observed in the human lung (Martin et al., 2000). These workers concluded that lung stiffness and resonant frequency are dominated by gas in the lungs, with a small contribution from the chest wall. This conclusion is consistent with the above conjecture that either marine mammal lungs, or other flexibly bounded air spaces such as the cranial air ways, will resonate largely as a function of the properties of their gaseous contents. Martin et al. (2000) also found the fundamental frequency of resonance to vary (i.e., increase) with depth. Sound stimulus was provided at frequencies between 20 and 500 Hz, with subjects exposed to depth pressures ranging from 0 to 36.6 m of seawater. Lung resonance frequencies ranged from 39 Hz at surface pressure, to 71 Hz at 36.6 m pressure. Cranial air spaces may not be an appropriate site to look for resonance in small-medium sized cetaceans. Larger air spaces (lungs, trachea) may be better candidates for resonance at the sorts of frequencies involved in military sonar and G&G activities.

d. *Implications for G&G Sources*

The observed cetacean mortalities (Frantzis, 1998; Balcomb and Claridge, 2000, 2001; Balcomb, 2001), and possible acoustically induced trauma to cranial tissues, with conjectured links to high powered sound sources (Balcomb, 2001; Tepley, 2001; NMFS, 2001), have thus far been ascribed to the deployment of military sonar equipment. There is, however, a fundamental difference between such sounds sources and the primary source of G&G noise – the airgun array. Whereas the sonars such as LFA may produce relatively long, multi-second tone bursts, or possibly swept frequency signals, seismic arrays produce relatively short impulse sounds. Sonars are likely to deliver more energy per pulse for an equivalent sound pressure level, due to the sustained pulse duration. If resonance were to be an issue, then such relatively sustained sound pulses could also sustain resonance in air filled structures for longer durations – presumably leading to greater potential of trauma. Seismic sources are not obvious contenders for inducing and maintaining resonance.

However, resonance effects with respect to seismic sources are still worthy of consideration. First, both the military sonar systems described above, and typical seismic sources, operate at sound pressure levels close to theoretical maxima, above which water will begin to cavitate. A sound pressure level of 220 dB re 1 µPa is equivalent to about one atmosphere of pressure, so in near surface waters a single transducer will tend to cavitate if attempts are made to operate it at much above this level. Effective source level for LFA has been quoted at 240 dB re 1 µPa and is achieved with multiple transducers. Seismic arrays operate at similar source levels and also employ multiple transducers. Second, it is possible for resonance to be induced with an impulsive sound stimulus. Two analogies were noted previously – the abstract analogy of the impulse driven resonant filter response, and the rather less abstract analogy of the church bell. Further, Tepley (2001) concludes during his resonance computations for LFA that it may take only five cycles for LFA and mid-frequency sonar sounds to drive the lung and cranial air spaces of cetaceans into their resonant state under the appropriate conditions.

At the LFA and mid-level sonar frequencies, Tepley (2001) computed the required stimulus durations to about 17 and 1.4 ms, respectively (i.e., for all intents and purposes, instantaneous with respect to the sonar pulse duration). Stimuli of 17 to 1.4 ms fall within the realm of seismic pulse durations (see **Figure G-4**). A seismic pulse is unlikely to be able to sustain the resonant state for very long, since there is a non-continuous input of energy, and damping will attenuate the resonant state when the energy source is removed. Damping occurs in all real world systems and prevents those systems from reaching unlimited amplitude oscillations and also from resonating indefinitely. However, even given the relatively short stimulus of a seismic pulse, Tepley (2001) further conjectures that resonance may theoretically amplify an incident sound level by 20 dB and that a mere few oscillations at resonance may be enough to cause physiological trauma to tissue structures.

e. *Airgun Pulses – Shape and Duration*

Airgun pulses are brief transients, and typical pulse specifications for an array will show a sharply rising and falling low frequency pressure spike within the space of 10 to 20 ms. However, measurements in the far-field at varying aspects to the array may show quite different pulse shapes. Further, the introduction of wideband recording reveals the complexity within seismic pulses, especially when recordings are made off the main transmission lobe (Goold and Fish, 1998; Sodal, 1999; Seiche, 2000). High frequency components are present, and if monitored through an appropriate audio high pass filter, set to roll off below approximately 3 kHz, the sound perceived under certain conditions may not be that dissimilar to the sound made by a pair of clashing cymbals, although it is somewhat less musical in quality (Goold, personal observation). **Figure G-4** illustrates the pulse shape and spectrogram of a seismic array pulse, recorded at 750-m horizontal range from source. The pulse is clearly complex and

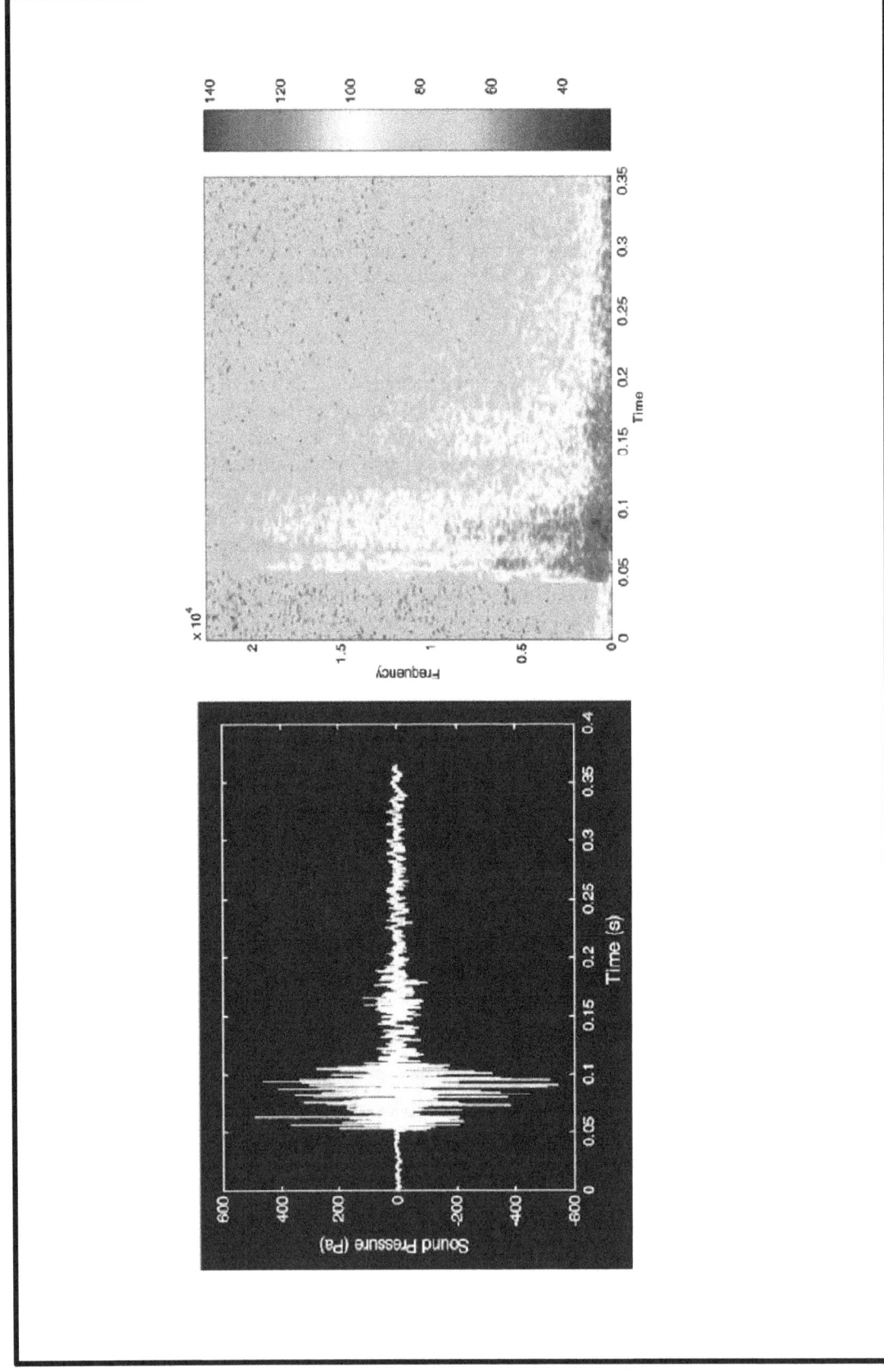

Figure G-4. Waveform trace of seismic pulse, left, and associated spectrogram of pulse, right, recorded at 750 m horizontal range. Spectrogram shows spectral density, dB re 1μPa/Hz. Frequency axis is x10,000 Hz (i.e., 0-22 kHz.)

has major energy over at least 50 ms with substantial reverberant energy beyond this. The spectrogram illustrates the wideband character of this pulse, albeit with a preponderance of low frequency energy.

The seismic pulse illustrated in **Figure G-4** was recorded in relatively "shallow" water, in this case approximately 100 m deep. Shallow water is commonly defined as water to a depth of about 200 m; however, in terms of multipath propagation, shallow water can become something of a relative concept (i.e., the horizontal range of the receiver from source, with respect to the water depth). As reflected in **Figure G-4**, horizontal range is approximately 7.5 times the water depth, hence multipath propagation tends to lengthen the pulse. Pulses may not only stretch but effectively become multiples if the propagation conditions are right. **Figure G-5** shows a similar seismic pulse with two major events and extensive reverberation with respect to the source signature. Low and mid-frequency reverberation has become stretched to approximately a quarter of a second (250 ms). Such stretched and reverberant pulse durations are far in excess of the 17 ms that Tepley (2001) conjectures is needed at low frequency to induce a resonant state.

Under the appropriate conditions, the duration of a seismic pulse can become greatly stretched, and it is even possible for its character to change completely (e.g., from a pulse to a chirp). Such a phenomenon has been documented in a waveguiding channel (Coates, 2001a; Goold and Coates, 2001). This can occur due to normal mode propagation of sound in a waveguide. The seabed acts as a node (i.e., the water cannot move up and down against the hard surface) and the sea surface creates an antinode (i.e., the water can move at a pressure release boundary). The water depth can therefore accommodate modes of vibration that are one-quarter, three-quarter, etc., wavelengths long. Wavelengths that fall between these fractions tend to create tilting at the wavefront, which in turn determines the angle of bounce at the channel top and bottom boundaries. Low frequencies are inclined at a steeper angle than high frequencies and hence travel a longer path and undergo more reflections by the time they are received at a distant point. This progressive increase in path length as frequency decreases means that the effective horizontal propagation of sound is frequency dependent and that high frequencies will be received first. The received sound is therefore a downwards chirp, turning an impulsive sound into a swept frequency signal. As frequency decreases beyond a certain point, sound incident at the seabed reaches the critical angle and is largely transmitted into the seabed rather than reflected. Energy at such low frequencies is rapidly lost from the horizontal transmission. **Figure G-6** depicts the modification of a seismic pulse to a chirp.

f. Vibroseis

Although frequency sweeps from a sonar may possibly be damaging to cetaceans, long range normal mode propagation of seismic pulses, changing their character from a pulse to a chirp, is likely to result in relatively low received levels. However, there is some relevance in the discussion of swept frequency signals in terms of G&G activities. Although not commonly deployed, vibroseis technology can produce swept frequency signals of several seconds duration and with source levels in the order of 220 dB re 1 µPa. The application of such sources has already raised questions in terms of the sustained pulse duration, possible masking effects on marine mammal sounds, and saturation of the integration time of the marine mammal ear. Peak amplitude is effectively traded off against pulse duration. Further, the sweep and duration of the signal makes it more like a low frequency sonar signal, hence it is possible that sustained resonance phenomena could be induced within cetacean air spaces. There is also a slight risk that pingers and transponders could emit sounds with appropriate properties to induce resonance.

2. Decompression Sickness (DCS)

The controversy surrounding the Bahamas mass stranding event, and subsequent beaked whale mass strandings in other parts of the world, has given rise to much debate. From this debate has arisen discussion of a previously unconsidered possibility in these unusual mortalities. The discussion centers

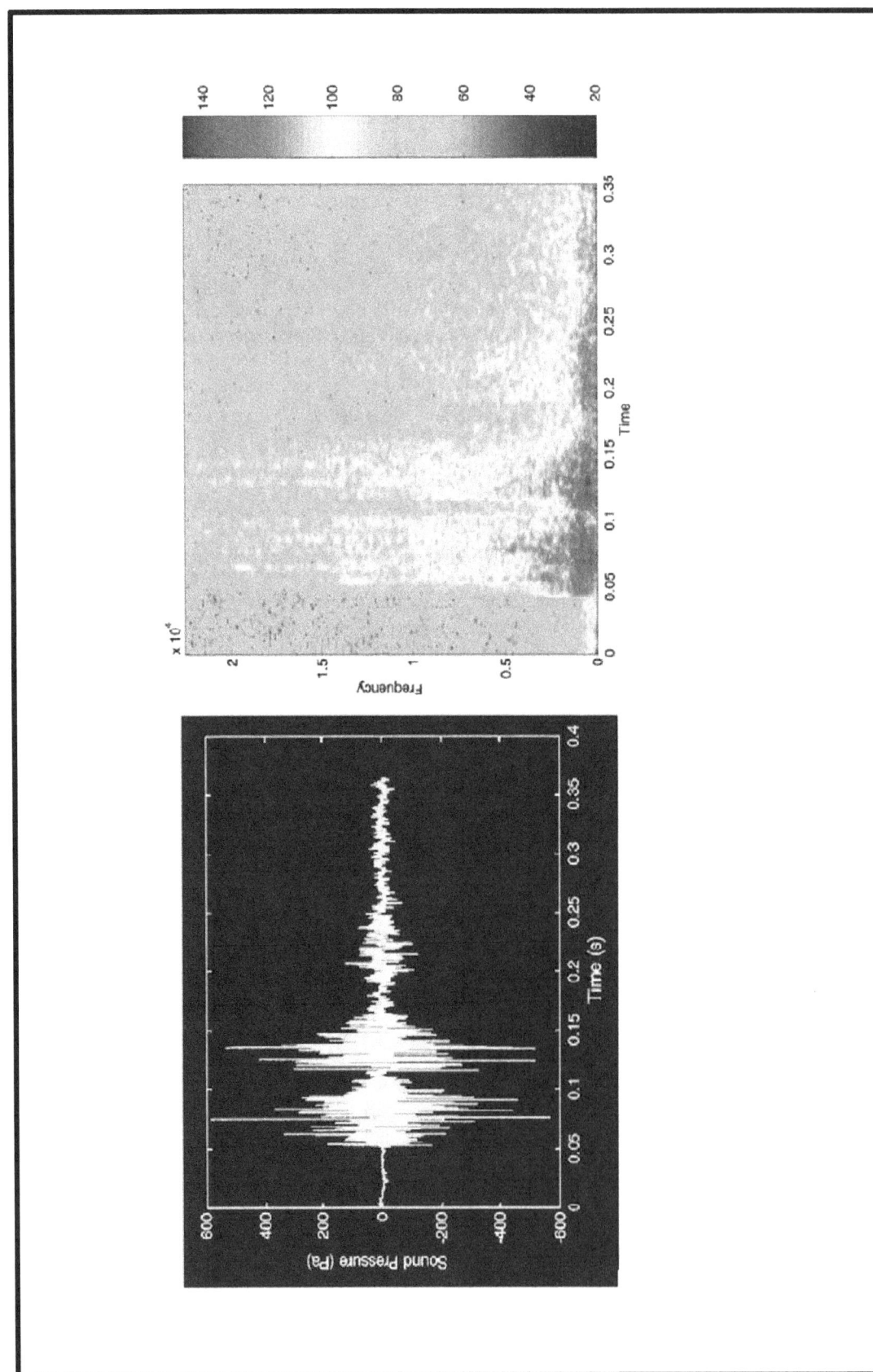

Figure G-5.   Waveform trace of seismic pulse at 1 km horizontal range after multi-path propagation, left, and associated spectrogram of pulse, right. Spectrogram shows spectral density, dB re 1 μPa/Hz. Frequency axis is x10,000 Hz (i.e., 0-22 kHz).

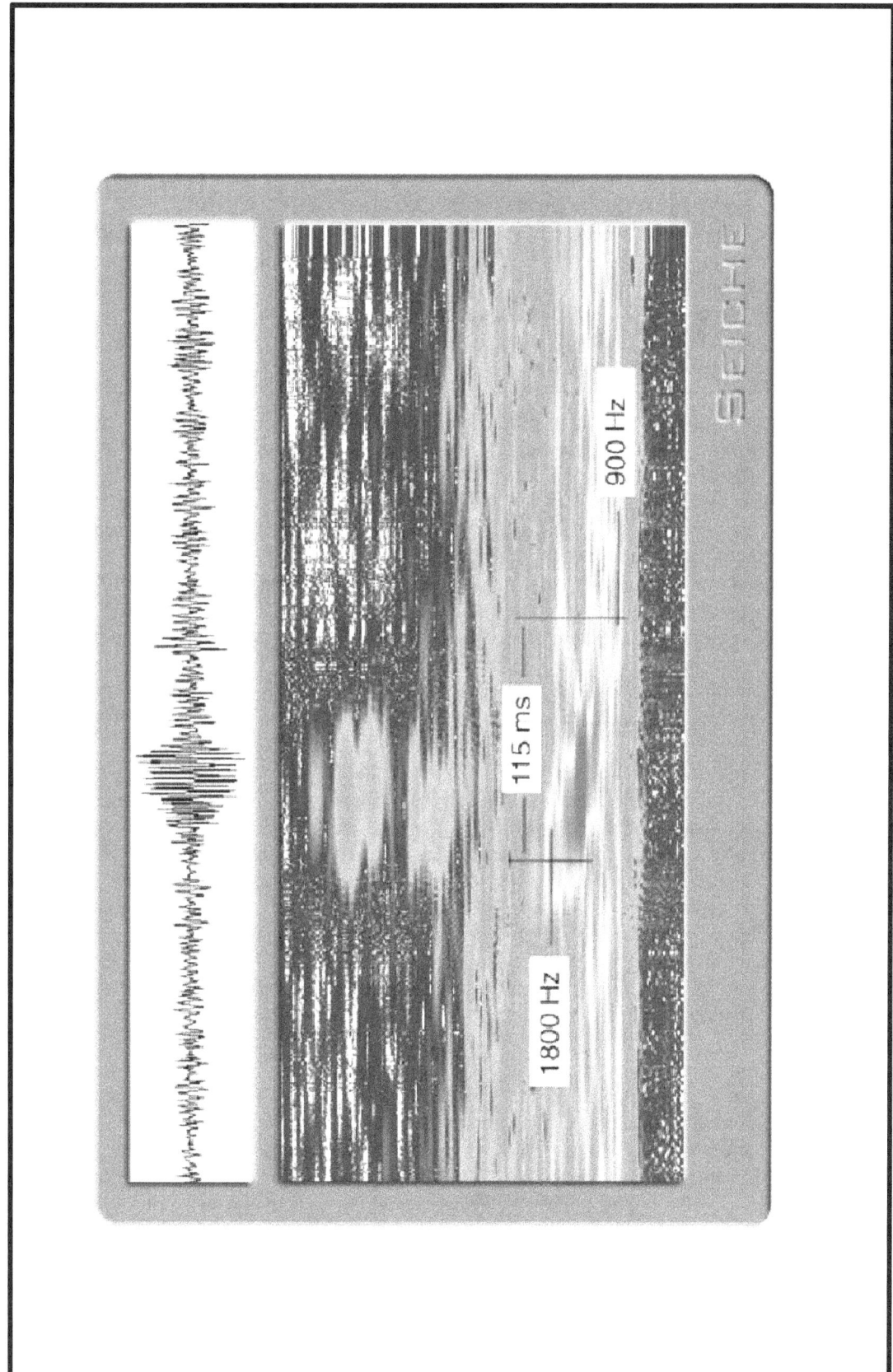

Figure G-6. Seismic pulse modified to a chirp via normal mode propagation in a waveguiding channel (From: Coates, 2001a, b).

around the possibility of acoustically induced decompression sickness (DCS) in deep diving marine mammals, particularly beaked whales.

The pathological findings from the Bahamas event indicated hemorrhages in the sub-arachnoid space, with blood clots in some inner ear spaces and in the lateral ventricals (i.e., bleeding around the inner ears and the brain) (NMFS, 2001). There was no evidence of gross fractures or contusions (i.e., from blows or blast shock waves), and the resonance hypothesis has been largely discounted as an agent in the Bahamas incident due to a mismatch between cavity volumes, resonant frequencies, incident sound frequencies, and probable particle displacements. Beaked whales involved in a subsequent mass stranding in the Canary Islands in September 2002 also were subject to detailed examination and showed similar hemorrhages. Indeed, hemorrhaging was found throughout the tissues and organs of the whole body (Degollada et al., 2003). Histology and detailed pathology indicates fat embolism, a secondary effect of air embolism (Degollada et al., 2003). This finding suggests that these unusual beaked whale mortalities are the result of DCS, more commonly known as "the bends" in human divers.

It is perhaps counter-intuitive to think that deep diving cetaceans should be susceptible to the bends. Cetaceans store the majority of their oxygen supply in vascular hemoglobin and muscular myoglobin, unlike a human diver who will take down a large volume of gaseous air compressed into a cylinder. However, nitrogen will be naturally absorbed in tissues, and even taking down just one lung full of air provides a reservoir for further absorption. Oxygen is only a small proportion of the gaseous component of air (~21%); nitrogen is by far the greatest component (~78%). Whereas oxygen is metabolized into the body, nitrogen is inert and becomes dissolved in solution or in tissues. Human divers saturate their body tissues with nitrogen under pressure, and decompression stops are needed when ascending from a deep dive, so as to gradually release this gas from solution without harmful effects. Cetacean body tissue can become supersaturated with nitrogen, especially in deep diving animals, since they descend to great depths that exert an enormous hydrostatic pressure on their bodies. This pressure effectively "squeezes" the nitrogen into solution in the body tissues.

Mammalian tissues appear to contain microbubbles naturally, even when saturated or supersaturated (Potter, 2003), which is crucial to the understanding of how DCS may come about. Absorption of nitrogen into the fluids and tissue of the body will initially take place via the lung alveoli, but after lung collapse this absorption also may occur through tissue boundaries in the tracheal spaces (Potter, 2003). Clearly marine mammals must have evolved to cope with diving without succumbing to DCS. However, in evolutionary terms, it would be sensible for deep diving cetaceans (e.g., sperm whales and beaked whales) to maximize their foraging time at depth, since deep dives are metabolically expensive. As a consequence of this, deep divers might operate close to their physiological limits in terms of foraging time at depth vs. oxygen depletion. An acoustic trigger that impacts on their physiology, or a modification of behavior brought about by an acoustic stimulus, could tip the balance.

It appears that naturally occurring microbubbles in mammalian tissues are stabilized by a mechanism that is not yet understood. This mechanism prevents the bubbles from completely disappearing into solution, which would normally be expected to occur in the case of a free bubble in a liquid, and also prevents the bubble from expanding to cause DCS in normal surfacing behavior. It is conjectured that such a mechanism could be a surfactant bubble coating of some kind (Potter, 2003). An acoustic pulse may be sufficient to stimulate the bubble and degrade such a coating, effectively "activating" the bubble, which could thereafter expand through the process of static diffusion until it causes air embolism. This is effectively an acoustic activation of DCS in a deep diving marine mammal. An alternative, or indeed parallel, situation may occur if there is behavioral modification (i.e., disturbance) in response to received sounds. The normal behavior of beaked whales after ascending from a deep dive is to remain as inactive as possible, whilst also undertaking a series of short, shallow dives. Remaining calm is important to avoid shock and cavitation events in the skeletal and cartilaginous joints. If tissues are supersaturated, shock and cavitation events can activate bubbles and cause them to expand, leading to DCS (Potter, 2003). Short dives are analogous to decompression dives, assuming the animal has ascended from depth to restore oxygen, but also needs to safely offload nitrogen from supersaturated tissues.

The surfacing behavior and subsequent shallow diving behavior of deep diving cetaceans may be critical to their physiology if they operate close to the DCS limit (Potter, 2003). Disturbance to the normal pattern of behavior may have a consequence for their physiology. Loud, anthropogenic sounds sources such as sonars and airguns might conceivably cause disturbance or even a panic response at depth, leading to excessively rapid surfacing without sufficient decompression time. In addition, a high degree of locomotory activity may be undesirable in terms of avoiding bubble activation at the joints. Further, if the subsurface waters are highly ensonified, the animals may be reluctant to descend back into the zone for decompression dives, thereby exacerbating DCS (Potter, 2003). If these situations were to occur in a deep diving cetacean operating at the limit of its physiology, DCS could result from the modified behavior. In addition, increased locomotory activity by the animal might exacerbate bubble growth and DCS in the body tissues, since cavitation events may occur in joints and cartilage. It is also conceivable that acoustic pulses could directly activate bubble growth in the tissues at much lower received levels than were conjectured for rectificied diffusion by Crum and Mao (1996). Current thinking is now that bubble activation and subsequent DCS might occur by static diffusion in body tissues, which once triggered by an initial acoustic pulse would not require continued ensonification to take effect (i.e., the effect would "run-away" by static diffusion once microbubbles present in marine mammal tissues had been "activated") (Potter, 2003).

The brain hemorrhages found in the Bahamas animals and the full body hemorrhaging and fat embolism found in the Canary Island animals are symptoms consistent with DCS. The fat embolism is a particularly significant find (Degollada et al., 2003). Air embolism is extremely difficult to see post mortem, but the fat embolism is the signpost to its occurrence. As nitrogen comes out of solution to expand microbubbles in the blood and tissues, capillaries may eventually block and create a back pressure that ruptures the capillary wall. Liquid fat in the tissues and interstitial spaces then enters the bloodstream. Fats are hydrophobic but will coagulate together, so eventually fat globules coalesce and ultimately block larger blood vessels, again creating back pressure and further ruptures in the vascular system. Fats can be stained in post mortem tissues, and were found in abnormal places in the brains, blood and organs of the Canary Island mass stranded animals (Degollada et al., 2003). This is a strong indication that DCS has occurred.

While the DCS hypothesis is currently taking shape, its implications to G&G activities, or even military sonar activities, is as yet unclear. However, as the hypothesis suggests, DCS may be induced either directly through acoustic stimuli or indirectly through behavioral modification.

3. Summary of Recent Theory

The subject of resonance phenomena as an agent in trauma of sensitive tissues surrounding air spaces within marine mammals has been hotly debated and is shrouded in conjecture and uncertainty. The scientific workshop on resonance phenomena (NMFS, 2002) concluded that resonance was unlikely to have been the cause of the Bahamas mass stranding, although the precise causative agent is still unclear. There is a near complete paucity of scientific study on the subject of resonance induced trauma in marine mammals, which renders any discussion largely hypothetical. The pathology of the Bahamas animals and the Canary Island animals indicated internal hemorrhaging around the inner ears and the brain. In addition, there was evidence of fat embolism throughout the body tissues of the Canary Island animals, indicative of decompression sickness. Acoustically induced trauma has been stated as the most likely cause of death (or at least precursor to a lethal stranding) in both the Bahamas mass stranding (NMFS, 2001) and the mass stranding in the Canaries (Degollada et al., 2003), but the precise mechanism by which acoustic trauma may be operating is a matter of debate at this time. Most of the debate has centered on lethal effects from military sonar sources, but given that similar effects may be emerging from G&G activities, it would be wise to exercise caution. It should be recognized that such evidence of acoustically induced trauma is extremely difficult to acquire, and high profile stranding events require the coincidence of activities the presence of animals, and favorable geography/topography. Detailed pathology of the Bahamas and Canary Island animals was a result of high profile incidents occurring under an unfortunate set of circumstances, in locations able to be serviced by appropriate specialists, with an urgent priority to conduct pathological examinations of whale carcasses. Such a concerted effort is not the norm with respect to individual cetacean strandings, and it is possible that trauma in other animals has been overlooked.

Given our limited knowledge at this time, plus a wide range of variables and uncertainties, it is not possible to draw any firm conclusions about either resonance phenomena or DCS in cetaceans as a result of G&G activities. It seems likely that resonance phenomena are not a primary candidate for generating acoustic trauma, at least in the ear spaces, and the short impulse nature of seismic survey sounds should mitigate to a certain extent. Acoustically induced DCS seems a more plausible candidate at this time and could account for lethal and sub-lethal effects seen in both the Bahamas mass stranding incident and the Canary Island mass stranding incident, both of which appear to have possibly been triggered by high level anthropogenic sound sources. It would be beneficial to have a more concerted effort to examine possible sites of acoustic trauma in stranded cetaceans (i.e., to conduct detailed ear and cranial air space pathology as standard procedure during necropsies). This should especially be the case if some correlation is perceived in space and time between G&G activities and strandings. Ascertaining such correlation requires coordination among Federal agencies (e.g., Minerals Management Service [MMS], NMFS strandings network coordinators) and G&G operators, such as is established in Notice to Lessees No. 2003-G10 (Vessel Strike Avoidance and Injured/Dead Protected Species Reporting). No conclusions or recommendations are offered other than to adopt precautionary measures, maintain vigilance of observation during G&G operations, and adopt all required mitigation measures to the highest standard.

II. MARINE MAMMAL IMPACT SENSITIVITY

Richardson et al. (1995) defined four zones of potential noise effects on marine mammals. In order of increasing severity, they are

- Audibility;
- Responsiveness (Behavioral Effects);
- Masking; and
- Hearing Loss, Discomfort, or Injury (Physical Effects).

The audibility of a sound depends largely on hearing threshold in the relevant frequency band and the level of ambient noise in that band. It seems unlikely that a sound that is just audible to a marine mammal (i.e., essentially a background sound) will have a major impact on an animal that spends its life in a perpetually noisy environment.

Responsiveness (behavioral effects) defines a condition whereby marine mammals exhibit changes in "normal" behavior (or in the case of captive animals "trained behavior") in response to certain levels of received sound. A number of studies have documented behavioral effects and will be considered.

In the case of masking, a sound signal that is of importance to a marine mammal (e.g., communication calls, echolocation, environmental sounds cues) is rendered undetectable due to the high noise to signal ratio in a relevant frequency band. In the case of G&G activities, where potential masking noise takes a pulsed form with a low duty cycle (~10%, or a 1-s disturbance in the sound field in every 10 s of ambient noise), the effect of masking is likely to be reduced relative to continuous sounds. In fact, Davis et al. (1998) considered masking to be of insignificant consequence in relation to possible impacts of seismic surveys on the Scotian Shelf, largely due to the low duty cycle of seismic pulses. Davis et al. (1998) suggest that only behavioral and physical effects rank with any consequence in relation to seismic surveys.

Physical effects (hearing loss, discomfort, and injury) are, undoubtedly, the most serious of all the above conditions. Physical effects can be both lethal and non-lethal, but by definition have a direct impact on the physiology of a marine mammal and its auditory system. At the lower end of this scale is a condition known as TTS. In TTS, the lower threshold of hearing in the relevant frequency band is increased (i.e., hearing becomes less sensitive) when exposed to a critical combination of sound intensity and duration. Moving up the physical effects scale, a phenomenon known as PTS occurs. In PTS, the hearing threshold shift is non-recoverable (i.e., permanent damage results to the auditory system). Repeated exposure to TTS levels without sufficient recovery time can lead to PTS. Finally, the most severe effects are physical damage, where organs and tissue structures rupture due to the high and rapid peak pressure fluctuations. Generally, such a severe event is only expected to occur to an organism in the near-field close to an airgun. However, in light of recent debate it may be considered that physical effects, especially decompression sickness, may be acoustically induced in deep diving cetaceans at received sound levels considerably lower than those required to produce TTS and PTS in auditory structures (NMFS, 2001; Potter, 2003).

Potential impacts in the Gulf of Mexico are likely to include behavioral effects, which could have extensive radii (kilometers) from airgun sources, and possibly physical effects extending from behavioral modification and acoustically induced DCS. Perceived wisdom from other studies suggests that audibility in itself is not likely to cause adverse impacts and that masking is not likely to pose a major problem due to the low duty cycle of seismic pulses. TTS and PTS are only likely to occur at close ranges (tens or perhaps hundred of meters from an airgun source). Therefore, physical damage to auditory structures is only likely in extreme proximity to airgun sources, although DCS might occur in response to sounds at much lesser proximity.

This discussion of impact sensitivity will address the species (or groups of species) most likely to be encountered in the Gulf of Mexico. As reviewed in **Appendix F, Section I.A. - Marine Mammals**, a wide variety of marine mammal species are present in northern Gulf of Mexico oceanic waters where they may come in contact with seismic survey operations. Species include members of Balaenopteridae, Physeteridae, Ziphiidae, and Delphinidae (primarily Bryde's whale, sperm whale, various beaked whales, dwarf and pygmy sperm whale, and a variety of dolphin species). Due to their coastal habitat preferences, sirenians (West Indian manatees) are highly unlikely to come in contact with seismic operations.

A. MYSTICETES

1. Behavioral Effects

The mysticetes (or baleen whales) have probably been the most studied group of marine mammals in the open ocean in terms of observations of behavioral changes in response to G&G operations and other high level sound sources. Bryde's whale is the only mysticete regularly occurring in the Gulf of Mexico (see **Appendix E**). Although there have been no studies of Bryde's whale reactions to seismic surveys, it is generally considered on the basis of vocalization frequencies and ear anatomy (Ketten, 1998) that the auditory abilities of all mysticete species are broadly similar. Limited data on Bryde's whale reactions to other anthropogenic disturbance suggest little response to slowly approaching boats (Watkins, 1981b), and also that this species, like others, appears to be easier to approach when feeding (Gallardo et al., 1983).

In terms of overall sensitivity to G&G activities, mysticetes are probably a relatively "high risk" category amongst the cetacea. Hearing sensitivity at low frequencies down to ~10 Hz is almost certainly good, and many of the vocalizations of mysticetes occur in the low tens to a few hundred Hertz (Thompson et al., 1990; Richardson et al., 1995; Crane and Lashkari, 1996; Rivers, 1997; Stafford et al., 1998, 1999), which implies, but does not prove, functional hearing in this range. As airgun arrays are configured to output maximal energy in the region of a few tens of Hertz, there is clearly a possible overlap between the expected frequencies of good hearing sensitivity (low threshold) in mysticetes and maximal airgun output at source.

Sound spreading loss significantly attenuates sound intensity with increasing distance from a source, which naturally reduces the received level of a sound at a distant point (e.g., a marine mammal ear) relative to source level. By way of example, using a 15log(R) cylindrical spreading model in shallow (~200 m shelf sea) water, spreading loss 10 km from source would reduce sound intensity by some 60 dB. However, low frequency sound in seawater is not further attenuated in seawater to any appreciable level by molecular absorption over the distances of interest, and hence low frequency sound will "reach" greater distance than will high frequency sounds. A signal at 100 Hz will only be attenuated by molecular absorption at a rate of ~0.001 dB per km, whereas a signal at 10 kHz will be attenuated at a rate of about 1 dB per kilometer. Therefore the 100-Hz signal would only lose a further 0.01 dB during a 10-km propagation, whereas a signal at 10 kHz would lose a further 10 dB, raising total attenuation from 60 to 70 dB at 10 kHz.

Given that no audiograms of mysticetes have been obtained, it is impossible to define what level of sound above hearing threshold may cause behavioral effects. For this reason, observations at sea have concentrated on relating received sound levels to observed behavioral changes (Malme et al., 1983, 1984, 1985, 1986, 1988; Reeves et al., 1984; Richardson et al., 1986; Ljungblad et al., 1988; Richardson and Malme, 1993; McDonald et al., 1993; Richardson, 1998; McCauley et al., 1998, 2000). It has been supposed, in the absence of audiogram data, that hearing threshold around 10 Hz may occur at a level of some 80 dB re 1 μPa (Ketten, 1998). If this were the case, exposure to sound at 10 Hz and at a level of, say, 160 dB re 1 μPa might represent a level 80 dB over hearing threshold in a mysticete.

The dynamic range of human hearing in air is ~130 dB. For air-borne human sensitivity measurements, using the in-air reference pressure of 20 μPa, 0 dB represents the quietest sound perceivable to a human subject (e.g., pin dropping), whereas 130 dB re 20 μPa is a level that induces pain. Health and safety standards have given guidelines for exposure to continuous sound sources, at a maximum of 80-85 dB re 20 μPa, using an A-weighted human hearing threshold criteria. There also is good reason to suppose, based on ear anatomy and physiology, that humans and marine mammals have similar dynamic hearing ranges, and hence similar criteria may apply.

Behavioral reactions (avoidance) have been noted in gray whales in response to received pulse levels of 164 dB re 1 μPa rms (Malme et al., 1984), equivalent to a sound exposure level of 158 dB re 1 μPa (Davis et al., 1998). Bowhead whales also have been shown to exhibit avoidance of an area of seismic surveying where received levels reached approximately 130 dB re 1 μPa rms (Richardson, 1998; Richardson, oral comm. to J.C. Goold). Such values stimulating behavioral avoidance are not inconsistent with the notion of low frequency hearing threshold of some 80 dB in mysticetes. The Australian seismic guidelines (Environment Australia, 2001) consider that "sounds heard by whales of over approximately 140 dB in feeding, breeding or resting areas may be considered likely to significantly disturb whales that are present. Sounds heard by whales of over 150 dB in other areas, such as migratory paths, may significantly disturb whales that are in the area." Such levels are not inconsistent with the behavioral reactions observed above.

It is sometimes assumed that because marine mammals produce high source level sounds in the course of their vocalizations, they are automatically tolerant of external, high level sounds. However, although emission of high level sounds implies tolerance, such emissions should not necessarily be taken as a measure of acceptable exposure levels for marine mammals to manmade sound. Cetacean ears have specializations that protect them from the animals' own high level vocalizations, but which would not necessarily protect them from an unanticipated, high level, externally generated sound (Ketten, 1998).

2. Physical Effects

Pressure pulses from airguns have the potential for damaging the hearing of marine mammals. Results of such an exposure could lead to TTS. For marine mammals, TTS could temporarily affect an individual's ability to hear calls, echolocation sounds, and other ambient sounds. Animals in closer proximity to a noise source could experience permanent hearing loss (i.e., PTS) under certain circumstances. Unfortunately, there are no data for TTS, PTS, or even hearing thresholds in mysticetes. However, there is evidence that blast injuries can be caused to mysticete ears, albeit by chemical explosives rather than G&G sources (Ketten et al., 1993). Since mysticetes are not typically deep divers, it is less likely they would suffer from acoustically and/or behaviorally induced DCS than would some of the odontocete families.

B. ODONTOCETES: SPERM WHALES

1. Behavioral Effects

Sperm whales have a fundamentally different lifestyle from that of mysticetes. Sperm whales are deep diving, pelagic predators that echolocate at depth using sonar clicks and feed on deep water cephalopods and fishes. Auditory thresholds of adult sperm whales have not been obtained, but it is reasonable to suppose, based on their vocalizations, that they are sensitive to a wide range of frequencies. Sperm whale sonar clicks are broadband, with energy ranging from a few hundred Hertz up to at least 30 kHz (Watkins, 1977). They also have distinct spectral components in their clicks at frequencies as low as 400 Hz (Goold and Jones, 1995). Mohl et al. (2000) recently evaluated sperm whale clicks as they pertain to directionality and source levels. Although the function of these low frequency components is not clear, their presence suggests functionality and implies the ability to perceive them through the auditory system. Source levels of sperm whale clicks have been quoted as high as 223 dB re 1 µPa rms (Mohl et al., 2000). Although this appears to be in the same region as source levels of some seismic sources, the duration of the p1 pulse in sperm whale clicks upon which this figure is based is an order of magnitude shorter than seismic pulses. Therefore the energy contained can be expected to be an order of magnitude lower. Further, as pointed out by Ketten (1998), it is misleading to assume that just because animals can produce high source level sounds, that they are immune to anthropogenic sounds of similar levels. Marine mammals have specializations that protect their auditory systems from high level, self-generated sounds, but which would not necessarily protect them from unanticipated, high level, externally generated sounds.

Possible sensitivity to low frequency sounds has been reported in sperm whales by Bowles et al. (1994), where sounds at 57 Hz with source levels of 209 to 220 dB re 1 µPa may have caused sperm whales to stop vocalizing and/or to leave the area of ensonification. Sperm whales are a highly vocal species under natural conditions (i.e., they click nearly continuously during dives), and interruption or cessation of their vocal activity has often been cited as a reaction to manmade noise. Watkins and Scheville (1975) showed that sperm whales interrupted click production in response to pinger (6 to

13 kHz) sounds. Mate et al. (1994) report decreased sperm whale abundance in an area of seismic operations in the Gulf of Mexico. Watkins et al. (1993) report interruption of vocal activity and immediate submergence by two sperm whales exposed to high level submarine sonar pulses. Andre et al. (1997) showed that 10 kHz pulses, with source levels of 180 dB re 1 μPa, induced startle reactions in sperm whales. Goold (1999) documents an event where six sperm whales in shallow water were actively "driven" through a narrow channel in a coherent formation, using ship noise and echosounder/fishfinder emissions from a flotilla of 10 vessels; sound characteristics and levels were not determined. One contradictory observation, however, reports no alteration in sperm whale vocal activity when exposed to received levels of 173 dB re 1 μPa rms from 1 g TNT detonators (Madsen and Mohl, 2000).

Sperm whales are most likely acoustically aware of their environment and can exhibit behavioral reactions in a number of ways, including interruption of vocal activity and locomotive avoidance. There are, as yet, insufficient data to assign thresholds for acoustic disturbance to sperm whales.

Although no behavioral audiograms have been obtained for adult sperm whales, Carder and Ridgway (1990) made spot measures of auditory threshold in a neonate sperm whale, using the auditory brainstem response procedure. Although not a comprehensive study, this revealed good auditory response at 5, 10, and 20 kHz; at 60 kHz auditory response was present but much weaker. This study suggests that sperm whales have relatively good hearing at low frequency.

2. Physical Effects

There are few documented data on physical effects of high level sound on sperm whales. As noted previously, pressure pulses from airguns have the potential for inducing TTS, which could temporarily affect an individual's ability to hear calls, echolocation sounds, and other ambient sounds. Animals in closer proximity to a noise source could experience permanent hearing loss (i.e., PTS) under certain circumstances. One study has examined the inner ears from two sperm whales killed by vessel collision in the waters around the Canary Islands. Sperm whales in these waters are heavily exposed to shipping noise, and the local "acoustic budget" may influence collision rates (Andre and Degollada, 2003). Analysis of the inner ear structures showed no overt evidence of impact, or ship strike related injuries. However, ears from both animals had reduced auditory nerve volumes. One animal also had patches of dense tissue in the inner ear. The findings were confirmed by histological analysis and are consistent with auditory nerve degeneration and fibrous growth in response to low frequency inner ear damage (Andre and Degollada, 2003).

An additional factor to consider in the case of sperm whales is their deep diving habit. Unlike mysticetes, which may remain close to the surface for long periods, sperm whales spend relatively little time at the surface during the course of feeding activity. This means they would be less likely to receive any surface shielding afforded by either the Lloyd mirror effect or refractive effects caused by near-surface hydrographic conditions, which can occur in some instances. In addition, the sperm whale dive takes them down to a depth where they could be passed over directly by an operating seismic vessel without their being visually detected. As airgun arrays are generally configured to produce a maximum, low frequency energy lobe directly downwards towards the seabed, sperm whales may enter a region of increased ensonification, relative to more near-surface species. As deep divers, sperm whales may be vulnerable to acoustically and/or behaviorally induced DCS.

C. ODONTOCETES: DWARF AND PYGMY SPERM WHALES, AND BEAKED WHALES

1. Behavioral Effects

Little is known of the acoustic abilities of either of these groups, although all the member species are deep diving (Willis and Baird, 1998) and almost certainly use echolocation clicks spanning the sonic and low ultrasonic frequency range. Pulsed sounds with peak frequencies below 13 kHz have been recorded from pygmy sperm whales (Caldwell and Caldwell, 1987), and the anatomy and physical properties of the dwarf sperm whale head have been shown to be consistent with production of echolocation clicks (Cranford et al., 1996; Goold and Clarke, 2000; Clarke, 2003). Thompson et al. (1990) reports a short duration (0.42 s) frequency modulated "cry" from a captive, stranded adult female, pygmy sperm whale, with start and finish frequencies of 1.36 kHz and 1.48 kHz, respectively. Hooker and Whitehead (1998) report echolocation clicks from diving bottlenose whales over a submarine canyon known as the Gully, on the Scotian Shelf. Clicks and frequency modulated whistles have been reported from beaked whales, with frequencies ranging between 300 Hz and 40 kHz (Lynn and Reiss, 1992; Dawson et al., 1998; Roger and Brown, 1999). Narrowband clicks, centered at frequencies between 15 and 17 kHz, have been identified from Cuvier's beaked whales (Frantzis et al., 2002).

Little, if anything, is known of detailed behavioral responses of these groups of animals to anthropogenic sound. However, their behavior may be linked to observed physical effects, and can only be extrapolated at this time.

2. Physical Effects

There are now several examples of possibly acoustically induced stranding events by beaked whales. In 1996, a mass stranding of Cuvier's beaked whales occurred along the coast of the Kyparissiakos Gulf, Greece, in a highly unusual manner (i.e., strandings were synchronous but spread out over 38.2 km of coastline) (Frantzis, 1998). Mass strandings of this species in the Mediterranean are very rare, and this anomalous event was highly correlated with nearby NATO military exercises using high powered sonar systems. There were no measurements of sound levels or characteristics, although source levels were expected to be in the range 215 to 230 dB re 1 µPa below 3 kHz. In March 2000, a mass stranding of beaked whales and other cetaceans occurred in the Bahamas. This stranding incident was highly coincident with naval activity in the area deploying mid-frequency tactical sonar (NMFS, 2001). Necropsies indicated hemorrhaging around the brain and inner ears. In September 2002, a mass stranding of beaked whales occurred in the Canary Islands, again correlated with naval exercises (Degollada et al., 2003). These animals showed clear indication of acoustically induced trauma and decompression sickness. The balance of evidence strongly suggests that these beaked whale stranding events are acoustically induced, and that they occur in response to received sound levels much lower than would be expected to give rise to "normal" physical trauma in marine mammals. Beaked whales appear to be particularly at risk from anthropogenic acoustic activity.

Beaked whales are deep divers, and as noted previously for sperm whales, this may increase their risk of being exposed to higher energy levels from downward-directed seismic pulses. There is evidence that beaked whales may be vulnerable to acoustically and/or behaviorally induced DCS. No audiograms have been obtained in pygmy sperm whales, dwarf sperm whales, or beaked whales, and there are insufficient data to ascribe avoidance thresholds. However, the sonar modeling from the Bahamas incident suggests that the whales affected may have only been exposed to a received level of ~165 dB re 1 µPa.

D. ODONTOCETES: DOLPHIN FAMILY (DELPHINIDAE)

1. Behavioral Effects

The Delphinidae is a diverse group including the true dolphins, the killer whales, and pilot whales. There have been few studies of the impact of seismic surveys on members of the Delphinidae. Indeed, Richardson et al. (1995) comments on an almost total lack of studies on effects of G&G activities on delphinid species. A few observations suggest tolerance by long finned pilot whales to drillship noise (Kapel, 1979), and common, Risso's, and bottlenose dolphin occurrences within 18 km of drillships (Sorenson et al., 1984). While it is true that delphinids have received less attention than mysticetes in terms of G&G effects, there have been some developments since the work of Richardson et al. (1995). There are still relatively few direct studies of the effects of seismic pulses on members of the Delphinidae, although there are now data on physiological and auditory impacts of other noise sources.

Variations in the amount of vessel traffic are not expected to cause significant impacts to delphinids in terms of noise signatures. However, there is evidence that vessel traffic can affect the acoustic behavior of dolphins (Van Parijs and Corkeron, 2001). These workers describe how the vocal behavior of Pacific humpback dolphins is modified in response to transiting vessel traffic. Dolphins significantly increased their rate of whistling immediately after a vessel had moved through the study area at ranges of 1.5 km or less. Effects were particularly noticeable for mother-calf groups, and it is suggested that the noise from transiting vessels affects dolphin group cohesion. Mother-calf pairs were the most disturbed, and have the greatest need to re-establish vocal contact following noise masking. This finding may have particular relevance for the disturbance of critical habitats (e.g., breeding, nursery areas). IAGC has noted that total vessel traffic in an area where a seismic vessel is operating may decrease as fishing and recreational vessels move out of the survey area to avoid vessel conflicts.

Historically, seismic survey operators have considered airguns to be sources of low frequencies (<200 Hz) only. As such, it was not considered that the emissions would be audible to dolphin species, given their high frequency biased hearing and their relatively poor sensitivity at low frequency. However, recent measurements of airgun sources at sea (Goold and Fish, 1998; Sodal, 1999) have demonstrated that, although airgun arrays are a source of primarily low frequency energy, there is also significant energy at higher frequencies. These energies encompass the entire audio frequency range of 20 Hz to 20 kHz (Goold and Fish, 1998) and extend well into the ultrasonic range up to 50 kHz (Sodal, 1999). This high frequency energy must be taken into account in considering seismic interactions with members of the Delphinidae. Further, and contrary to early perceptions, the low frequency components of airgun emissions are of sufficient level to exceed the dolphin auditory threshold curve at these low frequencies, even after considerable spreading loss.

Behavioral audiograms have been obtained for several delphinid species' including beluga, killer whale, bottlenose dolphin, Risso's dolphin and false killer whale (see reviews in Au, 1993 and Richardson et al., 1995; also see: Nachtigal et al., 1996; Ridgway and Carder, 1997; Popov and Supin, 1997; Popov et al., 1997; Erbe and Farmer, 1998; Sauerland and Dehnhardt, 1998; Szymanski et al., 1999). Although there is a degree of variability, the audiograms of small delphinids (the dolphins) follow a generic U-shaped pattern, with peak sensitivity (lowest threshold) of about 40 dB re 1 µPa in the mid-ultrasonic range at about 60 kHz. Hearing sensitivity decreases to about 110 dB re 1 µPa as frequency decreases to 200 Hz, and decreases to a similar level as frequency increases above the peak hearing range to about 150 kHz.

Larger delphinids, such as the killer whale, also exhibit a similar audiogram shape, but with peak hearing sensitivity in the region of 20 kHz (Szymanski et al., 1999), as might be expected from the body/ear scaling factor.

Since the delphinid auditory system has a relatively poor response at the low frequency end (about 110 dB re 1 μPa at 200 Hz) and increases in sensitivity towards the ultrasonic range, it is clear that a gradient of increasing sensitivity exists over a broad frequency range to the frequency of peak sensitivity. Further, although an airgun pulse will have maximal energy at a few tens of Hertz, with energy rolling off towards the higher kHz frequencies, the increase in dolphin hearing sensitivity in this region, in a sense, compensates for this. It may be considered, very generally, that the seismic pulse spectrum and the delphinid audiogram track one another over a bandwidth of tens of kHz.

Dolphin behavioral audiograms are obtained using pure tone bursts on captive animals that exhibit trained responses to signal presence/absence. In the assessment of perceived loudness of broadband airgun pulses to dolphins, a difficulty arises due to the measurement of seismic pulses and dolphin auditory thresholds. It is necessary to assess seismic pulses in terms of their power density spectrum. As dolphin thresholds are determined through tone burst sound responses, different units are used and the two are not directly comparable.

Goold and Fish (1998) recorded airgun pulses during a 2D seismic survey from a 2,120 cubic inch airgun array, across a frequency range 200 Hz to 22 kHz. Analysis of these pulses produced power spectral density measures, in dB re $1 \mu Pa^2/Hz$. In order to compare these spectra with the dolphin auditory curve, a system similar to the human A-weighting measure was adopted. An inverted representation of the dolphin audiogram, normalized to 0 dB at a reference frequency of 20 kHz, was used to weight the seismic spectrum, essentially attenuating the lower frequencies to compensate for the reduced hearing sensitivity at low frequency compared with the reference frequency of 20 kHz. Integration of the resultant seismic spectrum yielded a spot measure at the 20 kHz reference frequency that was directly comparable to the equivalent point on the behavioral audiogram curve.

Using the above method, it was determined that an equivalent narrow-band seismic signal level at 20 kHz was 133 dB re 1 μPa rms. This figure is some 78 dB above the equivalent point in the dolphin audiogram at 20 kHz, and is a figure obtained at 1 km from the airgun array. Observations of common dolphins during the seismic surveys from which these measurements were taken revealed that dolphins were able to tolerate seismic pulses at a distance of 1 km from the array (Goold, 1996), so it is assumed that a received level of 133 dB re 1 μPa rms at 20 kHz is tolerable. However, dolphins were not documented at ranges closer than 1 km, so it was considered that 133 dB re 1 μPa rms at 20 kHz might be representative of an upper level of tolerance. It should also be mentioned that the threshold in the dolphin audiogram used was increased by 5 dB to account for the fact that the seismic pulse power was concentrated into a duration of about 50 ms, a duration to which the dolphin auditory system would be less sensitive than a continuous tone. The adoption of a sound weighting method may well be necessary in future evaluations of G&G noise on delphinid behavior. Such research is, however, still in its infancy.

Seismic survey pulses also have been shown to modify the vocal behavior of common dolphins in the open sea. Wakefield (2001) demonstrated a shift in certain whistle parameters during airgun shooting, specifically: 1) there is an increase in the start, end, minimum and mean frequencies of whistles, and 2) the whistle contours become flatter. The significance of these changes is not clear, but they perhaps signify adaptation to the increased noise environment. Given the decreased airgun output at higher frequencies, there would seem a clear advantage in shifting vocalizations to higher frequencies to reduce noise masking. Whistle contours appear to become flatter as a consequence of the upper whistle frequency remaining largely unchanged.

Stone (1996, 1997ab, 1998) reported that common dolphins, white beaked dolphins, and whitesided dolphins were sighted in the vicinity of seismic surveys less often when the guns were firing than when they were not firing. These observations were statistically significant in the case of common dolphins.

Finneran et al. (2000) discuss a behavioral response paradigm to measure masked underwater hearing thresholds in two species of odontocete – bottlenose dolphin and beluga – before and after exposure to sound signatures representing distant underwater explosions. Their experiments produced no masked temporary threshold shift to simulated detonations as large as 500 kg HBX-1 charges, at ranges as close as 1.5 km. However, alterations in animals trained behavior began to occur at levels equivalent to 5 kg charges at 9.3 km and 5 kg charges at 1.5 km (two bottlenose dolphins) and 500 kg at 1.9 km (one beluga whale).

Morton and Symonds (2002) demonstrated convincingly that displacement of odontocetes can occur in response to anthropogenic sound disturbance. During a 15-year study of killer whales in Johnstone Strait and Broughton Archipelago, a significant decrease in killer whale abundance was observed during a 7-year period where acoustic harassment devices (AHDs) were installed in the Broughton Archipelago. Morton and Symonds (2002) reported on the use of 10-kHz Airmar AHDs with source levels of 194 dB re 1 μPa at 1 m, designed to cause physical pain to seals. The authors did not report on AHD signal duration or duty cycle. Killer whale abundance returned to baseline levels after the AHDs were removed. The abundance of killer whales in the control area, Johnstone Strait, which was not disturbed, remained stable throughout the study. Airmar AHDs also have been used as a marine mammal deterrent in at least one other study. Kraus et al. (1997) employed active alarms, which emitted a broadband signal with a fundamental frequency of 10 kHz and a source level of 132 dB re 1 μPa at 1 m, well within the hearing range of harbor porpoises and harbor seals. Signal duration was ~300 ms and was repeated every 4 s. Kraus et al. (1997) utilized a lower power setting to scare porpoises away (rather than a higher setting designed to cause pain to seals).

2. Physical Effects

Ridgway et al. (1997) have recently provided the first direct measurements of auditory system damage in toothed whales associated with impulsive noise. Working with four captive bottlenose dolphins, they measured TTS and observed behavioral responses to pure tones of one second duration. Ridgway et al. (1997) reported that TTS occurred in the range of 194 to 201 dB (re 1 μPa) at 3 kHz, 193 to 196 dB at 20 kHz, and 192 to 194 dB at 75 kHz. A conservative bound for sound pressure levels at which medium-sized odontocetes would not experience TTS was established at 192 dB re 1 μPa. Goold and Fish (1999) noted that the onset of TTS in the experiments of Ridgway et al. (1997) induced signs of distress in the captive dolphin subjects, and questioned whether TTS should be considered as an appropriate metric for setting "safe" exposure levels in marine mammals. High duty cycle, or quasi-continuous, noise exposure to TTS levels is likely to remain a controversial subject.

Au et al. (1999) observed 12-18 dB TTS in a bottlenose dolphin exposed to 50 minutes of octave-band noise centered at 7.5 kHz. Finneran et al. (2002) observed masked temporary threshold shift (MTTS) in a beluga when exposed to single underwater impulses produced by a seismic water gun. MTTS of 7 and 6 dB were observed at 0.4 and 30 kHz respectively, approximately 2 minutes following exposure to a pulse of peak-to-peak pressure of 226 dB re 1 μPa (total energy flux 186 dB re 1 μPa^2/Hz). Thresholds returned to within 2 dB of pre-exposure levels approximately 4 minutes after exposure. Finneran et al. (2002) noted that using MTTS may decrease the amount of TTS measured over non-masked procedures, but that the masking was a necessary consequence of the experimental design. The implication is that animals exposed in a "quiet" (i.e., non-masked) environment might develop greater levels of TTS to a similar exposure. Some members of the Delphinidae are very adept divers, such as pilot whales, which regularly dive to 500 m (Baird et al., 2003). It would therefore be prudent to consider that there is some risk of acoustically and/or behaviorally induced DCS. Indeed this may not pertain solely to the larger deeper diving members of the family, as some evidence of DCS type pathology has already been detected in Risso's dolphin, common dolphin, and harbor porpoise (Jepson et al., 2003).

E. SIRENIANS: WEST INDIAN MANATEE

Due to their coastal habitat preferences, manatees are highly unlikely to come in contact with seismic operations. The following discussion of manatee hearing is presented as background information only. There are no data on auditory or behavioral effects of seismic pulses or other low-frequency noise on sirenians.

The United States Fish and Wildlife Service (USFWS) (1996) notes that the external ear structure of manatees suggests that this species can hear low frequency sound within a relatively narrow, low frequency range. Earlier anatomical analyses suggested that manatee hearing is not acute and this species has difficulty localizing sound (Ketten, 1992). Gerstein et al. (1993, 1999) determined that the hearing sensitivity of the West Indian manatee ranges from 15 to 46 kHz, with best sensitivity between 6 and 20 kHz; peak sensitivity was 50 dB re 1 µPa between 16 and 18 kHz. Low frequency detection thresholds, of a possible vibrotactile origin, were measured between 15 and 200 Hz. Structural considerations aside, the USFWS (1996) indicates that results of electrophysiological testing suggest that the West Indian manatee has a greater low frequency (i.e., <3 kHz) sensitivity than other marine mammals (e.g., Gerstein, 1994).

III. LITERATURE CITED

Andre, M., M. Terada, and Y. Watanabe. 1997. Sperm whale (*Physeter macrocephalus*) behavioral response after the playback of artificial sounds. Reports on the International Whaling Commission 47(SC/48/NA13):499-504.

Andre, M., and E. Degollada. 2003. Effects of shipping noise on sperm whale populations, p. 109. *In*: Abstracts of the 17th Conference of the European Cetacean Society, Universidad de Las Palmas de Gran Canaria, 9-13 March 2003.

Aroyan, J.L., M.A. McDonald, S.C. Webb, J.A. Hildebrand, D. Clark, J.T. Laitman, and J.S. Reidenberg. 2000. Acoustic models of sound production and propagation. *In*: W.W.L. Au, A.N. Popper, and R.R. Fay (eds.), Hearing by Whales and Dolphins. Springer-Verlag, New York.

Au, W.W.L. 1993. The Sonar of Dolphins. Springer-Verlag, New York. 277 pp.

Au, W.W.L., R.W. Floyd, R.H. Penner, and A.E. Murchinson. 1974. Measurement of echolocation signals of the Atlantic bottlenose dolphin, *Tursiops truncatus* Montagu, in open waters. J. Acoust. Soc. Am. 56(4):1280-1290.

Au, W.W.L., M.O. Lammers, and R. Aubauer. 1999. A portable broadband data acquisition system for field studies in bioacoustics. Mar. Mamm. Sci. 15(2):526-531.

Awbrey, F.T., J.A. Thomas, W.E. Evans, and S. Leatherwood. 1982. The bioacoustics of the Dall porpoise-salmon drift net interaction. HSWRI Technical Report 79-120. Report prepared by Hubbs Sea World Research Institute, San Diego, CA for the National Marine Fisheries Service, Seattle, WA. 41 pp.

Awbrey, F.T., J.A. Thomas, and R.A. Kastelein. 1988. Low frequency underwater hearing sensitivity in belugas, *Delphinapterus leuca*s. J. Acous. Soc. Am. 84(6):2273-2275.

Backus, R.H., and W.E. Schevill. 1966. *Physeter* clicks, pp. 510-528. *In*: K.S. Norris (ed.), Whales, dolphins, and porpoises. University of California Press, Berkeley, CA. 789 pp.

Baird, R.W, D.J. McSweeney, M.R. Heithaus, and G.J. Marshall. 2003. Sub-surface and night-time behaviour of short-finned pilot whales in Hawaii: information from suction-cup attached time-depth recorders and video camera (crittercam) systems, p. 34. *In*: Abstracts of the 17th Conference of the European Cetacean Society, Universidad de Las Palmas de Gran Canaria, 9-13 March 2003.

Balcomb, K.C. 2001. Letter to Mr. J.S. Johnson, SURTASS LFA Sonar OEIS/EIS Program Manager, 901 North Stuart Street, Suite 708, Arlington, VA 22203, re: SURTASS LFA. 23 February 2001.

Balcomb, K., and D. Claridge. 2000. Summary report on Bahamas beaked whale strandings. Posting to MARMAM, 26 June 2000. Archived at http://www.escribe.com/science/marmam/

Balcomb, K.C., and D.E. Claridge. 2001. A mass stranding of cetaceans caused by naval sonar in the Bahamas. Bahamas Journal of Science 5:2-12.

Barham, E.G. 1973. Whales' respiratory volume as a possible resonant receiver for 20 Hz signals. Nature 245:220-221.

BBN Systems and Technologies. 1993. Assessment of the potential impact of experimental acoustic sources on marine animals and fisheries in the New York Bight. BBN Technical Memorandum No. W1182. Prepared for Advanced Research Projects Agency, Arlington, VA.

Beamish, P. 1979. Behavior and significance of entrapped baleen whales, pp. 291-309. *In*: H.E. Winn and B.L. Olla (eds.), Behavior of marine animals, current perspectives in research. Volume 3: Cetaceans. Plenum Press, New York, NY.

Beamish, P., and E. Mitchell. 1971. Ultrasonic sounds recorded in the presence of a blue whale *Balaenoptera musculus*. Deep-Sea Res. 18(8):803-809.

Beamish, P., and E. Mitchell. 1973. Short pulse length audio frequency sounds recorded in the presence of a minke whale (*Balaenoptera acutorostrata*). Deep-Sea Res. 20(4):375-386.

Bowles, A.E., B. Smultea, B. Würsig, D.P. DeMaster, and D. Palka. 1994. Relative abundance and behavior of marine mammals exposed to transmissions from the Heard Island Feasibility Test. J. Acous. Soc. Am. 96(4):2469-2484.

Buerki, C.B., T.W. Cranford, K.M. Langan, and K.L. Marten. 1989. Acoustic recordings from two stranded beaked whales in captivity, p. 10. *In*: Abstracts, 8th Biennial Conference on the Biology of Marine Mammals, Pacific Grove, CA, December 1989.

Busnel, R.-G., and A. Dziedzic. 1966a. Acoustic signals of the pilot whale *Globicephala melaena* and of the porpoises *Delphinus delphis* and *Phocoena phocoena*, pp. 607-646. *In*: K.S. Norris (ed.), Whales, dolphins, and porpoises. University of California Press, Berkeley, CA. 789 pp.

Busnel, R.-G., and A. Dziedzic. 1966b. Caractéristiques physiques de certains signaux acoustiques du delphinididé *Steno bredanensis*, Lesson. C.R. Acad. Sci. Paris, Ser. D:262-143-146.

Busnel, R.-G., and A. Dziedzic. 1968. Caractéristiques physiques des signaux acoustiques de *Pseudorca crassidens* Owen (Cetace Odontocete). Mammalia 32(1):1-5.

Busnel, R.-G., G. Pilleri, and F.C. Fraser. 1968. Notes concernant le dauphin *Stenella styx* Gray 1846. Mammalia 32:192-203.

Caldwell, D.K., and M.C. Caldwell. 1987. Underwater echolocation type clicks by captive stranded pygmy sperm whales, *Kogia breviceps*, p. 8. *In*: Abstracts, Seventh Biennial Conference on the Biology of Marine Mammals. Miami, Florida, Dec. 5-9, 1987.

Caldwell, D.K., M.C. Caldwell, and J.F. Miller. 1969. Three brief narrow-band sound emissions by a captive male Risso's dolphin, *Grampus griseus*. Bull. Soc. California Acad. Sci. 68(4):252-256.

Caldwell, M.C., and D.K. Caldwell. 1967. Intraspecific transfer of information via pulsed sound in captive odontocete cetaceans, pp. 879-937. *In*: R.G. Busnel (ed.). Animal Sonar Systems: Biology and Bionics II. Laboratorie de Physiologie Acoustique, Joue-en-Josas, France.

Caldwell, M.C., and D.K. Caldwell. 1968. Vocalization of naïve captive dolphins in small groups. Science 159(3819):1121-1123.

Caldwell, M.C., and D.K. Caldwell. 1969. Simultaneous but different narrow-band sound emissions by a captive eastern Pacific pilot whale, *Globicephala scammoni*. Mammalia 33:505-508 + plates.

Caldwell, M.C., and D.K. Caldwell. 1971a. Sounds produced by two rare cetaceans stranded in Florida. Cetology 4:1-6.

Caldwell, M.C., and D.K. Caldwell. 1971b. Underwater pulsed sounds produced by captive spotted dolphins, *Stenella plagiodon*. Cetology 1:1-7.

Caldwell, M.C., D.K. Caldwell, and N.R. Hall. 1973. Statistical evidence for individual signature whistles in the spotted dolphin, *Stenella plagiodon*. Cetology 16:1-21.

Caldwell, M.C., D.K. Caldwell, and P.L. Tyack. 1990. Review of the signature whistle hypothesis for the Atlantic bottlenose dolphin, pp. 199-234. *In*: S. Leatherwood and R.R. Reeves (eds.), The bottlenose dolphin. Academic Press, San Diego, CA. 653 pp.

Carder, D.A., and S. Ridgway. 1990. Auditory brainstem response in a neonatal sperm whale. J. Acous. Soc. Am. 88, Suppl. 1:S4.

Clark, C.W. 1982. The acoustic repertoire of the southern right whale, a quantitative analysis. Anim. Behav. 30(4):1060-1071.

Clark, C.W. 1983. Acoustic communication and behavior of the southern right whale (*Eubalaena australis*), pp. 163-198. *In:* R. Payne (ed.), Communication and behavior of whales. AAAS Sel. Symp. 76. Westview Press, Boulder, CO. 643 pp.

Clark, C.W. 1990. Acoustic behavior of mysticete whales, pp. 571-584. *In:* J.A. Thomas and R.A. Kastelein (eds.), Sensory ability of cetaceans: laboratory and field evidence. Plenum Press, New York, NY.

Clay, C.S., and H. Medwin. 1977. Acoustical Oceanography: Principles & Applications. Wiley, New York and London. ISBN 0471160415.

Coates, R.F.W. 2001a. The Sonar Course. www.seiche.com. ISBN 1-904055-00-1.

Coates, R.F.W. 2001b. The Advanced Sonar Course. www.seiche.com. ISBN 1-904055-01-X.

Crane, N.L., and K. Lashkari. 1996. Sound production of gray whales along their migration route, a new approach to signal analysis. J. Acous. Soc. Am. 100(3):1878-1886.

Cranford, T.W., M. Amundin, and K.S. Norris. 1996. Functional morphology and homology in the odontocete nasal complex: implications for sound generation. Journal of Morphology 228:223–285.

Cummings, W.C., and P.O. Thompson. 1971. Underwater sounds from the blue whale, *Balaenoptera musculus*. J. Acous. Soc. Am. 50(4) Part 2:1193-1198.

Cummings, W.C., and P.O. Thompson. 1994. Characteristics and seasons of blue and finback whale sounds along the U.S. west coast as recorded at SOSUS stations. J. Acous. Soc. Am. 95(5) Part 2:2853.

Cummings, W.C., J.F. Fish, and P.O. Thompson. 1972. Sound production and other behavior of southern right whales, *Eubalena* [sic] *glacialis*. Trans. San Diego Soc. Nat. Hist. 17(1):1-13.

Cummings, W.C., P.O. Thompson, and S.J. Ha. 1986. Sounds from Bryde's, *Balaenoptera edeni*, and finback, *B. physalus*, whales in the Gulf of California. Fish. Bull. 84(2):359-370.

Dahlheim, M.E., and D.K. Ljungblad. 1990. Preliminary hearing study on gray whales (*Eschrichtius robustus*), in the field, pp. 335-346. *In:* J. Thomas and R. Kastelein (eds.), Sensory abilities of cetaceans: laboratory and field evidence. Plenum Press, NY.

Davis, R.A., D.H. Thompson, and C.I. Malme. 1998. Environmental assessment of seismic exploration on the Scotian Shelf. Prepared by LGL Limited Environmental Research Associates, King City, Ontario for Canada/Nova Scotia Offshore Petroleum Board, Halifax, Nova Scotia. 181 pp. + app.

Dawson, S., J. Barlow, and D. Ljungblad. 1998. Sounds recorded from Baird's Beaked Whale, *Berardius bairdii*. Mar. Mamm. Sci. 14(2):335-344.

Department of the Navy. 2001. Final Overseas Environmental Impact Statement and Environmental Impact Statement for Surveillance Towed Array Sensor System Low Frequency Active (SURTASS LFA) Sonar. Prepared for the Department of the Navy, Chief of Naval Operations, Washington, DC. January 2001. 2 vols.

Degollada, E., M. Arbello, M. Andre, A. Blanco, and A. Fernandez. 2003. Preliminary ear analysis report of the 2002 Canary Islands *Ziphius* mass stranding, pp. 60-61. *In:* Abstracts of the 17th Conference of the European Cetacean Society, Universidad de Las Palmas de Gran Canaria, 9-13 March 2003.

Diercks, K.J. 1972. Biological sonar systems: a bionics survey. Applied Research Laboratories, Report ARL-TR-72-34. University of Texas - Austin, Austin, TX.

Diercks, K.J., R.T. Trochta, C.F. Greenlaw, and W.E. Evans. 1971. Recording and analysis of dolphin echolocation signals. J. Acoust. Soc. Am. 49(6) Part 1:1729-1732.

Edds, P.L. 1982. Vocalizations of the blue whale, *Balaenoptera musculus*, in the St. Lawrence River. J. Mammal. 63(2):345-347.

Edds, P.L. 1988. Characteristics of finback *Balaenoptera physalus* vocalizations in the St. Lawrence estuary. Bioacoustics 1(2/3):131-149.

Edds, P.L., D.K. Odell, and B.R. Tershy. 1993. Vocalizations of a captive juvenile and free-ranging adult-calf pairs of Bryde's whales, *Balaenoptera edeni*. Mar. Mamm. Sci. 9(3):269-284.

Environment Australia. 2001. Guidelines on the application of the Environmental Protection and Biodiversity Conservation Act to interactions between offshore seismic operations and larger cetaceans. Environment Australia, Canberra. October 2001. http://www.ea.gov.au/epbc/assessmentsapprovals/guidelines/seismic/pubs/seismic-guidelines-nomaps.pdf.

Erbe, C., and D.M. Farmer. 1998. Masked hearing thresholds of a beluga whale (*Delphinapterus leucas*) in icebreaker noise. Deep-Sea Res. II 45:1373-1388.

Evans, W.E. 1967. Vocalizations among marine mammals, pp. 159-186. *In:* W.N. Tavolga (ed.), Marine bio-acoustics. Pergamon Press, New York, NY.

Evans, W.E. 1973. Echolocation by marine delphinids and one species of freshwater dolphin. J. Acoust. Soc. Am. 54(1):191-199.

Evans, W.E., and J.H. Prescott. 1962. Observations on the sound production capabilities of the bottlenose porpoise: a study of whistles and clicks. Zoologica 47:121-128.

Finneran, J.J., C.E. Schlundt, R. Dear, D.A. Carder, and S.H. Ridgway. 2000. Masked temporary threshold shift (MTTS) in odontocetes after exposure to single underwater impulses from a seismic watergun. J. Acous. Soc. Am. 108(5), part 2, November 2000:2515.

Finneran, J.J., C.E. Schlundt, R. Dear, D.A. Carder, and S.H. Ridgway. 2002. Temporary shift in masked hearing thresholds in odontocetes after exposure to single underwater impulses from a seismic watergun. J. Acous. Soc. Am. 111(6):2929-2940.

Fish, J.F., and C.W. Turl. 1976. Acoustic sources levels of four species of small whales. NUC TP 547. U.S. Naval Undersea Center, San Diego, CA. NTIS AD-A037620. 14 pp.

Fleischer, G. 1978. Evolutionary principles of the mammalian middle ear. Springer-Verlag, Berlin, Heidelberg, New York.

Ford, J.K.B., and H.D. Fisher. 1983. Group-specific dialects of killer whales (*Orcinus orca*) in British Columbia, pp. 129-161. *In:* R. Payne (ed.), Communication and behavior of whales. AAAS Sel. Symp. 76. Westview Press, Boulder, CO. 643 pp.

Frantzis, A. 1998. Does acoustic testing strand whales? Nature 392:29.

Frantzis, A., J.C. Goold, E.K. Skarsoulis, M.I. Taroudakis, and V. Kandia. 2002. Clicks from Cuvier's Beaked Whales, *Ziphius cavirostris*. J. Acous. Soc. Am. 112 (1):34-37.

Gallardo, V.A., D. Arcos, M. Salamanca, and L. Pastene. 1983. On the occurrence of Bryde's whale in an upwelling area off central Chile. Reports of the International Whaling Commission 33:481-488.

Gerstein, E.R. 1994. The manatee mind: discrimination training for sensory perception testing of West Indian manatees (*Trichechus manatus*) pp. 10-21. *In:* Marine Mammals: Public Display and Research, Vol. 1.

Gerstein, E.R., L.A. Gerstein, S.E. Forsythe, and J.E. Blue. 1993. Underwater audiogram of a West Indian manatee (*Trichechus manatus*), p. 53(A). *In:* Proceedings, Tenth Biennial Conference on the Biology of Marine Mammals, Galveston, TX.

Gerstein, E.R., L. Gerstein, S.E. Forsythe, and J.E. Blue. 1999. The underwater audiogram of the West Indian manatee. J. Acous. Soc. Am. 105(6):3575-3583.

Glotov, V.P., P.A. Kolobaev, and G.G. Neuimin. 1962. Investigation of scattering of sound by bubbles generated by an artificial wind in seawater and the statistical distribution of bubble sizes. Sov. Phys. - Acoust. 7:341-345.

Goold, J.C. 1996. Acoustic assessment of populations of common dolphin, *Delphinus delphis*, in conjunction with seismic surveying. Journal of the Marine Biological Association of the United Kingdom 76(3):811-820.

Goold, J.C. 1999. Behavioral and acoustic observations of sperm whales in Scapa Flow, Orkney Islands. Journal of the Marine Biological Association of the United Kingdom 79:541-550.

Goold, J.C., and M.R. Clarke. 2000. Sound velocity in the head of the dwarf sperm whale, *Kogia simus*, with anatomical and functional discussion. Journal of the Marine Biological Association of the United Kingdom 80(3):535-542.

Goold, J.C., and P.J. Fish. 1998. Broadband spectra of seismic survey airgun emissions, with reference to dolphin auditory thresholds. J. Acous. Soc. Am. 103(4):2177-2184.

Goold, J.C., and P.J. Fish. 1999. Response to "Comments on Broadband spectra of seismic survey airgun emissions, with reference to dolphin auditory thresholds." J. Acous. Soc. Am. 105(3):2049-2050.

Goold, J.C., and S.E. Jones. 1995. Time and frequency domain characteristics of sperm whale clicks. J. Acous. Soc. Am. 98(3):1279-1281.

Goold, J.C., and R.F.W. Coates. 2001. Acoustic Monitoring of Marine Wildlife. www.seiche.com. ISBN 1-904055-02-8.

Hoff, L., P.C. Sontum, and J.M. Hovem. 2000. Oscillations of polymeric microbubbles: effect of the encapsulating shell. J. Acous. Soc. Am. 107(4):2272-2280.

Hooker, S.K., and H. Whitehead. 1998. Echolocation of northern bottlenose whales. Poster and Abstract, WWMSC, 1998.

Jepson, P.D., R. Deaville, T. Patterson, J.R. Baker, H.R. Ross, A. Pocknell, F. Howie, R.J. Reid, and A.A. Cunningham. 2003. Novel cetacean gas bubble injuries: acoustically induced decompression sickness, pp. 59-60. *In:* Abstracts of the 17th Conference of the European Cetacean Society, Universidad de Las Palmas de Gran Canaria, 9-13 March 2003.

Kamminga, C., and J.G. van Velden. 1987. Investigations on cetacean sonar. VIII. Sonar signals of *Pseudorca crassidens* in comparison with *Tursiops truncatus*. Aquat. Mamm. 13(2):43-49.

Kapel, F.O. 1979. Exploitation of large whales in west Greenland in the twentieth century. Reports of the International Whaling Commission 29:197-214.

Kargi, S.G., and K.L. Williams. 1998. Double monopole resonance of a gas-filled, spherical cavity in a sediment. J. Acous. Soc. Am. 103(1):265-274.

Ketten, D.R. 1992. The cetacean ear: form, frequency and evolution, pp. 53-75. *In:* Marine Mammal Sensory Systems, Plenum Press, NY.

Ketten, D.R. 1994. Functional analysis of whale ears: Adaptations for underwater hearing. IEEE Proceedings in Underwater Acoustics 11:264-270.

Ketten, D.R. 1998. Marine Mammal Auditory Systems: A summary of audiometric and anatomical data and its implications for underwater acoustic impacts. NOAA Technical Memorandum NMFS-SWFSC-256.

Ketten, D.R., J. Lien, and S. Todd. 1993. Blast injury in humpback whale ears: Evidence and implications. J. Acous. Soc. Am. 94(3):1849-1850.

Ketten, D.R., S. Ridgway, and G. Early. 1995. Apocalyptic hearing: Aging, injury, disease, and noise in marine mammal ears, p. 61. *In:* Abstracts of the 11th Biennial Conference on the Biology of Marine Mammals.

Knowlton, A.R., C.W. Clark, and S.D. Kraus. 1991. Sounds recorded in the presence of sei whales, *Balaenoptera borealis*, p. 40. *In:* Abstract, 9th Biennial Conference on the Biology of Marine Mammals, Chicago, IL, December 1991. 76 pp.

Kraus, S.D., A.J. Read, A. Solow, K. Baldwin, T. Spradlin, E. Anderson, and J. Williamson. 1997. Acoustic alarms reduce porpoise mortality. Nature 388:525.

Levenson, C. 1974. Source level and bistatic target strength of the sperm whale (*Physeter catodon*) measured from an oceanographic aircraft. J. Acoust. Soc. Am. 55(5):1100-1105.

Lilly, J.C., and A.M. Miller. 1961. Sounds emitted by the bottlenose dolphin. Science 133(3465):1689-1693.

Ljungblad, D.K., B. Würsig, S.L. Swartz, and J.M. Keene. 1988. Observations of the behavioral responses of bowhead whales to active geophysical vessels in the Alaskan Beaufort Sea. Arctic 41(3):183-194.

Lynn, S.K., and D.L. Reiss. 1992. Pulse sequence and whistle production by two captive beaked whales, *Mesoplodon* species. Mar. Mamm. Sci. 8(3):299-305.

Madsen, P.T., and B. Mohl. 2000. Sperm whales (*Physeter catodon* L.) do not react to sounds from detonators. J. Acous. Soc. Am. 107(1):668-671.

Malme, C.I., P.R. Miles, C.W. Clarke, P. Tyack, and J.E. Bird. 1983. Investigations of the potential effects of underwater noise from petroleum industry activities on migrating gray whale behavior. BBN Report 5366. Report by Bolt, Beranek, and Newman Inc, Cambridge, MA, for the U.S. Department of the Interior, Minerals Management Service, Anchorage, AK. NTIS PB86-174174.

Malme, C. I., P. R. Miles, C. W. Clark, P. Tyack, and J. E. Bird. 1984. Investigations of the potential effects of underwater noise from petroleum industry activities on migrating gray whale behavior/Phase II: January 1984 migration. BBN Report 5586. Report by Bolt, Beranek, and Newman Inc., Cambridge, MA, for the U.S. Department of the Interior, Minerals Management Service, Anchorage, AK. NTIS PB86-218377.

Malme, C.I., P.R. Miles, C.W. Clarke, P. Tyack, and J.E. Bird. 1985. Investigations of the potential effects of underwater noise from petroleum industry activities on feeding humpback whale behavior. BBN Report 5851. OCS Study MMS 85-0019. Report by BBN Laboratories Inc., Cambridge, MA, for the U.S. Department of the Interior, Minerals Management Service, Anchorage, AK. NTIS PB86-218385.

Malme, C.I., B. Würsig, J.E. Bird, and P. Tyack. 1986. Behavioral responses of gray whales to industrial noise: feeding observation and predictive modelling. BBN Report 6265. OCS Env. Assess. Progr. Final Rep. Princ. Invest., NOAA, Anchorage, AK 56:393-600. OCS Study MMS 88-0048; NTIS PB88-249008.

Malme, C.I., B. Würsig, J.E. Bird, and P. Tyack. 1988. Observations of feeding gray whale responses to controlled industrial noise exposure, pp. 55-73. *In:* W.M. Sackinger et al. (eds.), Port and ocean engineering under arctic conditions, vol II. Geophys. Inst., Univ. Alaska, Fairbanks. 111 pp.

Martin, J.S., P.H. Rogers, E.A. Cudahy, and E.L. Hanson. 2000. Low frequency response of the submerged human lung. J. Acous. Soc. Am. 107(5), part 2:2813.

Mate, B.R., K.M. Stafford, and D.K Ljungblad. 1994. A change in sperm whale (*Physeter macrocephalus*) distribution correlated to seismic surveys in the Gulf of Mexico. J. Acous. Soc. Am. 96(5):3268-3269.

McCauley, R.D., M.N. Jenner, C. Jenner, K.A. McCabe, and J. Murdoch. 1998. The response of humpback whales (*Megaptera novaeangliae*) to offshore seismic survey noise: preliminary results of observations about a working seismic vessel and experimental exposures. APPEA Journal 1998: 692-707.

McCauley, R.D., J. Fewtrell, A.J. Duncan, C. Jenner, , M-N. Jenner, J. Penrose, R.I.T. Prince, A. Adhitya, J. Murdoch, and K. McCabe. 2000. Marine seismic surveys - a study of environmental implications. APPEA Journal, May 2000:692-708.

McDonald, M.A., J.A. Hildebrand, S. Webb, L. Dorman, and C.G. Fox. 1993. Vocalizations of blue and fin whales during a mid-ocean ridge airgun experiment. J. Acous. Soc. Am. 94(3):1849.

McLeod, P.J. 1986. Observations during the stranding of one individual from a pod of pilot whales, *Globicephala malaena*, in Newfoundland. Can. Field-Nat. 100(1):137-139.

Mohl, B., M. Wahlberg, P.T. Madsen, L.A. Miller, and A. Surlykke. 2000. Sperm whale clicks: directionality and source level revisited. J. Acoust. Soc. Am. 107(1):638-648.

Moore, S.E., J.K. Francine, A.E. Bowles, and J.K.B. Ford. 1988. Analysis of calls of killer whales, *Orcinus orca*, from Iceland and Norway. Rit. Fisk. 11:225-250.

Morton, A.B., J.C. Gale, and R.C. Prince. 1986. Sound and behavioral correlations in captive *Orcinus orca*, pp. 303-333. *In:* B.C. Kirkevold and J.S. Lockard (eds.), Behavior biology of killer whales. Alan R. Liss, New York. 457 pp.

Morton, A.B., and H.K. Symonds. 2002. Displacement of *Orcinus orca* (L.) by high amplitude sound in British Columbia, Canada. ICES J. Mar. Sci. 59(1):71-80.

Mullin, K.D., L.V. Higgins, T.A. Jefferson, and L.J. Hansen. 1994. Sightings of the Clymene dolphin *Stenella clymene* in the Gulf of Mexico. Mar. Mamm. Sci. 10:464-470.

Nachtigal, P.E., W.W.L. Au, and J. Pawloski. 1996. Low frequency hearing in three species of odontocetes. J. Acous. Soc. Am. 100(4):2611.

National Marine Fisheries Service. 2001. Bahamas Marine Mammal Stranding Event of 15-16 March 2000. Joint Interim Report by National Marine Fisheries Service and Department of the Navy, December 2001.

National Marine Fisheries Service. 2002. Report of the workshop on acoustic resonance as a source of tissue trauma in cetaceans, April 24 & 25, 2002, Silver Spring, MD. Report by National Marine Fisheries Service and Department of the Navy, November 2002.

Norris, K.S., and W.E. Evans. 1967. Directionality of echolocation clicks in the rough-tooth porpoise, *Steno bredanensis* (Lesson), pp. 305-316. *In:* W.N. Tavolga (ed.), Marine bio-acoustics, volume 2. Pergamon, Oxford, UK. 353 pp.

Norris, K.S., B. Würsig, R.S. Wells, and M. Würsig, with S.M. Brownlee, C.M. Johnson, and J. Solow. 1994. The Hawaiian spinner dolphin. University of California Press, Berkeley, CA. 408 pp.

Orris, G.J., and M. Nicholas. 2000. Collective oscillations of fresh and salt water bubble plumes. J. Acous. Soc. Am. 107(2):771-787

Pace, N.G., A. Cowley, and A.M. Campbell. 1997. Short pulse acoustic excitation of microbubbles. J. Acous. Soc. Am. 102(3):1474-1479.

Payne, K., and R. Payne. 1985. Large scale changes over 19 years in songs of humpback whales in Bermuda. Z. Tierpsychol. 68(2):89-114.

Payne, K., P. Tyack, and R. Payne. 1983. Progressive changes in the songs of humpback whales (*Megaptera novaeangliae*): a detailed analysis of two seasons in Hawaii, pp. 9-57. *In:* R. Payne (ed.), Communication and behavior of whales. AAAS Selected Symposia Series 76. Westview Press, Boulder, CO.

Popov, V.V., and A.Y. Supin. 1997. Detection of temporal gaps in noise in dolphins: Evoked potential study. J. Acous. Soc. Am. 102(2):1169-1176.

Popov, V.V., A.Y. Supin, and V.O. Klishin. 1997. Frequency tuning of the dolphins hearing as revealed by auditory brain-stem response with notch-noise masking. J. Acous. Soc. Am. 102(6):3795-3801.

Potter, J.R. 2003. Acoustically triggered decompression sickness: A possible mechanism in deep-diving marine mammals? Unpublished Manuscript, National University of Singapore.

Pryor, T., K. Pryor, and K.S. Norris. 1965. Observations on a pygmy killer whale (*Feresa attenuata* Gray) from Hawaii. J. Mammal. 46(3):450-461.

Reeves, R.R., D.K. Ljungbald, and J.T. Clarke. 1984. Bowhead whales and acoustic seismic surveys in the Beaufort Sea. Polar Research 22(138):271-280.

Richardson, W.J. 1998. Reactions of Bowhead whales and Ringed seals to an open water seismic program in the Alaskan Beaufort Sea. *In:* Workshop Documentation, Seismic and Marine Mammals Workshop, 23-25 June, London.

Richardson, W.J., and C.I. Malme. 1993. Man made noise and behavioral responses, pp. 631-700. *In:* The Bowhead Whale. Spec. Publ. 2. Society for Marine Mammology, Lawrence, KS. 787 pp.

Richardson, W.J., C.R. Greene, Jr., C.I. Malme, and D.H. Thomson. 1995. Marine Mammals and Noise. Academic Press, San Diego. 576 pp.

Richardson, W.J., B. Würsig, and C.R. Greene. 1986. Reaction of bowhead whales to seismic exploration in the Canadian Beaufort Sea. J. Acous. Soc. Am. 79(4):1117-1128.

Ridgway, S.H., and D.A. Carder. 1997. Hearing deficits measured in some *Tursiops truncatus*, and discovery of a deaf/mute dolphin. J. Acous. Soc. Am. 101(1):590-594.

Ridgway, S.H., D.A. Carder, R.R. Smith, T. Kamolnick, C.E. Schlundt, and W.R. Elsberry. 1997. Behavioral responses and temporary shift in masked hearing threshold of bottlenose dolphins, *Tursiops truncatus*, to 1-second tones of 141 to 201 dB re 1 µPa. Technical Report 1751, July 1997. Naval Command, Control and Ocean Surveillance Center, RDT&E Division, San Diego, CA. 16 pp. + app.

Rivers, J.A. 1997. Blue whale, *Balaenoptera musculus*, vocalizations from the waters off central California. Mar. Mamm. Sci. 13(2):186-195.

Rogers, T.L., and S.M. Brown. 1999. Acoustic observations of Arnoux's beaked whale (*Berardius arnuxii*) off Kemp Land, Antarctica. Mar. Mamm. Sci. 15(1):192-198.

Santoro, A.K., K.L. Marten, and T.W. Cranford. 1989. Pygmy sperm whale sounds (*Kogia breviceps*), p. 59. *In:* Abstract, 8th Biennial Conference on the Biology of Marine Mammals, Pacific Grove, CA, December 1989. 81 pp.

Sauerland, M., and G. Dehnhardt. 1998. Underwater audiogram of the tucuxi (*Sotalia fluviatilis guianensis*). J. Acous. Soc. Am. 103(2):1199-1204.

Schevill, W.E., and W.A. Watkins. 1966. Sound structure and directionality in *Orcinus* (killer whale). Zoologica (N.Y.) 51(6):71-76 + plates.

Schevill, W.E., and W.A. Watkins. 1972. Intense low-frequency sounds from an Antarctic minke whale, *Balaenoptera acutorostrata*. Brevoria 388:1-8.

Schultz, K.W., and P.J. Corkeron. 1994. Interspecific differences in whistles produced by inshore dolphins in Moreton Bay, Queensland, Australia. Can. J. Zool. 72(6):1061-1068.

Schultz, K.W., D.H. Cato, P.J. Corkeron, and M.M. Bryden. 1995. Low frequency narrow-band sounds produced by bottlenose dolphins. Mar. Mamm. Sci. 11(4):503-509.

Seiche. 2000. Investigations into the high frequency emissions of geophysical air guns. Project Summary Report. Document Seiche 128 Issue 1, September 10, 2000.

Silber, G.K. 1986. The relationships of social vocalizations to surface behavior and aggression in the Hawaiian humpback whale (*Megaptera novaeangliae*). Can. J. Zool. 64:2075-2080.

Sodal, A. 1999. Measured underwater acoustic wave propagation from a seismic source. *In*: Proceedings of the Airgun Environmental Workshop, London, 6 July 1999.

Sorensen, P.W., R.J. Medved, M.A.M. Hyman, and H.E. Winn. 1984. Distribution and abundance of cetaceans in the vicinity of human activities along the continental shelf of the northwestern Atlantic. Marine Environmental Research 12(1):69-81.

Spero, D. 1981. Vocalizations and associated behavior of northern right whales *Eubalaena glacialis*, p. 108. *In*: Abstracts, 4th Biennial Conference on the Biology of Marine Mammals, San Francisco, CA. 127 pp.

Stafford, K.M., C.G. Fox, and B.R. Mate. 1994. Acoustic detection and location of blue whales (*Balaenoptera musculus*) from SOSUS data by matched filtering. J. Acoust. Soc. Am. 96(5) Part 2:3,250-3,251.

Stafford, K.M., C.G. Fox, and D.S. Clark. 1998. Long range acoustic detection and localization of blue whale calls in the northeast Pacific Ocean. J. Acous. Soc. Am. 104(6):3616-3625.

Stafford, K.M., S.L. Nieukirk, and C.G. Fox. 1999. Low frequency whale sounds recorded on hydrophones moored in the eastern tropical pacific. J. Acous. Soc. Am. 106(6):3687-3698.

Steiner, W.W. 1981. Species-specific differences in pure tonal whistle vocalizations of five western North Atlantic dolphin species. Behav. Ecol. Sociobiol. 9(4):241-246.

Steiner, W.W., J.H. Hain, H.E. Winn, and P.J. Perkins. 1979. Vocalizations and feeding behavior of the killer whale (*Orcinus orca*). J. Mammal. 60(4):823-827.

Stone, C.J. 1996. Cetacean observations during a seismic survey of the *M.V. Mintrop*, west of Shetland. Report to Conoco (UK) Limited.

Stone, C.J. 1997a. Cetacean and seabird observations in Tranche 52 during 1997. Report to Conoco (UK) Limited.

Stone, C.J. 1997b. Cetacean observations during seismic surveys in 1996. JNCC Reports, No. 228. Joint Nature Conservation Committee, Aberdeen, Scotland.

Stone, C.J. 1998. Cetacean observations during seismic surveys in 1997. JNCC Reports, No. 278. Joint Nature Conservation Committee, Aberdeen, Scotland.

Supreme Allied Command Atlantic, Undersea Research Centre – North Atlantic Treaty Organization (SACLANTCEN). 1998. Human Diver and Marine Mammal Environmental Policy and Risk Mitigation. Part II NATO Unclassified, from the SACLANTCEN Bioacoustics Panel, La Spezia, 15-17 June 1998.

Szymanski, M.D., D.E. Bain, K. Kiehl, S. Pennington., S. Wong, and K. R. Henry. 1999. Killer whale (*Orcinus orca*) hearing: Auditory brainstem response and behavioral audiograms. J. Acous. Soc. Am. 106(2):1134-1141.

Taruski, A.G. 1969. The whistle repertoire of the North Atlantic pilot whale (*Globicephala malaena*) and its relationship to behavior and environment, pp. 345-368. *In*: H.E. Winn and B.L. Olla (eds.), Behavior of marine mammals, volume 3: Cetaceans. Plenum Press, New York. 438 pp.

Tepley, L. 2001. Air-space resonances and other mechanisms which may cause tissue damage in cetaceans. On-line research paper, (http://home1.gte net/leetpley/resonance_p1_toc.html) also submitted to NMFS under the public comment period of the LFAS FEIS.

Thomas, J.A., and C.W. Turl. 1990. Echolocation characteristics and range detection threshold of a false killer whale (*Pseudorca crassidens*), pp. 321-3224. *In*: J.A. Thomas and R.A. Kastelein (eds.), Sensory abilities of cetaceans – laboratory and field evidence. Plenum Press, New York. 710 pp.

Thompson, P.O., W.C. Cummings, and S.J. Ha. 1986. Sounds, source levels, and associated behavior of humpback whales, southeast Alaska. J. Acoust. Soc. Am. 80(3):735-740.

Thompson, P.O., L.T. Findley, and O. Vidal. 1992. 20-Hz pulses and other vocalizations of fin whales, *Balaenoptera physalus*, in the Gulf of California, Mexico. J. Acous. Soc. Am. 92(6):3051-3057.

Thompson, T.J., H.E. Winn, and P.J. Perkins. 1979. Mysticete sounds, pp. 403-431. *In:* H.E. Winn and B.L. Olla (eds.), Behavior of marine animals, volume 3: Cetaceans. Plenum Press, New York. 438 pp.

Tyack, P. 1985. An optical telemetry device to identify which dolphin produces a sound. J. Acoust. Soc. Am. 78(5):1892-1895.

U.S. Fish and Wildlife Service. 1996. Florida manatee recovery plan (*Trichechus manatus latirostris*), second revision. Prepared by the Florida Manatee Recovery Team for the Southeast Region, U.S. Fish and Wildlife Service, Atlanta, GA. Approved January 29, 1996.

Van Parijs, S.M., and P.J. Corkeron. 2001. Boat traffic affects the acoustic behaviour of Pacific humpback dolphins, *Sousa chinensis*. Journal of the Marine Biological Association of the UK, 81:533-538.

Wakefield, E.D. 2001. The distribution and vocal behaviour of common dolphin (*Delphinus delphis* L.) in the Celtic sea and adjacent waters, with particular reference to the effects of seismic surveying. MSc. Thesis, University of Wales, Bangor.

Wang Ding, B. Würsig, and W. Evans. 1995. Comparisons of whistles among seven odontocete species. *In:* R.A. Kastelein, J.A. Thomas, and P.E. Nachtigall (eds.), Sensory systems of aquatic mammals. De Spil Publishers, Woerden, Netherlands.

Wartzok, D., and D. R. Ketten. 1999. Marine mammal sensory systems, pp. 117-175. *In:* J.E. Reynolds III and S.A. Rommel (eds.), Biology of marine mammals. Smithsonian Institution Press, Washington, DC. 590 pp.

Watkins, W.A. 1967. The harmonic interval: fact or artifact in spectral analysis of pulse trains, pp. 15-43. *In:* W.N. Tavolga (ed.), Marine bio-acoustics, volume 2. Pergamon, Oxford, UK. 353 pp.

Watkins, W.A. 1977. Acoustic behaviors of sperm whales. Oceanus 20:50-58.

Watkins, W.A. 1980a. Acoustics and the behavior of sperm whales, pp. 283-290. *In*: R.G. Busnel and J.F. Fish (eds.), Animal sonar systems. Plenum Press, New York. 1,135 pp.

Watkins, W.A. 1980b. Clicks sounds from animals at sea, pp. 291-297. *In*: R.G. Busnel and J.F. Fish (eds.), Animal sonar systems. Plenum Press, New York. 1,135 pp.

Watkins, W.A. 1981a. Activities and underwater sounds of fin whales. Sci. Rep. Whales. Res. Inst. 33:83-117.

Watkins, W.A. 1981b. Reaction of three species of whales to implanted radio tags. Deep-Sea Res. 28A(6):589-599.

Watkins, W.A., and W.E. Schevill. 1972. Sound source location by arrival-times on a non-rigid three-dimensional hydrophone array. Deep-Sea Res. 19(10):691-706.

Watkins, W.A., and W.E. Schevill. 1974. Listening to Hawaiian spinner porpoises, *Stenella* cf. *longirostris*, with a three-dimensional hydrophone array. J. Mammal. 55(2):319-328.

Watkins, W.A., and W.E. Schevill. 1975. Sperm whales react to pingers. Deep-Sea Res. 22:123-129.

Watkins, W.A., P. Tyack, K.E. Moore, and J.E. Bird. 1987. The 20-Hz signals of finback whales (*Balaenoptera physalus*). J. Acoust. Soc. Am. 82(6):1901-1912.

Watkins, W.A., M.A. Daher, K.M. Fristrup, T.J. Howald, and G.N. Disciara. 1993. Sperm whales tagged with transponders and tracked underwater by sonar. Mar. Mamm. Sci. 9:55-67.

Willis, P.M., and R.W. Baird. 1998. Status of the dwarf sperm whale, *Kogia simus*, with special reference to Canada. Can. Field Nat. 112:114-125.

Winn, H.E., and P.J. Perkins. 1976. Distribution and sounds of the minke whale, with a review of mysticete sounds. Cetology 19:1-12.

Winn, H. E., P. J. Perkins, and T. C. Poulter. 1970. Sounds of the humpback whale, pp. 39-52. *In:* Proc. Seventh Ann. Conf. on Biol. Sonar and Diving Mamm. Stanford Research Institute, Menlo Park, CA.

Wood, F.G., Jr. 1953. Underwater sound production and concurrent behavior of captive porpoises, *Tursiops truncatus* and *Stenella plagiodon.* Bull. Mar. Sci. Gulf Caribb. 3(2):120-133.

Würsig, B., C.W. Clark, E.M. Dorsey, M.A. Fraker, and R.S. Payne. 1982. Normal behavior of bowheads, pp. 33-143. *In:* W.J. Richardson (ed.), Behavior, disturbance responses, and feeding of bowhead whales *Balaena mysticetus* in the Beaufort Sea, 1980-81. Report prepared by LGL Ecological Research Associates for the U.S. Department of the Interior, Bureau of Land Management, Washington, DC. NTIS PB86-152170. 456 pp.

Würsig, B., C.R. Greene, and T.A. Jefferson. 2000. Development of an air bubble curtain to reduce underwater noise of percussive pile driving. Marine Environmental Research 49:79-93.

**Appendix H
Sea Turtle Hearing and Sensitivity to
Acoustic Impact**

Sea Turtle Hearing and Sensitivity to Acoustic Impact

This appendix describes the current state of knowledge regarding sea turtle hearing capabilities and sensitivity to acoustic impact. This summary serves as a basis for assessing the environmental impact of geological and geophysical (G&G) activities on sea turtle species that frequent outer continental shelf (OCS) waters of the Gulf of Mexico (see **Section III - Environmental Impacts** of the Programmatic Environmental Assessment [PEA]).

I. BACKGROUND

Until recently, little research has been performed on the auditory mechanism of sea turtles. Compared with the marine mammal literature, sea turtle sensory systems are virtually unstudied. Relatively little is documented or understood, for any sea turtle species, about their hearing ability or their dependency on sound, passive or active, for survival cues. Moreover, since sea turtles often have been reported to show a lack of response to even intense sounds (Wever, 1978), they were initially characterized as having insensitive hearing capabilities. A number of factors, including anatomy, electrophysiological response, and behavioral responses to sound, need to be considered when evaluating the hearing capabilities of sea turtles.

The anatomy of the sea turtle ear does not lend itself to aerial conduction but rather is structured for sound conduction through two media, bone and water (Békésy, 1948; Lenhardt, 1982; Lenhardt and Harkins, 1983). Neither mode of auditory reception is mutually exclusive; instead, the sea turtle appears to be able to process sound through both pathways. Moreover, from the structure and impedance of the tympanum, the sea turtle ear appears to be a poor aerial receptor (DeBurlet, 1934; Manley, 1970; Lenhardt et al., 1985). Consequently, the audiogram for the green sea turtle in air (Ridgway et al., 1969) provides information on frequency ranges but not the intensities to which the animal might respond.

Complicating this picture is the fact that each life stage of sea turtles is marked by exceptional differences in gross morphometry of auditory structures and in the physical parameters of their habitat. It is believed that physiological and behavioral adaptations may have evolved for the sea turtle on the basis of their selection of aquatic niches with each ontogenetic stage. For these three stages of life, the sensory environment is also changing. Shallow water habitats of the juvenile and adult stages are a much "noisier" world than the open ocean environment of the hatchling stage. Ambient noise in the inshore environment is heavily weighted to low frequency sound (Hawkins and Myrberg, 1983). In highly developed areas, low frequency noises associated with shipping lanes, recreational boat traffic, and biological organisms are prominent. This certainly applies to the Gulf of Mexico, which is habitat for Kemp's ridley (*Lepidochelys kempi*), loggerhead (*Caretta caretta*), green (*Chelonia mydas*), hawksbill (*Eretmochelys imbricata*), and leatherback (*Dermochelys coriacea*) sea turtles.

Differences in hearing capability (frequency selectivity and threshold levels) among species and life history stages have not been documented for sea turtles in the literature. In fact, only two species, loggerhead and green sea turtles, and one life history stage, juvenile, have undergone any auditory investigations.

II. MORPHOLOGY

Most of the research on the hearing capacity of sea turtles is limited to gross morphological dissections (Ridgway et al., 1969; Wever, 1978; Lenhardt et al., 1985). Most studies published on sea turtle hearing focus on the anatomy of the ear (Wever, 1978; Lenhardt et al., 1985) (**Figure H-1**). Sea

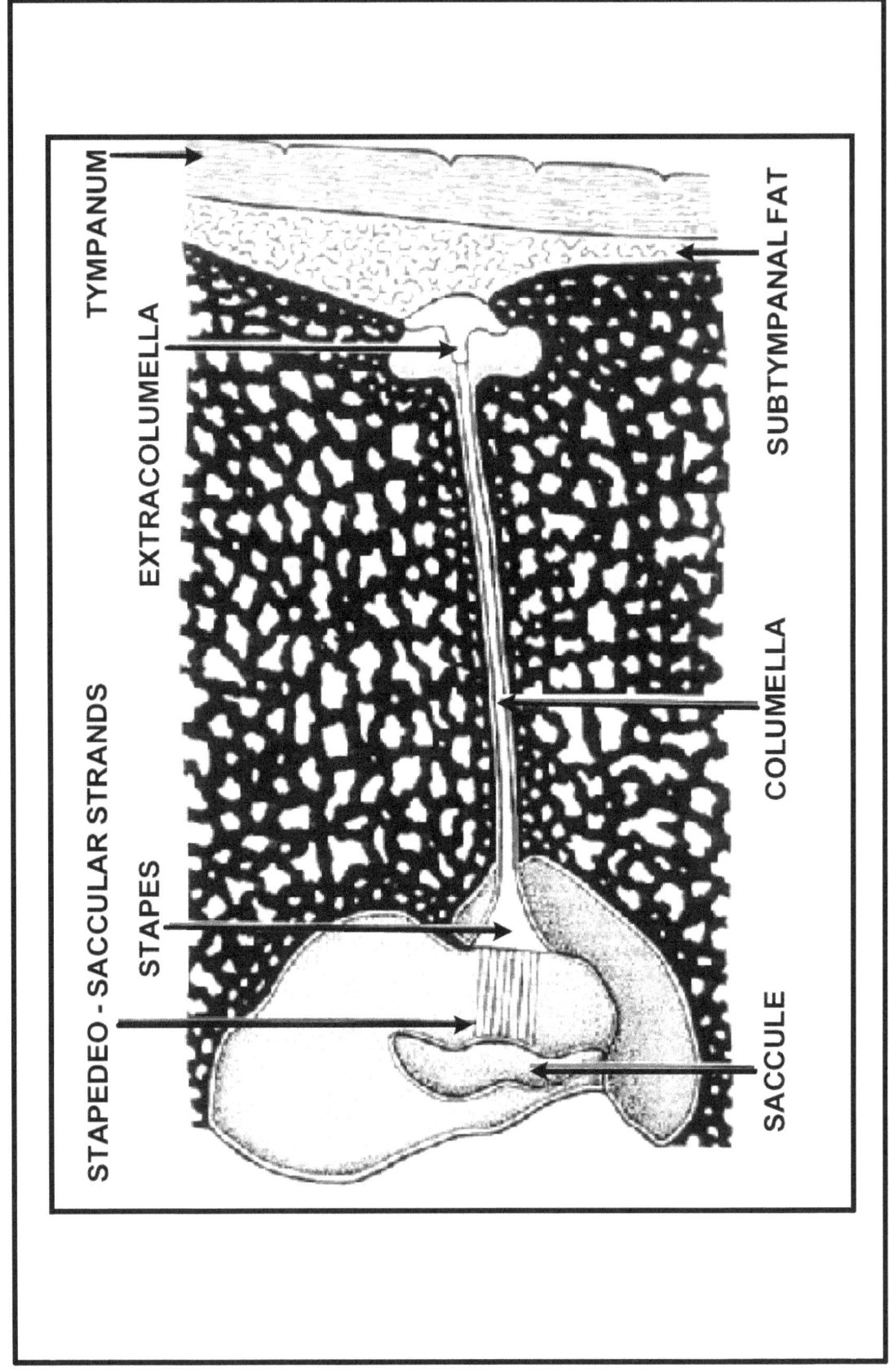

Figure H-1. Middle ear anatomy of the juvenile loggerhead sea turtle (Source: Moein, 1994).

turtles have a thick layer of subtympanal fat, a feature that distinguishes them from both terrestrial and semiaquatic species. There is no external ear, and the tympanum is a continuation of the facial tissue (i.e., there is no tympanic membrane *per se*). The tympanum of marine turtles is composed of layers of superficial tissues at the side of the head that are no more than a continuation of facial tissue over a depression in the skull that forms the middle ear cavity (Wever, 1978; Lenhardt et al., 1985; Bartol et al., 1999; Bartol and Musick, 2003). Removal of the tympanum in freshwater species produces only negligible change in the displacement of the columella (middle ear bone) to sound pressure, which suggests that the tympanum is a poorly adapted aerial receptor (Manley, 1970; Moffat and Capranica, 1978). The ossicular mechanism consists of two elements, the columella and the extracolumella. The extracolumella is a cartilaginous disk under the tympanic membrane, which is attached to the columella by ligaments. The columella, a long rod with the majority of the mass concentrated at each end, travels through a bone channel and expands within the oval window to form a funnel shaped stapes. Fibrous strands, the stapedo-saccular strands, connect the stapes and the oval window to the saccule, a unique feature of sea turtles. The stapedo-saccular strands presumably relay vibrational energy to the saccule (Wever and Vernon, 1956; Lenhardt et al., 1985).

Lenhardt et al. (1985) postulate from the morphology that the sea turtle ear is a poor aerial receptor. Instead, these researchers suggest that the shape of the columella and its interactions with the cochlea and saccule indicate that the sea turtle's middle ear is a compromise for sound conduction through two media, bone and water. When the turtle uses bone conduction to process sound, sound flows through the bones and soft tissue to stimulate the inner ear. The tympanum acts as a release mechanism rather than a sound receptor. Collectively, the thick tympanum does not inhibit overall sound reception, but rather acts as additional mass loading to the ear, allowing for reduction in the sensitivity of sound frequencies and increasing low frequency, bone conduction sensitivity (Lenhardt et al., 1985; Bartol et al., 1999; Bartol and Musick, 2003). However, when the turtle uses water conduction to process sound, the tympanum and subtympanal fat act as low impedance channels for underwater sound, resulting in columella displacement to stimulate the inner ear (Ketten, 1999). Neither sound reception mode need be mutually exclusive; instead, multiple pathways for stimulating the inner ear may exist.

III. HEARING STUDIES

A. ELECTROPHYSIOLOGICAL

Some attempts have been made to collect electrophysiological responses, specifically auditory brainstem responses (ABRs), to the stimulation of the sea turtle's hearing apparatus. Historically, ABRs have been used as a method for testing for audition and acoustic threshold in non-communicative species. ABRs are sequences of events originating in the brainstem and are generated by separate parts of the auditory pathway in the first 10 ms after stimulation. These recordings have been found to be consistent within species and similar across vertebrate classes in general form and origin, regardless of auditory apparatus (Corwin et al., 1982). Furthermore, these techniques are noninvasive and can be performed on awake subject animals (Bullock, 1981; Corwin et al., 1982).

Only two efforts to collect electrophysiological data from sea turtles have been published, one study performed on juvenile green sea turtles (Ridgway et al., 1969) and one study on juvenile loggerhead sea turtles (Bartol et al., 1999). Ridgway et al. (1969) used both aerial and vibrational stimuli to record cochlear potentials. Thresholds were not measured; instead, sensitivity curves to produce a cochlear potential of 0.1 µV were plotted. The frequencies tested on these turtles ranged from 50 to 2,000 Hz. The results revealed that green sea turtles detected limited sound frequencies (200 to 700 Hz) and displayed a high level of sensitivity at the low tone region of about 400 Hz. Moreover, with an increase in frequencies, their range of sensitivity declined by a rate of 35 dB per octave. Though this investigation examined two separate modes of sound reception (air conduction and bone conduction) by the green sea turtle, sensitivity curves were relatively similar. These results suggest that the inner ear is the main structure for determining frequency sensitivity (Ridgway et al., 1969).

Bartol et al. (1999) recorded auditory evoked potentials of bone conducted stimuli from juvenile loggerheads using subdermal platinum electrodes. Vibratory stimuli of known frequency were delivered directly to the sea turtle's tympanum. Evoked potentials were extracted from the electroencephalogram (EEG) by repeating and averaging single responses. The averaging of many responses time-locked to the stimulus isolated the single response from the biological noise associated with ongoing neural and muscular activity (Gelfand, 1990). Only responses that occurred within 10 ms of stimulation (ABRs) were recorded from the EEG. Thresholds were recorded for both tonal and click stimuli. Maximum sensitivity was in the low frequency region of 250 to 1,000 Hz. The decline in sensitivity was rapid after 1,000 Hz, and the most sensitive threshold tested was found to be at 250 Hz (the lowest frequency tested) with a mean threshold of –26.3 dB re: 1 gravity (g) rms (standard deviation $\pm$ 2.3 dB) (**Figures H-2** and **H-3**).

Vibratory stimuli (either click or tone burst) most likely provide valid threshold levels when used to evoke ABRs. However, the reference level for vibratory stimulation, dB of acceleration, cannot be readily compared to research with data presented in sound pressure levels (either in air or water). Thus, these data cannot be used to establish impact thresholds in the underwater environment but rather serve to reinforce the frequency range that is likely to affect sea turtles.

B. BEHAVIORAL

The relationship between electrophysiological measurements and behavioral responses is abundant in the mammal and bird literature. However, behavioral work with reptiles has proven to be more difficult to perform, and conditioning and/or identification of reliable responses with turtles has been elusive for researchers. Patterson (1966) successfully performed one conditioning study on *Chrysemys scripta* using head withdrawal as the conditioned response to sound stimuli. Close correlation was found between the behavioral thresholds and cochlear potential audiograms, both showing similar curves. Moreover, a behavioral study is currently underway at the New England Aquarium to train (i.e., operantly condition) an adult green sea turtle to respond to sound stimulation. Though this adult turtle appears responsive to training, recording threshold levels has not been possible, as yet.

Performing hearing studies for sea turtles in a tank environment is difficult. For hearing research, testing must be performed in a large tank, large enough to account for sound reflections along the air-water interface as well as harmonics set up by the tank walls. Dampening the tank walls with sound absorbing material is almost impossible when the stimuli are low frequencies (the material would have to be as thick as the dimensions of one wavelength of the sound) (Hawkins and Myrberg, 1983). As a consequence, underwater audiograms in a tank environment are yet to be reported.

IV. ACOUSTIC IMPACTS

Loud acoustic impulses can affect nearby animals anatomically, physiologically, and behaviorally. Experimental terrestrial animals have long been shown to have adverse reactions to acoustic overexposure, with the potential of damage occurring in all organ systems, including the auditory system (Saunders et al., 1985; Lenhardt, 1986; Yost, 1994). Little is known, however, of the effect of noise on sea turtles in the marine environment, and many problems arise when extrapolating research on terrestrial species to those in the marine environment. Certainly, the transmission of sound in air vs. water is quite different, and acoustic values that contribute to hearing damage to terrestrial animals cannot easily be

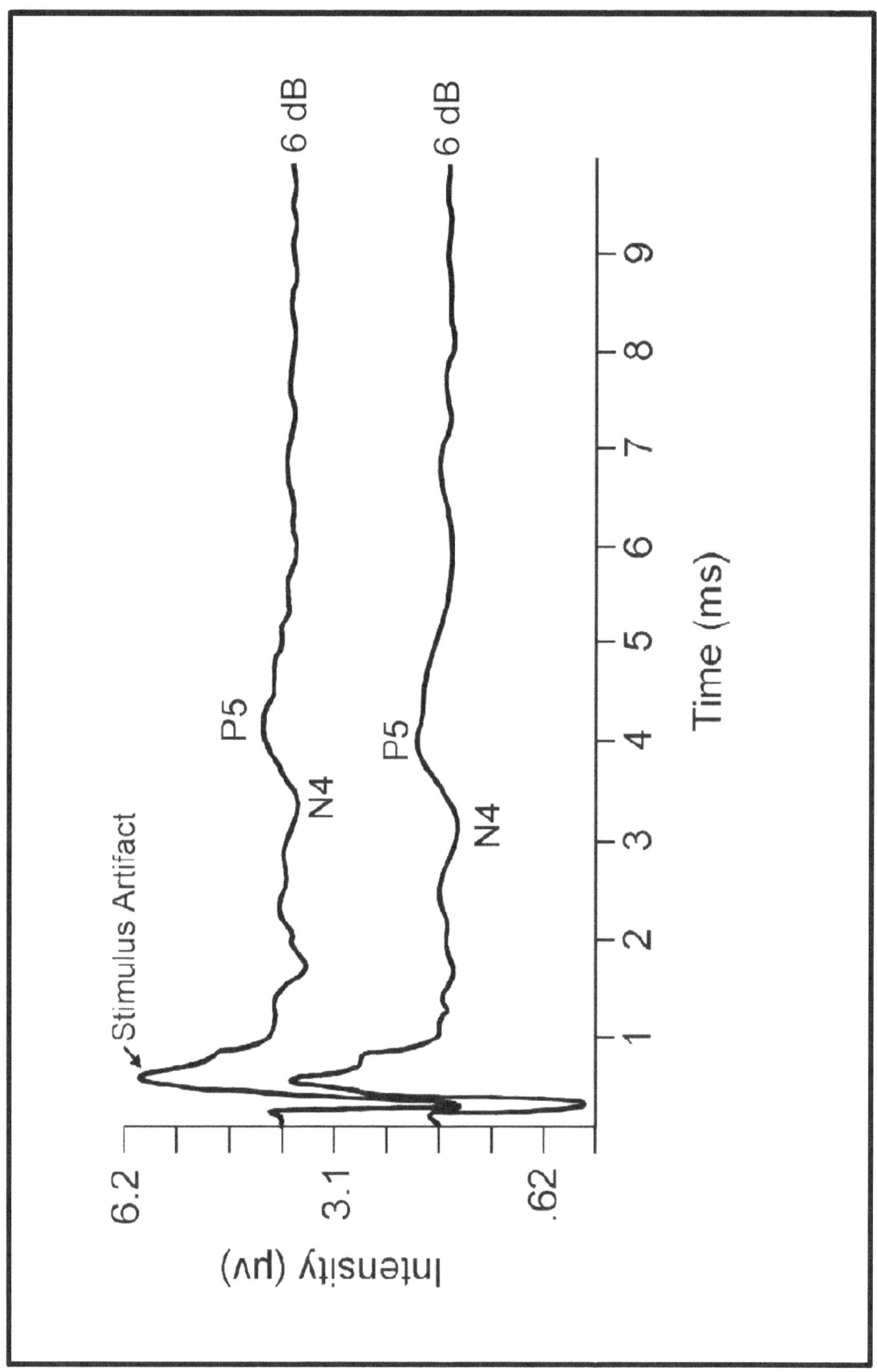

Figure H-2.　Juvenile loggerhead auditory brainstem responses evoked from broadband low frequency clicks.　Peak 5 (P5) was used to track the response of the loggerhead sea turtle to stimulus intensity.　Click intensities were measured in dB re: 1 gravity rms (Source: Bartol et al., 1999).

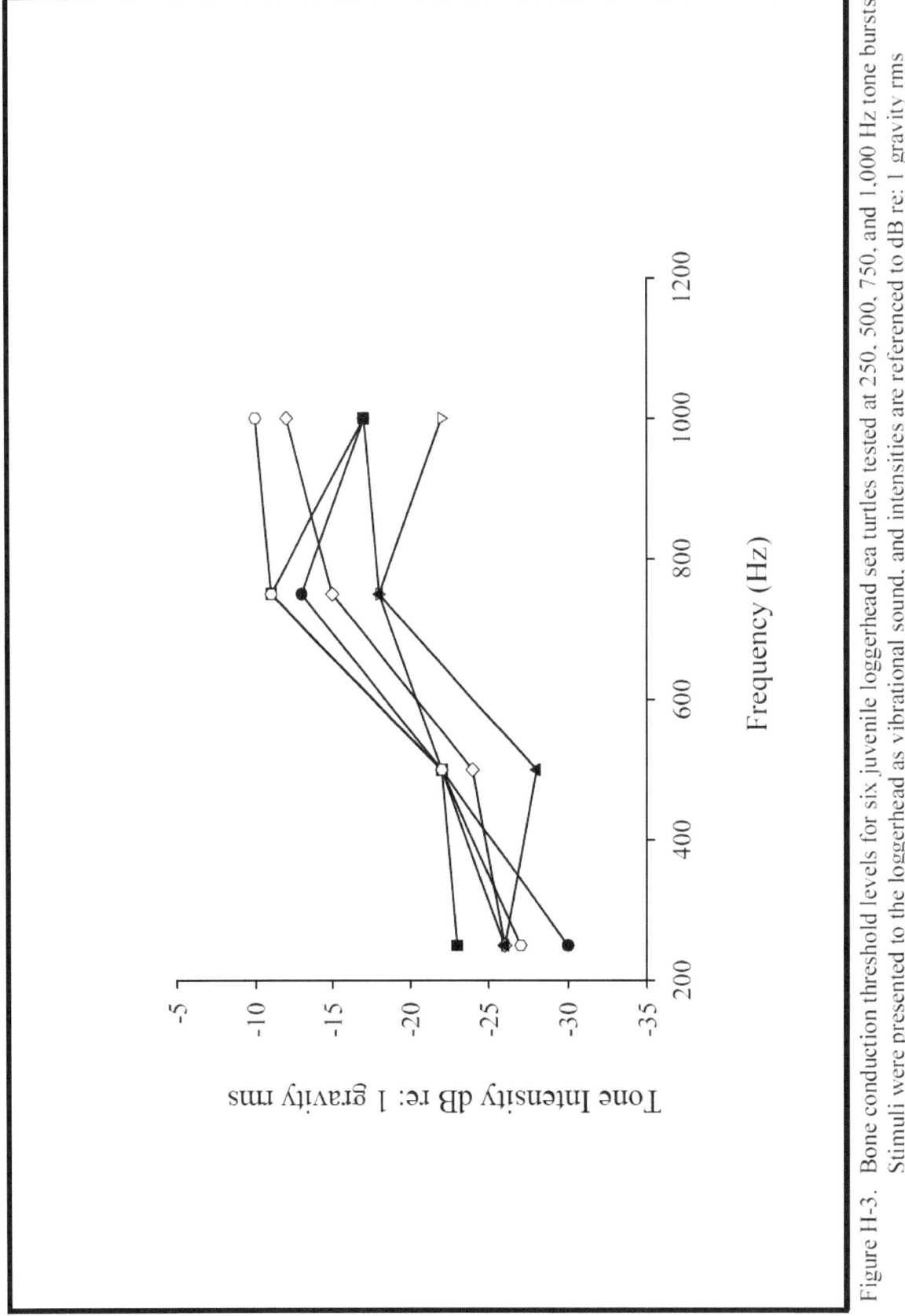

Figure H-3. Bone conduction threshold levels for six juvenile loggerhead sea turtles tested at 250, 500, 750, and 1,000 Hz tone bursts. Stimuli were presented to the loggerhead as vibrational sound, and intensities are referenced to dB re: 1 gravity rms (Source: Bartol et al., 1999).

translated to the aquatic environment. Moreover, comparison across species is also problematic, with many different pathways existing for sound reception (Lenhardt et al., 1985; Ketten, 1999).

A. BEHAVIORAL RESPONSES

Two U.S. research studies have examined the response of juvenile loggerheads to sound in their natural environment (O'Hara and Wilcox, 1990; Moein et al., 1995). In both cases, these studies were funded to assist in the development of an acoustic repelling device for sea turtles.

O'Hara and Wilcox (1990) attempted to create a sound barrier for loggerhead turtles at the end of a canal of Florida Power & Light, using seismic airguns. The test results indicated that at 140 kg/cm^2, the airguns were effective as a deterrent for a distance of about 30 m. The sound output of this system was characterized as approximately 220 dB re 1 µPa at 1 m in the 25 to 1,000 Hz frequency range. However, this study did not account for the reflection of sound by the canal walls. Consequently, the stimulus frequency and intensity levels are ambiguous (O'Hara and Wilcox, 1990).

Moein et al. (1995) investigated the use of pneumatic energy sources (airguns) to repel juvenile loggerhead sea turtles from hopper dredges. A net enclosure (approximately 18 m x 61 m x 3.6 m) was erected in the York River, Virginia, to contain the turtles, and an airgun was stationed at each end of the net. Using a float attached to the posterior of the carapace, the position of the turtle was noted as the airguns fired. Sound frequencies of the airguns ranged from 100 to 1,000 Hz (Zawila, 1995). Three source levels (175, 177, and 179 dB re 1 µPa at 1 m) were used. Beginning at the 175-dB source level, one airgun fired every 5 s for 5 min, followed by a 10-min resting period. Then, the airgun at the other end fired using the same source level. This was repeated using the successively higher source levels, resulting in exposure to a total of 30 min of airgun pulses. Avoidance to the airguns was observed upon first exposure for the juvenile loggerheads. However, these animals also appeared to habituate to the sound stimuli. After three separate exposures to the airguns, the turtles no longer avoided the stimuli (Moein et al., 1995). Another possible interpretation is that the turtles gave up trying to avoid the airguns since they could not truly escape the noise (J. Sigurdson, Department of the Navy, oral comm., 2001).

The most recent of the studies of sea turtles exposed to airgun pulses was completed by McCauley et al. (2000) off Western Australia. In this experiment, researchers exposed single caged individuals (i.e., green, loggerhead turtles) to pulses from a moving seismic source (i.e., a 20-in.^{3} airgun at 5-m depth operating at 1,500 psi; firing interval 10 s; two separate exposure periods: 1 and 2 h). Results indicated that at received levels above 166 dB re 1 µPa rms, the turtles noticeably increased swimming speed. Observed sea turtle behavior became increasingly erratic when received levels were >175 dB re 1 µPa rms. McCauley et al. (2000) suggested that the erratic behavior exhibited by caged sea turtles would be expressed as an avoidance response in free swimming individuals.

B. AUDITORY TRAUMA

There are two types of trauma for an animal from acoustic stimuli: lethal and sublethal. Typically, trauma that produces a lethal effect, trauma so severe that an animal in close proximity to an intense sound source dies within minutes, is not limited to impacts on the auditory system (i.e., the acoustic wave damages tissues other than those associated with hearing). This effect has been documented in association with use of underwater explosives (O'Keeffe and Young, 1984; Klima et al., 1988), but not seismic surveys using airguns. Sublethal impacts of acoustic impulses on the hearing system of sea turtles have been examined in only one study.

The extent of trauma to the auditory system is usually placed into two categories: permanent threshold shift (PTS) and temporary threshold shift (TTS) (see Saunders et al., 1985; Lenhardt, 1986 for review). Threshold levels of hearing are defined as the minimum intensity required for perception of a sound by an individual. Threshold levels will differ by frequency and are usually recorded from a subject animal without the interference of background noise. Loss of sensitivity, or an increase in the threshold level due to trauma, is a threshold shift. If this loss is irreversible, then PTS has occurred. This PTS could result from direct damage to the middle ear (i.e., the tympanum or columella) or inner ear (hair cells). If the sensitivity levels return to normal after a period of time, then the threshold shift is temporary (TTS). Many factors can determine the extent and duration of a TTS in an animal, including the frequency, intensity, and duration of the damaging sound. TTS can extend over a wide range of frequencies or be limited to a narrow band, and normal hearing can return over a period of days or months (Melnick, 1991; Lombarte et al., 1993; Hastings et al., 1996).

Possible response to threshold shifts by an animal can vary tremendously, and behavioral responses by sea turtles to TTS are unknown. Behavioral research on sea turtles in their natural environment is difficult, and only a few studies have attempted to measure reactions of healthy animals to sound stimuli, as noted previously. However, loss of hearing, either temporary or permanent, could potentially prevent an individual animal from detecting predators or prey, from successfully navigating along migration routes or to a nesting beach, or from avoiding manmade dangers, such as boating traffic, to name a few.

One study has produced evidence of possible TTS in juvenile loggerhead sea turtles (Caretta caretta) (Moein et al., 1995). The turtles in this study were used to test the efficacy of seismic airguns to repel sea turtles from the path of hopper dredges. As noted previously, three source levels (175, 177, and 179 dB re 1 μPa at 1 m) were used. Beginning at the 175-dB source level, one airgun fired every 5 s for 5 min, followed by a 10-min resting period. Then, the airgun at the other end fired using the same source level. This was repeated using the successively higher source levels (Moein et al., 1995; Zawila, 1995). Turtles were tested for stress levels and hearing thresholds before and after the airgun trials. These health tests not only ensured that all loggerhead turtles used in the net enclosure were healthy animals but also demonstrated the long-term effects of sound exposure on the turtles. A temporary alteration of blood chemistry values after exposure to the airguns indicated that these turtles might have been affected by exposure to repeated acoustic stimuli. Values indicated both an increase in the stress level of the animal as well as damage to tissues. However, the magnitude of the changes did not indicate significant injury to the turtle's organs, and levels returned to normal in approximately 2 weeks. Thus, this exposure to sound stimuli did cause minor, but reversible, changes to the turtle's tissues (Moein et al., 1995).

Hearing thresholds also were examined. Auditory evoked potentials were recorded from these juvenile loggerheads, and auditory thresholds were obtained before the first testing sequence in the net with the sound stimuli, within 24 h after each test, and approximately 2 weeks after the last exposure to the airgun. Half of turtles tested exhibited a shift in the latency of their auditory evoked potentials collected within 24 h of exposure (**Figure H-4**). This shift in Peak 5 (P5) latency, the peak used to track threshold from the recordings, was indicative of a change in the hearing physiology. However, in all five cases, hearing capabilities of each turtle returned to normal by the end of 2 weeks, and the effect was always temporary. Turtle position was not monitored in sufficient detail to calculate the received sound pressure levels to which any of the affected turtles were exposed.

V. CONCLUSIONS

Obviously, much more research needs to be conducted to estimate sound levels producing impacts in sea turtles. Very little is even known about their basic hearing mechanisms or the role of sound in their life cycle. Only one study has reported threshold levels for a species of sea turtle, the juvenile loggerhead

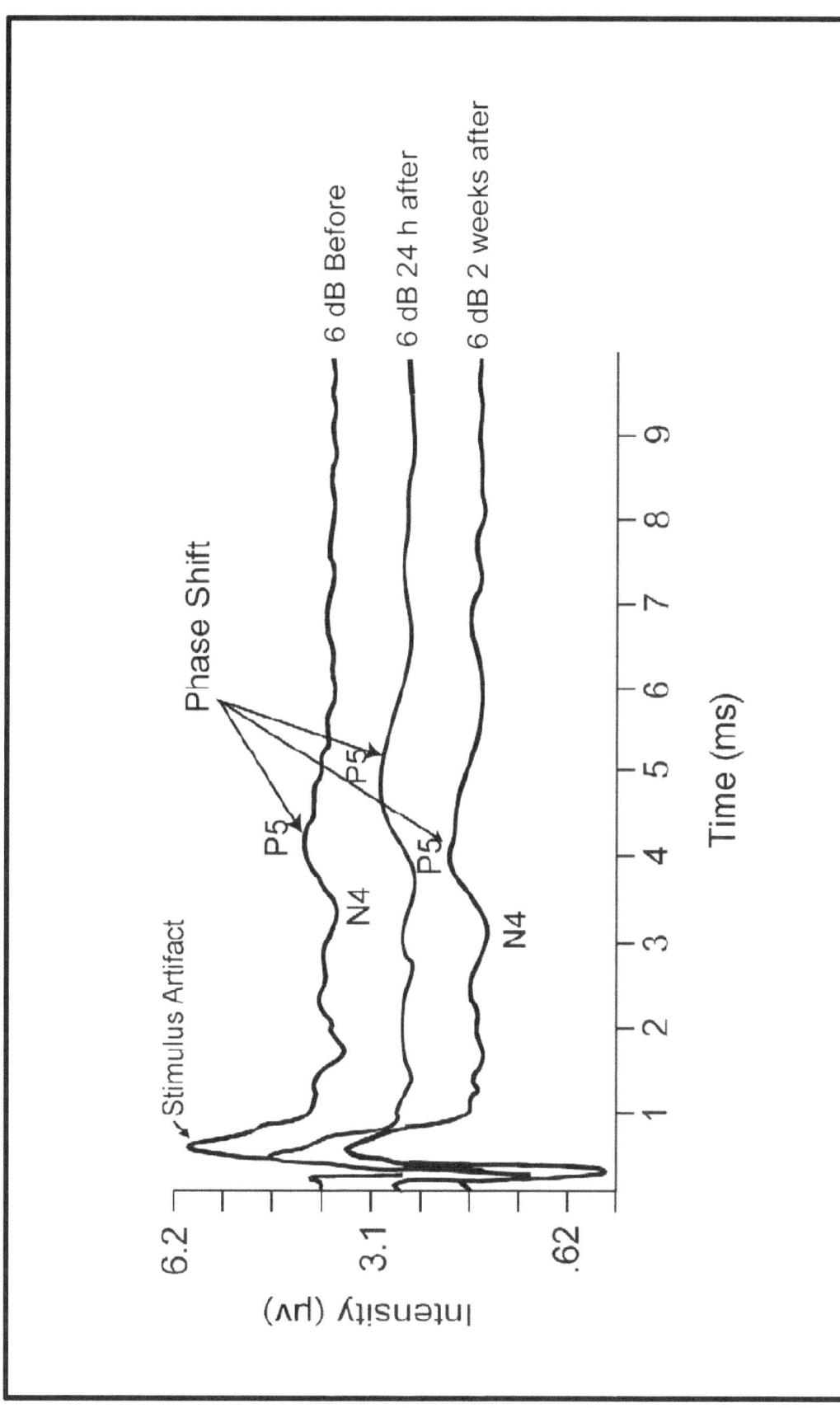

Figure H-4. Auditory brainstem responses (ABRs) obtained from the juvenile loggerhead sea turtles before the first testing sequence in the net with the sound stimuli (airgun), within 24 h after each test, and approximately 2 weeks after the last exposure to the airgun (Moein et al., 1995). A phase shift in the latency of Peak 5 (P5) was observed in half of the turtles tested within 24 h after exposure to the sound stimuli. In all cases, this phase shift was temporary, and the ABR returned to normal within 2 weeks.

turtle (Bartol et al., 1999). However, these thresholds were collected from vibratory stimulation, and the reference level, dB of acceleration, cannot be readily compared to sound pressure levels in water.

Auditory testing and behavioral studies show that turtles can detect and respond to low frequency sounds (250 to 1,000 Hz) from airguns. There is some indication of avoidance, at least initially, of airgun pulses with source levels of 175 to 179 dB re 1 µPa. The same study has produced possible evidence of a temporary effect on the auditory response of turtles exposed to a total of 30 min of airgun pulses (Moein et al., 1995). Source levels were as noted, above; unfortunately, the sound pressure levels received by the affected turtles are unknown.

VI. LITERATURE CITED

Bartol, S.M., and J.A. Musick. 2003. Sensory biology of sea turtles, pp. 79-102. *In:* P.L. Lutz, J.A. Musick, and J. Wyneken (eds.), The Biology of Sea Turtles, Volume II. CRC Press, Boca Raton, FL. 455 pp.

Bartol, S.M., J.A. Musick, and M. Lenhardt. 1999. Auditory evoked potentials of the loggerhead sea turtle (*Caretta caretta*). Copeia 99(3):836-840.

Békésy, G. 1948. Vibration of the head in a sound field, and its role in hearing by bone conduction. J. Acoust. Soc. Am. 20:749-760.

Bullock, T.H. 1981. Neuroethology deserves more study of evoked responses. Neuroscience 6:1203-1215.

Corwin, J.T., T.H. Bullock, and J. Schweitzer. 1982. The auditory brain stem response in five vertebrate classes. Electroenceph. Clin. Neurophysiol. 54:629-641.

DeBurlet, H.M. 1934. Vergleichende Anatomie des statoakustischen Organs, pp. 1,293-1,432. *In:* Handburch der vergleichenden Anatomie der Wirbeltiere. L. Bolk, E. Goppert, E. Kallius, and W. Lobosch (eds.). Urban an Schwarzenberg, Berlin.

Gelfand, S.A. 1990. Hearing: An Introduction to Psychological and Physiological Acoustics. Marcel Dekker, Inc., New York.

Hastings, M.C., A.N. Popper, J.J. Finneran, and P.J. Lanford. 1996. Effect of low frequency underwater sound on hair cells of the inner ear and lateral line of the teleost fish *Astronotus ocellatus*. J. Acoust. Soc. Am. 99:1759-1766.

Hawkins, A.D., and A.A. Myrberg, Jr. 1983. Hearing and sound communication under water, pp. 347-405. *In:* B. Lewis (ed.), Bioacoustics: A Comparative Approach. Academic Press, London.

Ketten, D.R. 1999. Acoustical fatheads: parallel evolution of underwater sound reception mechanisms in dolphins, turtles, and sea birds. J. Acoust. Soc. Am.

Klima, E.F., G.R. Gitchlag, and M.L. Renaud. 1988. Impacts of the explosive removal of offshore petroleum on sea turtles and dolphins. Marine Fisheries Review 50:33-42.

Lenhardt, E. 1986. Clinical Aspects of Inner Ear Deafness. Springer-Verlag, New York.

Lenhardt, M.L. 1982. Bone conduction hearing in turtles. J. Aud. Res. 22:153-160.

Lenhardt, M.L., and S.W. Harkins. 1983. Turtle shell as an auditory receptor. J. Aud. Res. 23:251-260.

Lenhardt, M.L., R.C. Klinger, and J.A. Musick. 1985. Marine turtle middle-ear anatomy. J. Aud. Res. 25:66-72.

Lombarte, A., H.Y. Yan, A.N. Popper, J.C. Chang, and C. Platt. 1993. Damage and regeneration of hair cell ciliary bundles in a fish ear following treatment with gentamicin. Hear. Res. 66:166-174.

Manley, G. 1970. Comparative studies of auditory physiology in reptiles. Z. Vergl. Physiol. 67:363-381.

McCauley, R.D., J. Fewtrell, A.J. Duncan, M.-N. Jenner, C. Jenner, R.I.T. Prince, A. Adhitya, K. McCabe, and J. Murdoch. 2000. Marine seismic surveys - a study of environmental implications. J. Australian Petrol. Product. Explor. Assoc. 40:692-708.

Melnick, W. 1991. Human temporary threshold shifts (TTS) and damage risk. J. Acoust. Soc. Am. 90:147-155.

Moein, S.M. 1994. Auditory evoked potentials of the loggerhead sea turtle (*Caretta caretta*). Master's Thesis. College of William and Mary, Williamsburg, VA.

Moein, S.E., J.A. Musick, J.A. Keinath, D.E. Barnard, M.L. Lenhardt, and R. George. 1995. Evaluation of seismic sources for repelling sea turtles from hopper dredges, pp. 90-93. *In:* L.Z. Hales (compiler), Sea Turtle Research Program: Summary Report. Prepared for U.S. Army Engineer Division, South Atlantic, Atlanta, GA, and U.S. Naval Submarine Base, Kings Bay, GA. Technical Report CERC-95-. 145 pp.

Moffat, A.J.M., and R.R. Capranica. 1978. Middle ear sensitivity in anurans and reptiles measured by light scattering spectroscopy. J. Com. Physiol. 127:97-107.

O'Hara, J., and J.R. Wilcox. 1990. Avoidance responses of loggerhead turtles, *Caretta caretta*, to low frequency sound. Copeia 1990(2):564-567.

O'Keeffe, D.J., and G.A. Young. 1984. Handbook on the environmental effects of underwater explosions. Naval Surface Weapons Center. NWC TR 83-240.

Patterson, W.C. 1966. Hearing in the turtle. J. Aud. Res. 6:453-464.

Ridgway, S.H., E.G. Wever, J.G. McCormick, J. Palin, and J.H. Anderson. 1969. Hearing in the giant sea turtle, *Chelonia mydas*. Proc. Nat. Acad. Sci. 64:884-890.

Saunders, J.C., S.P. Dear, and M.E. Schneider. 1985. The anatomical consequences of acoustical injury: a review and tutorial. J. Acoust. Soc. Am. 78(3):833-860.

Wever, E.G. 1978. The Reptile Ear: Its Structure and Function. Princeton University Press, Princeton.

Wever, E.G., and J.A. Vernon. 1956. Sound transmission in the turtle's ear. Proc. Nat. Acad. Sci. 42:229-299.

Yost, W.A. 1994. Fundamentals of Hearing: An Introduction. Academic Press, New York.

Zawila, J.S. 1995. Characterization of a seismic airgun acoustic dispersal technique at the Virginia Institute of Marine Science sea turtle test site, pp. 88-89. *In:* L.Z. Hales (compiler), Sea Turtle Research Program: Summary Report. Prepared for U.S. Army Engineer Division, South Atlantic, Atlanta, GA, and U.S. Naval Submarine Base, Kings Bay, GA. Technical Report CERC-95-. 145 pp.

**Appendix I
Fish Hearing and Sensitivity to Acoustic Impact**

Fish Hearing and Sensitivity to Acoustic Impact

This appendix describes the current state of knowledge regarding fish hearing capabilities and sensitivity to acoustic impact. This summary serves as a basis for assessing the environmental impact of geological and geophysical (G&G) activities on those fish resources found in outer continental shelf (OCS) waters of the Gulf of Mexico (see **Section III - Environmental Impacts** of the Programmatic Environmental Assessment [PEA]).

I. INTRODUCTION

This appendix summarizes and evaluates the literature on the effects of intense sound on fishes and speculates in cases where experimental data are lacking. The purpose of this appendix is to determine, to the extent presently possible, the known, probable, and possible impacts on fishes of the sorts of sounds produced during acoustic seismic exploration in aquatic environments. The general physiology of sound detection by fishes is relatively well understood (see Fay and Megela Simmons [1999], and Popper and Fay [1999] for recent reviews). In contrast, the usual acoustic behavior and uses of sound by fishes is less well documented. Finally, the effects of intense and potentially damaging sound on fish hearing and behavior is only poorly understood, with only a small number of studies published in the peer-reviewed literature. Thus, this appendix will be complete in summarizing general hearing mechanisms, less satisfying on the topic of acoustic behaviors and uses of sound, and generally lacking in substantive data on the adverse effects of sounds on fishes. Where data are limited, speculations will be made on the basis of a somewhat more complete literature on other vertebrate animals, including humans.

II. EFFECTS OF SOUND ON FISHES

A. SOURCES OF INTENSE SOUND UNDERWATER

In the case of seismic exploration in the Gulf of Mexico, the important source of sound to be considered is produced by pneumatic "airguns" commonly used in geological exploration of the seabed. Arrays of multiple airguns are used for seismic exploration and development (see **Appendix D**). The sound levels produced by airguns depend upon the number of guns employed and their size (gas volume). A single airgun produces a peak sound pressure between about 210 dB and 235 dB re 1 μPa at 1 m distance (source level), depending on source size and depth (Malme et al., 1986). Airguns in multiple arrays produce source levels as high as 255 dB. The levels reaching fish depend upon the distance between the fish and the source, water depth, and the type of distance attenuation taking place (spherical, cylindrical, or other types of spreading, which tend to be depth-dependent). The sounds produced are impulsive, damped sinusoids, generally with fundamental frequencies in the range between 10 and 200 Hz. All Gulf of Mexico fish species have best hearing frequencies within, or slightly above, this range, based on the results of the present analysis (see **Section II.E - Uses of Sound by Fishes** of this appendix), as well as prior summaries of hearing capabilities and sound use in fishes (e.g., Myrick, 1981; Zellick et al., 1999). Thus, airguns are likely to be audible by all fishes given sufficient received sound pressure levels.

B. HOW SOUND STIMULATES THE EARS OF FISHES

The acoustic portions of the ears of fishes are one or more of the otolith organs (saccule, lagena, and utricle). These organs are composed of a membranous sac that may or may not be bounded by bone. Within the sac is a sensory epithelium made up of support cells and hair cell mechanoreceptors that are overlain by an otolith, substantially calcium carbonate (density of about 3). The hair cells are similar in overall structure and function to those of the lateral line system and those of the inner ears of all vertebrates. At the hair cell's apical end, a bundle of stereovillae ("hairs") and a single, eccentrically located kinocilium project apically. Deflection of the stereovillae toward the kinocilium opens ion channels, causing the cell's depolarization and increasing the neurotransmitter release that stimulates input processes of auditory nerve fibers, resulting in an increased flow of action potentials (spikes) to the brain. Deflection of the stereovillae away from the kinocilium hyperpolarizes the hair cell and tends to reduce neurotransmitter release and spike probability in auditory nerve fibers. Sounds usually cause an oscillatory or back-and-forth deflection of the stereovillae, resulting in a modulation of spike probability according to the acoustic waveform. Deflection of the stereovillae results from a relative movement between the hair cell and the overlying otolith. This movement occurs as a result of two different stimulation routes of sound to the ears that operate to varying degrees for different organs and in different species.

The stimulation route common to all organs and species is known as the "direct" route. In this stimulation mode, acoustic particle motion or hydrodynamic motion accelerates the fish's soft tissues, including the sensory epithelium, with the surrounding water. The dense otoliths tend to remain at rest due to their inertia, and a relative motion is set up across the stereovillae. This mode of hearing is inherently directional since the effective stimulus is a vector quantity and the hair cells themselves are highly directional receivers (Hudspeth and Corey, 1977; Fay, 1984). In the region of 100 to 200 Hz, particle motion amplitudes of less than 0.1 nanometers (nm) are behaviorally detectable in several species tested, including cod (Chapman and Hawkins, 1973). In this primitive mode of hearing, sound pressure does not play a direct role, and species that apparently hear exclusively in this mode are known as "hearing generalists."

Some species have an additional mode of hearing that renders them sensitive to sound pressure. In these "hearing specialists," the swimbladder or other gas bubble is efficiently linked mechanically to the fluid systems of the ear, and thus to the otoliths. As sound is transmitted through the animal's body, sound pressure fluctuations cause the gas bladder to expand and contract, and the motions of its walls are transmitted to the sensory epithelia of, usually, one of the otolith organs. In most species, the saccule receives this motion most efficiently, but in clupeids (herring-like fishes), the utricle is stimulated in this way. Since sound pressure at a point is a scalar quantity, there is no directional information in the stimulus. The hearing specialists are particularly sensitive to sound, with best thresholds in the region of 50 dB re 1 µPa, and hear in a relatively wide frequency range (<100 to 3,000 Hz), with best sensitivity occurring between 200 and 1,000 Hz. Sensitivity to sound additionally depends on sound duration (with approximately equal energy detected up to sound durations of 400 ms), on the level of ambient or background noise in the frequency region of the signals (Fay, 1988), and on the physiological integrity of the auditory system.

C. PHYSIOLOGICAL AND ANATOMICAL EFFECTS OF INTENSE SOUND ON AUDITORY SYSTEMS OF FISHES

The study of sound-induced damage to hearing and the auditory system has been concentrated on vertebrate species other than fishes, especially on mammals. In every mammal species investigated, there are specific high sound levels that cause damage to the hair cells of the ears, including the shearing off of the stereovillae and eventual death of hair cells. In these cases, the effects of such damage are to raise sound detection thresholds (sometimes to the point of producing profound deafness) and to disrupt the normal frequency selectivity of the auditory organ. The effects on frequency selectivity arise from the highly frequency-selective nature of the mammalian cochlea, and presumably on hair cell mechanisms generally known to comprise "the cochlear amplifier" (Patuzzi, 1996) that contribute to this selectivity. While hair cell damage has also been observed in fishes, and the presumption of hearing loss can reasonably be made, the hair cells of fish ears are not especially frequency-selective and appear not to contain the cochlear amplifier mechanisms characteristic of mammals. In addition, hair cells of fish ears are known to regenerate after severe damage or death (Lombarte et al., 1993), while those of mammals are not. Recovery of function after such regeneration has not been investigated.

There are only two papers in the peer-reviewed literature specifically addressing the question of damage to the auditory endorgan following intense sound exposure (Enger, 1981; Hastings et al., 1996). Hastings et al. (1996) is the more comprehensive study, focusing on the oscar (*Astronotus ocellatus*), a freshwater cichlid and hearing generalist. In this study, fish were exposed to three sound levels (100, 140, and 180 dB re 1 μPa) at two frequencies (60 and 300 Hz) in a laboratory setting. Animals were stimulated for 1 hour, some with continuous tones, and others with intermittent tone bursts of 20% duty cycle (12 s on and 48 s off per minute). The sensory epithelia of all ear organs and lateral line organs were microscopically evaluated for structural damage after 1 to 4 days post-treatment. The only damage observed that could be attributed to sound exposure occurred in one or both ears of four of five animals stimulated at 300 Hz (continuous tone) that had been allowed to survive the exposure for 4 days. This damage was restricted to small portions (15%) of the lagena and utricle, but not the saccule. The saccule is thought to be the primary auditory organ in this species. Animals stimulated at lower sound pressure levels, using continuous tones or tone bursts, or allowed less than 4 days after exposure for examination, showed no hair cell damage. These results suggest that higher frequencies are likely to be more damaging than lower frequencies (given that both are detectable), but that damage is only slight following 180 dB continuous 300 Hz tone exposure for 1 hour. It was assumed that damage would be more likely at higher sound exposure levels, and possibly for longer exposure durations.

Enger (1981) exposed Atlantic cod (*Gadus morhua*) to pure tones of various frequencies between 50 and 400 Hz at 180 dB for from 1 to 5 hours. Fish were immediately evaluated for hair cell damage after stimulation. Under these circumstances, hair cell damage was observed in many animals in restricted regions of the saccular epithelium (other organs were not evaluated), the position of which was roughly dependent on stimulation frequency. These results show more robust damaging effects than in the Hastings et al. (1996) study on oscar. Recall that no saccular damage was observed in the oscar. The reasons for greater damage in the cod may be due to the higher hearing sensitivity in cod compared with oscar, and thus the higher level of the stimulating sound with respect to the hearing threshold. Hastings et al. (1996) speculate that levels of 220 to 240 dB at 300 Hz would be required to produce extensive hair cell damage in hearing generalists such as the oscar, and by extension, levels possibly 40 dB less than this to cause damage in the more sensitive hearing specialists.

In studies published only as abstracts but cited in Hastings et al. (1996), Cox et al. (1986a,b, 1987) exposed goldfish (hearing specialists) to pure tones between 250 and 500 Hz at levels of 182 to 204 dB re 1 μPa for 2 hours. In these studies, 100% of the fish stimulated at 204 dB (200 Hz) had damage to saccular hair cells, but none of the animals stimulated at any frequency at 182 dB (500 Hz) were damaged.

An analysis of the data acquired to date show that fish hearing capabilities are species-specific and varying degrees of sensitivity are evident. Underwater anthropogenic sounds (e.g., from underwater explosions, vessel traffic, military sonars, G&G operations) have the potential to mask the sounds normally used by fishes in their normal acoustic behaviors at levels as low as 60 to 80 dB (just above detection thresholds for many species). Levels as high as 160 dB may cause receiving fishes to change their behaviors and movements, which may temporarily affect the usual distribution of animals and commercial fishing. Continuous, long-term exposure to levels above 180 dB has been shown to cause damage to the hair cells of the ears of some fishes under some circumstances. These effects may not be permanent since damaged hair cells are repaired and/or regenerated in some fishes (Continental Shelf Associates, Inc., 2003). However, McCauley et al. (2003) found irrepairable hair damage for species exposed to seismic airgun noise levels of 180 dB re 1 μPa.

In summary, these data indicate that the hair cells of inner ear organs of fishes can be damaged by intense sound. The lowest levels causing documented damage are 180 dB for continuous, long-duration tones in a region of good hearing (200 to 500 Hz for many fish species). The effects of intermittent sound stimulation are not known except that a 20% duty cycle is clearly less effective in damaging hair cells than continuous sounds. It seems likely that most fishes exposed to airgun shots at a distance of 1 m or less could receive inner ear damage as a result of source levels in the range between 210 and 240 dB. However, as the distance between the fish and the airgun sources increased, the probability of hearing impairment would decrease according to the nature of distance attenuation taking place in the ensonified environment.

D. BEHAVIORAL EFFECTS OF INTENSE SOUND ON HEARING IN FISHES

The types of anatomical damage to the ears reviewed above is expected to cause elevated sound detection thresholds in the affected animals, and the defects would be expected to last for the period of time that the tissue damage occurred. This total or partial deafness would be termed permanent threshold shift (PTS) in an analogy with effects observed in terrestrial animals and humans. If the damage only extended to restricted areas of the sensory epithelium, as the several studies reviewed above demonstrated, the effect on hearing would be limited to those stimuli that selectively activate those areas. Thus, in cod for example, increased thresholds would be expected to occur for sounds in the frequency range served by the lesioned areas, but not necessarily for sounds of other frequencies. Since the hair cells of fishes can regenerate, beginning at about 7 days post-trauma (Lombarte et al., 1993), damage-induced threshold shifts may not be ultimately permanent but might last for only 1 or 2 weeks. However, there are no published functional data on PTS in fishes to evaluate these hypotheses.

More is known about transient (rapidly recovering) threshold shifts that could occur following exposures to sounds at lower sound pressure levels than those causing PTS. Popper and Clarke (1976) exposed goldfish to intense tones between 300 and 1,000 Hz at levels of 149 dB for 4 hours. Behavioral measures of sound detection thresholds showed that thresholds for an 800 Hz tone increased by as much as 28 dB as measured immediately following the intense tone stimulation. The effect on the detection of a 500 Hz tone was slightly less. Intense tone exposure at frequencies above and below 800 Hz produced less threshold shift (a minimum of 8 dB for 300 Hz exposures). Possible hair cell damage was not investigated. Sensitivity at all frequencies returned to normal (pre-exposure levels) within 24 hours. Thus, these effects were termed temporary threshold shifts (TTS) in an analogy with similar effects in mammals. The assumption here is that there were subtle disruptions of normal hair cell function that caused the threshold shifts, but that these rapidly recovered. It would be expected that hearing generalists having higher detection thresholds would require correspondingly higher sound pressure levels (20 dB or more) to exhibit similar TTS effects. In summary, it is likely that long duration, continuous sounds in the region of 150 to 170 dB at 800 Hz and below could cause transient hearing impairment (30 dB or so) in most fishes that could last as long as 24 hours.

Behavioral studies on several fish species show that the thresholds for tone detection can be raised by the presence of additional sounds. When the threshold for the detection of one sound (i.e., the "signal") is raised by the presentation of another sound (i.e., the "masker"), masking is said to occur. Masking is greatest when signal and masker are simultaneous (Fay and Coombs, 1988) but can also occur for brief periods of time (less than 1 second; typically less than 300 msec) before masker onset (backward masking) and following masker offset (forward masking) (Popper and Clarke, 1979). In both forms of masking, threshold shift depends upon masker level, and this effect is known to be essentially linear for the simultaneous masking case (Fay, 1974). Linearity means that the signal-to-noise (S/N) ratio at threshold is independent of masker level, and that noise levels can predict detection thresholds once these S/N ratios are known. In goldfish, S/N ratios for tones masked by broad band noise range between 13 and

24 dB, increasing linearly with log frequency between 100 and 1,200 Hz. S/N values for other species tested (marine catfish, tilapia, and pinfish; Tavolga, 1974) are slightly higher. In general, the auditory systems of fishes are crudely frequency-selective so that masking effects are restricted to signal frequencies in the frequency region of the masker (Hawkins and Chapman, 1975; Fay et al., 1978). Thus, for example, airguns would be expected to cause masking only within or near the limits of the airgun shot's spectral profile as it impinges on the fish, and only for up to 300 msec prior to and following the shot (non-simultaneous masking). The masking effect (threshold elevation) would be proportional to the received shot level. There are no masking data in the literature for intermittent, impulsive maskers similar to airgun shots.

Recent experiments on goldfish indicate that fish are capable of what has been called "auditory scene analysis" (Bregman, 1990). This means that a sound stream of interest can be "heard out" and analyzed for its informational content, independently of simultaneous, potentially interfering sounds (Fay, 1998). As demonstrated, this capacity depends on spectral and temporal pattern differences between the signal and the potentially interfering sounds. These studies were carried out using repetitive impulses or clicks as both signals and potentially interfering sounds. Thus, the presence of intermittent, audible airgun shots would not necessarily impair fishes in receiving and appropriately interpreting other biologically relevant sounds from the environment, so long as the shot sounds were not high enough in received level to cause PTS or TTS.

E. USES OF SOUND BY FISHES

Understanding of the uses of sound by fishes in normal behavior is very poor, with very few relevant published papers in the literature. Using the extant literature, common sense, and analogy with other vertebrate species studied, it is thought that fishes use sounds to aid in predator and prey detection and in general orientation to environmental sound sources and scatterers. In addition, as summarized above, it is thought that fishes are capable of resolving the general auditory scene, made up of the collection of sound sources and scatterers that normally occupy it. Thus, impaired hearing or excessive masking could have a general effect of rendering the affected fish generally less fit for survival and reproduction. In addition, some but not all species are known to make sounds used in communication. This topic has been reviewed recently by Zelick et al. (1999).

Sonic fishes are known to produce sounds during aggressive interactions, territorial defense, territorial advertisement, swimming, courtship, and mating. In general, sounds are best understood in the context of reproduction because they are associated with particular, stereotyped behaviors. Thus, it is likely that the major negative effects of impaired hearing or excessive masking would possibly be in disrupting the usual reproductive behavior, and thus the reproductive success, of some species. In addition, Myrberg (1981) has identified various categories of acoustic communication, some less obvious than others, that are used by fishes. These are startle or warning sounds that may help protect individuals and groups from predation; courting sounds used as part of the usual mating behaviors including advertisement; swimming sounds used in schooling and aggregation; aggressive sounds used when competing for mates; sounds used in other aggressive interactions (e.g., in territorial defense); sounds used by interceptor species to avoid predation or to locate prey; and sounds overheard and used to competitive advantage by competitors. This is a very wide range of sound types, some produced by specially adapted sound production organs, and some not. The uses of these sounds are very widespread, too, including listening species that produce sound using specialized mechanisms, and other species that do not produce sound, but only listen for them to some advantage. The families identified having species documented to use sound in behavior in some way, and that have species members in the Gulf, are presented in **Table I-1**. In general, this analysis of the potential use of sound in a communication context could potentially apply to most all fish species found in the Gulf of Mexico.

Table I-1
Fish Families in the Gulf of Mexico Having Species Members Known to Use Sound in Behavior
(From: Myrberg, 1981)

Family	Common Name(s)	Number of Species in the Gulf of Mexico
Albulidae	Bonefish	1
Alopidae	Thresher sharks	2
Ariidae	Catfish	2
Balistidae	Triggerfishes	12
Batrachoididae	Toadfish	3
Blenniidae	Blennies	9
Carangidae	Jack-like	25
Carcharhinidae	Requiem sharks	16
Clupeidae	Herring, menhaden	13
Dactylopteridae	Flying gurnards	1
Ephippidae	Spadefishes	1
Elopidae	Tarpons	2
Engraulidae	Anchovies	5
Gadidae	Cods	4
Gobiidae	Gobies	22
Holocentridae	Squirrelfishes	3
Kyphosidae	Sea chubs	1
Labridae	Wrasses	6
Lamnidae	Mackerel sharks	2
Lutjanidae	Snappers	13
Mullidae	Goatfishes	4
Myliobatidae	Eagle rays	3
Orectolobidae	Carpet sharks	1
Ostraciidae	Boxfishes	2
Polynemidae	Threadfins	1
Pomacentridae	Damselfishes	5
Haemulidae (formerly Pomadasyidae)	Grunts	9
Pomatomidae	Bluefishes	1
Scaridae	Parrotfishes	5
Sciaenidae	Croakers, drums	16
Scombridae	Mackerels, tuna	14
Serranidae	Basses	29
Sparidae	Porgies	10
Sphyraenidae	Barracudas	3
Sphyrnidae	Hammerhead sharks	5
Tetraodontidae	Puffers	7
Triglidae	Sea robins	12

In summary, some species of 37 families occurring in the Gulf of Mexico are known to use sound in behavior, according to the definitions of communication developed by Myrberg (1981). There have been no published reports on the effects of hearing impairment or excessive masking on the acoustic communication behavior of any species. Thus, we can only speculate that impairment has the potential to disrupt a wide range of behaviors in many Gulf species (up to 270 species). These behaviors include startle responses to predators, courtship and mate choice, territorial and individual advertisement, maintenance of schooling and aggregation, aggressive competition for mates and other resources, and overhearing or intercepting potential predators, prey, and competitors. In addition to these behaviors classified by Myrberg (1981) as communication, it is also likely that hearing is used to help form a general image of the auditory scene that may include both other fishes and abiotic sound sources and scatterers. Fishes unable to use the usual acoustic ambience for this scene formation would be expected to be less fit for survival, reproduction, and competition for resources. Since all fish species investigated have inner ear organs of essentially the same type, and have complex auditory brain pathways that are like most other vertebrates, it would be expected that hearing plays an important role in maintaining the fitness of all species.

F. POSSIBLE EFFECTS OF INTENSE SOUND ON FISH DISTRIBUTION AND COMMERCIAL CATCH

It is now well documented that intense sounds such as those produced by seismic airguns affect the spatial distribution of fishes during and following exposure, thus affecting the commercial catch by trawl or hook and line within the exposure area and for a certain period of time post-exposure. Klimley and Beavers (1998) have carried out the best controlled experimental study on this topic of fish movements, using intense, low frequency sounds of the type used in acoustic thermometry of ocean climate (ATOC; Munc et al., 1994). The goal of this study was to observe the behaviors of penned rockfish (*Sebastes flavidus*, *S. auriculatus*, and *S. mystinus*) during presentations of ATOC-like sounds (narrow-band signal centered at 75 Hz at peak sound pressure levels of 145 to 153 dB re 1 μPa in the vicinity of the fish). Sounds were 25 minutes in duration, including a 5-minute gradual "ramp-up" in level. Animals confined to sea pens were allowed to move up to 15 m away from the pen area closest to the source. Results showed remarkably few behaviors indicating that the fish found the intense sounds aversive. Prior to testing, many animals were stationed in the part of the pen closest to the transducer (where sound pressure levels would be greatest). During exposure, the animals tended to remain in these high exposure areas despite the opportunity to move 15 m away from the transducers, to a pen area where the sound level was more than 40 dB less. The authors concluded that the ATOC-like signals had no apparent effect on the distribution of rockfish within the pen, and thus would not be expected to cause animals to move from a location resembling a natural refuge. Of course, the major differences between the ATOC-like signal and airgun shots, are the brief, impulsive nature of the airgun sounds and their rapid rise time compared to the 5-minute ramp-up of the ATOC signal.

Pearson et al. (1992) investigated the effects of airgun sounds on rockfish (*S. mystinus*, *S. melanops*, *S. miniatus*, and *S. serranoides*) behavior. At levels of 180 dB re 1 μPa, animals appeared to be alarmed, and either aggregated more tightly, descended, or ascended in the water column. The lowest level causing any observed behavioral change was 161 dB.

Some species of sharks may be attracted to some sounds, and other species may move away from some sound sources. Klimley and Myrberg (1979) noted that lemon sharks (*Negaprion brevirostris*) were repulsed from a sound (noise) source at 123 dB but only when the signal was ramped up rapidly in amplitude (96 dB/sec). At slow ramp-up rates, no effect was observed at this pressure level. Nelson and Gruber (1963) observed that sharks were attracted by intermittent, low frequency sounds resembling those produced by struggling or injured fishes. Received levels in this experiment were not recorded.

Popper and Carlson (1998) have reviewed the literature on attempts to control and guide fish movements using sound. The major questions have concerned controlling some clupeids near water intakes at dams and power plants using very high frequency sounds, and controlling salmonid movements at dams and other manmade waterways using very low frequency sounds. Experiments using ultrasound to control clupeid behavior (e.g., Nestler et al., 1992) are not especially relevant to the question of airgun geophysical exploration since the airgun sounds are very low in frequency. It appears that ultrasound at high levels (110 to 140 kHz at 180 dB and above) can divert some herring species (e.g., American shad [*Alosa sapidissima*]) from sound sources, but this effect is restricted only to those species, like the American shad, which have developed ultrasonic hearing (Mann et al., 1997). Apparently, not all clupeids have this ability (Higgs and Popper, 2000). Knudsen et al. (1994) were successful in causing downstream migrating salmon to reverse their direction and swim away from intense, infrasound (<20 Hz) sources. This too may not be especially relevant with respect to seismic exploration in the Gulf of Mexico. However, some other species (e.g., Atlantic cod [*Gadus morhua*]) are known to detect efficiently very low frequency sound (0.1 to 30 Hz) (Sand and Karlsen, 1986), and it would not be surprising to find other Gulf species with such capabilities.

There have been a number of field studies investigating the effects of seismic airgun shots on cods and herrings distributions and catch successes in marine environments (Chapman and Hawkins, 1969; Matousek et al., 1988; Løkkeberg, 1991; Skalski et al., 1992; Engås et al., 1993, Løkkeberg and Soldal, 1993). Løkkeberg (1991) and Engås et al. (1993) both reported that the cod catch (by trawl) was reduced significantly (50%-80% reduction) during and following seismic shooting in the North Sea off the coast of Norway. The calculated sound pressure levels received by the fish were 191 and 160 dB, respectively. Remarkably, Engås et al. (1993) found that the reduced catch lasted for at least 5 days within a 33-km radius of the shooting. In the Pacific, off the coast of California, Skalski et al. (1992) found that calculated received levels of 161 dB caused rockfish (*Sebastes* sp.) to change behavior, to show alarm reactions at 180 dB, and to startle reactions at 200 to 205 dB.

Along the west coast of Scotland, Wardle et al. (2001) exposed several reef fish species in varying age classes (i.e., juvenile cod, *Gadus morhua*; adult pollack, *Pollachius pollachius*; juvenile saithe, *Pollachius virens*; and adult mackerel, *Scomber scombrus*) to airgun noise and recorded their reaction. Maximum seismic source levels from the three-airgun array were calculated at 218 dB re 1 μPa (peak to peak). While the fishes observed failed to move off the reef and airgun noise did not affect diurnal rhythms, involuntary reactions were elicited when firings occurred within 10 m of the fish being observed. Wardle et al. (2001) suggest that the visual cue of a firing airgun may elicit a more pronounced response than airgun noise alone.

In summary, the consensus is that seismic airgun shooting can result in reduced trawl and longline catch of several species when the animals receive levels as low as 160 dB. These effects would tend to be less in shallower water where sound propagation at these frequencies would be less efficient, and where sound levels would be lower due to the use of smaller airgun arrays. Reduced catch by trawl probably reflects temporary movement of the target species away from the ensonified area. Reduced catch by hook-and-line could be caused by fish moving away, or changing feeding behaviors. In any case, there are sufficient careful observations in the literature to conclude that airgun shooting may cause a temporary reduction in the commercial fish catch within at least several kilometers of the ensonified area.

G. CONCLUSIONS

All fish species investigated can hear, with varying degrees of sensitivity, within the frequency range of sound produced by seismic airguns. These sounds have the potential to cause masking of the sounds normally used by fishes in their usual acoustic behaviors at levels as low as 60 to 80 dB (just above detection thresholds for many species). Levels as high as 160 dB may cause receiving fishes to change their behaviors and movements that may temporarily affect the usual distribution of animals and commercial fishing. Continuous, long-term exposure to levels above 180 dB has been shown to cause damage to the hair cells of the ears of some fishes under some circumstances. These effects may not be permanent since damaged hair cells are repaired and/or regenerated in fishes. The effects of intermittent sound stimulation (such as airgun pulses) are not known, except that a 20% duty cycle is clearly less effective in damaging hair cells than continuous sounds. It seems likely that most fishes exposed to airgun shots at a distance of 1 m or less could receive inner ear damage as a result of source levels in the range between 210 and 240 dB. As the distance between the fish and the airgun sources increases, the probability of hearing impairment would decrease according to the nature of distance attenuation taking place in the ensonified environment.

Fishes of the Gulf of Mexico are generally representative of the species for which quantitative hearing data are available. It is thought that most, if not all, species of fishes in the Gulf of Mexico (and elsewhere) use the sense of hearing to contribute to their general fitness for survival and reproduction. The sense of hearing is thought to be used not only in intra-specific sound communication in some species, but also in many more general contexts in which most or all fishes monitor their acoustic environment by simply listening to the ambient sounds produced by other fishes and by abiotic sources and sound scatterers. There are no experimental studies investigating the effects of excessive masking or hearing impairment on the usual behaviors of fishes. However, several studies indicate that seismic shooting can temporarily alter the behaviors and movements of several fish species when received sound pressure levels are sufficiently high. The long-term effects of these hearing-related behavioral disruptions on fish populations are unknown.

III. LITERATURE CITED

Bregman, A.S. 1990. Auditory Scene Analysis: The Perceptual Organization of Sound. MIT Press, Cambridge, MA.

Continental Shelf Associates, Inc. 2003. Explosive removal of offshore structures - information synthesis report. Final proof copy. Prepared for the U.S. Department of the Interior, Minerals Management Service, Gulf of Mexico OCS Region, New Orleans, LA. Prepared under MMS Contract 85237. 190 pp. + apps.

Chapman, C.J., and A.D. Hawkins. 1969. The importance of sound in fish behaviour in relation to capture by trawls. FAO Fish. Rept. 62:717-729.

Chapman, C.J., and A.D. Hawkins. 1973. A field study of hearing in the cod, *Gadus morhua*. L.J. Comp. Physiol. 85:147-167.

Cox, M., P. Rogers, A. Popper, W. Saidel, and R. Fay. 1986a. Frequency regionalization in the fish ear. J. Acoust. Soc. Am., Suppl. 1, 79:S80 (abstract).

Cox, M., P. Rogers, A. Popper, and W. Saidel. 1986b. Anatomical effects of intense tone stimulation in the ear of bony fishes. J. Acoust. Soc. Am., Suppl. 1, 80:S75 (abstract).

Cox, M., P. Rogers, A. Popper, W. Saidel, R. Fay, and S. Coombs. 1987. Anatomical effects of intense tone stimulation in the goldfish ear. J. Acoust. Soc. Am., Suppl. 1, 89:S7 (abstract).

Enger, P.S. 1981. Frequency discrimination in teleosts - central or peripheral?, pp. 243-255. *In:* W.N. Tavolga, A.N. Popper and R.R. Fay (eds.), Hearing and Sound Communication in Fishes. Springer-Verlag, New York.

Engås, A., S. Lokkeborg, E. Ona, and A.V. Soldal. 1993. Effects of seismic movements on catches and availability of cod and haddock. Fisken Og Havet 3 (March 1993):1-111.

Fay, R.R. 1974. Masking of tones by noise for the goldfish (*Carassius auratus*). J. Comp. Physiol. Psychol. 87:708-716.

Fay R.R. 1984. The goldfish ear codes the axis of acoustic particle motion in three dimensions. Science 225:951-954

Fay, R.R. 1988. Hearing in Vertebrates: A Psychophysics Databook. Hill-Fay Associates, Winnetka, IL.

Fay, R.R. 1998. Auditory stream segregation in goldfish (*Carassius auratus*). Hear. Res. 120:69-76.

Fay, R.R., and S.L. Coombs. 1988. Psychophysics and neurophysiology of frequency selectivity and masking in the goldfish, pp. 169-176. *In:* H. Duifhuis, W. Horst, and H. Wit (eds.), Basic Issues in Hearing. Groningen University Press, Groningen, The Netherlands.

Fay, R., and A. Megela Simmons. 1999. The sense of hearing in fishes and amphibians, pp. 269-318. *In:* R. Fay and A. Popper (eds.), Comparative Hearing: Fish and Amphibians. Springer Handbook of Auditory Research, Vol. 11. Springer-Verlag, New York.

Fay, R.R., W.A. Ahroon, and A.A. Orawski. 1978. Auditory masking patterns in the goldfish (*Carassius auratus*): Psychophysical tuning curves. J. Exp. Biol. 74:83-100.

Hastings, M., A. Popper, J. Finneran, and P. Lanford. 1996. Effects of low-frequency underwater sound on hair cells of the inner ear and lateral line of a teleost fish. J. Acoust. Soc. Am. 99:1759-1766.

Hawkins, A.D., and C.J. Chapman. 1975. Masked auditory thresholds in the cod, *Gadus morhua* L. J. Comp. Physiol. A103:209-226.

Higgs, D., and A. Popper. 2000. Possible morphological correlates to ultrasound detection by the American shad (*Alosa sapidissima*), p. 280. *In:* Abstracts of the Association for Research in Otolaryngology, Mid-Winter meeting, #971.

Hudspeth, A.J., and D.P. Corey. 1977. Sensitivity, polarity and conductance change in the response of vertebrate hair cells to controlled mechanical stimuli. Proc. Nat. Acad. Sci. USA 74:2407-2411.

Klimley, A., and S. Beavers. 1998. Playback of acoustic thermometry of ocean climate (ATOC) – like signal to bony fishes to evaluate phonotaxis. J. Acoust. Soc. Am. 104:2506-2511.

Klimley, A., and A. Myrberg. 1979. Acoustic stimuli underlying withdrawal from a sound source by adult lemon sharks, *Negaprion brevirostris* (Poey). Bull. Mar. Sci. 29:447-458.

Knudsen, F., P. Enger, and O. Sand. 1994. Avoidance responses to low-frequency sound in downstream migrating Atlantic salmon smolt, *Salmo salar*. J. Fish. Biol. 45:227-233.

Løkkeberg, S. 1991. Effects of geophysical survey on catching success in longline fishing. ICES CM 40:1-9.

Løkkeberg, S., and A. Soldal. 1993. The influence of seismic exploration with airguns on cod (*Gadus morhua*). ICES Marine Science Symposium 196:62-67.

Lombarte, A., H. Yan, A. Popper, J. Chang, and C. Platt. 1993. Damage and regeneration of hair cell ciliary bundles in a fish ear following treatment with gentamycin. Hear. Res. 66:166-174.

Mann, D., Z. Lu, and A. Popper. 1997. Ultrasound detection by a teleost fish. Nature (London) 389:341.

Malme, C., P. Smith, and P. Miles. 1986. Characterization of geophysical acoustic survey sounds. U.S. Department of the Interior, Minerals Management Service, Los Angeles, CA. Rep. OCS/MMS-86/0032.

Matousek, J., A. Wells, and P. McGroddy. 1988. Field testing of behavioral barriers for fish exclusion at cooling-water intake systems. Electric Power Research Institute, Project # 221406. Report No. EPRI CS-5995.

McCauley, R.D., J. Fretwell, and A.N. Popper. 2003. High intensity anthropogenic sound damages fish ears. J. Acoust. Soc. Am. 113(1): 638-642.

Munc, W., R. Spindel, A. Baggeroer, and T. Birdsall. 1994. The Heard Island feasibility test. J. Acoust. Soc. Am. 96:2330-2342.

Myrberg, A. 1981. Sound communication and interceptions by fishes, pp. 395-426. *In:* W.N. Tavolga, A.N. Popper, and R.R. Fay (eds.), Hearing and Sound Communication in Fishes. Springer-Verlag, New York.

Nelson, D., and S. Gruber. 1963. Sharks: attraction by low-frequency sounds. Science 142:975-977.

Nestler, J., G. Ploskey, J. Pickens, J. Menezes, and C. Schildt. 1992. Responses of blueback herring to high-frequency sound and implications for reducing entrainment at hydropower dams. N. Amer. J. Fish. Man. 12:667-683.

Patuzzi, R. 1996. Cochlear micromechanics and macromechanics, pp. 186-257. In: P. Dallos, A. Popper, and R. Fay (eds.), The Cochlea. Springer Handbook of Auditory Research, Vol. 8. Springer-Verlag, New York.

Pearson, W., J. Skalski, and C. Malme. 1992. Effects of sounds from a geophysical survey device on behavior of captive rockfish. Can. Fish. Aquatic Sci. 49:1343-1356.

Popper, A., and T. Carlson. 1998. Application of sound and other stimuli to control fish behavior. Trans. Amer. Fish. Soc. 127:673-707.

Popper, A.N., and N.L. Clarke. 1976. The auditory system of the goldfish (*Carassius auratus*): Effects of intense acoustic stimulation. Comp. Biochem. Physiol. 53A:11-18.

Popper A.N., and N.L. Clarke. 1979. Non-simultaneous auditory masking in the goldfish *C. auratus*. J. Exp. Biol. 83:145-158.

Popper, A., and R. Fay. 1999. The auditory periphery in fishes, pp. 43-100. *In:* R. Fay and A. Popper (eds.), Comparative Hearing: Fish and Amphibians. Springer Handbook of Auditory Research, Vol. 11. Springer-Verlag, New York.

Sand, O., and H. Karlsen. 1986. Detection of infrasound by the Atlantic cod. J. Exp. Biol. 125:197-204.

Skalski, J., W. Pearson, and C. Malme. 1992. Effects of sounds from a geophysical survey device on catch per unit effort on a hook-and-line fishery for rockfish (*Sebastes* spp.). Can. J. Fish. Aquatic Sci. 49:1357-1365.

Tavolga W.N. 1974. Signal/noise ratio and the critical band in fishes. J. Acoust. Soc. Am. 55: 1323-1333.

Wardle, C.S., T.J. Carter, G.G. Urquhart, A.D.F. Johnstone, A.M. Ziolkowski, G. Hampson, and D. Mackie. 2001. Effects of seismic air guns on marine fish. Cont. Shelf Res. 21(2001):1005-1027.

Zelick, R., D. Mann, and A. Popper. 1999. Acoustic communication in fishes and frogs, pp. 363-412. *In:* R. Fay and A. Popper (eds.), Comparative Hearing: Fish and Amphibians. Springer Handbook of Auditory Research, Vol. 11. Springer-Verlag, New York.

Appendix J
Cumulative Activity Scenario and Associated Impacts

Cumulative Activity Scenario and Associated Impacts

This appendix describes the current level of maritime activity within the Gulf of Mexico, exclusive of geological and geophysical (G&G) operations. This summary of cumulative activities serves as a basis for assessing the relative contribution (of environmental impact) of G&G activities to the cumulative impacts of other maritime activities in the Gulf of Mexico.

I. COMMERCIAL SHIPPING

The northern Gulf of Mexico supports an extensive maritime industry transporting goods between the major Gulf ports and ports located outside the Gulf. Gulf waters and adjacent waters, including the Gulf Intracoastal Waterway (GIWW) and large rivers and canals, are heavily used for domestic and foreign waterborne commerce within the Gulf. Cargo ships, bulk freighters, and tankers of various sizes and flags ply its waters, transporting a large range of commodities including crude oil, fuel, petroleum products, raw materials, manufactured goods, food and farm products, and chemicals.

In addition to being the leader in domestic offshore oil production, the Gulf Coast region has developed into the major U.S. oil-handling center for crude and refined petroleum products originating from and destined for both domestic and foreign markets. The Gulf offers extensive refinery capacity, easy port access, and a well-developed transportation system. The only deep-water crude-oil terminal in the country, the Louisiana Offshore Oil Port (LOOP), and the National Strategic Petroleum Reserve are located within the Gulf. Crude oil is brought by tankers into Gulf refineries from domestic production occurring into the Atlantic and Pacific Oceans and crude oil is also transported by barges among Gulf terminals to reach terminals and onshore transportation routes. Petroleum products from refinery complexes in the Gulf are then transported by tanker or barge to other states.

The Department of Energy (DOE) projects that tanker imports and exports into the Gulf will continue to increase, projected to double during the 20-year period from 1995-2015. In 2000, approximately 2.08 billion barrels of oil (BBO) and 1.09 BBO of petroleum products moved through the 13 largest Gulf ports, representing 38 and 13 percent of the U.S. total, respectively (U.S. Department of the Interior, Minerals Management Service [USDOI, MMS], 2002a). MMS projects that importation of crude oil and other petroleum products moving through Gulf ports will increase 35 and 60 percent, respectively, by 2020. The actual rate of increase will be dependent upon the rate of growth of domestic petroleum consumption. Imports of refined petroleum products are also expected to increase as domestic refinery capacity approaches maximum utilization rates by 2015 (USDOI, MMS, 1998). A large volume of crude oil will continue to be transported by tanker from Alaska, California, and Atlantic production into the Gulf for refining.

The LOOP, located in Grand Isle Block 59, approximately 30 km from shore, handles a significant portion of the crude oil transported into the Gulf from domestic (60,000 bbl/d in 1996; 350,000 bbl/d in 2001) and foreign (823,000 bbl/d in 1996; 1.1 million bbl/d in 2001) sources. LOOP is the only U.S. port capable of offloading deep draft tankers (e.g., ultra large crude carriers, ULCC; very large crude carriers, VLCC; LOOP also offloads smaller tankers. In 2001, more than 300 tanker calls occurred at LOOP (i.e., >600 tanker trips) (U.S. Commission on Ocean Policy, 2002). Crude oil offloaded at the LOOP facility is transported to shore via submarine pipeline, then pumped to refineries in Louisiana, along the Gulf coast, and in the U.S. midwest.

Data for 1999 vessel activity levels (i.e., tanker and other freight traffic) compiled by the U.S. Army Corps of Engineers and summarized by USDOI, MMS (2002a) are outlined in **Table J-1**.

Louisiana, Alabama, Texas, and Mississippi realized approximately 140,000, 50,000, 28,000, and 18,000 trips (involving tankers and other freight carriers) into and out of major ports, respectively, during 1999. Tanker and freight vessel traffic through the GIWW comprised nearly 43,000 vessel trips during the same period. Gulfwide, 1999 tanker and freight traffic amounted to about 279,000 vessel trips, including use of both Gulf waters and the GIWW. Of these totals, tanker traffic contributed one quarter to more than one half of all vessel trips (i.e., tankers, other freight traffic) through the Gulf. Additional information pertinent to current Gulf vessel traffic and associated noise characteristics is provided in **Appendix F**, **Section IV - Noise Environment** (e.g., see **Table F-14**).

Table J-1
Summary of Tanker and Other Freight Vessel Trip Activity into Major Gulf Ports
(Adapted from: U.S. Department of the Interior, Minerals Management Service, 2002a)

Waterway	Upbound			Downbound		
	Total	Tanker	Other	Total	Tanker	Other
GIWW	21,636	18,747	14,909	21,264	18,784	14,530
TX	14,517	9,965	15,116	13,650	9,808	14,479
LA	70,128	34,823	78,411	69,845	34,923	86,972
MS	10,650	3,907	6,743	10,540	3,887	6,653
AL	24,774	5,772	22,573	24,833	5,751	23,482

GIWW= Gulf Intracoastal Waterway.

II. OIL AND GAS SUPPORT VESSEL OPERATIONS

Service vessels that support various requirements of offshore oil and gas activities of about 24 types are categorized into supply, crew, and utility vessels. Large supply boats (50 to 70 m in length) with a capacity of 300 tons and draft of 3.5 m when loaded make up a large proportion of service vessels in the Gulf of Mexico. Crew and utility boats are about 30 m in length and draw 2 to 2.8 m and 3 to 3.7 m of water, respectively. Service vessels utilized in deep water include offshore supply vessels, fast supply vessels, and anchor-handling towing supply/mooring vessels; vessels employed in deep-water operations typically are larger and/or faster than those usually supporting oil and gas operations in shallower water closer to shore (USDOI, MMS, 2000a). Supply and utility boats carry water, cement, drilling fluids, equipment, food, and miscellaneous supplies, while crew boats are active during change-outs of personnel. Service vessels produce noise above and underwater, discharges, and air emissions; make waves that erode channel banks; and disturb the sea bottom with anchors. As the number of deep-water development facilities located greater distances from shore increases, larger supply vessels with greater cargo carrying capacities and fast crew boats are being used (USDOI, MMS, 2001a).

Using some assumptions about the number of vessel crew members per boat, number of trips to existing as well as projected platforms, the number of development wells, the number of trips per well per week, transit times, and distances to sites from service bases, etc., the average annual number of service vessel trips in each of the planning areas has been estimated by MMS. Trips are also assumed to originate from service bases and new vessels are assumed to be built to handle transport to distant oil and gas deep-water platforms. Service vessel trips are greatest during exploration, drilling, and the development phase and are significantly reduced during the production phase.

On an annual basis, Gulf-wide service vessel trips are expected to range from 304,807 to 319,921 annually (**Table J-2**) in support of the OCS Program (USDOI, MMS, 2002a), although the distribution of this activity is disproportionate between planning areas and water depths (i.e., most oil and gas production occurs in nearshore or shelf waters, with increasing interest in deep-water development). In the Western Planning Area, 31,359 to 36,923 service vessel trips per year are estimated, representing ~10 to 12 percent of the Gulf-wide projected service vessel trips per year. Using similar assumptions, MMS projections of service-vessel activity for the Central Planning Area are estimated to be 272,923 to 281,948 trips annually, representing ~88 to 90 percent of the Gulf-wide service vessel trips. In the Eastern Planning Area, only 525 to 1,050 service vessel trips are expected annually, representing 0.2 to 0.3 percent of the Gulf-wide total (USDOI, MMS, 2002a).

Table J-2
Summary of Annual Service Trip Activity by Planning Area (From: U.S. Department of the Interior,
Minerals Management Service, 2002a)

Planning Area	Low Projection (trips/year)	High Projection (trips/year)
Western	31,359	36,923
Central	272,923	281,948
Eastern	525	1,050

III. MILITARY ACTIVITIES

The surface and airspace of large areas of the Gulf of Mexico are used extensively by the military for testing, evaluation, training, and qualification of aircraft, vessels, weapon systems, and personnel. Military activities in the Gulf of Mexico have been extensively summarized in USDOI, MMS (1997, 1998, 1999, 2002a). These activities normally consist of various air-to-air, air-to-surface, and surface-to-surface fleet-training and air force exercises. In addition to carrier operations, air-to-air, air-to-surface, and subsurface operations, the Navy uses the Gulf for shakedown cruises on newly built ships, and for ships completing overhaul or extensive repairs in Gulf shipyards such as those located in Pascagoula, Mississippi. No aircraft carriers or submarines are currently home-ported or stationed in the Gulf but may from time to time conduct operations there. Military Warning Areas (MWAs) have been designated by the Federal Aviation Administration (FAA) in the Gulf of Mexico and are used by various Air Force, Navy, and Air National Guard units. The airspace over MWAs is FAA Special Use Airspace reserved for military use, and FAA air traffic controllers may prohibit unauthorized aircraft from entering MWAs. The specific boundaries, authorized activities, and military command contacts for each MWA are available in appropriate U.S. Air Force Flight Information Publications and Navy Fleet Training Area Range Manuals.

Under Department of Defense (DOD) and DOI agreements, lease activities in MWAs may be under Military Areas Stipulations, which include hold and save harmless, electromagnetic emissions, and operational sections to reduce potential multi-use conflicts. The stipulations promote safety and reduce interference between operations but do not reduce or eliminate the actual physical presence of oil and gas operations in areas where military operations are conducted. Except for a few areas and activities, detailed information on the level of present and future levels of military activity in the MWAs is not available. Because of the lease stipulations and Memorandums of Agreement (MOAs) between DOD and MMS, multi-use conflicts will be minimized.

Some MWAs are designated by the U.S. Air Force for the conduct of various testing and training missions, and others are designated by the Navy. Air Force operations are controlled out of the Eglin Military Complex located in northwest Florida, Macdill Air Force Base in southwest Florida, and Offutt Air Force Base in Omaha, Nebraska and through Air National Guard bases located in Corpus Christi and Houston, Texas; New Orleans, Louisiana; and Gulfport, Mississippi. Naval operations are controlled and coordinated through Corpus Christi, Texas; New Orleans, Louisiana; and Pensacola and Key West, Florida.

In the Western Gulf Planning Area, all or a portion of four MWAs are used for military operations, per Department of the Navy (1998) and USDOI, MMS (2002a):

Warning Area	Defense Operations Conducted
W-59A	Air National Guard flight training (W-59A predominantly occurs in the Central Planning Area)
W-147A-E	Air National Guard flight training (W-147A and B predominantly occur in the Central Planning Area)
W-228A-D	Mine warfare training, air-to-air gunnery, rocket firing, aircraft operations, submarine operations
W-602	High altitude training

Mine warfare testing and training activities are conducted by the Navy in several Mustang Island (MI) Area blocks that are permanently deferred to Navy activities and are excluded from leasing. In addition, an MMS and Navy formal agreement requires that some MI blocks carry a multi-use mitigation stipulation if leased. The stipulation will eliminate potential impacts from multi-use conflicts on these blocks.

In the Central Planning Area, all or a portion of five designated warning areas and portions of two of the six Eglin Water Test Area (EWTA) are used for military operations, per Department of the Navy (1998) and USDOI, MMS (2002a):

Warning Area	Defense Operations Conducted
W-59A-C	Air National Guard flight training (small segment of W-59A occurs in the Western Planning Area)
W-92	Air-to-air gunnery, rocket-firing, sonar buoy operations
W147A-B	Air National Guard flight training
W-453	Air National Guard flight training, live fire air-to-air gunnery
W-155A-C	Carrier operations, carrier pilot training, live fire air-to-air and air-to-surface operations, subsurface operations (W-155A-C occur predominantly in the Eastern Planning Area)
EWTA-1 and -3	Rocket and missile testing and research (EWTA-1 and -3 occur predominantly in the Eastern Planning Area)

The Eastern Planning Area contains five partial or complete MWAs and the Eglin Water Test Area, per Department of the Navy (1998) and USDOI, MMS (2002a). The EWTA and Areas W-151 and W-470 together are referred to as the Eglin Gulf Test Area.

Warning Area	Defense Operations Conducted
W-151A-F	Air combat training, air-to-air training, air-to-surface training, live fire air-to-air activities
W-155A-C	Carrier operations, carrier pilot training, live fire air-to-air and air-to-surface operations, subsurface operations (W-155A-C occur predominantly in the Eastern Planning Area)
W-168A	Air combat training, live fire air-to-air gunnery, anti-submarine warfare operations, Air National Guard flight training
W-174A-G	Air Force and Navy flight training, weapons testing
W-470A-D	Air-to-air training, air combat training, live fire air-to-air and surface-to-air operations
EWTA	Rocket and missile testing and research (EWTA-1, 2A-C, -3, -4, -5, -6; EWTA-1 and -3 occur predominantly in the Eastern Planning Area)

The Eglin Gulf Test Range may also be used for testing the Theatre Missile Defense (TMD) system, which requires a large area crucial to the development and deployment of TMD missiles. A supplemental Environmental Impact Statement states that "up to 24 testing and training events" could occur each year, requiring "up to 48 interceptor missiles be launched each year from land sites in Eglin AFB and/or from Navy ships in the Gulf of Mexico." However, TMD officials have stated at public meetings that this level of testing is considered to be the extreme "worst case" and that it is more likely that there would be three to four tests each year once TMD testing becomes operational (USDOI, MMS, 1999).

Although live ordnance air-to-surface training is currently accomplished on the Eglin land ranges, the Air Force has made attempts to reopen the offshore ranges for high explosive (HE) gunnery training. If approved, it is estimated that 14 missions per year will be required to accomplish the required over-water, air-to-surface gunnery and test operations (USDOI, MMS, 1997; Department of the Navy, 1998).

IV. IMPACTS ASSOCIATED WITH THE CUMULATIVE SCENARIO

Results of the preliminary screening (see **Section III - Environmental Impacts** of the Programmatic Environmental Assessment [PEA]) indicate that G&G activities (i.e., primarily seismic surveys) have the potential to adversely affect several marine resources within the Gulf of Mexico – marine mammals, sea turtles, fishes, and commercial and recreational fisheries, as well as coastal and marine birds and benthic communities. The estimated cumulative impacts on these resources from these other maritime-related activities identified under the cumulative scenario are outlined below, utilizing the impact classification scheme (i.e., **significant adverse**, **adverse but not significant**, **negligible**) and resource-specific significance criteria outlined in **PEA Section III**. Impact determinations arising from the cumulative scenario (summarized below) provide the basis for an evaluation of the incremental impact projected from G&G activities; resource-specific incremental impacts are evaluated and summarized in **PEA Section III**. Pertinent reviews of cumulative impacts in the Gulf of Mexico region also may be found in USDOI, MMS (1999, 2001a,b, 2002a,b) and Department of the Navy (2001).

A.   MARINE MAMMALS

Marine mammals may be affected by several impact producing factors (i.e., vessel traffic, noise) resulting from commercial shipping, oil and gas support vessel operations, and military activities under the cumulative scenario. Vessel traffic produces the potential for shipstrike, whereas noise from vessels may result in behavioral response and/or short term displacement.

In terms of vessel traffic, oil and gas support vessel activity accounts for approximately one quarter to one third of a million transits per year in Gulf waters, with tanker and other freight vessels >10,000 DWT contributing another ~279,000 trips in the waters of the Gulf and Gulf Intracoastal Waterway (see **Appendix F**). Military exercises and routine fleet operations also contribute to vessel traffic levels in the Gulf. For some marine mammal species, low frequency noise from vessel engines and propeller cavitation may serve to warn of an approaching vessel, thereby reducing the potential for shipstrike. For other species (i.e., delphinids), vessel noise is a cue to approach a transiting vessel and bowride. In its assessment of cumulative activities occurring on the Gulf of Mexico OCS, USDOI, MMS (2001a) notes that the potential for short term disruption of movement and behavior patterns from service vessel traffic is unlikely to affect survival or productivity of marine mammal populations in the Gulf of Mexico, unless such disruptions occur frequently. Long term displacement is not expected, but is possible. Considering the significance criteria noted in **PEA Section III**, impacts to marine mammals from vessel traffic under the cumulative scenario are considered to be potentially **adverse but not significant**.

Noise from commercial shipping, oil and gas support vessel activity, and military operations may mask acoustic signals in the marine environment, or elicit a startle response or avoidance behavior from marine mammals. USDOI, MMS (1999, 2001a) notes that noise from service vessel traffic may also result in active avoidance behavior or displacement of individuals or groups, with reactions varying by species, age, sex, and psychological status. Displacement towards a transiting vessel may also occur. Short term disruptions in movement or behavior patterns resulting from vessel activity are unlikely to adversely affect population survival or productivity. The potential for long term displacement remains to be determined. On the basis of the significance criteria provided in **PEA Section III**, impacts to marine mammals from noise under the cumulative scenario (i.e., commercial shipping, oil and gas support vessel operations, and military activities) are considered to be potentially **adverse but not significant**.

B.   SEA TURTLES

Sea turtles may be affected by several impact producing factors (i.e., vessel traffic, noise) resulting from commercial shipping, oil and gas support vessel operations, and military activities, as well as other activities not considered in this cumulative scenario.[1] As noted previously, oil and gas support vessel activity accounts for the vast majority of vessel transits within the Gulf each year, with commercial vessels and military operations also contributing to current traffic levels. As with marine mammals, sea turtles may respond either visually or acoustically (i.e., recognizing the low frequency noise from vessel engines and propeller cavitation) to an approaching vessel. This short term behavioral response, or startle reaction, increases the stress level of the individual turtle exposed (NRC, 2000). Such responses are not expected to affect population survival or productivity adversely. Vessel traffic is estimated to cause approximately nine percent of all sea turtle deaths in the southeastern U.S. (USDOI, MMS, 2001a).

---

[1] Impacts to sea turtles or their nesting or foraging habitats have also been attributed to structure installation (e.g., pipelines), dredging operations, water quality and habitat degradation, presence of trash and debris in marine and coastal environments, explosive platform removal operations, commercial and recreational fishing, beach lighting, and power plant entrainment.

Regions of greatest concern for vessel strikes are those with the highest vessel activity levels, particularly recreational boat traffic (e.g., within coastal bays and nearshore waters). Consideration of the significance criteria provided for sea turtles in **PEA Section III**, indicates that impacts from vessel traffic under the cumulative scenario are considered to be potentially **adverse but not significant**.

Noise from vessel activities occurring under the cumulative scenario may elicit a startle response, as noted previously. USDOI, MMS (2001a) notes that noise from cumulative vessel traffic sources may result in short term disruptions in sea turtle movement or behavior patterns. Such responses are unlikely to adversely affect survival or productivity. Of greater concern are nearshore activities (e.g., dredging, pipelaying operations) that affect nearshore water quality, degrade nursery habitat, and alter nesting beaches. Using the significance criteria provided in **PEA Section III**, an analysis shows impacts to sea turtles from commercial shipping, oil and gas support vessel operations, and military activities to be potentially **adverse but not significant**.

C.   FISHES

Fishes may be affected by several impact producing factors (i.e., vessel traffic, noise) resulting from commercial shipping, oil and gas support vessel operations, and military activities, as well as other activities not considered in this cumulative scenario. Impacts to fishes from these cumulative activities are expected to very limited, consisting primarily of short-term avoidance behavior. Considering the significance criteria provided for fishes in **PEA Section III**, an analysis shows that impacts from cumulative vessel traffic and noise under the cumulative scenario are considered to be **negligible**.

D.   COMMERCIAL AND RECREATIONAL FISHERIES

Commercial and recreational fisheries may be affected by several impact producing factors (i.e., vessel traffic, noise) resulting from commercial shipping, oil and gas support vessel operations, and military activities, as well as other activities not considered in this cumulative scenario (e.g., installation of offshore structures [platforms, subsea completions, pipelines], oil spills, oil spill response and cleanup activities, alteration or destruction of essential fish habitat [EFH]). USDOI, MMS (2001a) notes that activities such as seismic surveys and pipeline trenching operations result in negligible impacts to commercial fisheries, while structure emplacement, underwater OCS impediments, explosive removal activities, and oil spills result in slightly higher impacts. Impacts to EFH may be attributed to any activity that has a deleterious effect on water or sediment quality or benthic resources. Vessel traffic and noise impact factors are not likely to produce any impact on EFH within the Gulf of Mexico. Similarly, anchoring by oil and gas support vessels will be precluded around sensitive benthic communities, including EFH (see **Subsection F - Benthic Communities**, below), under existing or proposed lease stipulations. For recreational fisheries, no adverse impacts from cumulative vessel activity are expected. On the basis of the significance criteria noted in **PEA Section III**, an analysis shows that impacts to commercial and recreational fisheries and EFH from commercial shipping, oil and gas support vessel operations, and military activities are considered **negligible**.

E.   COASTAL AND MARINE BIRDS

Coastal and marine birds may be affected by several impact producing factors (i.e., vessel traffic, noise) resulting from commercial shipping, oil and gas support vessel operations, and military activities, as well as other activities not considered in this cumulative scenario (e.g., oil spills, oil spill response and cleanup activities, alteration or destruction of preferred and/or critical coastal habitat). Impacts from the cumulative activities noted are expected to be sublethal, although some displacement attributed to vessels may occur to local individuals or groups, particularly in coastal habitats used for nesting or foraging (USDOI, MMS, 2001a). Consideration of the significance criteria provided for coastal and marine birds in **PEA Section III**, shows impacts from vessel traffic and noise under the cumulative scenario to be **negligible**.

F.   BENTHIC COMMUNITIES

Benthic communities may be affected by one impact producing factor (i.e., anchoring impacts) resulting from oil and gas support vessel operations. Military activities and commercial shipping in surface and near surface waters are unlikely to affect benthic communities. In addition, MMS has either implemented or proposed a series of lease stipulations (e.g., Live Bottom Stipulation, Pinnacle Trend Stipulation) designed to protect sensitive benthic resources from anchoring and other bottom founded activities. Such measures should provide adequate protection for sensitive benthic resources. Therefore, impacts to benthic communities from commercial shipping, oil and gas support vessel operations, and military activities under the cumulative scenario are considered **negligible**.

V.   LITERATURE CITED

Department of the Navy. 1998. Fleet Training Area/Range Directory. Naval Warfare Assessment Station, Coronado, CA. 4105 Ser SE 40/122 (CD-ROM).

Department of the Navy. 2001. Final environmental impact statement: Shock trial of the WINSTON S. CHURCHILL (DDG 81). Southern Division, Naval Facilities Engineering Command, North Charleston, SC. February 2001. 229 pp. + apps.

National Research Council. 2000. The decline of sea turtles: Causes and prevention. National Academy Press, Washington, D.C. 183 pp.

Turner, R.E., and D.R. Cahoon. 1988. Causes of wetland loss in the coastal Central Gulf of Mexico. Prepared under Contract 14-12-0001-30252 for the U.S. Department of the Interior, Minerals Management Service, Gulf of Mexico OCS Region, New Orleans, LA. OCS Study 87-0119. 3 volumes.

U.S. Commission on Ocean Policy. 2002. Meeting notes - Gulf of Mexico Regional Meeting, March 7-8, 2002, New Orleans, LA. http://www.oceancommission.gov/meetings/mar7_8_02/mar7_8_02 html

U.S. Department of the Interior, Minerals Management Service. 1997. Gulf of Mexico OCS Oil and Gas Lease Sales 169, 172, 175, 178, and 182, Central Planning Area. Final Environmental Impact Statement. U.S. Department of the Interior, Minerals Management Service, Gulf of Mexico OCS Region, New Orleans, LA.

U.S. Department of the Interior, Minerals Management Service. 1998. Gulf of Mexico OCS Oil and Gas Lease Sales 171, 174, 177, and 180, Western Planning Area. Final Environmental Impact Statement. U.S. Department of the Interior, Minerals Management Service, Gulf of Mexico OCS Region, New Orleans, LA.

U.S. Department of the Interior, Minerals Management Service. 1999. Destin Dome 56 Unit Development and Production Plan and Right-of-Way Pipeline Application. Draft Environmental Impact Statement. U.S. Department of the Interior, Minerals Management Service, Gulf of Mexico OCS Region, New Orleans, LA.

U.S. Department of the Interior, Minerals Management Service. 2000. Gulf of Mexico Deepwater Operations and Activities - Environmental Assessment. U.S. Department of the Interior, Minerals Management Service, Gulf of Mexico OCS Region, New Orleans, LA. OCS EIS/EA MMS 2000-001. 267 pp.

U.S. Department of the Interior, Minerals Management Service. 2001a. Gulf of Mexico OCS Oil and Gas Lease Sale 181, Eastern Planning Area. Final Environmental Impact Statement. U.S. Department of the Interior, Minerals Management Service, Gulf of Mexico OCS Region, New Orleans, LA. OCS EIS/EA MMS 2001-055. 2 vols.

U.S. Department of the Interior, Minerals Management Service. 2001b. Proposed Use of Floating Production, Storage, and Offloading Systems on the Gulf of Mexico Outer Continental Shelf – Western and Central Gulf Planning Areas. Final Environmental Impact Statement. U.S. Department of the Interior, Minerals Management Service, Gulf of Mexico OCS Region, New Orleans, LA. OCS EIS/EA MMS 2000-090.

U.S. Department of the Interior, Minerals Management Service. 2002a. Gulf of Mexico OCS Oil and Gas Lease Sales: 2002-2007. Central Planning Area Sales 185, 190, 194, 198, and 201. Western Planning Area Sales 187, 192, 196, and 200. Final Environmental Impact Statement. U.S. Department of the Interior, Minerals Management Service, Gulf of Mexico OCS Region, New Orleans, LA. OCS EIS/EA MMS 2002-052. 2 vols.

U.S. Department of the Interior, Minerals Management Service. 2002b. Gulf of Mexico OCS Oil and Gas Lease Sales 189 and 197. Eastern Planning Area. Draft Environmental Impact Statement. U.S. Department of the Interior, Minerals Management Service, Gulf of Mexico OCS Region, New Orleans, LA. OCS EIS/EA MMS 2002-056. 2 vols.

Appendix K
Risk of Vessel Accidents

Risk of Vessel Accidents

This appendix describes the risk of vessel accidents associated with geological and geophysical (G&G) operations. This summary serves as a basis for assessing the environmental impact of G&G activities on resources of the Gulf of Mexico (see **Section III - Environmental Impacts** of the Programmatic Environmental Assessment [PEA]).

A. ACCIDENT CHARACTERIZATION

Vessels conducting G&G operations in the Gulf of Mexico vary in size from about 37 to 93 m long. A description of the size of the vessels performing the various activities is presented in **PEA Section II - Alternatives**. **Table II-4** of **PEA Section II** estimates the number of crews (i.e., "crew[s]" refer[s] to a seismic data acquisition operation involving one or more vessels) needed to perform the surveying activities associated with the Proposed Action. As can be seen from the table, it is estimated that there will be between 12 and 61 crews operating to perform the proposed surveying.

These vessels have the potential to become involved in incidents that can result in damage to property, damage to the environment, injury, and loss of life. The type of incidents that have the potential to occur include the following:

- Spills during fueling activities;
- Jet fuel releases from streamers;
- Groundings;
- Collisions with other vessels;
- Allisions (running into fixed objects such as platforms)[1];
- Damage to subsea pipelines;
- Structural/equipment failure; and
- Weather induced incidents (severe weather conditions can increase the potential of the other incidents occurring).

The U.S. Coast Guard (USCG) vessel database Port State Information eXchange (PSIX) System classifies these vessels as research vessels. The Marine Safety Information System (MSIS) was queried to examine incidents involving research vessels. The database includes incidents for the calendar years 1996 and 1997 and includes all incidents in U.S. waters for which a report was filed. The following summarizes incidents involving research vessels during 1996 and 1997.

- There was a total of 12 incidents, all of which had a casualty classification of *routine*. This classification is based on factors such as amount of damage done and seaworthiness of the vessel. Routine represents the lowest level of casualty classification.
- Three of the incidents occurred in the Gulf of Mexico. One of these three was listed as being in coastal waters and one in the intracoastal waterway.
- Two of the three incidents were a result of collisions and one was a result of an allision.
- One of the three incidents required that a pollution report be filed. The severity of the pollution incident was classified as *minor*.

---

[1] Defined as the collision between one ship in motion with another stationary vessel, structure (e.g., bridge), or facility (e.g., offshore platform). This is considered distinct from collision, where two vessels, both of which are in motion, collide.

The USCG reports on pollution incidents in and around U.S. waters. Recent incidents noted in the USCG (2003) database were accessed to examine spills involving research vessels during the 7 years from 1994 through 2000. These incidents were not broken down by cause and were only broken down by location for the years 1999 and 2000. The results are summarized in **Table K-1**, below. As can be seen from **Table K-1**, there have been approximately 20 spills per year from research vessels. In 1999 and 2000, there were five and two spills, respectively, in the Gulf of Mexico from research vessels. It is noted here that the spills from 1994 through 1998 were not necessarily in the Gulf of Mexico, nor did any of the spills necessarily involve G&G vessels.

Table K-1
Summary of Research Vessel Spills in U.S. Waters, 1994-2000
(From: U.S. Coast Guard, 2003)

Year	Number of Spills	Spill Volume (gal.)	Average Spill Size (gal.)	Median Spill Size (gal.)	Maximum Spill Size (gal.)
2000	18	2,429	135	1	1,500
1999	17	71	4	1	20
1998	21	32	2	1	5
1997	19	273	14	1	138
1996	19	959	50	1	500
1995	23	2,997	130	1	2,200
1994	21	1,442	69	20	500
Total	138	8,203	59	Unknown	2,200

Prelease G&G permit activity is an indicator of the number of G&G surveys being conducted annually in the Gulf of Mexico, as outlined in further detail in **Appendix D, Section V - Level of Activity**. No data are currently available for postlease survey or other survey (e.g., pipeline surveys) activity levels in the Gulf of Mexico. Minerals Management Service (MMS) data show that prelease G&G seismic survey activity in the Gulf of Mexico has been variable over the 1995-2000 period, ranging from 3,000 to approximately 6,500 blocks surveyed annually. A review of MMS 3D permit records for 2002 and 2003 indicates that G&G operators complete 30+ surveys per year (R. Brinkman, MMS, oral comm., 2003), each of which typically encompasses operations across multiple outer continental shelf (OCS) blocks. While G&G survey vessel operations (i.e., acquiring seismic data, deploying/retrieving equipment) are of variable duration (i.e., from a few days to several weeks or more), supply and service vessels operate on a daily basis, supporting offshore (non-OCS and OCS) oil and gas exploration, development and production, and facility abandonment activities. MMS has analyzed current and future service vessel activity in the three Gulf of Mexico OCS planning areas as part of its multiple lease sale environmental analyses. The majority of current activity occurs within the Central Planning Area, followed by the Western Planning Area and, to a very limited extent, the Eastern Planning Area. Total service vessel activity Gulf-wide is expected to range from 304,807 to 319,921 trips annually. This estimate is based on projected future lease sales in the Gulf of Mexico, future exploration and development activity, and support of existing offshore exploration and production operations (U.S. Department of the Interior, Minerals Management Service, 2002a,b). On a Gulf-wide basis, more than 300,000 service vessel trips are expected to occur annually (with ~90% occurring in the Central Planning Area) as compared with approximately 20 G&G surveys annually.

Each of the potential types of incidents listed above is discussed below.

Spills During Fueling Activities – All of the vessels involved in these activities burn diesel fuel. The amount of fuel that can be carried by one of these vessels depends on the design of the vessel and most importantly, on its size. The larger vessels can carry up to 1,514,000 L of fuel, stored in four to six tanks, while the smaller vessels have fuel capacities of around 473,125 L. Vessels can be refueled in port, either from dockside hoses or from fuel barges that pull alongside, or at sea from fuel supply vessels using hoses. Fuel transfer at sea is more dangerous than fuel transfer in port for several reasons. First, sea motion and wind can cause the two vessels to move, which in turn can lead to fuel hoses being pulled apart. This potential risk is mitigated by the use of procedures that limit fuel transfer operations during adverse weather and/or sea conditions. Second, spill response equipment is readily available in port. Fueling spills, both in port and at sea, are rare, and when they do occur, the amounts released are small. Because of the small potential for fueling spills, the small amounts of fuel released when such spills do occur, and the mitigation measures employed during fueling operations, the potential impact from fueling spills is considered negligible.

Jet Fuel Releases from Streamers – Many of the streamers pulled behind the survey vessels contain jet fuel. There is a trend for the newer streamers to be solid and not contain jet fuel. The streamers are approximately 2 inches in diameter and come in 23-m sections. Thus, a section of streamer can contain approximately 45 L of jet fuel, which could be released into the ocean if the streamer is damaged. This small release would not result in significant impacts. It is highly unlikely that multiple streamers would release their jet fuel simultaneously. Vessels that pull streamers containing jet fuel may also store up to several hundred gallons of jet fuel in tanks located on the vessel for the purpose of repairing damaged streamers.

Groundings – As noted previously, none of the incidents in the Gulf of Mexico involved groundings. There was one grounding incident that occurred outside the Gulf of Mexico. It did not result in a spill. In general, the G&G vessels are equipped with state-of-the-art navigation equipment including depth finders and/or other devices to measure water depth. In most cases, a grounding would result there in a scraping of the vessel bottom or possibly in the vessel becoming grounded. It would be unlikely that a grounding would result in a spill unless the bottom were rocky.

Collisions – As noted previously, there were two collisions in the Gulf of Mexico involving research vessels during the 2-year period from 1996 through 1997. One of the groundings resulted in torn plating, flooding, and minor pollution. As mentioned previously, the G&G vessels are equipped with state-of-the-art navigation equipment including radars that should mitigate the potential for collisions. If a collision were to occur, the consequences could be anywhere from minor to catastrophic. A catastrophic collision could result in the vessel sinking, spilling some or all of its fuel, and in injury and/or fatalities. Similar types of consequences could result from the other vessel involved in the collision. None of the incidents listed previously resulted in injuries or fatalities.

Allisions – Allisions are impacts between a moving object and a fixed object. Allisions most often involve vessels running into platforms or bridges. One of the incidents occurring in the Gulf of Mexico was an allision. Again, the G&G vessels are equipped with radars. The consequences of an allision could be similar to that of a collision.

Damage to Subsea Pipelines – Subsea pipelines run between various platforms and from many platforms to shore. Subsea pipelines located in less than 60 m of water are required to be buried. Subsea pipelines carry various hazardous materials such as oil and gas. If one of these pipelines were broken, oil or gas could be released. A gas release could become ignited, resulting in a fire. An oil release could result in significant pollution and possibly a fire. It is possible that a vessel can scrape a pipeline with its keel or drag an anchor or towed object that snags the pipeline, resulting in damage to the pipeline. Survey vessels normally use dynamic positioning and not anchors to stay in place, especially when working in areas containing pipelines. For vessels requiring anchoring, mapped geohazards information is also employed to avoid existing pipelines. None of the incidents listed previously involved damage to a subsea pipeline.

Structural/Equipment Failure – It is possible for structural elements of vessels to fail due to improper construction techniques (e.g., bad welds), faulty material, flexing over time, wearing out (e.g., valves), etc. Equipment failure can include steering failures that can result in other types of incidents such as collisions and allisions. These type of incidents are rare but could result in consequences similar to collisions.

Weather Induced Incidents – Extreme weather such as hurricanes or severe storms can result in structural failures or flooding. An offshore support vessel capsized in the Gulf of Mexico due to a severe storm in 1983 (National Transportation Safety Board, 1984). Five of the six crew members were killed. This vessel was approximately 50 m in length. Proper procedures could have prevented the vessel from sinking. While these types of events are rare, they can occur.

B. SPILL RESPONSE

Following a spill of diesel fuel, several physical, chemical, and biological processes, collectively called weathering, interact to change the physical and chemical properties of the spill and thereby influence its harmful effects on marine organisms and ecosystems. The most important weathering processes for spilled oil (including refined diesel fuel) are spreading, evaporation, dissolution, dispersion into the water column, formation of water-in-oil emulsions, photochemical oxidation, microbial degradation, adsorption to suspended particulate matter, and stranding on shore or sedimentation to the seafloor (Boehm et al., 1987; Payne et al., 1987; Neff et al., 2000). In the case of diesel fuel, stranding and sedimentation are feasible only in very shallow and/or sheltered environments.

The more toxic, light aromatic and aliphatic hydrocarbons are lost rapidly by evaporation and dissolution. Evaporated hydrocarbons are degraded rapidly by sunlight. Biodegradation of spilled fuel on the water surface and in the water column by marine bacteria and fungi initially removes the n-alkanes and subsequently the light aromatics. Other components are biodegraded more slowly. Photooxidation attacks mainly the medium and high molecular weight polycyclic aromatic hydrocarbons of a spill.

Diesel fuel is toxic to fishes and other wildlife in high concentrations. It is lighter than water and tends to floats on the surface, with dissolution or mixing into the water column being a function of physical factors (e.g., waves, wind chop) present on the sea surface. Diesel fuel evaporates fairly rapidly and does not leave a tarball residue like crude oil. Diesel fuel is classified as a combustible material and is not easily ignited once it is released on the water, although it is possible that spilled diesel can ignite under certain circumstances.

Sorbent pads or other sorbent materials are normally used to collect small diesel spills. The use of containment booms (i.e., booming) routinely occurs when oceanographic and wind conditions allow for effective herding of a spill. Booming may be used to protect sensitive resources or to contain diesel fuel if the release occurs near shore. Booming would not normally be used for releases of diesel some distance from the shoreline. Sorbent boom may be pulled through large open water releases to recover some of the released fuel. Skimmers are not very effective at recovering released diesel fuel and would not normally be used. After recovering as much diesel as possible using sorbent materials, the remaining diesel is often left to evaporate and disperse naturally.

C. SUMMARY

The table previously presented shows that the largest spill that occurred from a research vessel over the time period covered was 2,200 gallons (52 barrels), with the median size less than 1 gallon for 4 of the 7 years. The average spill size was 59 gallons, slightly larger than 1 barrel. The maximum amount of fuel that could be released would be the total fuel capacity of the vessel. As stated above, the largest survey vessels can hold up to 400,000 gallons of diesel fuel. However, since the diesel fuel is stored in multiple tanks, it is unlikely that the contents of all the tanks would be released.

Because data are limited on the number of surveys conducted, number of survey vessel trips, number of survey vessel days at sea, or survey vessel miles traveled during the time period used in **Table K-1**, it is not possible to derive a quantitative estimate for the probability of an incident involving a survey vessel from the data presented in the table. It can be concluded that the probability of such incidents occurring is quite low, with the potential for a pollution incident even lower. It can be concluded that an event involving a survey vessel can have significant consequences (loss of life and/or release of diesel fuel), but that such an event has an extremely remote probability of occurring.

D. LITERATURE CITED

Boehm, P.D., M.S. Steinhauer, D.R. Green, B. Fowler, B. Humphrey, D.L. Fiest, and W.J. Cretney. 1987. Comparative fate of chemically dispersed and beached crude oil in subtidal sediments of the arctic nearshore. Arctic 40 (Supplement):133-148.

National Transportation Safety Board. 1984. Marine accident report - capsizing of the U.S. offshore vessel *Laverne Hebert*, Gulf of Mexico, November 9-10, 1983. National Transportation Safety Board, Bureau of Accident Investigation, Washington, D.C. 48 pp.

Neff, J.M., S. Ostazeski, W. Gardiner, and I. Stejskal. 2000. Effects of weathering on the toxicity of three offshore Australian crude oils and a diesel fuel to marine animals. Environmental Toxicology and Chemistry 19(7):1809-1821.

Payne, J.R., C.R. Phillips, and W. Hom. 1987. Transport and transformations: water column processes, pp. 175-231. *In:* D.F. Boesch and N.N. Rabalais (eds.), Long-term environmental effects of offshore oil and gas development. Elsevier Applied Science Publishers, London.

U.S. Coast Guard. 2003. Pollution incidents in and around U.S. waters. Internet version - www.uscg.gov.

U.S. Department of the Interior, Minerals Management Service. 2002a. Gulf of Mexico OCS Oil and Gas Lease Sales: 2002-2007. Central Planning Area Sales 185, 190, 194, 198, and 201. Western Planning Area Sales 187, 192, 196, and 200. Final Environmental Impact Statement. U.S. Department of the Interior, Minerals Management Service, Gulf of Mexico OCS Region, New Orleans, LA. OCS EIS/EA MMS 2002-052. 2 vols.

U.S. Department of the Interior, Minerals Management Service. 2002b. Gulf of Mexico OCS Oil and Gas Lease Sales 189 and 197. Eastern Planning Area. Draft Environmental Impact Statement. U.S. Department of the Interior, Minerals Management Service, Gulf of Mexico OCS Region, New Orleans, LA. OCS EIS/EA MMS 2002-056.

Appendix L
Potential Incidental Harassment and Take

Potential Incidental Harassment And Take

I.   BACKGROUND

Incidental take estimates have been calculated for a single event, or a limited number of sequential events (e.g., Navy ship shock trials; see Department of the Navy 1998, 2001), as well as for smaller scale, site- or area-specific operations (e.g., National Science Foundation and Lamont-Doherty Earth Observatory (LDEO) seismic surveys in the Gulf of Mexico (GOM); see LGL Ltd., environmental research associates, 2003a,b). In these assessments, the total area potentially ensonified (to a level of interest) was calculated as a product of impact radii (e.g., energy flux density; sound exposure levels [dB]) and regional or area-specific population densities. In the case of a moving sound source (e.g., seismic array), total transect length also was integrated into an estimate of total area potentially affected. However, only recently has accommodation been made for the effectiveness of visual monitoring or the detectability of marine mammal species (e.g., see LGL Ltd., environmental research associates, 2003b).

Significant data gaps remain in regards to our understanding of species- or group-specific recognition and avoidance of seismic noise, as well as other noted data deficiencies (e.g., hearing capabilities of Gulf species). Additional limitations and assumptions regarding the following incidental take methodology and associated calculations are outlined in **Table L-1**.

Recent estimates of the incidental take of marine mammal species routinely utilize, at a minimum, the following factors or data sources, including:

- Number of line miles (or line kilometers) traversed;
- Estimated radial distance to the edge of a safety, impact, or exclusion zone; and
- Densities of marine mammals present.

In the case of the most recent Navy shock trial environmental impact statement (EIS) (Department of the Navy, 2001), considerable discussion was devoted to estimating probability of sighting vs. non-sighting for aerial and shipboard monitoring. Shipboard monitoring criteria are directly applicable to geological and geophysical (G&G) operations working under Notice to Lessees and Operators (NTL) requirements. The Navy's analysis gives attention to the problems of submergence (i.e., availability bias, as outlined in Marsh and Sinclair, 1989) and underscores one of the inherent limitations of visual monitoring. Given the similarities between Navy mitigation (i.e., pre-detonation clearance of a moving, 2-nmi safety zone) and the requirements of exclusion zone clearance under NTL No. 2004-G01, the following analysis draws upon existing Navy EIS information to support the present calculation of take numbers.

II.   APPROACH

Several elements provide the basis for this determination of incidental take, including:

- Summary of seismic survey activity levels, with an emphasis on the distribution of activity by planning area and water depth, derived from recent Minerals Management Service (MMS) seismic survey permit records;
- Abundance and distribution of marine mammal species in the GOM, derived from newly revised depth and planning area-specific estimates of marine mammal abundance; and
- Quantification of effects of detection (availability) bias for shipboard observations (visual monitoring), using a recent marine mammal mitigation program.

Table L-1

Summary of Identified Limitations and Assumptions to be Considered in Calculations of Incidental Take

Limitation or Assumption	Comment
Simplified Calculation of Radial Distance to Isopleths of Interest	A simplifying assumption in the calculation of distance to the isopleths of interest is that the ensonified zone is circular or radial. A reasonable accommodation is also made in the calculations for the array effect. As noted in **Appendices C (Section III - Seismic Source Levels)** and **D (Section II.A - Airguns)**, the seismic signal from an array is directed in a downward, vertical direction, towards the seafloor. Array geometry (e.g., airgun separation and size) and aspect (i.e., in-line versus cross-line array axes) also affect the distribution of the sound field created. As a consequence, the area of ensonification surrounding an array is elliptical when viewed from above, indicative of the fact that there is a difference in the sound attenuation rate along the water surface fore and aft of the array versus port and starboard of the array (see **Appendix C, Figure C-12** and **Appendix D, Figure D-4**). Consideration of the array effect takes this fact into account. For the purpose of this analysis, the greatest radial distance to the target isopleth is calculated; this provides a maximum estimate of distance to a particular isopleth. However, calculating the total area ensonified on the basis of this maximum radial distance produces an overestimate. A review of a recent model validation study (LGL Ltd., environmental research associates, 2003c) and field measurements of seismic arrays (e.g., see Goold and Fish, 1998) indicates that the actual elliptical area ensonified by either single or multiple pulses is approximately one-half of that calculated using a radial geometry. Therefore, a 50% reduction in area ensonified has been adopted in the following harassment and take analysis to account for conversion from a radial to an elliptical zone of ensonification.
Recognition and Avoidance of Seismic Noise	National Oceanic and Atmospheric Administration Fisheries (NOAA-F) has established 160 and 180 dB re 1 μPa (root mean squared [rms]) as possible exposure levels to which marine mammals may exhibit behavioral response or physiological injury. These levels are subject to change as more data regarding sound exposure levels and marine mammal sensitivity and response become available. Specific information about the reactions of some baleen and odontocete whales to low frequency noise pulses has come from observing responses of bowhead whales in the Beaufort Sea, migrating gray whales off California, and migrating or lingering humpback whales off western Australia to pulses from airguns and other non-explosives (Richardson et al., 1986; Malme et al., 1983, 1984; McCauley et al., 1998, 2000). Avoidance reactions to seismic sounds at received levels of about 160 to 170 dB re 1 μPa (rms) and, in some cases, somewhat lower levels, have been documented. Humpback whales showed avoidance at a mean received sound level of 140 dB re 1 μPa (rms). About 50% of feeding gray whales will cease feeding at an average peak pressure level of 173 dB (Malme et al. 1988), and about 50% of migrating gray whales will avoid a seismic source at about the same average received sound levels (Malme and Miles, 1985). Similar observations for Gulf of Mexico (GOM) marine mammals exposed to seismic are generally lacking. Further, the question of acclimation to seismic noise (i.e., use of seismic arrays in a mature oil producing region such as the GOM) remains unanswered. In addition, marine mammal researchers have suggested that the orientation of the marine mammal at risk relative to the sound source may also affect how much an animal may be affected (e.g., whether temporary threshold shift or permanent threshold shift results from the exposure). Currently, there is no provision in the incidental take calculation methodology to accommodate the potential for recognition and avoidance of the seismic source by marine mammals in the general area (e.g., beyond the 160-dB [rms] isopleth).

Table L-1

Summary of Identified Limitations and Assumptions to be Considered in Calculations of Incidental Take
(Continued)

Limitation or Assumption	Comment
Seismic Pulses and the Range of Marine Mammal Hearing Sensitivities	Previously, there was no accommodation (within the incidental take calculation methodology) to account for the predominant frequencies of seismic signals and the perceived overlap (or, more importantly, the lack thereof) between seismic signals and the vocalization/estimated hearing frequencies of species of concern. As outlined in **Appendix G (Section I.E.1.b - Biological Context)**, hearing threshold data for marine mammals are very limited, particularly for the larger whale species. Hearing thresholds have been inferred from vocalization data. In general, most toothed cetaceans are good mid- to high-frequency vocalizers (i.e., sonic to ultrasonic specialists, 100 Hz to 100 kHz), while baleen whales generally vocalize in the lower frequency bands (i.e., sensitive to infrasonic sounds, 10 Hz and upwards; see **Appendix G, Section I.A – Marine Mammal Hearing**). One major consideration in a thorough incidental take analysis should be the output frequencies of the seismic source and the hearing/frequency sensitivities of the species of concern. An incidental take analysis should take into consideration the differential nature of hearing and vocalization among mysticetes and odontocetes. Mysticetes exhibit inferred hearing thresholds of 10 to 31,000 Hz, with dominant frequencies of 16 to 25,000 Hz; odontocete hearing thresholds are highly species-specific, with dominant frequencies of 120 to 60,000 Hz in delphinids (**Appendix G, Tables G-1 and G-2**; see Ketten, 1998). For this geological and geophysical (G&G) Programmatic Environmental Assessment, take estimations consider the possibility of frequency overlap between the predominant components of a seismic pulse and the inferred hearing capabilities of GOM marine mammal species.
Intermittent vs. Continuous Noise Exposure	Inherent in the simplified approach to calculating incidental take is the assumption that marine mammals would be continuously exposed to seismic noise - along the entire transect or series of transects surveyed. This is certainly not the case. Airgun arrays are fired intermittently, every 12 to 16 s, and in some cases slightly longer. The duration of each pulse is measured in terms of milliseconds, as are the reflections of the pulse off the seafloor or sea surface. In addition, both the sound source and the marine mammal potentially at risk are moving; the relative movement of vessel and whales either away from or towards one another over a period of minutes may be significant. Both the type of exposure (i.e., intermittent, pulsed from G&G seismic sources) and the duration of the exposure (i.e., due to marine mammal proximity and location in the water column [surface vs. submerged], ability to move away from a potentially adverse sound source, as noted above) are important elements in estimating exposure and incidental take.
Effect of Biotic and Abiotic Factors in Marine Mammal Distribution	Oceanographic and other conditions may strongly influence the distribution and numbers of marine mammals present in an area. Therefore, for some species, the density estimates used may not be fully representative of the densities that may be encountered during future seismic operations. In addition, only limited life history and ecological data are available for certain species while others are better known. The results of this assumption may either underestimate or overestimate calculations of incidental take. This is particularly problematic for species that often congregate (i.e., show clumped distribution patterns). To represent the potential presence of a particular species in shallow or deep water more accurately, within one of three GOM Planning Areas, refined density estimates have recently been developed from survey efforts completed by the National Marine Fisheries Service. These revised density estimates, categorized by water depth (i.e., <200 m; ≥200 m) and planning area, have been integrated into this analysis. The problem persists, however, for smaller scale oceanographic features (e.g., rings), which may attract select species, or cross depth or planning area boundaries.

Table L-1
Summary of Identified Limitations and Assumptions to be Considered in Calculations of Incidental Take
(Continued)

Limitation or Assumption	Comment
Effectiveness of Existing Mitigation Measures	MMS NTL No. 2004-G01 was implemented to provide assurance that marine mammals and turtles (at water depths $\geq$200 m throughout the GOM; all outer continental shelf [Federal] waters in the Eastern Planning Area [EPA]) will not be subjected to relatively high sound levels in close proximity to an array. The use of ramp-up procedures, coupled with protected species observation (visual monitoring) and reporting and establishment of a 500-m (radial) zone (exclusion zone) around the center of the towed airgun array and the area within the immediate vicinity of the survey vessel, are intended to 1) provide confidence to the G&G operator that the initiation of ramp-up will occur without marine mammals in close proximity, and 2) provides an audible warning to marine mammals in the general area that seismic operations are being initiated and to allow sufficient time for those animals to leave the immediate vicinity.
Other Assumptions	• Assumption that 3D seismic survey activity levels realized in 2002 will, in general, remain relatively constant and distributed between water depths and planning areas in similar fashion. Limited accommodation is made in this analysis for increased activity in select planning areas where no activity occurred in 2002 (i.e., EPA).
	• Utilization of 0.3 and 3.0 km as appropriate maximum radial distances to the 180- and 160-dB (rms) isopleths is based on a "typical" seismic array (i.e., 4,550-in.^{3} airgun array; 240 dB re 1 μPa [zero-to-peak], 230-dB re 1 μPa [rms]) in use in the GOM (See **Appendix C, Section III - Seismic Source Levels**). Actual array output varies by operator and can be lower than the typical system employed in this analysis. Lower output systems produce smaller zones of ensonification. For example, for an array with a specified source level 10 dB less than the typical array, the range to the 180- and 160-dB isopleths would be on the order of 100 and 1,000 m, respectively.

- Calculation of potential take in shallow and deep water under two scenarios - a single 30-minute ramp-up of a typical array (i.e., 4,550 in.[3]; 240 dB re 1 µPa [zero-to-peak], 230 dB re 1 µPa (root mean squared [rms]) and cumulative survey ramp-up and restart activity (i.e., approximately 20 seismic surveys per year in Federal waters of the GOM). The following elements will be integrated into the analysis of take:

  ➢ Planning area and depth delineated marine mammal abundance data;
  ➢ Planning area and depth delineated seismic survey activity;
  ➢ Integration of ~300- and ~3,000-m (radial) distance estimates for the 180- and 160-dB rms isopleths, respectively, and subsequent reduction of total area ensonified by 50% to account for the elliptical shape of the ensonification zone;
  ➢ Visual monitoring and its effectiveness;
  ➢ Species detection characteristics and estimates of detected and non-detected individuals;
  ➢ Hearing capabilities and documented marine mammal presence in the GOM;
  ➢ Calculations of exposure under both scenarios - single 30-minute ramp-up and cumulative survey ramp-up - to estimate uncorrected (i.e., first order) take; and
  ➢ Calculations of final take estimates, considering hearing capabilities and presence in the Gulf.

Enhancement of and revision to the site-specific approach to calculating incidental take are warranted given the programmatic nature of this environmental assessment (i.e., a programmatic EA, or PEA). Incidental take calculations found within recent site-specific assessments have routinely modeled known seismic arrays and specific areas of operation; these modeling results may not be representative of typical G&G systems being used in commercial applications by operators in the U.S. GOM. For example, the summer 2003 surveys conducted by LDEO in the northern GOM employed several different seismic arrays, the largest of which was comprised of 20 airguns (8,580 in.[3], 255 dB re 1 µPa [zero-to-peak], source output) and was considerably larger than those currently in commercial use. None of the remaining LDEO arrays were representative of typical G&G systems being used in the GOM. Further, recent LDEO seismic survey modeling assumptions (which continue to be evaluated) indicate that model predictions in shallow water significantly underestimate the range to target isopleths, whereas model predictions in deeper water are reasonable approximations (see LGL Ltd., environmental research associates, 2003c).

A second issue of concern lies with the detectability of marine mammal species. Recent take estimates associated with Navy shock trials integrated probabilities of detection vs. non-detection of individuals or groups to refine take estimates further. However, none of the recent take estimates have accounted for species recognition of a sound field and avoidance. This incidental take analysis integrates the probability of detection/non-detection into the species- and planning area-specific calculations of potential take.

The purpose of presenting two take scenarios is to characterize the range of potential seismic sound exposure, from the total annual G&G activity level (i.e., cumulative ramp-up or system start up, with estimated shutdown and restart; by water depth and by planning area) to a single event (i.e., single ramp-up or similar system start-up). This approach recognizes that several marine mammal species exhibit avoidance reactions to seismic sounds at received levels of about 160 to 170 dB re 1 µPa (rms) and, in some cases, somewhat lower levels (e.g., bowhead whales in the Beaufort Sea - Richardson et al., 1986; migrating gray whales off California - Malme et al., 1983, 1984; humpback whales off western Australia - McCauley et al., 1998, 2000). Estimating exposure and "uncorrected" incidental take on the basis of total line kilometers does not account for either the potential for avoidance or the effects of mitigation (e.g., shutdown when a marine mammal is observed within the exclusion zone). In contrast, take estimates based on a single 30-minute ramp-up represent a single event. With nearly 20 large scale surveys conducted annually in the GOM and considering the possibility of routine array shutdown on several occasions during a single survey, it is suggested that marine mammals may potentially be exposed to these levels on several occasions annually. For the purpose of this analysis, uncorrected take numbers from a single event will be multiplied by a factor of 120 (i.e., approximately 20 surveys annually, six start-ups during each survey) to approximate cumulative, annual potential exposure more closely.

The following calculations outline the potential for incidental harassment and take that may result from G&G seismic survey activities in Federal waters of the GOM. Marine mammal densities and seismic survey activity levels (e.g., by depth, by planning area) are reasonably well known, whereas other important elements necessary for take calculations must be estimated in the absence of available data. For example, the precise characteristics of the entire suite of seismic arrays cannot be readily evaluated; therefore, a typical seismic array has been considered in this analysis. As another example, the detectability of marine mammals at the surface has been determined in previous analyses; this factor has been utilized in the take calculations to more closely approximate the percentage of a marine mammal population that may remain undetected, and thus potentially at risk. Further, the ability of marine mammals to recognize and avoid approaching sound sources cannot be fully quantified at this time, although results of limited field observations indicate that marine mammals (potentially sensitive to seismic noise) will move to avoid continued exposure. In the absence of available data, reasonable estimates or assumptions have been employed. Emphasis must be placed on the methodology and associated rationale regarding the potential for injury arising from G&G seismic operations; take resulting from G&G activities may occur but are not a certainty (i.e., exposure is not necessarily equivalent to take).

III.   FACTORS CONSIDERED IN ESTIMATING INCIDENTAL TAKE

A.   ESTIMATED 3D SEISMIC ACTIVITY

The number of line kilometers traversed annually has previously been described from a review of MMS 3D permit data for 2002 and 2003 (see **Appendix D, Section V.C. - Correlation of Factors to Activity**). Details have been summarized in **Table L-2**.

Table L-2
Summary of Permitted 3D Seismic Activity in the Three Planning Areas of the Gulf of Mexico, 2002, by Water Depth

2002 3D Seismic Surveys, Gulf of Mexico	Shallow (<200 m)			Deep (≥200 m)		
	WPA	CPA	EPA	WPA	CPA	EPA
Total Tracklines (km)	1,931	33,287	0	60,257	116,299	21,532
% of Total	0.8	14.3	0	25.8	49.8	9.2

CPA = Central Planning Area.
EPA = Eastern Planning Area.
WPA = Western Planning Area.

There was considerable variability in the size and location of 3D surveys conducted in the GOM in 2002. A total of 233,306 line km were surveyed during 17 3D surveys conducted in outer continental shelf (OCS) waters of the Gulf under MMS permit during that year. Deep water (≥200 m) seismic activities in the Central Planning Area accounted for nearly 50% of the total 3D seismic activity in 2002, with 116,299 line km traversed. Deep waters of the Western Planning Area and shallow (shelf) waters of the Central Planning Area contributed 26% and 14% to the total, respectively. Nearly 80% of the surveys conducted in 2002, if done post-June 2003, would be required to comply with applicable NTL requirements (i.e., NTL No. 2003-G08; superceded by NTL No. 2004-G01) and attendant ramp-up and visual monitoring requirements.

B.   CURRENT ESTIMATES OF MARINE MAMMAL SPECIES DENSITY

Recent survey efforts undertaken by the National Marine Fisheries Service (NMFS), have developed species-specific density estimates for GOM cetaceans. NMFS data have been organized by MMS OCS planning area and by depth regime (i.e., continental shelf, 20 to 200 m: deeper water, ≥200 m to the edge of the U.S. Exclusive Economic Zone [EEZ]). **Table L-3** summarizes group size, density, and abundance estimates for predominant species in continental shelf waters 20 to 200 m deep (K. Mullin, NMFS, written comm., 2003).

Table L-3
Group Size, Density, and Abundance Estimates of Cetaceans in Northern Gulf of Mexico Continental Shelf Waters 20 to 200 m Deep by Minerals Management Service Planning Area (From: K. Mullin, National Marine Fisheries Service, Written Comm., 2003)

Species by Planning Area	n	S	$CV(S)$	D	N	$CV(N)$	95% CI
Bottlenose dolphin (*Tursiops truncatus*)							
Eastern	27	10.9	0.18	11.21	15,432	0.36	7,544-31,566
Central	21	10.5	0.26	10.01	4,697	0.41	2,106-10,475
Western	20	9.4	0.25	8.19	5,034	0.39	2,364-10,720
Atlantic spotted dolphin (*Stenella frontalis*)							
Eastern	32	24.3	0.19	19.48	26,823	0.30	14,867-48,394
Central	7	13.3	0.28	2.79	1,305	0.47	511-3,336
Western	4	19.8	0.33	2.26	1,391	0.80	313-6,179
Bottlenose dolphin/Atlantic spotted dolphin (*T. truncatus/S. frontalis*)							
Eastern	8	2.4	0.22	0.67	926	0.57	305-2,813
Central	2	2.5	0.60	0.21	99	0.82	10-961
Western	6	5.5	0.30	1.34	820	0.56	274-2,454
Rough-toothed dolphin (*Steno bredanensis*)							
Eastern	1	11.0	-	0.42	577	0.85	117-2,860
Central	0	-	-	0	0	-	-
Western	2	14.0	0.43	1.13	696	0.96	118-4,099

CI = confidence interval.
CV = coefficient of variation.
D = animals/100 km^2.
N = number of animals.
n = number of groups sighted.
S = mean group size.

Table L-4 provides similar parameters for predominant cetacean species in deeper waters (≥200 m to the edge of the EEZ) of the northern GOM. While data presented in these tables have been derived from survey efforts conducted in the GOM, understanding of marine mammal distribution patterns remains highly variable on a species by species basis.

Cetaceans exposed to airgun sounds with received levels ≥160 dB re 1 µPa (rms) might be sufficiently disturbed to be "taken by harassment." For the purpose of incidental take calculations, water depth and planning area distinctions were maintained, as this allowed for comparison of mitigated (NTL) and non-mitigated operations.

Table L-4

Group Size, Density, and Abundance Estimates of Cetaceans in Northern Gulf of Mexico Oceanic Waters
≥200 m Deep (to the seaward boundary of the U.S. Exclusive Economic Zone) by Minerals Management
Service Planning Area (From: K. Mullin, National Marine Fisheries Service, Written Comm., 2003)

Species by Planning Area	*n*	*S*	*CV(S)*	*D*	*N*	*CV(N)*	*95% CI*
Bryde's whale (*Balaenoptera edeni*)							
Eastern	4	2.0	0.35	0.03	56	0.71	15-205
Central	0	-	-	0	0	-	-
Western	0	-	-	0	0	-	-
Sperm whale (*Physeter macrocephalus*)							
Eastern	29	2.1	0.23	0.29	467	0.35	240-908
Central	26	2.3	0.15	0.44	567	0.43	249-1,293
Western	12	1.7	0.15	0.25	222	0.47	91-542
Pygmy and dwarf sperm whales (*Kogia* spp.)							
Eastern	38	1.5	0.08	0.21	351	0.37	172-716
Central	15	2.3	0.14	0.21	275	0.51	105-721
Western	5	1.6	0.25	0.08	73	0.71	20-264
Cuvier's beaked whale (*Ziphius cavirostris*)							
Eastern	4	1.8	0.27	0.03	56	0.46	23-140
Central	1	4.0	-	0.03	40	0.70	11-145
Western	0	-	-	0	0	-	-
Beaked whales (*Mesoplodon* spp.)							
Eastern	2	2.0	-	0.02	32	0.86	6-179
Central	2	1.0	-	0.02	20	0.68	6-70
Western	4	1.3	0.20	0.07	58	0.63	18-189
Unidentified ziphiid (beaked whales)							
Eastern	3	2.3	0.57	0.03	56	0.80	11-287
Central	3	2.0	0.29	0.05	61	0.55	21-176
Western	1	1.0	-	0.01	12	0.59	4-36
Pygmy killer whale (*Feresa attenuata*)							
Eastern	3	7.7	0.36	0.12	188	0.60	58-616
Central	1	15.0	-	0.12	154	0.98	30-805
Western	0	-	-	0	0	-	-
False killer whale (*Pseudorca crassidens*)							
Eastern	3	40.7	0.31	0.62	1,014	0.70	281-3,660
Central	0	-	-	0	0	-	-
Western	0	-	-	0	0	-	-
Killer whale (*Orcinus orca*)							
Eastern	1	3.0	-	0.02	25	0.78	6-98
Central	3	2.3	0.38	0.06	73	0.71	19-276
Western	1	2.0	-	0.03	24	1.04	4-136
Short-finned pilot whale (*Globicephala macrorhynchus*)							
Eastern	0	-	-	0	0	-	-
Central	2	56.0	0.52	0.90	1,169	0.99	192-7,103
Western	4	23.3	0.30	1.28	1,120	0.47	434-2,891
Melon-headed whale (*Peponocephala electra*)							
Eastern	0	-	-	0	0	-	-
Central	2	80.0	0.13	1.35	1,740	0.60	567-5,339
Western	1	125.0	-	1.79	1,568	0.82	365-6,735
Risso's dolphin (*Grampus griseus*)							
Eastern	23	8.6	0.13	0.87	1,417	0.37	704-2,850
Central	11	9.3	0.21	0.71	921	0.48	369-2,298
Western	4	8.5	0.26	0.40	354	0.83	82-1,528

Table L-4
Group Size, Density, and Abundance Estimates of Cetaceans in Northern Gulf of Mexico Oceanic Waters
≥200 m Deep (to the seaward boundary of the U.S. Exclusive Economic Zone) by Minerals Management
Service Planning Area (From: K. Mullin, National Marine Fisheries Service, Written Comm., 2003)
(Continued)

Species by Planning Area	n	S	CV(S)	D	N	CV(N)	95% CI
Bottlenose dolphin (*Tursiops truncatus*)							
Eastern	18	16.2	0.40	1.48	2,426	0.52	898-6,555
Central	2	11.5	0.39	0.19	240	0.76	55-1,043
Western	3	27.0	0.58	1.11	975	0.74	177-5,374
Rough-toothed dolphin (*Steno bredanensis*)							
Eastern	2	4.0	0.50	0.04	67	0.88	12-369
Central	2	21.5	0.16	0.35	449	0.69	127-1,583
Western	1	15.0	-	0.21	181	1.04	32-1,029
Fraser's dolphin (*Lagenodelphis hosei*)							
Eastern	1	117.0	-	0.62	1,014	0.87	227-4,529
Central	0	-	-	0	0	-	-
Western	0	-	-	0	0	-	-
Atlantic spotted dolphin (*Stenella frontalis*)							
Eastern	0	-	-	0	0	-	-
Central	1	15.0	-	0.12	117	1.17	24-1014
Western	0	-	-	0	0	-	-
Spinner dolphin (*Stenella longirostris*)							
Eastern	8	191.4	0.35	8.11	13,268	0.49	5,111-34,443
Central	4	69.5	1.02	2.34	3,025	1.14	165-55,393
Western	0	-	-	0	0	-	-
Pantropical spotted dolphin (*Stenella attenuata*)							
Eastern	85	54.4	0.11	24.79	40,559	0.21	27,366-60,113
Central	55	69.1	0.15	32.65	41,838	0.28	24,429-71,651
Western	15	24.7	0.30	5.37	4,700	0.67	1,373-16,084
Clymene dolphin (*Stenella clymene*)							
Eastern	0	-	-	0	0	-	-
Central	11	97.1	0.26	8.98	11,617	0.52	4,388-30,757
Western	3	100.0	0.25	4.30	3,764	0.83	869-16,305
Striped dolphin (*Stenella coeruleoalba*)							
Eastern	4	21.3	0.41	0.45	737	0.72	194-2,793
Central	4	64.0	0.24	2.15	2,784	0.57	960-8,071
Western	4	64.3	0.50	3.69	3,225	1.09	528-19,697
Stenella spp.							
Eastern	3	3.8	0.14	0.06	99	0.44	43-231
Central	2	13.5	0.78	0.23	294	1.02	20-4,330
Western	1	15.0	-	0.22	188	1.04	33-1,070
Unidentified dolphin							
Eastern	21	6.5	0.26	0.40	662	0.37	326-1,344
Central	13	6.4	0.52	0.39	503	0.60	159-1,594
Western	2	4.5	0.78	0.07	63	0.95	2-1,990
Unidentified small whale							
Eastern	4	1	-	0.02	32	0.53	12-86
Central	3	1.7	0.40	0.04	51	1.03	9-281
Western	1	1.0	-	0.01	12	0.83	3-51
Unidentified large whale							
Eastern	5	1.2	0.17	0.03	42	0.47	17-102
Central	1	2.0	-	0.01	18	1.14	3-111
Western	0	-	-	0	0	-	-

Table L-4
Group Size, Density, and Abundance Estimates of Cetaceans in Northern Gulf of Mexico Oceanic Waters
≥200 m Deep (to the seaward boundary of the U.S. Exclusive Economic Zone) by Minerals Management
Service Planning Area (From: K. Mullin, National Marine Fisheries Service, Written Comm., 2003)
(Continued)

Species by Planning Area	n	S	CV(S)	D	N	CV(N)	95% CI
Unidentified odontocete							
Eastern	4	1.5	0.33	0.03	48	0.54	17-140
Central	2	1.0	-	0.02	20	0.71	6-73
Western	2	2.5	0.50	0.07	58	0.84	13-258

CI = confidence interval.
CV = coefficient of variation.
D = animals/100 km².
N = number of animals.
n = number of groups sighted.
S = mean group size.

C. DETECTION BIAS AND OTHER FACTORS

The effectiveness of visual monitoring is limited by the ability of an observer or observers to detect animals at or near the surface. To estimate the probability of detecting animals on the surface, an Effective Strip Width (ESW) was estimated for shipboard observations, using the approach most recently cited by Department of the Navy (2001). ESW is a quantity derived from analysis of aerial survey data (e.g., see Buckland et al., 1993), as described below, and defines the width of sea surface in front of and lateral to a sighting platform (e.g., aircraft, surface vessel) that can be effectively surveyed for the presence of marine mammals (**Figure L-1**). The following discussion has been derived and modified from the *Winston S. Churchill* EIS analysis of mitigation (pre-detonation monitoring) effectiveness.

During marine mammal and turtle surveys, sighting probability varies with perpendicular distance from the aircraft or ship. A sighting probability function is developed to describe this variation. Densities of sightings are estimated using f(0), which is the sighting probability function evaluated at zero perpendicular distance. The half-strip width (μ) is the reciprocal of f(0), and it is the width such that, if all groups were detected out to this distance to either side of the aircraft or ship and none were detected beyond it, then the expected number of groups would be the same as for the actual survey. This provides a strip width that can be used to estimate the area effectively monitored. As noted by Department of the Navy (2001), data to estimate ESW are available from aerial and shipboard surveys that produced site-specific or regional marine mammal density data. However, since individual surveys often have only a few sightings of each species, it is preferable to estimate ESW from a larger data set (Buckland et al., 1993). Estimates produced during the GulfCet II program (Mullin and Hoggard, 2000) were used as described below.

For shipboard monitoring, the ESW values for shipboard GulfCet II surveys were used, as summarized in **Table L-5**. The half-strip width (μ=ESW/2) values were used to define a radius of 100% detection around the ship (see **Figure L-1**). Note that the GulfCet numbers were estimated using two observers, comparable to the visual monitoring teams being required under NTL No. 2004-G01. However, slight differences may exist between observation techniques used during GulfCet vessel surveys and the various G&G vessels operating in the GOM under the current NTL (e.g., use of "big-eye" binoculars during marine mammal surveys vs. use of standard, hand-held binoculars during seismic mitigation).

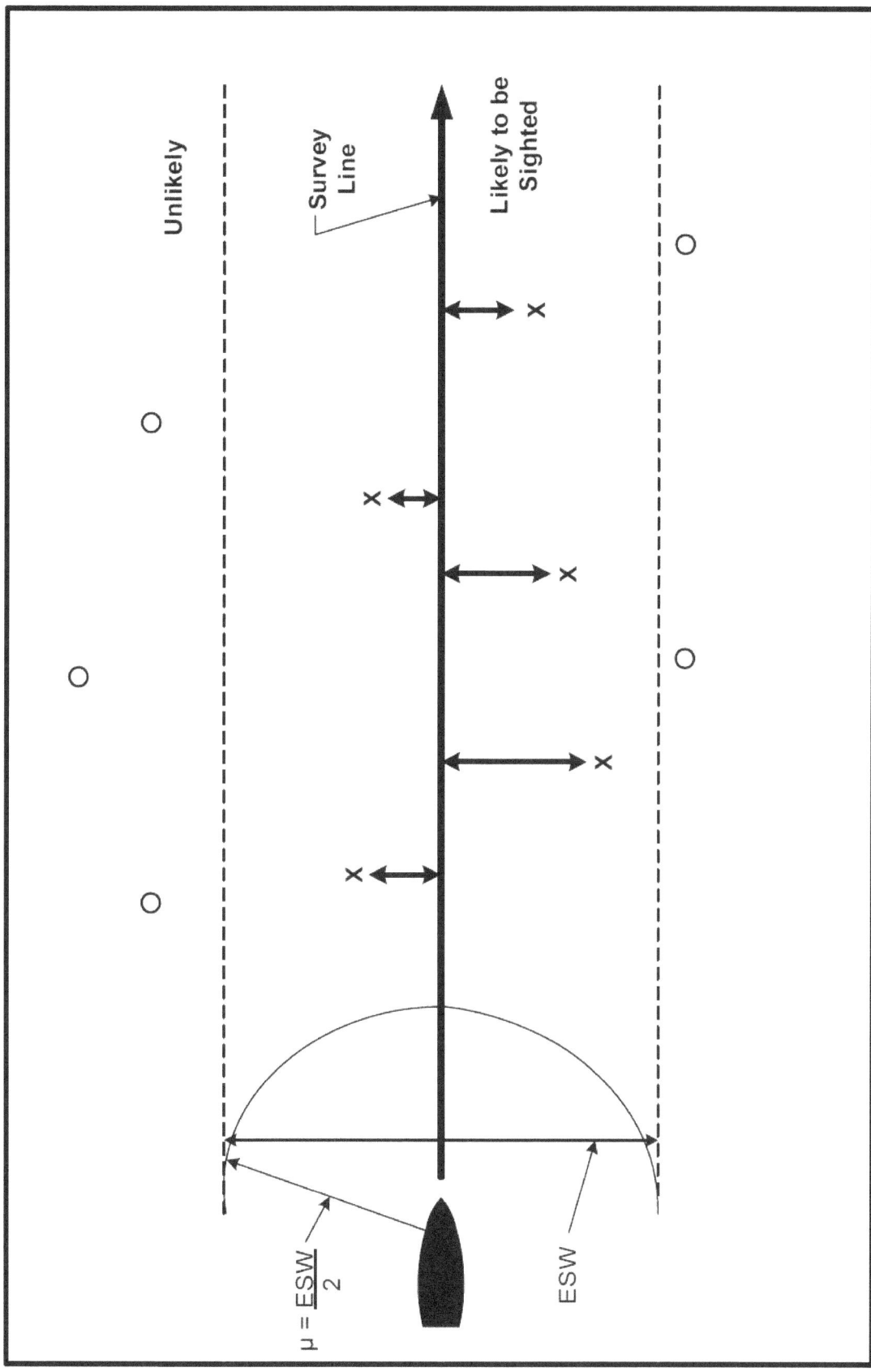

Figure L-1.   Diagram of seismic vessel traversing a survey line, indicating effective strip width (ESW), half-strip width (µ), and sighted (x) and unsighted (o) individuals (Adapted from: Arvanitis and Portier, 1997).

Table L-5
Estimated Mitigation Effectiveness of Visual Shipboard Observations (Adapted from: Department of the Navy, 2001)

Group	Effective Half-Strip Width[1] (km)	Area of Strip Width (km^2)[1]	Total 180-dB Isopleth Area[2] (km^2)	Proportion of 180-dB Zone Covered	Total 160-dB Isopleth Area[3] (km^2)	Proportion of 160-dB Zone Covered
Large whales	2.141	14.40	0.14	1.00	14.14	1.00
Cryptic species	1.235	4.79	0.14	1.00	14.14	0.34
Small whales/ Large dolphins	1.799	10.17	0.14	1.00	14.14	0.72
Small dolphins	1.672	8.78	0.14	1.00	14.14	0.62
Dolphins, unid.	2.222	15.51	0.14	1.00	14.14	1.00

[1] Effective half-strip width, shipboard observations, from Department of the Navy (2001), Appendix C. Half-strip widths were used to define a radius (circular zone) of 100% detection around a seismic survey vessel (when marine mammals are available for sighting); area calculated as $A = \pi \times r^2$.

[2] 0.3-km radius for the 180-dB (rms) isopleth, based on typical 4,550 in.^{3} array (240 dB re 1 μPa [zero-to-peak], 230 dB re 1 μPa [rms]); see **Appendix C (Section III - Seismic Source Levels)** of this Programmatic Environmental Assessment (PEA); total area calculated as $A = \pi \times r^2$, where r = 0.3 km; total area subsequently multiplied by 0.5 to account for the elliptical nature of the 180-dB (rms) ensonification zone (see **Figure L-2**).

[3] 3.0-km radius for the 160-dB (rms) isopleth, based on typical 4,550 in.^{3} array (240 dB re 1 μPa [zero-to-peak], 230 dB re 1 μPa [rms]); see **Appendix C (Section III - Seismic Source Levels)** of this PEA; total area calculated as $A = \pi \times r^2$, where r = 3 km; total area subsequently multiplied by 0.5 to account for the elliptical nature of the 160-dB (rms) ensonification zone (see **Figure L-2**).

Table L-5 also identifies the calculated areas of 160- and 180-dB (rms) ensonification, allowing comparison to the area calculated from the half-strip width (i.e., area of 100% detection of individuals or groups on or near the surface). The area ensonified to 180 dB is well within the area covered by the half-strip width, meaning that all individuals at or near the surface within this zone will be detected. However, the area ensonified to 160 dB extends well beyond the area covered by the half-strip width and varies by marine mammal group (e.g., large whales, cryptic whales, small dolphins, etc.), meaning that some proportion of the marine mammals present will remain undetected simply because of the limitations of the ESW; the relative proportion remaining undetected is group-specific.

The ESW values refer only to detection of animals present on the surface (i.e., those that are potentially detectable by observers). A certain proportion of the population is submerged and is therefore unavailable for visual detection. This is referred to as "availability bias" (Marsh and Sinclair, 1989). To correct for availability bias, the probability of detecting a marine mammal on the surface must be multiplied by the probability of the species being present on the surface. Similarly, observed densities can be corrected for availability bias by dividing them by the proportion of time the species is present on the surface within the viewing range. For example, if a species is on the surface 10% of the time, the observed density would be divided by 0.1 (i.e., multiplied by 10). For species with long dive times, the simplest estimate is the mean proportion of time the animal spends at the surface. However, for species with short dive times and surface intervals, the time within the range of the aerial (or shipboard) observer becomes a significant factor. The probability of being on surface during this time interval (t) is given by Eberhardt et al. (1979) and Barlow et al. (1988) as follows:

$$S_t = (s + t)/(s + d) \qquad\qquad [Equation\ 1]$$

where S_t = probability of being on surface within the viewing interval (t); s = mean surface time; and d = mean dive time. The equation also can be expressed as:

$$S_t = [s/(s + d)] + [t/(s+d)]$$ [*Equation 2*]

i.e., S_t = (proportion of time on surface) + [t/(s+d)]

The viewing interval (t) can be estimated based on the ESW values and, in this case, vessel speed. For shipboard monitoring, a speed of 4.5 kn (8.334 km/h) is assumed. This is the maximum speed that a G&G survey vessel will be moving during visual monitoring, prior to ramp-up. The ESW for large whales would be covered in about 15.4 minutes. The ESW for cryptic whales would be covered in about 9 minutes. The ESW for small whales/large dolphins and small dolphins would be covered in about 13 to 16 minutes, respectively.

Sperm Whales. For sperm whales, a wide range of dive times is evident. For sperm whales in the GOM, median dive time (d) and surface time (s) are estimated to be about 40-50 minutes and 10-14 minutes, respectively (Jochens and Biggs, 2003). Based on analysis of shipboard detection, the viewing interval is 15.5 minutes, and Equation 1 predicts that the probability of being on the surface during the viewing interval would be, on average, approximately 0.48 (i.e., longest surface interval, shortest dive interval = 0.546; shortest surface interval, longest dive interval = 0.425). While other combinations of dive and surface times produce slightly higher probabilities of being on the surface (e.g., shortest surface interval, shortest dive; longest surface interval, longest dive), 0.48 represents a conservative estimate of sighting probability for GOM sperm whales.

Other Large Whales. Mysticetes (the only large whales in the GOM, other than sperm whales) are estimated to be on the surface about 10% to 40% of the time (Leatherwood et al., 1982; Winn, 1982; Calambokidis et al., 1989; Stern, 1992; Kopelman and Sadove, 1995). Using 20% as a starting point and assuming a typical surface plus dive time of 10 minutes or less, the probability of being on the surface during visual observations should be close to one. The shipboard viewing interval of 15.5 minutes and surface plus dive times typically of 10 minutes or less, the probability of being on surface at least once during the shipboard viewing interval should be close to 1.0, based on Equation 2. A conservative estimate of 0.75 is used.

Cryptic Whales. Based on analysis of dive interval studies by Barlow et al. (1997), beaked whales and dwarf/pygmy sperm whales are expected to be on the surface about 10% of the time. Anecdotal information from the GOM indicates that beaked whales and dwarf/pygmy sperm whales exhibit minimal surface times, perhaps only 5% of the time. With a shipboard viewing interval of about 9 minutes and surface plus dive times assumed to be 20 to 30 minutes, the probability of being on surface at least once during the shipboard viewing interval is estimated to be 0.10.

Small Whales, Large Dolphins, and Small Dolphins. Small odontocetes typically have short dive intervals (a few seconds to a few minutes) and are on the surface 30% to 80% of the time (Evans, 1974; Barlow et al., 1988; Mate et al., 1994). The shipboard viewing interval is 13 to 16 minutes, making the probability of being on the surface at least once during this interval essentially 1.0. Some of the species in this category form large groups in which individual dive times are not synchronized, increasing the likelihood of at least one individual being on the surface at a given time (thereby allowing detection of the rest of the group). A conservative value of 0.90 is used.

Table L-6 estimates the proportion of marine mammal populations that may remain undetected during visual monitoring. Because 100% of the 180-dB area is effectively covered, the proportion of the population expected to be detected is the same as the proportion of the population available for detection.

For the 160-dB area, however, the proportion of the zone covered varies (by group), ranging from 17% to 55%. Consequently, the proportion of the population expected to be detected in this range is quite variable. The proportion of the population expected to remain undetected is determined by subtracting the proportion of the population expected to be detected from unity. For example, for large whales, the estimate for the proportion of large whale groups expected to be available for detection is 0.48. For the 180-dB zone, 100% of the zone is covered during visual monitoring, indicating that 48% of the large whale groups will be detected, or 52% of the large whale groups will remain undetected. For the 160-dB zone, the same proportion of the large whale groups will be available for detection (0.48). However, only 51% of the 160-dB zone will be covered during visual monitoring, indicating that only 24% of the large whale groups will be detected (i.e., 76% will remain undetected during visual monitoring).

Table L-6
Estimated Proportion of Marine Mammal Populations that May Be Expected to Remain Undetected
During Visual Monitoring and Ramp-Up

Group	Proportion Available for Detection[1]	180 dB (rms)			160 dB (rms)		
		Proportion of Zone Covered[2]	Proportion Expected to be Detected[3]	Proportion Expected to be Undetected[4]	Proportion of Zone Covered[2]	Proportion Expected to be Detected[3]	Proportion Expected to be Undetected[4]
Large whales	0.48	1.00	0.48	0.52	1.00	0.48	0.52
Cryptic species	0.10	1.00	0.10	0.90	0.34	0.03	0.97
Small whales/ Large dolphins	0.90	1.00	0.90	0.10	0.72	0.65	0.35
Small dolphins	0.90	1.00	0.90	0.10	0.62	0.56	0.44
Dolphins, unid.	0.90	1.00	0.90	0.10	1.00	0.90	0.10

[1] Proportion of marine mammal population (by group category) available for detection from Department of the Navy (2001), Appendix C, supplemented by several Gulf of Mexico sources; see text for further explanation.
[2] From **Table L-5**.
[3] Calculated by multiplying the proportion available for detection by the proportion of the 180-dB (rms) zone (or 160-dB [rms] zone) covered by visual observers; see **Table L-5** and text for further explanation.
[4] Calculated by subtracting the proportion expected to be detected from unity; see text for further explanation.

D.   RADIAL DISTANCE TO 160- AND 180- dB rms ISOPLETHS

A simplified calculation of radial distance to the appropriate dB contours (isopleths) has been developed using a typical GOM G&G seismic array (see **Appendix C, Section III - Seismic Source Levels**) – 4,550 in.³ array; 240 dB re 1 µPa (zero-to-peak), 230 dB re 1 µPa (rms). Assumptions and limitations to these estimates are detailed in **Appendix C, Section III**. Estimated maximum range to each isopleth is as follows:

- *Actual range to 180-dB re 1 µPa (rms) target isopleth*   =   ~300 m
- *Actual range to 160-dB re 1 µPa (rms) target isopleth*   =   ~3,000 m

A simplistic application of each target isopleth has been used, in conjunction with the length of ramp-up transect, to calculate area ensonified to each respective level.

As outlined in **Table L-1**, a simplifying assumption in the calculation of distance to the 160- and 180-dB isopleths is that the ensonified zone is circular or radial. As noted in **Appendices C** (**Section III - Seismic Source Levels, Figure C-12**) and **D** (**Section II.A - Airguns**), the seismic signal from an array is directed in a downward, vertical direction, towards the seafloor. As a consequence, the area of ensonification is elliptical when viewed from above (**Figure L-2**), indicative of the fact that there is a difference in the sound attenuation rate along the water surface fore and aft of the array versus port and starboard of the array (see also **Appendix C**, **Figure C-12** and **Appendix D**, **Figure D-4**). For the purpose of this analysis, the greatest radial distance to the target isopleth is calculated; this provides a realistic estimate of distance to a particular isopleth, particularly in those regions lateral to the array (see **Figure L-2**). However, calculating the total area ensonified on the basis of this maximum radial distance produces an overestimate. A review of a recent model validation study (LGL, Ltd., environmental research associates, 2003c) indicates that the actual elliptical area ensonified by either single or multiple pulses is approximately one-half of that calculated using a radial geometry. Therefore, a 50% reduction in area ensonified has been adopted in the following harassment and take analysis to account for conversion from a radial to an elliptical zone of ensonification.

In addition, the frequency spectra of a seismic signal has not been accounted for in the distance calculations (i.e., radial distance to a target isopleth). It is known that a seismic signal is comprised primarily of low frequency components (i.e., significant energy at 10 to 1,000 Hz, peak frequency at 50 to 60 Hz, with lesser contributions from both mid- and high frequency components). Although some data regarding differential frequency attenuation of seismic sound have been published over the past few years (e.g., Goold and Fish, 1998; Caldwell and Dragoset, 2000; Fontana, 2002), insufficient information is available to estimate accurately and integrate frequency components into the isopleth calculations within the framework of a PEA (e.g., due to limited field measurements, variations in seismic array composition and geometry, etc.). As a consequence, frequency spectra cannot readily be accounted for in the current take analysis.

E. MARINE MAMMAL PRESENCE IN GULF WATERS AND CONSIDERATION OF HEARING VS. SEISMIC FREQUENCY OVERLAP

Table L-7 revisits the findings pertinent to vocalization and perceived hearing frequencies of marine mammals found in the GOM, as previously detailed in **Appendix G**, **Table G-1**. **Table L-7** also characterizes species presence (e.g., common, uncommon, rare, or extralimital) in the GOM, based on analyses from Würsig et al. (2000) as previously summarized in **Appendix F, Table F-1**. **Table L-7** is intended to reflect the potential for overlap of species-specific marine mammal vocalization (and by inference, hearing) frequencies with the highest output from a typical seismic array (i.e., 10 to 1,000 Hz, with a recognized lower contribution from the higher frequencies).

IV. TAKE ESTIMATES

Estimates of the number of potential animals affected by seismic operations are outlined in **Tables L-8** through **L-11**. Two different calculation scenarios are outlined - for a single start-up (30-min duration) and cumulative start-up and restart for all surveys conducted during a year. For operations in Gulf waters <200 m deep, the requirements of NTL No. 2004-G01 are not in force except within the Eastern GOM Planning Area. **Tables L-8** and **L-9** calculate first order take estimates for 1) a single, shallow water (<200 m), 30-minute start-up (i.e., with no ramp-up except in the Eastern GOM, per NTL 2004-G01); and 2) a single, deep water (≥200 m) ramp-up where NTL No. 2004-G01 is in force in all planning areas.

Total area ensonified considers system start-up at a low output level (e.g., 160 dB [rms]), with incremental increases to full power within approximately 30 min. The area ensonified is equal to the length of a transect covered in 30 min (i.e., 4.17 km, using vessel speed of 4.5 kn (nmi/h) = 8.34 km/h), multiplied by twice the isopleth radius (i.e., r = 0.3 or 3.0 km, for 180 and 160 dB [rms], respectively).

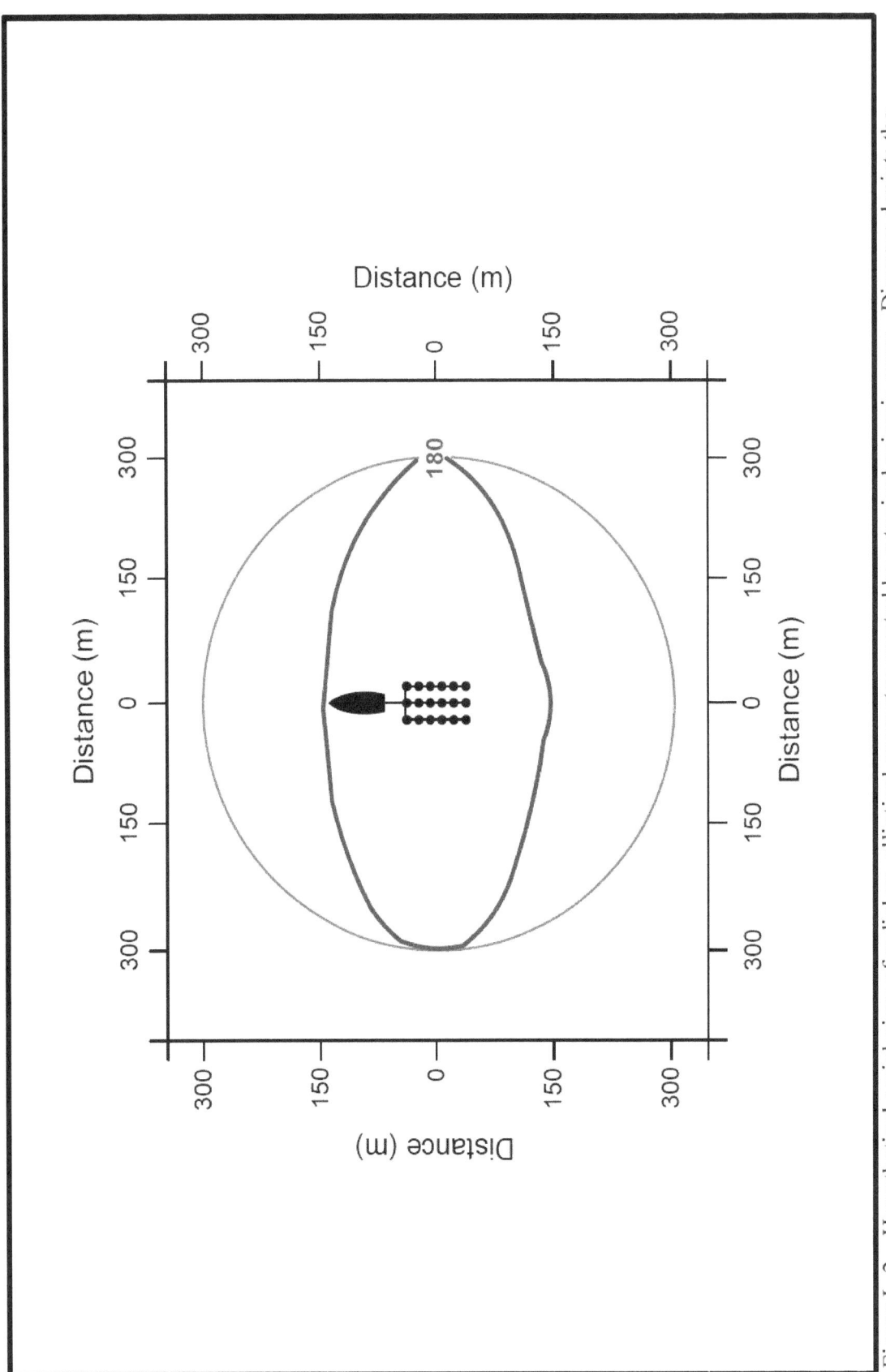

Figure L-2.   Hypothetical aerial view of radial vs. elliptical geometry created by a typical seismic array.   Diagram depicts the 180-dB re 1 μPa (rms) isopleth for a 4,550 in.^{3} array. Radial geometry based on array output and assumes spherical spreading (see **Appendix C**). Elliptical geometry derived from model and model validation (LGL Ltd., environmental research associates, 2003a,c), field measurements (e.g., Goold and Fish, 1998), and near-field/far-field spreading loss assumptions. Radial and elliptical isopleths have been drawn to scale, while vessel and towed array have not.

Table L-7

Vocalization Characteristics of Baleen (Mysticete) and Toothed (Odontocete) Whales[1] (Including Frequency Ranges and Source Levels) Compared to Species Presence in the Gulf of Mexico and Potential Overlap of Vocalization Characteristics with Maximum Seismic Output

Species	Signal Type	Frequency Range (Hz)	Dominant Frequencies (Hz)	Source Level[2] (dB)	Gulf of Mexico Presence[3]	Overlap of Vocalization Characteristics with Seismic Output[4]
Baleen Whales:						
Right whale[5]	tonal	30-1,250	160-500	172-187	Extralimital	Yes
	pulsive	30-2,200	50-500	181-186		
	call	<400	<200	-		
	moans	<400	-	-		
Fin whale	moans, downsweeps	14-118	20	160-186	Rare	Yes
	constant call	20-40	-	-		
	moans, tones, upsweeps	30-750	-	155-165		
	rumble	10-30	-	-		
	whistles?, chirps?	1,500-5,000	1,500-2,500	-		
	clicks?	16,000-28,000	-	160-190		
	moans	16-750	20	-		
	pulse	40-70	20	-		
	pulse	18-25	20	-		
	ragged pulse	<30	-	-		
	rumble	-	<30	-		
Blue whale	moans	12-400	12-25	188	Extralimital	Yes
	clicks?	6,000-8,000, 21,000-31,000, 16-18	6,000-8,000, 25,000, 16 5	130, 159		
Sei whale	FM sweeps	1,500-3,500	-	-	Rare	Possible
Humpback whale	song components	30-8,000	120-4,000	144-174	Rare	Yes
	shrieks	-	750-1,800	179-181		
	horn blasts	-	410-420	181-185		
	moans	20-1,800	35-360	175		
	grunts	25-1,900+	-	190		
	pulse trains	25-1,250	25-80	179-181		
	underwater blows	100-2,000	-	158		
	fluke and flipper slap	30-1,200	-	183-192		
	clicks	2,000-8,200	100-4,000	144-186		
	songs	30-8,000	<3,000	-		
	social	50-10,000	-	-		
Minke whale	downsweeps	60-130	-	165	Rare	Yes
	moans, grunts	60-140	60-140	151-175		
	ratchet	850-6,000	850	-		
	clicks	3,300-20,000	<12,000	151		
	thump trains	100-2,000	100-200	-		
	sweeps, moans	60-140	-	151-175		

Table L-7

Vocalization Characteristics of Baleen (Mysticete) and Toothed (Odontocete) Whales[1] (Including Frequency Ranges and Source Levels) Compared to Species Presence in the Gulf of Mexico and Potential Overlap of Vocalization Characteristics with Maximum Seismic Output

(Continued)

Species	Signal Type	Frequency Range (Hz)	Dominant Frequencies (Hz)	Source Level[2] (dB)	Gulf of Mexico Presence[3]	Overlap of Vocalization Characteristics with Seismic Output[4]
Bryde's whale	moans	70-245	124-132	152-174	Uncommon	Yes
	pulsed moans	100-930	165-900	-		
	discrete pulses	700-950	700-900	-		
Toothed Whales:						
Sperm whale	clicks	100-30,000	2,000-4,000, 10,000-16,000	160-180	Common	Possible
	clicks in coda	16,000-30,000				
Pygmy sperm whale	clicks	60,000-200,000	120	-	Common	No
Dwarf sperm whale	Unknown	Unknown	Unknown	-	Common	Unknown
Blainville's beaked whale	chirps/short whistles	<1,000-6,000	-	-	Rare-Common	Yes
	whistles	2,600-10,700				
Cuvier's beaked whale	Unknown	Unknown	Unknown	Unknown	Rare-Common	Unknown
Gervais' beaked whale	Unknown	Unknown	Unknown	Unknown	Uncommon	Unknown
Sowerby's beaked whale	Unknown	Unknown	Unknown	Unknown	Extralimital	Unknown
Killer whale	whistles	1,500-18,000	6,000-12,000	-	Uncommon	Possible
	pulsed calls	500-25,000	1,000-6,000	160		
	echolocation clicks	100-35,000	12,000-25,000	180		
	clicks	250-500	-	-		
	scream	2,000				
False killer whale	echolocation clicks	-	25,000-30,000 95,000-130,000	220-228	Uncommon	No
	whistles	-	4,000-9,500	-		
Pygmy killer whale	growls, blats			-	Uncommon	Unknown
Long-finned pilot whale	whistles	1,000-8,000	1,600-6,700	-	Common	Yes
	clicks	1,000-18,000	-	-		
	echolocation clicks	-	6,000-11,000			
Short-finned pilot whale	whistles	500-20,000+	2,000-14,000	180	Common	Yes
	echolocation clicks	-	30,000-60,000	180		
Bottlenose dolphin	echolocation clicks	-	110,000-130,000	218-228	Common	Yes
	whistles	800-24,000	3,500-14,500	125-173		
	low frequency narrow band	<2,000	300-900	-		
	rasp/grate/mew/bark/yelp					
	click	200-150,000	30,000-60,000	-		
	bark	200-16,000				
	whistle	4,000-20,000				
Risso's dolphin	whistles		3,500-4,500	-	Common	Possible
	rasp/pulse burst	100-8,000+?	2,000-5,000	-		
	echolocation clicks	-	65,000	~120		
Fraser's dolphin	whistles	7,600-13,400	-	-	Common	No

Table L-7

Vocalization Characteristics of Baleen (Mysticete) and Toothed (Odontocete) Whales[1] (Including Frequency Ranges and Source Levels) Compared to Species Presence in the Gulf of Mexico and Potential Overlap of Vocalization Characteristics with Maximum Seismic Output

(Continued)

Species	Signal Type	Frequency Range (Hz)	Dominant Frequencies (Hz)	Source Level[2] (dB)	Gulf of Mexico Presence[3]	Overlap of Vocalization Characteristics with Seismic Output[4]
Spotted dolphin[6]	whistles	5,000-19,800	6,700-17,900	-	Common	Possible
	clicks	1,000-8,000	-	-		
	squawk, barks, growls	100-3,000	-	-		
	chirps	4,000-8,000	-	-		
	echolocation clicks	-	-	-		
Pantropical spotted dolphin	whistles	3,100-21,400	6,700-17,800	-	Common	No
	whistles	-	-	-		
	pulse	-	-	-		
Striped dolphin	whistles	to 150,000	8,000-12,500	-	Common	No
Spinner dolphin	echolocation clicks	6,000-24,000+	-	-	Common	Possible
	whistles (= squeals?)	low-65,000+	6,800-16,900	109-125?		
	pulse bursts	1,000-22,500	5,000-60,000	108-115		
	screams	very wide	-	-		
Clymene dolphin	whistles	6,300-19,200	-	-	Common	No
Rough-toothed dolphin	echolocation clicks	-	5,000-32,000	-	Common	No
	whistles	-	4,000-7,000	-		

Notes and Footnotes:

[1] Indicates data infrequently recorded and/or questionable correlation of sound with species, per Richardson et al. (1995)

[1] Data provided for species whose presence has been documented in the U.S. Gulf of Mexico; vocalization data from Richardson et al. (1995) and Wartzok and Ketten (1999)

[2] Decibel level (dB) using an underwater/in-water reference pressure unit of one microPascal at one meter distance (re 1 µPa at 1 m), rms

[3] Gulf of Mexico presence derived from Würsig et al. (2000), as presented in **PEA Section III, Table III-3**

[4] Potential overlap based on vocalization characteristics (frequency range, dominant frequencies) and maximum power output of a seismic array (i.e., significant energy at 10 to 1,000 Hz, peak frequency at 50 to 60 Hz, with lesser contributions from both mid- and high frequency components)

[5] Both northern and southern right whale results presented

[6] Spotted dolphin (*Stenella plagiodon*) is a congener to Atlantic spotted dolphin (*S. frontalis*), the latter of which is common in the Gulf of Mexico; no sound or vocalization characterization data are available for Atlantic spotted dolphin

Table L-8

First Order Calculations of Potential Incidental Take for Cetaceans in Northern Gulf of Mexico Continental Shelf Waters 20 to 200 m Deep by Planning Area. Exposure Estimates Calculated on the Basis of a *Single Seismic System Activation* With No Ramp-Up and No Visual Monitoring (except in the Eastern Planning Area), No Recognition of Seismic Sound and Avoidance by Marine Mammals, and No Consideration of Overlap (or Lack of Overlap) Between Marine Mammal Vocalization Characteristics and Seismic Output

Species by Planning Area	D^1	180 dB (rms)			160 dB (rms)		
		Total Area Ensonified (km²)²	Proportion Expected to be Undetected³	First Order Calculated Take	Total Area Ensonified (km²)²	Proportion Expected to be Undetected³	First Order Calculated Take
Bottlenose dolphin (*Tursiops truncatus*)							
Eastern	11.21	0.6255	0.10	<0.01	6.255	0.44	0.31
Central	10.01	0.6255	1.00	0.06	6.255	1.00	0.63
Western	8.19	0.6255	1.00	0.05	6.255	1.00	0.51
Atlantic spotted dolphin (*Stenella frontalis*)							
Eastern	19.48	0.6255	0.10	0.01	6.255	0.44	0.54
Central	2.79	0.6255	1.00	0.02	6.255	1.00	0.17
Western	2.26	0.6255	1.00	0.01	6.255	1.00	0.14
Bottlenose dolphin/Atlantic spotted dolphin (*T. truncatus/S. frontalis*)							
Eastern	0.67	0.6255	0.10	<0.01	6.255	0.44	0.02
Central	0.21	0.6255	1.00	<0.01	6.255	1.00	0.01
Western	1.34	0.6255	1.00	<0.01	6.255	1.00	0.08
Rough-toothed dolphin (*Steno bredanensis*)							
Eastern	0.42	0.6255	0.10	<0.01	6.255	0.44	0.01
Central	0	0.6255	1.00	0	6.255	1.00	0
Western	1.13	0.6255	1.00	<0.01	6.255	1.00	0.07

[1] D = number of animals/100 km²; taken from **Table L-3**.

[2] Total area ensonified considers system start-up at a low output level (e.g., 160 dB [rms]), with incremental increases to full power within approximately 30 min. Area ensonified = length of transect covered in 30 min (4.17 km, using vessel speed of 4.5 kn (nmi/h) = 8.34 km/h) x twice the isopleth radius (r = 0.3 or 3.0 km, for 180 and 160 dB [rms], respectively) x 0.50 (i.e., to account for gradual increase in airgun output) x 0.50 (i.e., conversion from circular to elliptical area of ensonification).

[3] Proportion expected to be undetected equals 1.00 (100% undetected) for species in the Central and Western Planning Areas in water depths <200 m because NTL 2004-G01 is not in effect in these areas; the requirement for visual monitoring (and ramp-up) in OCS waters <200 m deep in the Eastern Planning Area prompts use of the estimates of the proportion of small dolphins expected to be undetected (see **Table L-6**). Historical seismic activity is extremely low in the Eastern Planning Area (see **Table L-2**).

Table L-9

First Order Calculations of Potential Incidental Take for Cetaceans in Northern Gulf of Mexico Oceanic Waters $\geq$200 m Deep (to the Seaward Boundary of the U.S. Exclusive Economic Zone) by Planning Area. Exposure Estimates Calculated on the Basis of a *Single Seismic System Activation* With Ramp-Up and Visual Monitoring (Per NTL No. 2004-G01), but No Recognition of Seismic Sound and Avoidance by Marine Mammals and No Consideration of Overlap (or Lack of Overlap) Between Marine Mammal Vocalization Characteristics and Seismic Output

Species by Planning Area	D^1	180 dB (rms)			160 dB (rms)		
		Total Area Ensonified (km²)²	Proportion Expected to be Undetected	First Order Calculated Take	Total Area Ensonified (km²)²	Proportion Expected to be Undetected	First Order Calculated Take
Bryde's whale (*Balaenoptera edeni*)							
Eastern	0.03			<0.001			<0.001
Central	0	0.6255	0.52	0	6.255	0.52	0
Western	0			0			0
Sperm whale (*Physeter macrocephalus*)							
Eastern	0.29			<0.001			0.009
Central	0.44	0.6255	0.52	0.001	6.255	0.52	0.014
Western	0.25			<0.001			0.008
Pygmy and dwarf sperm whales (*Kogia* spp.)							
Eastern	0.21			0.001			0.013
Central	0.21	0.6255	0.90	0.001	6.255	0.97	0.013
Western	0.08			<0.001			0.005
Cuvier's beaked whale (*Ziphius cavirostris*)							
Eastern	0.03			<0.001			0.002
Central	0.03	0.6255	0.90	<0.001	6.255	0.97	0.002
Western	0			0			0
Beaked whales (*Mesoplodon* spp.)							
Eastern	0.02			<0.001			0.002
Central	0.02	0.6255	0.90	<0.001	6.255	0.97	0.002
Western	0.07			<0.001			0.008
Unidentified ziphiid (beaked whales)							
Eastern	0.03			<0.001			0.002
Central	0.05	0.6255	0.90	<0.001	6.255	0.97	0.003
Western	0.01			<0.001			<0.001
Pygmy killer whale (*Feresa attenuata*)							
Eastern	0.12			<0.001			0.003
Central	0.12	0.6255	0.10	<0.001	6.255	0.35	0.003
Western	0			0			0

Table L-9

First Order Calculations of Potential Incidental Take for Cetaceans in Northern Gulf of Mexico Oceanic Waters ≥200 m Deep (to the Seaward Boundary of the U.S. Exclusive Economic Zone) by Planning Area. Exposure Estimates Calculated on the Basis of a *Single Seismic System Activation* With Ramp-Up and Visual Monitoring (Per NTL No. 2004-G01), but No Recognition of Seismic Sound and Avoidance by Marine Mammals and No Consideration of Overlap (or Lack of Overlap) Between Marine Mammal Vocalization Characteristics and Seismic Output (Continued)

Species by Planning Area	D^1	180 dB (rms)			160 dB (rms)		
		Total Area Ensonified $(km^2)^2$	Proportion Expected to be Undetected	First Order Calculated Take	Total Area Ensonified $(km^2)^2$	Proportion Expected to be Undetected	First Order Calculated Take
False killer whale (*Pseudorca crassidens*)							
Eastern	0.62			<0.001			0.014
Central	0	0.6255	0.10	0	6.255	0.35	0
Western	0			0			0
Killer whale (*Orcinus orca*)							
Eastern	0.02			<0.001			<0.001
Central	0.06	0.6255	0.10	<0.001	6.255	0.35	0.001
Western	0.03			<0.001			<0.001
Short-finned pilot whale (*Globicephala macrorhynchus*)							
Eastern	0			0			0
Central	0.90	0.6255	0.10	<0.001	6.255	0.35	0.020
Western	1.28			<0.001			0.028
Melon-headed whale (*Peponocephala electra*)							
Eastern	0			0			0
Central	1.35	0.6255	0.10	<0.001	6.255	0.35	0.030
Western	1.79			0.001			0.039
Risso's dolphin (*Grampus griseus*)							
Eastern	0.87			<0.001			0.019
Central	0.71	0.6255	0.10	<0.001	6.255	0.35	0.016
Western	0.40			<0.001			0.009
Bottlenose dolphin (*Tursiops truncatus*)							
Eastern	1.48			<0.001			0.032
Central	0.19	0.6255	0.10	<0.001	6.255	0.35	0.004
Western	1.11			<0.001			0.024
Rough-toothed dolphin (*Steno bredanensis*)							
Eastern	0.04			<0.001			<0.001
Central	0.35	0.6255	0.10	<0.001	6.255	0.35	0.008
Western	0.21			<0.001			0.005

Table L-9

First Order Calculations of Potential Incidental Take for Cetaceans in Northern Gulf of Mexico Oceanic Waters ≥200 m Deep (to the Seaward Boundary of the U.S. Exclusive Economic Zone) by Planning Area. Exposure Estimates Calculated on the Basis of a *Single Seismic System Activation* With Ramp-Up and Visual Monitoring (Per NTL No. 2004-G01), but No Recognition of Seismic Sound and Avoidance by Marine Mammals and No Consideration of Overlap (or Lack of Overlap) Between Marine Mammal Vocalization Characteristics and Seismic Output (Continued)

Species by Planning Area	D^1	180 dB (rms)			160 dB (rms)		
		Total Area Ensonified (km²)²	Proportion Expected to be Undetected	First Order Calculated Take	Total Area Ensonified (km²)²	Proportion Expected to be Undetected	First Order Calculated Take
Fraser's dolphin (*Lagenodelphis hosei*)							
Eastern	0.62			<0.001		0.44	0.017
Central	0	0.6255	0.10	0	6.255		0
Western	0			0			0
Atlantic spotted dolphin (*Stenella frontalis*)							
Eastern	0			0			0
Central	0.12	0.6255	0.10	<0.001	6.255	0.44	0.003
Western	0			0			0
Spinner dolphin (*Stenella longirostris*)							
Eastern	8.11			0.005		0.44	0.223
Central	2.34	0.6255	0.10	0.001	6.255		0.064
Western	0			0			0
Pantropical spotted dolphin (*Stenella attenuata*)							
Eastern	24.79			0.016		0.44	0.682
Central	32.65	0.6255	0.10	0.020	6.255		0.899
Western	5.37			0.003			0.148
Clymene dolphin (*Stenella clymene*)							
Eastern	0			0			0
Central	8.98	0.6255	0.10	0.006	6.255	0.44	0.247
Western	4.30			0.003			0.118
Striped dolphin (*Stenella coeruleoalba*)							
Eastern	0.45			<0.001		0.44	0.012
Central	2.15	0.6255	0.10	0.001	6.255		0.059
Western	3.69			0.002			0.102
Stenella spp.							
Eastern	0.06			<0.001		0.44	0.002
Central	0.23	0.6255	0.10	<0.001	6.255		0.006
Western	0.22			<0.001			0.006

Table L-9

First Order Calculations of Potential Incidental Take for Cetaceans in Northern Gulf of Mexico Oceanic Waters ≥200 m Deep (to the Seaward Boundary of the U.S. Exclusive Economic Zone) by Planning Area. Exposure Estimates Calculated on the Basis of a *Single Seismic System Activation* With Ramp-Up and Visual Monitoring (Per NTL No. 2004-G01), but No Recognition of Seismic Sound and Avoidance by Marine Mammals and No Consideration of Overlap (or Lack of Overlap) Between Marine Mammal Vocalization Characteristics and Seismic Output (Continued)

Species by Planning Area	D^1	180 dB (rms)			160 dB (rms)		
		Total Area Ensonified $(km^2)^2$	Proportion Expected to be Undetected	First Order Calculated Take	Total Area Ensonified $(km^2)^2$	Proportion Expected to be Undetected	First Order Calculated Take
Unidentified dolphin							
Eastern	0.40			<0.001			0.003
Central	0.39	0.6255	0.10	<0.001	6.255	0.10	0.002
Western	0.07			<0.001			<0.001
Unidentified small whale							
Eastern	0.02			<0.001			<0.001
Central	0.04	0.6255	0.10	<0.001	6.255	0.35	<0.001
Western	0.01			<0.001			<0.001
Unidentified large whale							
Eastern	0.03			<0.001			0.001
Central	0.01	0.6255	0.52	<0.001	6.255	0.52	<0.001
Western	0			0			0
Unidentified odontocete							
Eastern	0.03			<0.001			0.002
Central	0.02	0.6255	0.90	<0.001	6.255	0.97	0.001
Western	0.07			<0.001			0.004

[1] D = number of animals/100 km²; taken from **Table L-4**.

[2] Total area ensonified considers system start-up at a low output level (e.g., 160 dB [rms]), with incremental increases to full power within approximately 30 min. Area ensonified = length of transect covered in 30 min (4.17 km, using vessel speed of 4.5 kn (nmi/h) = 8.34 km/h) x twice the isopleth radius (r = 0.3 or 3.0 km, for 180 and 160 dB [rms], respectively) x 0.50 (i.e., to account for gradual increase in airgun output) x 0.50 (i.e., conversion from circular to elliptical area of ensonification).

Table L-10

Calculation of Potential Incidental Take for Cetaceans in Northern Gulf of Mexico Continental Shelf Waters 20 to 200 m Deep by Planning Area. Exposure Estimates Calculated on the Basis of a Single Event and Cumulative Annual Seismic With No Ramp-Up and No Visual Monitoring (except in the Eastern Gulf of Mexico). Species Presence and Susceptibility to Seismic Sources Have Been Factored in to Determine Potential for Take

Species by Planning Area	180 dB (rms)		160 dB (rms)		Susceptible to Seismic Frequencies[1]	Gulf of Mexico Presence[2]	Potential Maximum Take[3]	
	Single Event – First Order Calculated Take	Cumulative Annual Seismic – First Order Calculated Take	Single Event – First Order Calculated Take	Cumulative Annual Seismic – First Order Calculated Take			180 dB (rms)	160 dB (rms)
Bottlenose dolphin (*Tursiops truncatus*)								
Eastern	<0.01	1.0	0.31	37.2			1	38
Central	0.06	7.2	0.63	75.6	Yes	Common	8	76
Western	0.05	6.0	0.51	61.2			6	62
Atlantic spotted dolphin (*Stenella frontalis*)								
Eastern	0.01	1.2	0.54	64.8			2	65
Central	0.02	2.4	0.17	20.4	Possible	Common	3	21
Western	0.01	1.2	0.14	16.8			2	17
Bottlenose dolphin/Atlantic spotted dolphin (*T. truncatus/S. frontalis*)								
Eastern	<0.01	1.0	0.02	2.4			1	3
Central	<0.01	1.0	0.01	1.2	Possible	Common	1	2
Western	<0.01	1.0	0.08	9.6			1	10
Rough-toothed dolphin (*Steno bredanensis*)								
Eastern	<0.01	1.0	0.01	1.2			0	0
Central	0	0	0	0	No	Common	0	0
Western	<0.01	1.0	0.07	8.4			0	0

[1] Susceptibility to seismic noise based on **Table L-7**.

[2] Gulf of Mexico presence derived from Würsig et al. (2000), as presented in **PEA Section III, Table III-3**.

[3] Potential maximum take estimates consider a single event (e.g., ramp-up, start-up) and cumulative annual seismic activity (i.e., single event multiplied by a factor of 120; see text for explanation). First order take estimates shown at the 180-dB and 160-dB (rms) levels - e.g., Central Gulf of Mexico bottlenose dolphin single and cumulative take estimates are 0.06 and 7.2 individuals, respectively, using the 180-dB (rms) threshold; at the 160-dB (rms) threshold, single and cumulative take estimates for this species in the Central Gulf are 0.63 and 75.6 individuals, respectively. Potential maximum take estimates use the cumulative first order take numbers at the 180- and 160-dB (rms) exposure levels. For example, a total of 8 bottlenose dolphin may be taken annually in the Central Gulf as a result of exposure to 180 dB (rms); using the 160-dB (rms) exposure level, a total of 76 bottlenose dolphin may realize take. Gulf of Mexico presence (i.e., common) and possible frequency overlap with vocalization (i.e., yes) also are considered in the determination.

Table L-11

Calculation of Potential Incidental Take for Cetaceans in Northern Gulf of Mexico Oceanic Waters ≥200 m Deep (to the Seaward Boundary of the U.S. Exclusive Economic Zone) by Planning Area. Exposure Estimates Calculated on the Basis of System Ramp-Up Per Existing Requirements of NTL No. 2004-G01. Species Presence and Susceptibility to Seismic Sources Have Been Factored in to Determine Potential for Take

Species by Planning Area	180 dB (rms)		160 dB (rms)		Susceptible to Seismic Frequencies[1]	Gulf of Mexico Presence[2]	Potential Maximum Take[3]	
	Single Event – First Order Calculated Take	Cumulative Annual Seismic – First Order Calculated Take	Single Event – First Order Calculated Take	Cumulative Annual Seismic – First Order Calculated Take			180 dB (rms)	160 dB (rms)
Bryde's whale (*Balaenoptera edeni*)								
Eastern	<0.001	<0.1	<0.001	<0.1			1	1
Central	0	0	0	0	Yes	Uncommon	0	0
Western	0	0	0	0			0	0
Sperm whale (*Physeter macrocephalus*)								
Eastern	<0.001	<0.1	0.009	1.1			1	2
Central	0.001	0.1	0.014	1.7	Possible	Common	1	2
Western	<0.001	<0.1	0.008	1.0			1	1
Pygmy and dwarf sperm whales (*Kogia* spp.)								
Eastern	0.001	0.1	0.013	1.6			0	0
Central	0.001	0.1	0.013	1.6	Unknown-No	Common	0	0
Western	<0.001	<0.1	0.005	0.6			0	0
Cuvier's beaked whale (*Ziphius cavirostris*)								
Eastern	<0.001	<0.1	0.002	0.2			1	1
Central	<0.001	<0.1	0.002	0.2	Unknown	Rare-Common	1	1
Western	0	0	0	0			0	0
Beaked whales (*Mesoplodon* spp.)								
Eastern	<0.001	<0.1	0.002	0.2			1	1
Central	<0.001	<0.1	0.002	0.2	Unknown	Extralimital-Common	1	1
Western	<0.001	<0.1	0.008	1.0			1	1
Unidentified ziphiid (beaked whales)								
Eastern	<0.001	<0.1	0.002	0.2			1	1
Central	<0.001	<0.1	0.003	0.4	Unknown	Extralimital-Common	1	1
Western	<0.001	<0.1	<0.001	<0.1			1	1
Pygmy killer whale (*Feresa attenuata*)								
Eastern	<0.001	<0.1	0.003	0.4			1	1
Central	<0.001	<0.1	0.003	0.4	Unknown	Uncommon	1	1
Western	0	0	0	0			0	0

Table L-11

Calculation of Potential Incidental Take for Cetaceans in Northern Gulf of Mexico Oceanic Waters ≥200 m Deep (to the Seaward Boundary of the U.S. Exclusive Economic Zone) by Planning Area. Exposure Estimates Calculated on the Basis of System Ramp-Up Per Existing Requirements of NTL No. 2004-G01. Species Presence and Susceptibility to Seismic Sources Have Been Factored in to Determine Potential for Take
(Continued)

Species by Planning Area	180 dB (rms)		160 dB (rms)		Susceptible to Seismic Frequencies[1]	Gulf of Mexico Presence[2]	Potential Maximum Take[3]	
	Single Event – First Order Calculated Take	Cumulative Annual Seismic – First Order Calculated Take	Single Event – First Order Calculated Take	Cumulative Annual Seismic – First Order Calculated Take			180 dB (rms)	160 dB (rms)
False killer whale (*Pseudorca crassidens*)								
Eastern	<0.001	<0.1	0.014	1.7			0	0
Central	0	0	0	0	No	Uncommon	0	0
Western	0	0	0	0			0	0
Killer whale (*Orcinus orca*)								
Eastern	<0.001	<0.1	<0.001	<0.1			1	1
Central	<0.001	<0.1	0.001	0.1	Possible	Uncommon	1	1
Western	<0.001	<0.1	<0.001	<0.1			1	1
Short-finned pilot whale (*Globicephala macrorhynchus*)								
Eastern	0	0	0	0			0	0
Central	<0.001	<0.1	0.020	<0.1	Yes	Common	1	1
Western	<0.001	<0.1	0.028	<0.1			1	1
Melon-headed whale (*Peponocephala electra*)								
Eastern	0	0	0	0			0	0
Central	<0.001	<0.1	0.030	3.6	Unknown	Common	1	4
Western	0.001	0.1	0.039	4.7			1	5
Risso's dolphin (*Grampus griseus*)								
Eastern	<0.001	<0.1	0.019	2.3			1	3
Central	<0.001	<0.1	0.016	1.9	Possible	Common	1	2
Western	<0.001	<0.1	0.009	1.1			1	2
Bottlenose dolphin (*Tursiops truncatus*)								
Eastern	<0.001	<0.1	0.032	3.8			1	4
Central	<0.001	<0.1	0.004	0.5	Possible	Common	1	1
Western	0.001	0.1	0.024	2.9			1	3
Rough-toothed dolphin (*Steno bredanensis*)								
Eastern	<0.001	<0.1	<0.001	<0.1			0	0
Central	<0.001	<0.1	0.008	1.0	No	Common	0	0
Western	<0.001	<0.1	0.005	0.6			0	0

Table L-11

Calculation of Potential Incidental Take for Cetaceans in Northern Gulf of Mexico Oceanic Waters ≥200 m Deep (to the Seaward Boundary of the U.S. Exclusive Economic Zone) by Planning Area. Exposure Estimates Calculated on the Basis of System Ramp-Up Per Existing Requirements of NTL No. 2004-G01. Species Presence and Susceptibility to Seismic Sources Have Been Factored in to Determine Potential for Take (Continued)

Species by Planning Area	180 dB (rms)		160 dB (rms)		Susceptible to Seismic Frequencies[1]	Gulf of Mexico Presence[2]	Potential Maximum Take[3]	
	Single Event – First Order Calculated Take	Cumulative Annual Seismic – First Order Calculated Take	Single Event – First Order Calculated Take	Cumulative Annual Seismic – First Order Calculated Take			180 dB (rms)	160 dB (rms)
Fraser's dolphin (*Lagenodelphis hosei*)								
Eastern	<0.001	<0.1	0.017	2.0			0	0
Central	0	0	0	0	No	Common	0	0
Western	0	0	0	0			0	0
Atlantic spotted dolphin (*Stenella frontalis*)								
Eastern	0	0	0	0			0	0
Central	<0.001	<0.1	0.003	0.4	Possible	Common	1	1
Western	0	0	0	0			0	0
Spinner dolphin (*Stenella longirostris*)								
Eastern	0.005	0.6	0.223	26.8			1	27
Central	0.001	0.1	0.064	7.7	Possible	Common	1	8
Western	0	0	0	0			0	0
Pantropical spotted dolphin (*Stenella attenuata*)								
Eastern	0.016	1.9	0.682	81.8			0	0
Central	0.020	2.4	0.899	107.9	No	Common	0	0
Western	0.003	0.4	0.148	17.8			0	0
Clymene dolphin (*Stenella clymene*)								
Eastern	0	0	0	0			0	0
Central	0.006	0.7	0.247	29.6	No	Common	0	0
Western	0.003	0.4	0.118	14.2			0	0
Striped dolphin (*Stenella coeruleoalba*)								
Eastern	<0.001	<0.1	0.012	1.4			0	0
Central	0.001	0.1	0.059	7.1	No	Common	0	0
Western	0.002	0.2	0.102	12.2			0	0
***Stenella* spp.**								
Eastern	<0.001	<0.1	0.002	0.2			1	1
Central	<0.001	<0.1	0.006	0.7	Possible-No	Common	1	1
Western	<0.001	<0.1	0.006	0.7			1	1

Table L-11

Calculation of Potential Incidental Take for Cetaceans in Northern Gulf of Mexico Oceanic Waters ≥200 m Deep (to the Seaward Boundary of the U.S. Exclusive Economic Zone) by Planning Area. Exposure Estimates Calculated on the Basis of System Ramp-Up Per Existing Requirements of NTL No. 2004-G01. Species Presence and Susceptibility to Seismic Sources Have Been Factored in to Determine Potential for Take (Continued)

Species by Planning Area	180 dB (rms)		160 dB (rms)		Susceptible to Seismic Frequencies[1]	Gulf of Mexico Presence[2]	Potential Maximum Take[3]	
	Single Event – First Order Calculated Take	Cumulative Annual Seismic – First Order Calculated Take	Single Event – First Order Calculated Take	Cumulative Annual Seismic – First Order Calculated Take			180 dB (rms)	160 dB (rms)
Unidentified dolphin								
Eastern	<0.001	<0.1	0.003	0.4			1	1
Central	<0.001	<0.1	0.002	0.2	Possible	Common	1	1
Western	<0.001	<0.1	<0.001	0.1			1	1
Unidentified small whale								
Eastern	<0.001	<0.1	<0.001	<0.1			1	1
Central	<0.001	<0.1	<0.001	<0.1	Possible	Extralimital-Common	1	1
Western	<0.001	<0.1	<0.001	<0.1			1	1
Unidentified large whale								
Eastern	<0.001	<0.1	0.001	0.1			1	1
Central	<0.001	<0.1	<0.001	<0.1	Possible	Extralimital-Uncommon	1	1
Western	0	0	0	0			0	0
Unidentified odontocete								
Eastern	<0.001	<0.1	0.002	0.2			1	1
Central	<0.001	<0.1	0.001	0.1	Possible	Extralimital-Common	1	1
Western	<0.001	<0.1	0.004	0.5			1	1

[1] Susceptibility to seismic noise based on **Table L-7**.

[2] Gulf of Mexico presence derived from Würsig et al. (2000), as presented in **PEA Section III, Table III-3**.

[3] Potential maximum take estimates consider a single event (e.g., ramp-up, start-up) and cumulative annual seismic activity (i.e., single event multiplied by a factor of 120; see text for explanation). First order take estimates shown at the 180-dB and 160-dB (rms) levels - e.g., Central Gulf of Mexico spinner dolphin single and cumulative take estimates are 0.001 and 0.1 individuals, respectively, using the 180-dB (rms) threshold; at the 160-dB (rms) threshold, single and cumulative take estimates for this species in the Central Gulf are 0.064 and 7.7 individuals, respectively. Potential maximum take estimates use the cumulative first order take numbers at the 180- and 160-dB (rms) exposure levels. For example, a total of one spinner dolphin may be taken annually in the Central Gulf as a result of exposure to 180 dB (rms); using the 160-dB (rms) exposure level, a total of 8 spinner dolphin may realize take. Gulf of Mexico presence (i.e., common) and possible frequency overlap with vocalization (i.e., possible) also are considered in the determination.

This product is subsequently multiplied by 0.50 to account for the gradual increase in airgun output to maximum power; see **Figure L-3**. Finally, the product is again multiplied by 0.50 to allow for conversion from a circular to elliptical area of ensonification. **Figure L-3** depicts, in a simplified form, the gradual increase in output from a typical seismic array realized between initial start-up and full power. Over the course of this 30 min-start-up period, the survey vessel travels >4 km with an increasing number of airguns shooting every 12 to 16 sec (i.e., 112 to 150 shots total). In spite of the fact that the duration of each seismic pulse lasts only a few milliseconds, there remains considerable overlap between zones of ensonification created by the first shot and the next (i.e., sound fields overlap as the seismic vessel and array move forward 28 to 37 m between shots). Concurrently, zones of ensonification expand proportionally as more airguns come on line to reach full power. **Figure L-3** does not depict all of the potential shots and their associated radii, nor does it reflect the elliptical nature of the sound field surrounding an operational airgun array.

First order take estimates also factor in the detectability of Gulf marine mammal species (see **Table L-6**). In shallow water, the proportion of animals expected to remain undetected is 100% for shallow water surveys because visual monitoring (and ramp-up) are not required, except in the Eastern GOM Planning Area, where a detectability factor has been integrated into the first order take calculations. In deep water, where operations are uniformly conducted under NTL No. 2004-G01, the probabilities of detection and non-detection have been incorporated, using the same approach outlined in Department of the Navy (2001). Further, a factor has been added to account for the effectiveness of surface (shipboard) visual monitoring. No quantitative estimates are currently available to account for noise avoidance resulting from ramp-up effectiveness or species recognition of a sound source (at some low level, e.g., 160 dB or less) and subsequent avoidance of that sound source. These limitations are noteworthy, as is the potential overlap between vocalization characteristics (and, by inference, hearing capabilities) and the maximum output of seismic noise.

Summary take calculations presented in **Table L-10** and **L-11** are estimates of potential take in differing water depths and within different planning areas. Calculations are provided for the 160- and 180-dB (rms) contours (i.e., indicative of potential Level B and Level A harassment distances, respectively) and are differentiated on the basis of recent marine mammal density estimates (i.e., by planning area, by water depth - <200 m and ≥200 m). These potential maximum take estimates are based on the first order calculated take associated with a single event (e.g., ramp-up, start-up), multiplied by an estimate of the number of times annually that such an event may occur (i.e., factor of 120). Maximum take estimates are provided for each species potentially exposed to 180-dB and 160-dB (rms) levels. The potential for frequency overlap (i.e., know/inferred hearing/vocalization vs. seismic array output) and species presence also are considered in the determinations. For example, in **Table L-10**, a total of 8 bottlenose dolphin in shallow waters of the Central Gulf may be exposed annually to seismic noise at the 180-dB (or greater) threshold, while the estimate of take at the 160-dB threshold is 76 individuals per year (all dB values as rms).

Estimates of maximum incidental take in all shallow OCS waters of the GOM range from 1 to 76 individuals for bottlenose dolphin and 2 to 65 individuals for Atlantic spotted dolphin (**Table L-10**). Both of these species are common shallow water inhabitants with likely to possible susceptibility to seismic frequencies. While there was a lack of seismic activity in the shallow waters of the Eastern Gulf in 2002, it is possible that future seismic activity may occur there; consequently, first order take estimates were calculated for this species in this region. Rough-toothed dolphin, also a shallow water inhabitant, exhibited moderate densities in the Western and Eastern Gulf and was absent from the shallow waters of the Central Gulf. This species vocalizes in frequency ranges that far exceed those characteristic of seismic operations; consequently, a qualified estimate of zero take has been calculated for this species. This rule was applied to all species that exhibit vocalization characteristics that do not overlap with seismic output (i.e., assignment of zero take to species with no overlap between vocalization and seismic

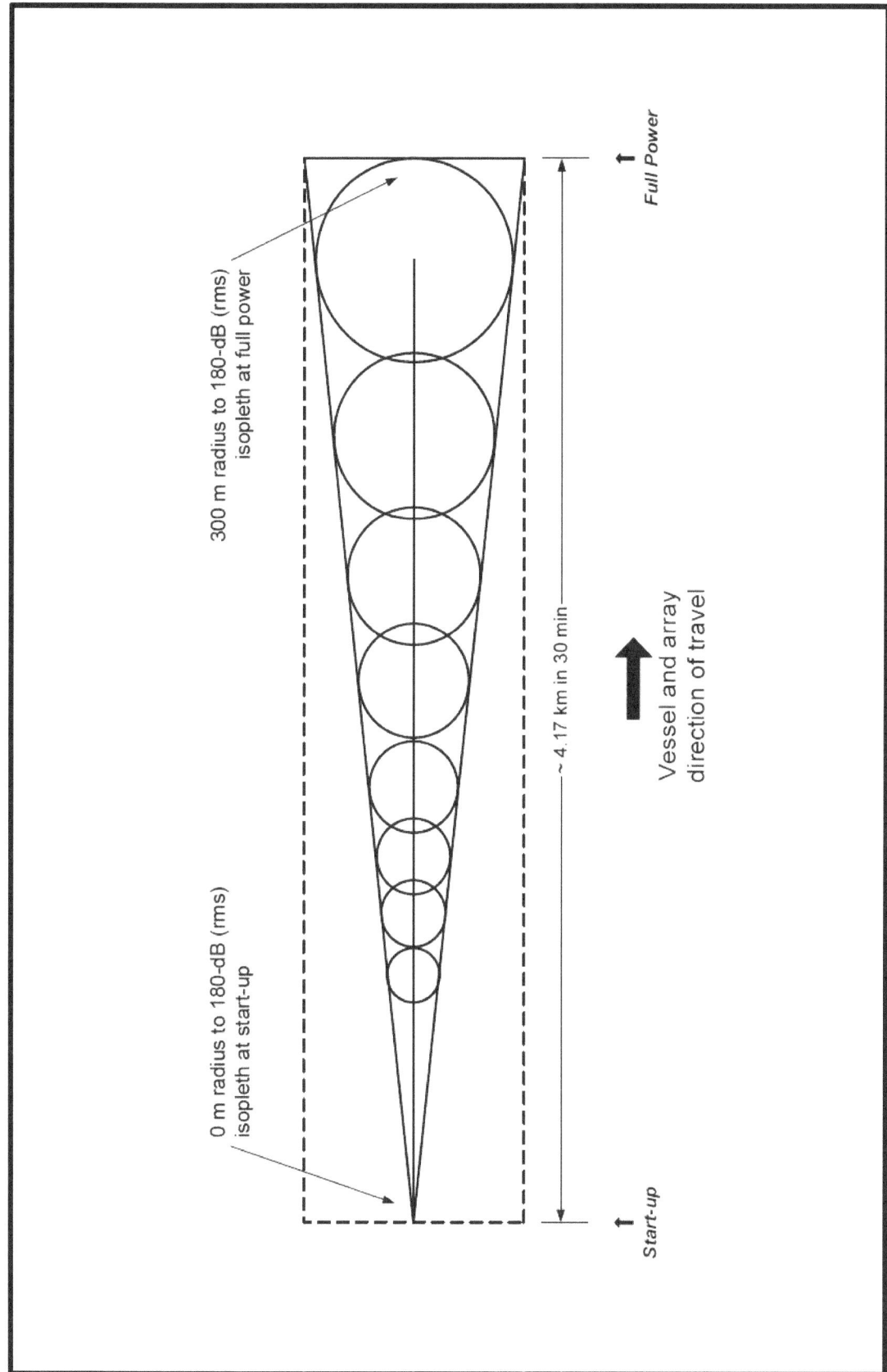

Figure L-3.   Schematic of a hypothetical sound field generated between seismic system start-up and full power. Vessel and array travel 4.17 km during 30 min start-up.   Between start-up and full power, the radial distance (from the array to the 180-dB [rms] isopleth; shown as radial distance) increases from 0 to 300 m. Example derived from a 4,550 in.³ array.

output frequencies). In those instances where vocalization characteristics are unknown, take estimates were developed from the cumulative first order take calculations.

In deeper water (≥200 m), a total of 17 cetacean species and several broader groups (e.g., *Stenella* spp.) were identified from recent NMFS surveys (**Table L-11**). Lack of overlap between vocalization characteristics and seismic output is suggested for several species, including false killer whale and rough-toothed, Fraser's, pantropical spotted, Clymene, and striped dolphins; zero take estimates were determined for these species in deeper waters of the Gulf. Most remaining species exhibit possible to definite overlap of vocalization characteristics with the frequencies of seismic output, with the exception of those for which vocalization is unknown. Notable determinations of incidental take associated with G&G operations in deeper Gulf waters include the following:

- Bryde's whale: <0.1 individual potentially taken annually at both the 180 and 160 dB (rms) exposure levels, resulting in potential maximum annual take estimates of 1 and 1. Take estimates were applicable only to the Eastern Gulf, with a high probability of frequency overlap between vocalizations and seismic noise and a relatively low seismic survey activity level.
- Sperm whale: a maximum of 0.1 and 1.7 individuals potentially taken annually at 180 and 160 dB (rms), respectively. Potential maximum take estimates were established at 1 and 1 or 1 and 2, depending upon the planning area. Take estimates were developed for all planning areas, with possible frequency overlap between vocalizations and seismic noise.
- Cuvier's beaked whale: <0.1 to 0.2 individual potentially taken annually at 180 and 160 dB (rms), respectively. Take estimates were applicable only to the Eastern and Central Gulf, with unknown frequency overlap between vocalizations and seismic noise. Other beaked whales reflected take estimates of <0.1 to 1.0 individuals (at 180 and 160 dB [rms], respectively). Cuvier's beaked whales and other beaked whale species are cryptic, indicative of the limited life history information that is currently available for this marine mammal group. In addition, beaked whales have shown an extreme sensitivity to mid- and high frequency noise sources (e.g., sonars). Although G&G seismic exposure levels have been calculated based on the best available survey data and density estimates, beaked whales may be more prevalent in the GOM than currently believed. Caution regarding the calculations of potential incidental take for members of this group is strongly suggested.
- Short-finned pilot whale: <0.1 individual potentially taken annually at both the 180 and 160 dB (rms) exposure levels. Potential maximum take estimates were established at 1 and 1 for these exposure levels. Take estimates were applicable only to the Central and Western Gulf, with a high probability of frequency overlap between vocalizations and seismic noise.
- Highest calculated annual take estimates were noted for bottlenose and spinner dolphins and melon-headed whales, with possible to unknown frequency overlap between vocalizations and seismic noise.

These estimates of incidental take consider current density estimates of marine mammal species found within shallow and deeper waters of the three GOM Planning Areas, as well as recent seismic survey activity by water depth and planning area. Measures of ESW (from historic survey data) and availability bias have been integrated, as appropriate, into these estimates. Preliminary calculations of radial distance to isopleths of interest (i.e., 160 and 180 dB [rms]) are based on a typical GOM G&G seismic array, as outlined in **Appendix C, Sections III** and **IV**. Limitations to isopleth calculations have been outlined in **Table L-1**; appropriate adjustments also have been integrated, with proper justification, into the analysis (e.g., adjustment to radial calculations to account for the elliptical shape of the zones of ensonification). In addition, the area ensonified during a single event (e.g., ramp-up or seismic system start-up) and cumulative seismic surveys in the Gulf were determined. Because ramp-up or system start-up may occur multiple times during a single survey, and approximately 20 surveys may be completed in any given year, maximum potential take calculations were developed as a product of a single event multiplied by 120 (i.e., the cumulative case). Additional data regarding susceptibility to seismic noise (i.e., potential for overlap between seismic noise and vocalization frequencies) and the potential for recognition and avoidance of mobile sound sources are necessary to develop more complete estimates of incidental take, and to identify further those marine mammal species at greatest potential risk from seismic operations.

V. LITERATURE CITED

Arvanitis, L.C., and K.M. Portier. 1997. Natural resource sampling – line transect method. University of Florida, Gainesville, FL. Available at www.ifasstat.ufl.edu/nrs98/LTS htm.

Barlow, J., C.W. Oliver, T.D. Jackson, and B.L. Taylor. 1988. Harbor porpoise, *Phocoena phocoena*, abundance estimation for California, Oregon, and Washington: Volume II, aerial surveys. Fish. Bull. 86(3):433-444.

Barlow, J., K. Forney, A. Von Saunder, and J. Urban-Ramirez. 1997. A report of cetacean acoustic detection and dive interval studies (CADDIS) conducted in the southern Gulf of California, 1995. National Marine Fisheries Service Technical Memorandum. NOAA-TM-NMFS-SWFSC-250. 48 pp.

Buckland, S.T., D.R. Anderson, K.P. Burnham, and J.L. Laake. 1993. Distance Sampling: Estimating Abundance of Biological Populations. Chapman and Hall, London. 446 pp.

Calambokidis, J., G.H. Steiger, J.C. Cubbage, K.C. Balcomb, and P. Bloedel. 1989. Biology of humpback whales in the Gulf of the Farallones. Final report for contract CX-8000-6-0003 to Gulf of the Farallones National Marine Sanctuary, NOAA, Fort Mason Center, San Francisco, CA. 93 pp.

Caldwell, J., and W. Dragoset. 2000. A brief overview of seismic air-gun arrays. The Leading Edge 19:898-902.

Department of the Navy. 1998. Final environmental impact statement: Shock testing the SEAWOLF submarine. Southern Division, Naval Facilities Engineering Command, North Charleston, SC.

Department of the Navy. 2001. Final environmental impact statement: Shock trial of the WINSTON S. CHURCHILL (DDG 81). Southern Division, Naval Facilities Engineering Command, North Charleston, SC. February 2001. 229 pp. + apps.

Eberhardt, L.L., D.G. Chapman, and J.R. Gilbert. 1979. A review of marine mammal census methods. Supplement to J. Wildl. Manage. 43(1). Wildlife Monograph No. 63.

Evans, W.E. 1974. Radio-telemetric studies of two species of small odontocete cetaceans, pp. 386-394. *In:* W.E. Schevill (ed.), The whale problem: a status report. Harvard Univ. Press, Cambridge, MA.

Fontana, P. 2002. Acoustic emissions produced by seismic exploration activities. Paper presented at the Boston Marine Mammal Workshop, International Association of Geophysical Contractors, Boston, MA, 16-17 May 2002.

Goold, J.C., and P.J. Fish. 1998. Broadband spectra of seismic survey airgun emissions, with reference to dolphin auditory thresholds. J. Acoust. Soc. Am. 103(4):2177-2184.

Jochens, A.E., and D.C. Biggs (eds). 2003. Sperm whale seismic study in the Gulf of Mexico: Annual Report: Year 1. U.S. Department of the Interior, Minerals Management Service, Gulf of Mexico OCS Region, New Orleans, LA. OCS Study MMS 2003-069. 139 pp.

Ketten, D.R. 1998. Marine Mammal Auditory Systems: A summary of audiometric and anatomical data and its implications for underwater acoustic impacts. NOAA Technical Memorandum NMFS-SWFSC-256. National Oceanic and Atmospheric Administration, National Marine Fisheries Service, Southwest Fisheries Science Center, La Jolla, CA. 74 pp.

Kopelman, A.H., and S.S. Sadove. 1995. Ventilatory rate differences between surface-feeding and non-surface-feeding fin whales (*Balaenoptera physalus*) in the waters off eastern Long Island, New York, U.S.A., 1981-1987. Mar. Mamm. Sci. 11(2):200-208.

Leatherwood, S., K. Goodrich, A.L. Kinter, and R.M. Truppo. 1982. Respiration patterns and 'sightability' of whales. Rep. Int. Whal. Comm. 32:601-613.

LGL, Ltd., environmental research associates. 2003a. Environmental assessment of marine seismic testing conducted by the *R/V Maurice Ewing* in the northern Gulf of Mexico, May – June 2003. Prepared for Lamont-Doherty Earth Observatory and the National Science Foundation by LGL Ltd., environmental research associates, King City, Ontario. 7 March 2003. LGL Report TA2822-3. 99 pp.

LGL, Ltd., environmental research associates. 2003b. Request by Lamont-Doherty Earth Observatory for an incidental harassment authorization to allow the incidental take of marine mammals during marine seismic testing in the northern Gulf of Mexico, April 2004. Prepared for Lamont-Doherty Earth Observatory by LGL Ltd., environmental research associates, King City, Ontario. November 2003. LGL Report TA2822-17. 111 pp.

LGL, Ltd., environmental research associates. 2003c. Marine mammal monitoring during Lamont-Doherty Earth Observatory's Acoustic Calibration Study in the northern Gulf of Mexico, 2003. Prepared for Lamont-Doherty Earth Observatory and National Marine Fisheries Service by LGL Ltd., environmental research associates, King City, Ontario. September 2003. LGL Report TA2822-12. var pp.

Malme, C.I., and P.R. Miles. 1985. Behavioral responses of marine mammals (gray whales) to seismic discharges, pp. 253-280. *In:* Proc. Workshop on Effects of Explosive Use in the Marine Environment, January 1985, Halifax, Nova Scotia. Technical Report No. 5, Canadian Oil and Gas Lands Administration, Environmental Protection Branch, Ottawa, Ontario. 398 pp.

Malme, C.I., P.R. Miles, C.W. Clarke, P. Tyack,. and J.E. Bird. 1983. Investigations of the potential effects of underwater noise from petroleum industry activities on migrating gray whale behavior. BBN Rep. 5366. Report from Bolt Beranek & Newman Inc., Cambridge MA, for U.S. Department of the Interior, Minerals Management Service, Anchorage, AK. NTIS PB86-174174.

Malme, C.I., P.R. Miles, C.W. Clarke, P. Tyack, and J.E. Bird. 1984. Investigations of the potential effects of underwater noise from petroleum industry activities on migrating gray whale behavior/Phase II: January 1984 migration. BBN Rep. 5586. Report by Bolt Beranek & Newman Inc., Cambridge, MA, for the U.S. Department of the Interior, Minerals Management Service, Anchorage, AK. NTIS PB86-218377.

Malme, C.I., B. Würsig, J.E. Bird, and P. Tyack. 1988. Observations of feeding gray whale responses to controlled industrial noise exposure, pp. 55-73. *In:* W.M. Sackinger et al., Port and ocean engineering under arctic conditions, vol II. Geophys. Inst., Univ. Alaska, Fairbanks. 111 pp.

Marsh, H., and D.F. Sinclair. 1989. Correcting for visibility bias in strip transect aerial surveys of aquatic fauna. J. Wildl. Manage. 53:1,017-1,024.

Mate, B.R., K.M. Stafford, R. Nawojchik, and J.L. Dunn. 1994. Movements and dive behavior of a satellite-monitored Atlantic white-sided dolphin (*Lagenorhynchus acutus*) in the Gulf of Maine. Mar. Mamm. Sci. 10(1):116-121.

McCauley, R.D., M.N. Jenner, C. Jenner, K.A. McCabe, and J. Murdoch. 1998. The response of humpback whales (*Megaptera novaengliae*) to offshore seismic survey noise: preliminary results of observations about a working seismic vessel and experimental exposures. APPEA Journal 1998:692-707.

McCauley, R.D., J. Fewtrell, A.J. Duncan, C. Jenner, M.N. Jenner, J. Penrose, R.I.T. Prince, A. Adhitya, J. Murdoch, and K. McCabe. 2000. Marine seismic surveys - A study of environmental implications. APPEA Journal 40:692-708.

Mullin, K.D., and W. Hoggard. 2000. Visual surveys of cetaceans and sea turtles from aircraft and ships, chapter 4. *In:* R.W. Davis, W.E. Evans, and B. Würsig (eds.), Cetaceans, Sea Turtles and Seabirds in the Northern Gulf of Mexico: Distribution, Abundance and Habitat Associations. Volume II: Technical Report. Prepared by Texas A&M University at Galveston and the National Marine Fisheries Service. U.S. Department of the Interior, U.S. Geological Survey, Biological Resources Division, USGS/BRD/CR-1999-005 and Minerals Management Service, Gulf of Mexico OCS Region, New Orleans, LA. OCS Study MMS 2000-003.

Richardson, W.J., B. Würsig, and C.R. Greene. 1986. Reaction of bowhead whales to seismic exploration in the Canadian Beaufort Sea. J. Acoust. Soc. Am. 79(4):1117-1128.

Richardson, W.J., C.R. Greene, Jr., C.I. Malme, and D.H. Thomson. 1995. Marine Mammals and Noise. Academic Press, San Diego. 576 pp.

Stern, S.J. 1992. Surfacing rates and surfacing patterns of minke whales (*Balaenoptera acutorostrata*) off central California, and the probability of a whale surfacing within the visual range. Rep. Int. Whaling Commn. 42:379-386.

Wartzok, D., and D.R. Ketten. 1999. Marine mammal sensory systems, pp. 117-175. *In:* J.E. Reynolds III and S.A. Rommel (eds.), Biology of marine mammals. Smithsonian Institution Press, Washington, DC. 590 pp.

Winn, H.E. (ed.). 1982. A characterization of marine mammals and turtles in the mid- and north Atlantic areas of the U.S. outer continental shelf. Final report of the Cetacean and Turtle Assessment Program. University of Rhode Island, Kingston, RI. Prepared for U.S. Department of the Interior, Bureau of Land Management, Washington, DC. Available through National Technical Information Service, Springfield, VA. PB83-215855.

Würsig, B., T.A. Jefferson, and D.J. Schmidley. 2000. The marine mammals of the Gulf of Mexico. Texas A&M University Press, College Station, TX. 232 pp.

The Department of the Interior Mission

As the Nation's principal conservation agency, the Department of the Interior has responsibility for most of our nationally owned public lands and natural resources. This includes fostering sound use of our land and water resources; protecting our fish, wildlife, and biological diversity; preserving the environmental and cultural values of our national parks and historical places; and providing for the enjoyment of life through outdoor recreation. The Department assesses our energy and mineral resources and works to ensure that their development is in the best interests of all our people by encouraging stewardship and citizen participation in their care. The Department also has a major responsibility for American Indian reservation communities and for people who live in island territories under U.S. administration.

The Minerals Management Service Mission

As a bureau of the Department of the Interior, the Minerals Management Service's (MMS) primary responsibilities are to manage the mineral resources located on the Nation's Outer Continental Shelf (OCS), collect revenue from the Federal OCS and onshore Federal and Indian lands, and distribute those revenues.

Moreover, in working to meet its responsibilities, the **Offshore Minerals Management Program** administers the OCS competitive leasing program and oversees the safe and environmentally sound exploration and production of our Nation's offshore natural gas, oil and other mineral resources. The MMS **Minerals Revenue Management** meets its responsibilities by ensuring the efficient, timely and accurate collection and disbursement of revenue from mineral leasing and production due to Indian tribes and allottees, States and the U.S. Treasury.

The MMS strives to fulfill its responsibilities through the general guiding principles of: (1) being responsive to the public's concerns and interests by maintaining a dialogue with all potentially affected parties and (2) carrying out its programs with an emphasis on working to enhance the quality of life for all Americans by lending MMS assistance and expertise to economic development and environmental protection.